This is the first full-length historical study of Gestalt psychology – an attempt to advance holistic thought *within* natural science. Holistic thought is often portrayed as a wooly minded revolt against reason and modern science, but this is not necessarily so. On the basis of rigorous experimental research and scientific argument as well as on philosophical grounds, the Gestalt theorists Max Wertheimer, Wolfgang Köhler, and Kurt Koffka opposed conceptions of science and mind that equated knowledge of nature with its effective manipulation and control. Instead, they attempted to establish dynamic principles of inherent, objective order and meaning – in current language, principles of self-organization – in human perception and thinking, in human and animal behavior, and in the physical world. The impact of their work ranged from cognitive science and theoretical biology to film theory.

Based on exhaustive research in primary sources, including archival material cited here for the first time, this study illuminates the multiple social and intellectual contexts of Gestalt theory and analyzes the emergence, development, and reception of its conceptual foundations and research programs in Germany from 1890 to 1967. The book challenges stereotypical dichotomies between modern and antimodern, rational and irrational, democratic and proto-Nazi thinking that have long dominated the history of German science and culture. It also contributes to the debate on continuity and change in German science after 1933 with a new look at Wolfgang Köhler's effort to resist Nazism, at the work of Gestalt theorists who remained in Nazi Germany after the founders emigrated, and at the impact of the Cold War and the professionalization of psychology in Germany on the reception of Gestalt theory after 1945.

T0234011

Gestalt psychology in German culture, 1890–1967

Cambridge Studies in the History of Psychology

GENERAL EDITORS: MITCHELL G. ASH AND WILLIAM R. WOODWARD

This new series provides a publishing forum for outstanding scholarly work in the history of psychology. The creation of the series reflects a growing concentration in this area by historians and philosophers of science, intellectual and cultural historians, and psychologists interested in historical and theoretical issues.

The series is open both to manuscripts dealing with the history of psychological theory and research and to work focusing on the varied social, cultural, and institutional contexts and impacts of psychology. Writing about psychological thinking and research of any period will be considered. In addition to innovative treatments of traditional topics in the field, the editors particularly welcome work that breaks new ground by offering historical considerations of issues such as the linkages of academic and applied psychology with other fields, for example, psychiatry, anthropology, sociology, and psychoanalysis; international, intercultural, or gender-specific differences in psychological theory and research; or the history of psychological research practices. The series will include both single-authored monographs and occasional coherently defined, rigorously edited essay collections.

Also in the series

Constructing the subject: Historical origins of psychological research
KURT DANZIGER

Metaphors in the history of psychology
edited by DAVID E. LEARY

Crowds, psychology, and politics, 1871–1899
JAAP VAN GINNEKEN

The professionalization of psychology in Nazi Germany
ULFRIED GEUTER

Gestalt psychology in German culture, 1890–1967

Holism and the quest for objectivity

Mitchell G. Ash

University of Iowa

CAMBRIDGE
UNIVERSITY PRESS

CAMBRIDGE UNIVERSITY PRESS
Cambridge, New York, Melbourne, Madrid, Cape Town, Singapore, São Paulo

Cambridge University Press
The Edinburgh Building, Cambridge CB2 8RU, UK

Published in the United States of America by Cambridge University Press, New York

www.cambridge.org
Information on this title: www.cambridge.org/9780521475402

© Cambridge University Press 1998

First published 1995
First paperback edition 1998

A catalogue record for this publication is available from the British Library

ISBN 978-0-521-47540-2 hardback
ISBN 978-0-521-64627-7 paperback

Transferred to digital printing 2007

Contents

List of illustrations   *page* vii
Preface   ix

Introduction   1

Part I   The social and intellectual settings

1 The academic environment and the establishment of
   experimental psychology   17
2 Carl Stumpf and the training of scientists in Berlin   28
3 The philosophers' protest   42
4 Making a science of mind: Styles of reasoning in sensory
   physiology and experimental psychology   51
5 Challenging positivism: Revised philosophies of mind and science   68
6 The Gestalt debate: From Goethe to Ehrenfels and beyond   84

Part II   The emergence of Gestalt theory, 1910–1920

7 Max Wertheimer, Kurt Koffka, and Wolfgang Köhler   103
8 Laying the conceptual and research foundations   118
9 Reconstructing perception and behavior   135
10 Insights and confirmations in animals: Köhler on Tenerife   148
11 The step to natural philosophy: *Die Physischen Gestalten*   168
12 Wertheimer in times of war and revolution: Science for the
   military and toward a new logic   187

Part III   The Berlin school in Weimar Germany

13 Establishing the Berlin school   203
14 Research styles and results   219
15 Theory's growth and limits: Development, open systems, self
   and society   247

v

16 Variations in theory and practice: Kurt Lewin, Adhemar Gelb,
    and Kurt Goldstein   263
17 The encounter with Weimar culture   284
18 The reception among German-speaking psychologists   307

Part IV Under Nazism and after: Survival and adaptation

19 Persecution, emigration, and Köhler's resistance in Berlin   325
20 Two students adapt: Wolfgang Metzger and Kurt Gottschaldt   342
21 Research, theory, and system: Continuity and change   362
22 The postwar years   382

Conclusion   405

Appendix 1: Tables   413
Appendix 2: Dissertations   419
List of unpublished sources   427
Notes   431
Index   501

Illustrations

Figures

1	Subjective grouping and contour (Schumann 1900)	page 92
2	Müller-Lyer figure, as employed by Benussi (1904)	94
3	Setups for Wertheimer's motion experiments (1912)	126
4	Setups for Koffka and Kenkel's motion experiments (1913)	140
5	"Genuine achievements" (a) and "imitations of chance" (b) (Köhler 1917)	153
6	Figure and ground (Rubin 1915/1921)	180
7	Uniform magnetic field distributed by an electric current in a straight conductor (Maxwell 1873, as cited in Köhler 1920)	183
8	Maximum–minimum distribution: String on soap film (from Ernst Mach, *The Science of Mechanics,* as cited in Köhler 1920)	185
9	Diagram of the Wertheimer-Hornbostel "Directional Listener" (Patented 1915)	189
10	Solving the external angles of a polygon (Wertheimer 1920)	193
11	Gestalt "laws" – self-organizing tendencies in vision (Wertheimer 1923)	225
12	Leveling and sharpening in figures drawn from memory (Wulf 1921)	227
13	Figural "belongingness" and brightness contrast (Benary 1924)	228
14	Spatial effects in two-dimensional drawings (Kopfermann 1930)	232
15	Phenomenal identity (Ternus 1926)	233
16	*Prägnanz* in motion perception (Metzger 1934)	234
17	Sander parallelogram (1928)	319

18 Diagramming heredity (Metzger 1937)   348
19 Phenomenal similarity (Goldmeier 1937)   367
20 Model of "interior psychophysics" (Metzger 1941, as revised
    in the 1960s)   377

Photos

1   Carl Stumpf in rectoral robes   36
2   Max Wertheimer circa 1912   104
3   Kurt Koffka   109
4   Wolfgang Köhler   112
5   Frankfurt University circa 1910   119
6   Max Wertheimer with modified Schumann tachistoscope   128
7a Apes stacking boxes   154
7b Apes stacking boxes   155
8   Sultan with double stick   157
9   Max Wertheimer, Albert Einstein, and Max Born.
    Photomontage 1918   197
10 Imperial Palace, Berlin   206
11 Experiment in progress on Berlin *Ganzfeld* setup   230
12 Erich von Hornbostel   237
13 Karl Duncker   239
14 Kurt Lewin   264
15 Adhemar Gelb   276
16 Experiment with brain-injured patient, ca. 1917   278
17 Rudolf Arnheim (photo by Wilfried Basse)   300
18 "Conversations in Germany" by Wolfgang Köhler – *Deutsche
    Allgemeine Zeitung,* 28 April 1933   329
19 Setup for Turhan's experiment on brightness and depth
    perception (1935). Photo from Metzger (1936)   365

Preface

Holistic thought is often portrayed as a wooly minded revolt against reason, an attempt to escape the constraints on both thought and action imposed by modern science. This book is a historical study of an attempt to advance holistic thought *within* natural science, one that emerged in Germany at a time supposedly dominated by a widespread revolt against scientistic categories in intellectual life. In the second decade of this century, Max Wertheimer, Wolfgang Köhler, and Kurt Koffka created what they believed to be both a new approach in psychology and the germ of a scientific worldview, which they called "Gestalt theory." By the 1920s they, their students, and their co-workers had applied this version of Gestalt principles in fields ranging from experimental psychology and theoretical biology to film theory. The influence of their work is still evident today, both inside and outside academic psychology.

Despite growing interest in holistic thought in recent years, there has been no historical monograph on the so-called Berlin school of Gestalt theory. Beyond filling that gap, the overall aim of the book is to illuminate for one important case the complex social and cultural relations of experimental psychology, and thus contribute to the ongoing debate in the history and philosophy of science on the role of society and culture in the development of scientific thought and research. The book also considers the work of students of the Berlin school who remained in Germany and attempted to continue Gestalt-theoretical research after the Nazi takeover of power, even though their teachers had emigrated to the United States. It thus examines the issue of continuity and change in one segment of German science in greatly altered political and social circumstances.[1]

The term "Gestalt psychology" has had multiple referents. It has been used as a designation for research on the experience of form or wholes in general, and hence also for the work of psychologists and philosophers on these topics who were not part of the Berlin school. Contemporaries sometimes employed the term when speaking of two other groups, the Graz or Austrian school of Alexius Meinong

ix

and Christian von Ehrenfels, and the Leipzig school headed by Felix Krueger, who called his doctrine "holistic psychology" (*Ganzheitspsychologie*). Both approaches will be discussed here briefly; so will the work of Kurt Lewin and his students, who worked closely with Wertheimer and Köhler in Berlin, as well as that of Fritz Heider, Adhemar Gelb, and Kurt Goldstein, who were associated more loosely with the Berlin school. The center of this text is nonetheless the work of Wertheimer, Köhler, Koffka, and their students. It was they who coined the term "Gestalt theory" for their ideas, and to whom others initially assigned the term. More important, they drew the most far-reaching and radical conclusions from their work on the Gestalt problem.

The group is called the Berlin school for several reasons. Though Wertheimer received his degree elsewhere, all three of Gestalt theory's founders as well as Kurt Lewin and Adhemar Gelb learned to be experimenting psychologists by studying with the philosopher and psychologist Carl Stumpf at the University of Berlin. Köhler succeeded his teacher Stumpf as professor of philosophy and director of the Psychological Institute in Berlin from 1923 to 1935; Wertheimer taught there from 1916 to 1929. Wertheimer first elaborated the basic propositions of Gestalt theory in Frankfurt between 1910 and 1914; he was a full professor there from 1929 until his dismissal in 1933. But it was during the time that he and Köhler were in Berlin together, in the 1920s, that Gestalt theory became a school in the ordinary sense of the word, achieved international prominence, and first intersected publicly with broader trends in German culture.

To grasp Gestalt theory's aims, history, and significance in German culture, this book employs approaches taken from history and sociology of science, general intellectual history, history of philosophy, and history of psychology. I have tried to combine these varied styles of scholarship without letting any one of them dominate the book. Rather, the text is a theoretically informed but not theory-driven narrative. The parts of the book are divided roughly chronologically, whereas the discussions within each part are arranged thematically. This makes it possible to address individually a variety of specific issues – such as the development of Gestalt theory and research over time, or its interactions with changing cultural milieux – while still giving shape to the whole. A more practical problem of interdisciplinary writing is the use of technical terms. It has not been possible to avoid these completely. Many of the issues at stake were technical, and so was the language of the actors; one scholar's jargon is another's ordinary language. Technical terms are generally defined where they appear first in the text.

One consequence of limiting the book's purview to Germany is that the opposition to behaviorism for which the Gestalt theorists were and are well known in North America is not central here. With the single exception of the German edition of Köhler's *Gestalt Psychology*, which first appeared in the United States, that motif is largely absent from the Gestalt theorists' German writings and those of their students until after 1945. Very few German psychologists took behavior-

ism seriously until then in any case. Another consequence of the book's geographical focus is that there is no discussion of Gestalt therapy here. Gestalt therapy was
founded in America in the 1950s, and had only tenuous connections with Gestalt
theory.

This is a historian's, not a psychologist's, history. Psychologists may feel a
certain tension brought about by a reluctance to weigh and measure the overall
contribution of that theory by current standards. My aim, instead, is to place
Gestalt theory in historical context – to discover how and why it developed as it
did. In this respect I have learned from the Gestalt theorists themselves, who
consistently argued that theories or other judgments about a state of affairs should
be based on a thorough understanding of the structure of the situation at a given
time. Nonetheless, it will be clear that I hold some of the criticisms that have been
made of Gestalt theory to be cogent. It will also be clear that I think its creators
asked important questions that still tug at the heart of psychological and philosophical thinking, and that they pursued their attempt to reconcile holism and
science with brilliance, boldness, and engagement.

The first two of the book's four parts are condensed, considerably revised, and
updated from my 1982 dissertation on the emergence of Gestalt theory. Parts III
and IV, which treat the subsequent history of Gestalt theory in Weimar, Nazi, and
postwar Germany, have been researched and written for this volume. Because this
book has taken so long to prepare, I have incurred more than the usual number of
debts. I can name here only some of those who have helped me along the way.
Research and writing were supported by National Science Foundation grant SES
85–11230, as well as a summer fellowship, a developmental assignment, and a
Faculty Scholar Award from the University of Iowa. The facilities, the delightful
combination of solitude and interdisciplinary conversation, and the friendly professionalism of Jay Semel and Lorna Olson at the University of Iowa Center for
Advanced Studies were indispensable. Mary Strottman and Rex Strottman helped
with word processing and printing out the manuscript; a subvention from the
office of the Vice President for Research of the University of Iowa helped to cover
the costs of publishing it. Alex Holzman, Helen Wheeler, and Cynthia Benn of
Cambridge University Press saw the book through the various stages of acquisi-
tion and production with extraordinary professionalism.

The staffs of the archives named in the list of unpublished sources at the end of
the book all provided expert assistance. Special thanks also go to the staff at the
City and University Library in Frankfurt am Main, to the staff of the Psychology
Library at the Free University of Berlin, to Eckhardt Henning and his co-workers
at the Library and Archive of the Max Planck Society in Berlin, and last but not
least to the staffs of the Psychology Library and the Photo Service at the University of Iowa. Molly Harrower gave generously of her time and knowledge in
discussions about Kurt Koffka. Michael Wertheimer allowed me to work with the
papers of his father, Max Wertheimer, and helped in many other ways. Students of

the Gestalt theorists, whose names appear in the list of unpublished sources, shared their memories with me.

Historians Fritz Ringer and Allan Megill, psychologists Kurt Danziger, David Leary, and Michael Wertheimer, and philosophers Evan Fales and Barry Smith read all or substantial portions of the manuscript. Berlin psychologists Ulfried Geuter, Siegfried Jaeger, Lothar Sprung, Helga Sprung, and Irmingard Staeuble have been discussion partners and friends. Carl-Friedrich Graumann and Wolf Lepenies helped put me in touch with historically interested psychologists and sociologists of science in Germany. My advisor and mentor, Donald Fleming, gave all manner of support to this project, and to me, long after it ceased to be a dissertation and I was no longer his student. These people, and others too numerous to name here, offered helpful suggestions and criticisms of many kinds, but responsibility for the text as it stands remains, of course, my own. Except where otherwise indicated, all translations from the German are also my own.

To Christiane Hartnack, my wife and most rigorous critic: thanks for patience, gentle but firm pressure, and loving support.

Acknowledgment is made to the following for permission to quote from copyright material: Emory University, for excerpts from my article, "Academic Politics in the History of Science: Experimental Psychology in Germany, 1871–1941," *Central European History,* 13 (1980), 255–286; Springer-Verlag, for excerpts in translation from my article, "Ein Institut und eine Zeitschrift. Zur Geschichte des Berliner Psychologischen Instituts und der Zeitschrift 'Psychologische Forschung' vor und nach 1933," in Carl-Friedrich Graumann (ed.), *Psychologie im Nationalsozialismus* (Berlin, Heidelberg, New York, Tokyo, 1985), 113–137; The Albert Einstein Archives, the Hebrew University of Jerusalem, for excerpts from letters by Albert Einstein; Professor Michael Wertheimer, for quotations from the papers of Max Wertheimer; and Molly Harrower, Ph.D., for quotations from the papers of Kurt Koffka.

Introduction

What was Gestalt psychology? Max Wertheimer, Wolfgang Köhler, and Kurt Koffka did not claim that the whole is *more* than the sum of its parts. Rather, they maintained, there are experienced objects and relationships that are *fundamentally different* from collections of sensations, parts, or pieces, or "and-sums," as Wertheimer called them. The Gestalt theorists opposed the assumption that sensory "elements" are the basic constituents of mental life then characteristic of psychological theory and research in Germany and elsewhere. In 1890, one of Wertheimer's teachers, Christian von Ehrenfels, had already tried to reform elementism in psychology by introducing the notion of additional "Gestalt qualities" given in and with sensory elements. As paradigmatic examples he cited melodies, which sound the same in any key because they have such qualities. Going beyond von Ehrenfels, the Gestalt theorists asserted that dynamic structures in experience *determine* what will be wholes and parts, figure and background, in particular situations.[1]

Going further still, Wertheimer introduced the principle of *Prägnanz* in 1914, according to which experienced structures spontaneously take on the "best" or simplest arrangement possible in given conditions. In 1920, Köhler stated that the brain events underlying perception follow the same dynamic, self-organizing principle that Wertheimer had enunciated for perception. This he likened to the tendency of physical systems to approach maximum order, or equilibrium, with minimum expenditure of energy. Since dynamic self-organizing processes occur in both inorganic and organic nature, he argued, they are not structures of "the understanding" imposed on experience, but are properties of both mind and nature. In opposition to what Köhler called "machine theory," that is, to technological conceptions of science, life, and mind that equated knowledge of nature with its effective manipulation and control, the Gestalt theorists attempted to introduce an aesthetic dimension of inherent order, meaning, and simplicity into the evaluation of scientific theories, and into the fabric of experience and nature itself.

1

Gestalt psychology has had a major impact on research in perception and the psychology of art, and has also contributed significantly to studies on problem solving and thinking. In addition, Köhler was one of the first to introduce the distinction between "closed" and "open" systems into theoretical biology. Most fundamental, however, were the philosophical implications of Gestalt theory. The Gestalt theorists asserted the primacy of perception over sensations in the constitution of consciousness, and advanced a conception of the subject as involved in, rather than separated from the world – ideas that had an important impact on phenomenology and existentialism.[2] In terms more familiar to analytical philosophers, they challenged the empiricist assumption that "sense data" are the "atomic facts" of experience by arguing that there are no such unambiguous "data." Rather, they maintained, the objects we perceive are always located in what would now be called self-organizing systems – constantly changing dynamic contexts or situations, of which our phenomenal selves, too, are parts. Against both Kant and the constructivist mode of thought dominant in cognitive science then and now, they claimed that form and order are not imposed on otherwise chaotic sensory "material," or constructed on a foundation of hypothetical sensory information according to fixed cognitive schemata or logical rules. According to the Berlin school, Gestalten are experienced because they *are* such structures, not because they have them.[3] One of the Gestalt theorists' favorite words was *sachlich,* meaning objective; theirs was a quest for objective order that lies not behind, but *within* the flux of experience.

Why did this school of thought emerge when it did, and why did contemporaries consider it important? Psychologist Mary Henle, a student and for many years a close colleague of the Gestalt theorists, suggests that psychologists in Germany were "deeply involved" in what was then called "the crisis of science." To many, science seemed incapable of dealing with the most significant human problems. Rather than abandoning natural science, "Wertheimer and Köhler proposed that the difficulty was not with science itself, but with the current conception of natural science among psychologists."[4] They and Koffka offered a radically reformed conception of scientific psychology that would do justice to the intrinsic meaning and value in human experience and thus overcome the divide between the natural and the human sciences. Historian Fritz Ringer argues that talk of a "crisis of science" was part of a larger problem of modernity faced by many educated Germans at the turn of the century and exacerbated after the First World War. The perception that traditional values were in crisis was widespread both in Europe and the United States at that time. Gestalt theory, in Ringer's view, was a "modernist" attempt to bring together the demands of science and the hopes of humanism, one of very few to do so through the medium of holistic thought.[5] In a similar vein, Martin Leichtman argues that the Gestalt theorists were part of the "revolt against positivism" in European thought after 1890, and advanced a "liberal humanistic" or "liberal democratic world view" rooted in Enlightenment values that was opposed to the prevailing conservatism of the German academic elite.[6]

This book combines and expands these interpretations in an attempt to develop a synthetic account of Gestalt psychology's history in Germany. Taking Ringer's view still further, and applying it to the vicissitudes of both the Gestalt category and the careers of the Berlin school's remaining adherents under Nazism and thereafter, it challenges stereotypical dichotomies between modern and anti-modern, rational and antirational, democratic or progressive and conservative or proto-Nazi thinking that have been matters of course in German intellectual history for decades. Jeffrey Herf's book, *Reactionary Modernism,* and the recent outpouring of literature on eugenics or "race hygiene" in Germany have shown how particular versions of biomedical science and conceptions of technology could interact with supposedly antimodern cultural nationalism and racism before 1933, and either resonate with or become part of Nazi ideology and practice thereafter.[7] This study contributes to that discussion by showing how one incarnation of holistic thought had more complex, both supportive and refractory relationships with Weimar, Nazi, and postwar German cultures than conventional dualisms would predict. Like organicist thinking, Gestalt discourse was ideologically *multivalent,* or heterogeneous. Though that category was widely associated with political conservatism and later with Nazism, the history of the Berlin school indicates that such linkages were contested and contingent rather than necessary.[8]

Further, the book shows that Gestalt theory was not only, or not simply, a revolt against positivism. The Gestalt theorists vigorously opposed mechanistic assumptions about perception, as well as sensationist, elementist, empiricist, and associationist conceptions of consciousness. In all these respects they challenged scientific beliefs generally considered central to positivism. But they remained firmly committed to rigorous natural-scientific method in psychology and to the hope of creating a science-based philosophy of mind. In their efforts to achieve this aim they did not cast aside, but rather reconstructed concepts, theoretical models, and research practices encountered in the laboratory of their common teacher, Berlin philosopher and psychologist Carl Stumpf. They also drew upon conceptual resources provided by contemporary developments in philosophy of mind, theoretical biology, physics, physical chemistry, and sensory physiology, only some of which were antipositivistic. Indeed, their chief intellectual alliance in the Weimar period outside psychology was to the Society for Scientific Philosophy founded by physicist and logical empiricist philosopher Hans Reichenbach.

Finally, this study indicates the impact of both political and practical changes on Gestalt theory under Nazism and in the postwar period. The history of psychology as a science and that of the psychological profession in this century are inseparable. In Germany, formal professionalization came during the Nazi period, driven mainly by demand for personality assessment and diagnostics as part of officer selection in the Wehrmacht as well as in the German Labor Front. Psychologists were in a position to offer such services not because of their subservience to Nazi ideology – though some engaged in that – but due to a shift of emphasis that had already begun in the Weimar period from questions of philosophical concern, such as the problem of knowledge, to more immediately applicable though still culturally resonant topics like character and personality.[9] That shift had powerful feedback effects on the reception of conceptions of science founded on aesthetic rather than technological categories, like Gestalt theory. Considering this aspect of Gestalt theory's history thus exemplifies some of the problems and tensions of a profound change in the status and self-concept of many educated Germans, including scientists, from bearers of culture to members of professional society.[10]

Redefining the contexts of science

The literature on the Berlin school of Gestalt theory is enormous.[11] A chapter on the subject is de rigeur in textbooks for "history and systems" or "systems and theories" courses in psychology. Few such accounts consider its emergence, development, and reception over time, or set any of these aspects in a broader context. Instead, they generally present systematic summaries of what Gestalt theory allegedly stood for and assess its contributions in the light of current thinking. This may suffice for the intended pedagogical purposes, but such discussions foreclose the possibility of asking how or why Gestalt theory changed over time.[12] Many standard accounts also contain significant errors. These are not merely misunderstandings, but examples of how taking current presuppositions in a science for granted can distort, marginalize, or domesticate alternative approaches. A prime example is the routine assertion that Gestalt theory is a "nativist" position, that is to say, one that discounted the role of learning in perception. Parallel to this is the common claim that the Gestalt theorists attributed perceived wholes and structures to inherited structures in the brain. The operative assumption seems to be that a theory that challenges certain empiricist claims must therefore be a descendent of the allegedly "Kantian," more accurately Cartesian, doctrine of innate ideas.[13] Students of Gestalt theory have long pointed out that its aim was to overcome such dualistic thinking by discovering not whether perception is primarily learned or not, but just what it is that people perceive, and what the experiences of perception and learning are.[14]

The point of view of this study is that of the contextualist scholarship in history of science and history of psychology that has been emerging for the past twenty years. This history is no longer one of linear progress, but of parallel, differing, and diverging lines of development. It is less a story of the discovery of more and more truths about (human) nature with ever greater technical precision, than of the construction and reconstruction of social, technological, and cultural artifacts. And it is less a tale of insights by individuals, though they are very much in evidence, than of knowledge generated by agreements – or arguments – in scientific communities, according to discursive norms, methodological rules, and societal goals that change with time and place.[15] Whether the results of such processes also correspond to "objective" knowledge is a question for philosophers of science, though even the answers to such questions – like scientists' own arguments for the objectivity of their knowledge – are not immune from historical change.[16]

This book reinforces contextualist accounts of science by locating experimental psychology and Gestalt theory in the changing social and institutional milieux of German academic and cultural life from the 1890s to the 1960s. It also places Gestalt theory in a complex intellectual setting of competing claims about science and mind, which changed over time as well. To construct such a multilevel interpretation, the account proceeds from the assumption that scientists, like other professionals, belong simultaneously to multiple sociocultural groupings, each of which defines itself by constructing its own discursive and practical universe. Relevant for Gestalt psychology are three such groupings, corresponding to their own multiple memberships: the working group in a single laboratory or the scientific school encompassing one or more such groups; the discipline or subdiscipline, encompassing parallel or rival groups of scientists competing for position in both epistemic or conceptual and institutional space; and the broader cultures and societies in which disciplines and laboratories in turn locate themselves.[17]

In science studies, these multiple contexts of social life and discourse are often portrayed as mutually constraining. Thus, institutional situations are said to constrain problem choice and method selection; methodological constraints are said to feed back on problem choice; and broader cultural or political contexts are said to determine both problem choice and the interpretation of results. These social settings are portrayed differently here – not as rigid constraints, but as frames or boundary conditions within which a range of motion occurs, as challenges to which a variety of responses is not only possible but inevitable.[18] Most important, this book argues that it is insufficient to consider only one of these social realms at a time – to study only the politics of science on the one hand, or the histories of disciplines or laboratories on the other. Scientists live simultaneously in *all* of these social realms. Focusing on a single important group of scientists, such as the Gestalt theorists, is a way of showing how their multiple social identities and their discourses interact.

This study devotes considerable attention to German psychological laboratories and the scientific schools that arose within them. In doing so it confronts some of the issues discussed in the growing literature on scientific practice.[19] This is necessary because institutional frameworks set the conditions for deciding what scholarship and science are. Scientific institutions are the loci of socialization into particular sets of research practices which then become privileged modes of meaning-generation – "forms of life," as Timothy Lenoir calls them, following

Wittgenstein.[20] As Helen Longino writes, "One 'enters into a world' and learns how to live in that world from those who already live there."[21] In the hierarchically organized world of German academic science at the turn of the century, would-be scientists encountered these practices and their conceptual underpinnings during their training not as alternatives from which they could choose, but as embodiments of science per se in a given discipline. They acquired in such settings not only a set of tools for problem-solving, but assumptions and methodological rules that gave meaning to and justified the use of these tools. Due to the amount of time and effort invested and the personal relationships created in such apprenticeships, the commitments involved often carried emotional as well as intellectual weight for young scientists.[22] In such circumstances, responses to problems perceived as threatening to normal science are likely to be created and presented as innovations that restructure thinking and practices within existing institutions, rather than overthrowing them. Gestalt theory began as such a revolt from within.

The members of subcultures that emerge within laboratories generally try to advance acceptance of their knowledge in disciplinary communities. However, limiting consideration to a single discipline or to its institutions alone will not suffice to grasp the complexity of the situation in which Gestalt theory arose. Wolf Lepenies has pointed out that emerging disciplines like experimental psychology and sociology faced two difficult problems simultaneously: establishing the autonomy of their subject matter as well as the scientific legitimacy of their methods in relation, often in competition, with those of neighboring disciplines; and specifying the relations of these claims to the common sense – the explicit or implicitly shared discursive norms – of a culture at a given time.[23] Precisely because of its ambiguous status as both a would-be natural science and a subfield of philosophy, and because many experimenting psychologists had training in both fields, experimental psychology in Germany interacted with theoretical assumptions and research practices in sensory physiology, neurophysiology, developmental biology, and philosophy of mind, all of which had subtle cultural or ideological resonances. For the same reason, innovators like the Gestalt psychologists had a rich fund of conceptual and methodological resources – and culturally resonant metaphors – from which to draw.

Relations within or among disciplinary subcultures and their broader sociocultural settings cannot be fully separated from one another. Pierre Bourdieu has characterized academic disciplines as systems of agents acting as a more or less cohesive subfield located within but in a refractive relationship to larger fields defined by class and cultural identity. To the extent that they become social microsystems that organize and restrict inquiry, they achieve relative autonomy, which means in Bourdieu's terms that they become centers of power and authority in their own right. But this happens only to the degree to which societies value what they do, for reasons that may or may not have much to do with the motives of the practitioners. Thus, neither universities nor their constituent disciplines are separate from the discursive fields constituted by shared ideologies, cultural presuppositions, and organizing metaphors, or from the societal matrixes in which these are enacted.[24]

Gestalt theory's location in this sociocultural field changed over time. At first, the Gestalt theorists neither understood their innovation nor cast their discourse in terms of the alleged "crisis of science." Instead they couched their position in language specific to the scientific community into which they had been initiated. But Wertheimer's use of the culturally resonant term "Gestalt" effectively linked the disciplinary and sociocultural realms. The pressure-cooker atmosphere of Weimar culture heightened both the intensity and the ideological resonance of their claims. It also shaped the reception of their ideas, in which both ideology and intradisciplinary competition played important roles. From the 1930s onward, however, the development and reception of Gestalt theory in Germany depended on three other factors – the forced emigration of Wertheimer, Köhler, and many of their students after 1933, the professionalization of psychology under Nazism, and the restructuring of scientific and professional life under occupation and in the two German states after 1945. All of these developments posed challenges to the students of Gestalt theory who remained in Germany that proved difficult to overcome.

The arguments in outline

Gestalt theory originated within a discipline constructing itself first as a science and then as a science-based profession in German culture. Its history was therefore shaped as much by that of Germany as a whole and of psychology in that context as by purely conceptual considerations. The book's four parts try to capture that interaction by considering all of the social settings discussed above within a broadly chronological framework, in order to bring out the dimensions of continuity and change in each. The first two parts discuss the emergence of Gestalt theory to 1920. Parts III and IV describe the establishment of Gestalt theory as a scientific school in the 1920s and trace its subsequent elaboration and reception.

Part I examines the social, institutional, and intellectual situations within which Gestalt theory emerged. A common term for the process leading to the formation of new disciplines is "differentiation"; but this makes the process seem to be a natural product of knowledge growth, rather than a site of bitter intellectual and academic-political struggles.[25] Even though a community of experimenting psychologists was fully formed by 1910, academic psychology did not "differentiate" from philosophy in Germany until 1941. That ambiguous disciplinary location was of central importance for the history of Gestalt psychology. The role of philosophy in teacher training was the entry point for psychology into German academic life. Knowledge of "empirical psychology" was specifically mentioned as part of the philosophy or "general culture" requirement in the statutes for Prussian state teachers' examinations. Whether such knowledge should be acquired by experimental methods was not specified. Between 1890 and 1910, a community of experimenting psychologists formed, with a scientific society, journals, and laboratories. Most of the leading members of that community continued to be professors of philosophy, and they believed that their research was relevant to philosophical problems, especially in logic and the theory of knowledge. For this reason, among others, their work focused on topics in sensory psychology and cognition. Seen in this light, experimental psychology's challenge was to incorporate innovative teaching and research methods while preserving the field's traditional function in the university system as part of the philosophical propaedeutic.

The Gestalt theorists were trained in rigorous experimental method and acquainted with the broader issues at stake by philosopher-scientists such as Christian von Ehrenfels in Prague, Oswald Külpe in Würzburg, and above all Carl Stumpf in Berlin. These professors accepted experimental psychology's double identity as a natural science and a part of philosophy. Indeed, though Külpe began as a positivist, all three of them came to see their work as a way of establishing an alternative philosophical standpoint between Neo-Kantian idealism and positivism. But other leading philosophers of the time vehemently denied that empirical research could contribute to philosophy at all, let alone address the truly pressing social and moral issues of the time. Shortly after Koffka, Köhler, and Wertheimer began their careers, the controversy took on a threatening dimension, when 107 university teachers signed a petition against appointing additional experimenting psychologists to chairs of philosophy.

Complicating this tense institutional situation was the sheer difficulty of constructing psychology as an autonomous intellectual and research domain at a time of rapid industrialization, when concepts of science and of mind were in flux. Legitimacy for psychology as a natural science in Germany came at first by adapting methods and mechanical modes of explanation from physiology and physics to the study of sensation, perception, and memory. The epistemic space was a shared one at first; for the parallel dualisms of sensation and perception, peripheral and central events in the nervous system, and physiological and psychological processes seemed congenial to both physiologists and psychologists. But tension grew due to the problems of extending to perception the assumptions applied to sensation by Hermann von Helmholtz. This created a conflict between the normative demand for mechanical explanation on the one hand, and the call for construction of an autonomous psychical realm with its own laws on the other. For some the procedures and theoretical models proposed by physiologist Ewald Hering offered an attractive alternative.

At the same time, new theories of mind challenged the elementistic assumptions about consciousness shared by most experimenting psychologists. Philosophers both inside and outside Germany who sought a middle way between ideal-ism and positivism, such as William James, Henri Bergson, Wilhelm Dilthey, and Edmund Husserl, rejected sensationalistic and atomistic concepts of conscious-ness and described mind as process, a view also taken in psychology by Oswald Külpe's Würzburg school. The controversy over the status of the Gestalt category, introduced from aesthetics into psychology by Mach and Ehrenfels, impinged on all of these issues. The Gestalt problem became linked with the mechanism–vitalism controversy when neovitalist Hans Driesch and psychovitalist Erich Becher maintained that the "psychological reality" of perceived form justified the notion of an independent "psychical causality." Related controversies in neu-rophysiology began when Johannes von Kries and Becher criticized the prevailing view of nerve transmission by isolated conducting pathways as inadequate to account for the perception of ordinary, complex objects. Thus, new conceptual and research issues of obvious philosophical, scientific, and wider cultural signifi-cance had come to the fore, but the descriptive categories and philosophies of science then employed by experimenting psychologists seemed incapable of deal-ing with them satisfactorily. Such difficulties underlined philosophers' claims that experimental psychology could not solve philosophical problems or believably articulate the higher values appropriate to bearers of culture.

Part II shows how Koffka, Köhler, and Wertheimer responded to this complex challenge by radically reconstructing psychology's conceptual framework be-tween 1910 and 1920. The core of that reconstruction was the fundamental revi-sion of the Gestalt concept explained earlier, first articulated by Wertheimer in 1912 and 1913. In his famous 1912 paper on the seeing of motion, he provided, with the so-called *phi* phenomenon – perception of motion without a moving object – what he took to be experimental evidence for the existence of essentially dynamic mental realities that cannot be composed of, or built up from, elements. In the same paper he conjectured that there are "structured whole processes" in the brain corresponding to these psychical events.

Ironically, in view of Thomas Kuhn's use of the term "Gestalt switch" in his theory of scientific revolutions, the creation of Gestalt theory involved not only the sudden change of perspective proposed by Wertheimer, but also an elaborate set of conceptual transformations that took almost ten years to work through.[26] First, Köhler disputed Helmholtz's assumption, shared by Stumpf, that sensations are strictly determined by physical stimuli. Rejecting this so-called "constancy hypothesis" opened the way to making perception and its objects – things in relation to one another – rather than sensations the primary foci of psychological research. Koffka then extended the Gestalt category from perception to action, claiming that behavior is not a bundle of reflexes connected by association, but an achievement released by the interaction of structured environmental events with organisms in particular states. This brought Gestalt theory close to American functionalist psychology and pragmatist philosophy. However, instead of making perception the product of action, as John Dewey and others did, Gestalt theory placed them on the same plane, saying that both conformed to the same structural laws. Köhler's research on animal problem solving on the island of Tenerife from 1914 to 1918 supported this extension of the Gestalt concept to behavior. In further studies, he attempted to show that animals can perceive relations directly, and claimed that such "structural functions" are therefore a property of living matter, not of mind alone.

This set the stage for Köhler's declaration, in 1920, that inanimate nature also contains Gestalten, physical systems that cannot be described as simple summations of isolated events. That step introduced Gestalt theory as a philosophy of nature, as well as a theory of knowledge, perception, and behavior. In the same year, Wertheimer extended his new epistemology to thought, a step for which he had prepared the way in a 1912 paper on the number concepts of so-called primitive peoples. Traditional logic might be suitable for ordering what is already known, he argued, but it is an inadequate guide to what happens in innovative thinking. Such thinking is characterized by a fundamental "recentering" or restructuring of problem situations – rather like the "Gestalt switches" later described by Kuhn. Research on such transformations, he suggested, might lead to a new "Gestalt logic." The Gestalt theorists accepted their teachers' belief that psychological research could contribute to philosophy. But they had seen during their training that dualistic presuppositions about mind were inadequate guides to psychological reality. They accepted the institutional situation in which they worked, and tried to resolve the intellectual and practical problems involved with fundamental innovations.

Part III analyzes the establishment, development, and reception of the new approach from 1920 to 1933. The Gestalt theorists were rewarded for their boldness with career advancement. This, along with the journal *Psychologische Forschung,* founded in 1921, assured the institutional anchorage of their new approach. The Gestalt theorists and their students employed a variety of research styles, consistent with their own rather different personalities. Wertheimer was both a philosopher-scientist and a Jewish intellectual loosely associated, like Franz Kafka, with the Prague circle. His 1923 paper laying down "Gestalt laws" of perceptual grouping presented an open-ended research program – to discover the principles of perceptual organization in both its static and dynamic aspects. Many Berlin school students tried to unpack Wertheimer's suggestive formulations or apply them to already existing research in form, color, and depth perception, creating what Dudley Shapere has called a conceptual and research domain by metaphorical and analogical extension.[27] Köhler, the son of a Prussian Gymnasium director, had studied both philosophy and physics, as well as experimental psychology. His study of successive comparison, also published in 1923, served as an exemplar for employing perceptual research to develop and test hypotheses about underlying brain events.

The terms "Gestalt," "whole," "field," and "system" thus functioned as what Kurt Danziger has called "generative metaphors," organizing research and acquiring increasingly rich new meanings within particular laboratory settings while simultaneously connecting them with wider scientific and cultural discursive fields. But these metaphors were enacted as components of Kuhnian exemplars – unities of theory and practice that embodied science per se for young experimenters in Berlin, Giessen, or Frankfurt.[28] Precisely this multiplicity of experimental designs in the service of a single group of generative metaphors accounts for the productivity of the Gestalt approach in the 1920s. The coherence of the Berlin school's open-ended research program – and of the Berlin school as a group – thus came both from the fundamental theoretical commitment to the Gestalt concept as reformulated by Wertheimer, and from shared allegiance to a common research style.[29] The Gestalt theorists and their students continued the experimenter–subject relationship established in Wilhelm Wundt's Leipzig, even as they revised the conceptual foundations of Wundt's psychology. Compared with psychological research practices in the United States at the time, Berlin-school research defended a conception of psychology as a science of subjectivity rather than of behavior and social control. This was consistent with the elite social makeup of the student population in Berlin school laboratories.[30]

Gestalt theory's principles continued to evolve in this period, primarily by extending Wertheimer's and Köhler's metaphors, but also by means of new conceptual transformations. First chronologically was Koffka's attempt in 1921 to lay the foundations for a Gestalt theory of psychical development that would incorporate both evolutionary history and the role of society and culture. Next and most original was Köhler's extension of his physical Gestalt concept to biology in 1924. In response to a challenge by neovitalist Hans Driesch, he then elaborated the concept of self-regulating, "open" systems, which he distinguished from those described by analytical mechanics, thus helping to introduce the systems concept into theoretical biology. Finally, building on earlier remarks by Wertheimer, Köhler tried to extend Gestalt theory to emotion and to social psychology by arguing that our perception of other people and their emotional expressions is as immediate as any experience we have. Provocative as these metaphorical and theoretical extensions were, it was not clear whether it would be possible to develop the same rigorous research agendas for these issues as had been done for perception, or how Köhler's conception of brain action could account for the enormous range of phenomena Gestalt theory now encompassed.

In related lines of theory and research, scientists associated with the Berlin school indicated the variety that was possible even in the orbit closest to that of the inner circle. Kurt Lewin, who worked closely with Köhler and Wertheimer in Berlin, developed a pluralistic theory of science that allowed him to apply metaphors from physics, biology, and Gestalt theory to the psychology of action and emotion, with the aim of humanizing societal practice in factories and schools.

Adhemar Gelb and Kurt Goldstein's work with brain-damaged soldiers supported Wertheimer's claim that dynamic self-organizing structures were primary in experience; for that was what was lacking in pathological cases. Goldstein, however, developed an "organismic" approach that differed in important respects from Gestalt theory.

The Gestalt theorists did not hesitate to bring out the implications of their thinking for the ideological debates of the Weimar period. In that tense era, calls for a comprehensive worldview were ubiquitous, and conservative intellectuals and academics often employed holistic vocabulary in their efforts to counter the perceived threats of urbanization, industrialization, and democratization. At the same time, proponents of so-called humanistic psychology (*geisteswissenschaftliche Psychologie*) and popular outsiders like Ludwig Klages presented alternative approaches to psychological research and professional practice based on intuitive diagnoses of character types rather than experimental methods. In this situation, Wertheimer and Köhler employed holistic terminology to support a worldview that would overcome the opposition of mind and nature without sacrificing natural science or succumbing to the popular irrationalism of the day. Wertheimer's student Rudolf Arnheim's theory of film and his remarks on the politics of art in the influential journal, *Die Weltbühne,* expressed a similarly radical middle position in aesthetics. He opposed easy equations of the "new objectivity" or radical modernist aesthetics with progressive politics, saying that only films that realized the medium's full expressiveness by accepting the medium's objective constraints, its laws of form, would have the desired social impact.

The reception of Gestalt theory among German-speaking psychologists also reflected the general cultural situation, but with complications posed by intra-disciplinary competition. That the Gestalt theorists were in tune with the times was clear from the many efforts to show that their holism did not go far enough. Felix Krueger's so-called Leipzig school of holistic, or *Ganzheitspsychologie,* for example, claimed that the Berlin school paid insufficient attention to the role of feeling and will in the constitution of experience. William Stern and his students focused on the Gestalt theorists' alleged underemphasis on the role of the person. Karl Bühler complained that the lack of attention to the category of language in Gestalt theory made it inapplicable to a general theory of culture. All of these theoretical objections also pointed to central components of the ideology or self-concept of educated middle-class Germans that the Berlin school had apparently not addressed satisfactorily. The Gestalt theorists' failure to address practical issues – with the conspicuous exception of Lewin – and their refusal to engage in disciplinary politics assured their cohesiveness as a school, but also contributed to their relative isolation.

Part IV treats Gestalt theory's German trajectory from 1933 to the postwar period, ending with the death of Köhler in 1967. The Nazi takeover of power suddenly reshaped the institutional situation of Gestalt theory. Wertheimer, who shared the humanistic, progressive convictions of his friend Albert Einstein, was one of the first academics dismissed in 1933. Lewin departed for the United States in the same year before meeting the same fate, and in 1935 Köhler, the only non-Jewish founder of Gestalt theory, resigned in protest against the political firing of his assistants. New evidence shows that student support and pressure from abroad aided Köhler's struggle to maintain the integrity of the Berlin institute against denunciations by Nazi students and erstwhile colleagues. In the end it proved vain to hope, as some students and perhaps Köhler also did, that it would be possible to create a special sphere where science might be pursued without outside interference.

But it was easier to dismantle previously productive centers of psychological research than to replace them with Nazified versions. Contrary to widespread impressions, important students of Gestalt theory remained in Germany during the Nazi period. In Frankfurt, Wolfgang Metzger continued the research he had begun in Berlin, culminating in a monograph on vision (1936) and a broader synthesis of theoretical psychology (1941), both based on Gestalt principles. This work earned him a professorship in Münster in 1942. Metzger also tried to show that Gestalt theory was compatible with National Socialism – a position he reversed later. Rather different was the story of Kurt Gottschaldt, a former student and assistant of Köhler, who combined ideas and methods of Gestalt theory with then-current theories of personality in research on the psychological development of twins at the Kaiser Wilhelm Institute for Anthropology, Human Heredity, and Eugenics from 1935 to 1945. Because he was working on an issue central to Nazism's aims, he had no difficulty citing the work of his former teachers. However, his published results did not fully support the claim that psychological abilities and functions were inherited; and his suggestion that additional research and better methodologies were needed to unravel the complex interaction of heredity and environment in psychological development did not support the aims of Nazi eugenicists seeking to breed a "master race" overnight.

In contrast to ideological and pragmatic adaptations of other German psychologists in the Nazi era, the research conducted under Metzger in Frankfurt showed remarkable continuity, and some of it substantially advanced Gestalt theory. Most notable were studies on the role of Gestalt factors in depth perception, Erich Goldmeier's 1937 dissertation on similarity, and Edwin Rausch's work of the same year on summative and nonsummative categories. This work was not imbued with obviously ideological vocabulary, though later work under Metzger did respond to criticism from other psychologists. But neither Gestalt theory nor experimental psychology in general was dominant in Germany at the time. Research in personality or "characterology" was more in line with the discipline's emerging practical functions – officer selection for the Wehrmacht, occupational assessment for the German Labor Front, and child diagnosis for the National Socialist Welfare organization.

In the postwar period, Metzger, Rausch, and Gottschaldt reconstructed Gestalt theory's institutional bases in Münster, Frankfurt, and the Humboldt University in East Berlin, respectively. There they fostered a subtle blend of continuing studies on perception and cognition begun in the 1920s and new themes. These included Metzger's work on the problem of "creative freedom" in pedagogy and Gottschaldt's studies on juvenile delinquency, the developmental psychology of twins, and the social psychology of small groups. Prominent in all of this, as elsewhere in postwar German psychology, was an effort to preserve institutional and intellectual continuity despite the growing impact of American research styles in West Germany and of Soviet viewpoints in East Germany. That effort had at best mixed results. Only a few of the students trained by Metzger, Rausch, and Gottschaldt could properly be called Gestalt psychologists.

The dominant vocabulary in cognitive science today is not that of Gestalt theory. Indeed, current talk of information being processed by minds or brains working like computers sounds rather like the thinking against which the Gestalt theorists originally rebelled. But their questions and concepts continue to stimulate research, and to function as productive irritants, in that field, as well as in theoretical biology and philosophy. So does the fact that what we see are things in relation to one another, not pieces of information. The gap between the discourse of cognitive science and that of lived experience still yawns. That is why the odyssey of Gestalt psychology is of continuing interest.

Part I

The social and intellectual settings

1

The academic environment and the establishment of experimental psychology

The institutionalization of experimental psychology in nineteenth-century Germany has long been portrayed as a continuous success story. Standard histories of the field refer to an all-encompassing Zeitgeist, or spirit of the age, that was somehow favorable to the rise of laboratory research, and thus to the establishment of psychological laboratories.[1] Sociologists of science speak of the initial institutionalization of a scientist's role in Germany in chemistry and physiology, and then describe later developments in biology, physics, and engineering as imitations of, or as the results of intellectual differentiation or personnel transfer from, the leading disciplines.[2] At first glance, the founding in the 1870s of the world's first continuously operating psychological laboratory by Wilhelm Wundt, a physiologist turned philosopher, seems to fit this model. In an often-cited study, Joseph Ben David and Randall Collins assert that this instance of "role hybridization," as they call it, marked experimental psychology's "take-off into sustained growth" as a scientific discipline.[3]

Whatever we may think of historicisms like Zeitgeist or sociologisms like "role hybridization," experimental psychology in Germany had no "take-off into sustained growth." By Wundt's own count there were "only four or five very modestly equipped psychological laboratories" in Prussia, the largest German state, by 1913, and some of these were "hardly more than collections of instruments for demonstration purposes."[4] The director of one of these laboratories, Georg Elias Müller, complained in his address as chair of the German Society for Experimental Psychology in 1914 that only about 140 Marks from his modest annual budget of 1,200 Marks were available to buy apparatus. "Even a fully uninformed layman recognizes," he concluded, "that with such a complete dependence upon uncertain or unexpected supplementary funds from the state or from private sources, the normal, regulated development and activity of a scientific institute is impossible."[5]

Thirty-five years after the founding of Wundt's institute, the combined budgets

17

of the seven best-funded psychological laboratories for the academic year 1913–1914, 17,600 Marks, amounted to somewhat more than one-fourth of the figure for the physiological institute at the University of Berlin alone in the same year – 63,1116 Marks (see Appendix 1, Table 2).[6] The hybrid status of experimental psychology had not brought it the level of sustained support enjoyed by other fields of experimental research. When the scientists who later created Gestalt theory decided to devote their lives to experimental psychology in the first decade of this century, they thus aligned themselves not with an already established, institutionally independent discipline, but with a tenuously supported subspecialty of philosophy. To understand what this meant for their identity as scientists in German culture, it is necessary to describe the historical situation of experimental psychology in Wilhelmian Germany in greater detail.

The academic environment

The twenty-one universities in the German Empire retained a nearly complete monopoly of the means of basic research there in the nineteenth century.[7] Their primary function, however, was the training of Germany's educated elite. Both the classification of disciplines and the state support awarded them depended on their promise to fulfill some aspect of this function. The methods of philosophy and classical philology, not those of natural science, were the original models for the German concept of *Wissenschaft* (systematic scholarship), and thus for the professionalization of academic teaching and research in Germany. Philosophy had already acquired a secure position in this general scheme in 1810, when the Prussian government required candidates for state teachers' examinations to attend lectures and seminars in the field as part of the university reform program developed by Wilhelm von Humboldt. Proclaiming the ideal of the unity of teaching and research, the reformers prescribed the study of philosophy for future Gymnasium teachers to convey to these purveyors of the high ideal of humanistic *Bildung* (general cultivation) the rudiments of *Wissenschaft*.[8]

Psychology was traditionally regarded as a part of philosophy, and the reformed teachers' examination statute of 1866 – later valid for half the Empire – specifically required knowledge of "the main points of empirical psychology." Even the revised statute of 1887, codified in 1898, which greatly reduced the space accorded to general "cultivation" in favor of greater mastery of a particular subject, still specified a philosophy seminar to acquaint teaching candidates with "the tasks of psychology and pedagogy."[9] Herbert Schnädelbach aptly describes the change in the culture of science in Germany during the nineteenth century as a shift "from culture through science to science as a vocation."[10] Psychology's role in teacher preparation meant that it combined both identities.

Max Weber may have exaggerated, but only somewhat, when he wrote in 1908 that "The 'freedom of science' exists in Germany within the limits of ecclesiastical and political acceptability. Outside these limits there is none."[11] Since full professors were state civil servants with permanent tenure, government officials took care when appointing them to ensure that their political and religious views were congenial to the heads of the German state in question, or, at least publicly, nonexistent. Because of their strategic role in teacher training, philosophers were especially likely to be subjected to such scrutiny. The Berlin philosopher and psychologist Carl Stumpf, who taught all of the founders of Gestalt theory, showed that he was fully aware of this when he wrote to the Prussian official in charge of university affairs, Friedrich Althoff, on behalf of his friend and colleague Edmund Husserl in 1895: "Despite his earlier Judaism and his current Austrian citizenship, I believe him to be reliable."[12] In the vast majority of cases, official caution was unnecessary. Professors tended to be sincere patriots who claimed to view practical politics with contempt. Yet they looked to the state as the guarantor both of their social and economic status and also of their right to pursue their own research as they wished.[13]

Despite continuing proclamations of the autonomy and prestige of scientific research, the institutionalization of new disciplines was not a simple process. University faculties enjoyed a certain degree of independence from state intervention, but budgetary decisions and professorial appointments remained in the hands of the educational and financial officials of the various German states. In the recognition of new disciplines, which meant establishing new professorships, financial and other extrascientific considerations often held back innovation.[14] For their part, university faculties often tried to accommodate new fields by granting their representatives temporary teaching contracts or nonbudgetary associate professorships, which expired with the appointee's departure. Especially during the period of rapid economic growth after 1890, a variety of disciplines, from meteorology and physical chemistry to the romance languages, either established or extended their institutional bases. But consistent financial support in the form of full professorships and their accompanying seminars or institutes still came only after long and difficult struggles.[15]

One way to circumvent such problems was to use the traditional privilege known as *Lehrfreiheit,* the freedom to lecture in any field one wished within certain broadly defined limits. This strategy helped prepare the way for the emergence of experimental psychology in Germany. Philosopher Rudolph Hermann Lotze, for example, taught courses in empirical psychology to two generations of students at Göttingen (from 1844 to 1881).[16] To reach the level of prestige and security necessary to carry out such innovations, however, one had first to become a full professor. For younger scientists, there was no guarantee of a regular income during the period between the completion of the dissertation and that of the *Habilitationsschrift,* a second, more extensive piece of research required to earn the right to teach. Even after that, the only official source of support until the appointment to a professorship was income from student lecture fees; but full professors tended to monopolize the better attended lectures. Since the number of full professorships did not increase as rapidly as enrollment, the waiting time between the doctorate and a chair lengthened from an average of twelve years in the 1860s to sixteen years by 1909. Less than half of those who began academic careers advanced so far at all. In the academic community, as contemporaries realized, men from economically secure backgrounds had the best chance of success. Women, who were not even admitted as full-time students until the early 1900s, had no chance at all until the 1920s.[17]

Two features of the situation, one traditional and one more modern, could mitigate these harsh facts. The traditional mode was patronage of younger scholars by their professorial masters. Such procedures also led to the formation of schools of thought such as the Marburg and Southwest German schools of Neo-Kantian philosophy. The more modern mode was the gradual introduction of the position of "assistant" (*Assistent*), which enabled some young academics to pursue their research as salaried employees and thus avoid complete dependence on personal or family wealth. In humanities seminars, assistants conducted the exercises linked to general lecture courses and ran the library. In the natural sciences they carried out demonstrations during the lectures, conducted laboratory exercises, and trained beginners in the use of equipment. However, control of assistants' appointments remained firmly in the hands of the full professors, who often ran their institutes as personal fiefdoms. Thus, although sociologists describe the assistantship as part of a pattern of differentiation and rationalization in the German universities, it was seen at the time as an extension of paternalistic patronage.[18]

An important part of this developing mix of traditional and modern social attitudes and institutions was the establishment of experimental methods in chemistry, physiology, and physics before and during the Wilhelmian period. The entry of laboratory research into the German universities and their subsequent growth and development followed no single pattern. In each case complex negotiations were needed to convince university colleagues that such methods deserved the hoary title *Wissenschaft,* and to persuade nonscientists of their practical value. It was in these pursuasive efforts that scientific and sociocultural discourse intersected.

In the case of chemistry, social and intellectual factors interacted most tightly. After Justus Liebig demonstrated the practical potential of the laboratory for reorganizing large theoretical issues into solvable research problems, entrepreneurs, many of them trained chemists, successfully applied the techniques, organizational forms, and results involved to soil assaying, dye manufacture, and pharmaceutical production. It was at this point that state officials became interested. The resulting cooperation among science, industry, and the state – pushed by Liebig's propaganda with early success in Baden and Bavaria, against initial resistance in Prussia – led to the creation of professorships in chemistry to help meet the rapidly growing need for trained scientists. By the turn of the century, as Heinrich Mann's novel *Der Untertan* mordantly conveys, a doctorate in this field was becoming a ticket to upward mobility for children of the so-called new middle classes.[19]

In the case of physiology, both the theoretical and the clinical utility of the new methods could be shown in the university clinics where experimenting physiologists worked. Through their refusal to accept other than physical explanations for biological phenomena or to be content with anatomical description alone, scientists like Hermann Helmholtz, Theodor Schwann, and Carl Ludwig set themselves apart from both the philological disciplines and from anatomy. At the same time, they shared their opponents' distaste for applied science, arguing that their methods had an equal if not greater right to the name *Wissenschaft*. Nonetheless, the reform of medical education in Baden (1858) and Prussia (1861), requiring laboratory instruction alongside clinical work, was the result of belief in the pedagogical value of experimental method and the critical thinking it allegedly fostered for medical practice, even though that value had been demonstrated mainly in diagnosis rather than in therapy. When he became professor in Heidelberg in 1862, Helmholtz spoke confidently in his inaugural lecture of the scientific enterprise as an effort to achieve "the intellectual mastery of nature."[20] Such rhetoric enhanced the appeal of science to the state without sacrificing the ideal of pure research. At least as important as such intellectual militance, however, was that victory in the institutionalization struggle came from within the established patronage system, by campaigning in state ministries of education for new professorships and then filling them with students trained in the founders' own laboratories. Within a single generation after 1848, every German university had at least one chair of physiology.[21]

Still a third route to institutionalization for experimental methods was that of physics. In the first step, leaders such as Wilhelm Weber in Göttingen and especially Franz Ernst Neumann in Königsberg instituted a four-semester sequence of lectures in theoretical physics alongside the traditional one-semester service course in experimental physics. Next they took over the seminar model from classics; physical institutes and experimental practica came only later, along with associate professorships in theoretical physics and assistantships. The resulting tension between the ideology of humanistic *Bildung* (general cultivation), and *Ausbildung* (occupational training) was embodied in the effort to combine the training of future science teachers and that of future scientists. Even after physics institutes were founded in the 1860s and later, the primacy of pedagogy over professionalization and of theoretical over experimental physics remained in place. The discipline's student audience, unlike that of chemistry, consisted mainly of future teachers, students in other sciences, and humanists in search of a general background in science. Still later in the century, the needs of science-based industries for trained personnel and standardized measurement became increasingly influential. Electricity became a central research topic, and the results fueled the expansion of the electrical industry in the 1860s and after; but training in electrotechnics occurred mainly in the technical academies. For students, experimental physics was often a route to upward mobility, while theoretical physics was a preserve of the children of the already educated.[22]

As such transformations multiplied in the Wilhelmian period, a constituency arose in the universities that favored establishing experimental methods wherever they might prove useful, even in philosophy. Since the natural sciences were not then separated from the humanities in most German philosophical (arts and sciences) faculties, this constituency was in a position to influence professorial appointments, given the opportunity. But the word "useful" could have a variety of meanings. On the one hand, natural scientists and humanists alike proclaimed the priority of theoretical over applied knowledge. If experimental methods were to gain even a foothold in philosophy, their applicability to philosophically relevant problems would have to be demonstrated. On the other hand, the social function of the universities remained what it had been, the training of Germany's educated elites, especially civil servants. Large-scale state support in the form of new chairs and laboratories would be available only if it could be shown that the new methods were useful in this sense. The German university reformers had already squared this circle in the case of philosophy by making that subject a practical requirement for Gymnasium teachers. How could experimental psychology hope to find a legitimate place in such a situation?

The institutionalization of experimental psychology

By the 1870s, experimental methods were well on the way to establishment in medicine and the natural sciences, where both their usefulness in research and the practical value of their results were being demonstrated. Physiologist Wilhelm Wundt then proposed to bring such new, potentially costly ways of doing science into philosophy, the very homeland of the humanistic ideal of "pure" science, when he moved in 1875 to a chair of philosophy at Leipzig, then the largest university in Germany.

Wundt was well aware of the delicate social and intellectual situation he was getting into. For many years he had been Helmholtz's assistant in Heidelberg, where, as he remembered it, scientists of the stature of Gustav Kirchoff and Robert Bunsen were regarded by the historians and philosophers as "mere apothecaries" meddling in humanists' affairs. The situation was better at Leipzig. There natural scientists like Ernst Heinrich Weber and Gustav Theodor Fechner, the founders of experimental psychophysics, had long been looked upon as equals, and philosophy itself was represented by Moritz Wilhelm Drobisch, a former mathematician. Moreover, the faculty member who pushed most strongly for his appointment was an astronomer, Carl Friedrich Zöllner.[23] This would not be the last time that support for experimental psychology in the philosophical faculty would come from natural scientists.

Nonetheless, Wundt was careful to secure his position at Leipzig by the traditional method of achieving high enrollments in his lecture courses before establishing his psychological institute on a private basis in 1879, and then founding the journal *Philosophische Studien* in 1881.[24] Since one of the journal's purposes was to publish the research of the institute, historians of psychology point to it with pride as the world's "first effective organ for experimental psychology." This it was, but it also contained numerous essays in the theory of knowledge, philosophy of science, and other philosophical topics, mostly by Wundt himself. In the 1880s, he anchored his institutional innovation by placing his psychology within a comprehensive system of the sciences. For him the human sciences (*Geisteswissenschaften*) are based upon "immediate" (*unmittelbare*) experience, and the natural sciences upon "mediate" (*mittelbare*) experience, abstracted from the former. Since "inner or psychological experience is not a special area of experience alongside others, but immediate experience itself," psychology is thus "the most general" human science. "Physiological" psychology – that is, experimental psychology in Wundt's sense – is a mediating link between the two groups of disciplines; but psychology as a whole is the foundation of all the humanistic fields, which in turn provide data for psychological analysis. Wundt dedicated his own work in Leipzig increasingly to systematic philosophy, ethics, natural philosophy, and social, or ethnological psychology (*Völkerpsychologie*). He later remarked that his journal's title was meant to be "a call to battle," but the aim of the struggle was to show only "that this new psychology had the claim to be a subdiscipline [*Teilgebiet*] of philosophy."[25] Wundt's career from the 1880s on could be described as an attempt to gain a secure, if carefully delimited, conceptual and institutional location for experimental psychology while demonstrating his own worthiness to belong to the philosophers' guild.

This goal was consistent with Wundt's conception of psychological research, which he had outlined in a comprehensive, hierarchically ordered program as early as 1862. Wundt placed higher psychological processes, particularly thought, beyond the reach of experiment, and limited the role of experimental methods to the classification and measurement of phenomena that could be treated "physiologically," or psychophysically, such as sensation, reaction time, and attention span.[26] Even so, such limits did not prevent Wundt from organizing what amounted to a knowledge factory to produce the results that those methods could deliver. According to his own description of his institute's operation, he announced the research topics himself at the beginning of each semester, taking the wishes of older members into account "when possible." Each topic was assigned to a separate research group led by a senior member, who assembled the results and prepared them for publication. The data were considered "the property of the institute, whether the investigation is published or not."[27]

This procedure had some of the features commonly attributed to large-scale scientific research today, particularly hierarchical organization and the institutional ownership of results. However, closer examination of the experimentation actually done in Wundt's laboratory reveals one important difference. The aim of Wundt's psychology was to discover not the principles of behavior but those of "psychical causality," which he postulated alongside physical causality. The proper subjects for the experimental portion of such a project could only be normal adult human beings, preferably with practice in psychophysical observation. This he limited to very simple judgments; in experiments with tones, for example, a single word such as "higher" or "lower" would suffice. Both the stimulus conditions and the time required to make the judgments could be measured with recording instruments like the kymograph and the chronoscope. The records made in Wundt's institute consisted almost entirely of such measurements, not of verbal introspective reports. This procedure appears similar to modern, apparatus-oriented, data-driven cognitive science. In Leipzig, however, not the number of subjects, but the number of observations was important. The membership of the institute was seldom more than twenty-five, and the investigators, including Wundt, often served as subjects in one another's experiments. One purpose of the careful organization of research in Leipzig was to ensure that the participants in this collaborative enterprise were properly trained to carry it out.[28]

The Leipzig institute's output, combined with Wundt's own articles, was more than sufficient to fill the annual numbers of the *Philosophische Studien*. The total of 186 dissertations produced at Leipzig under Wundt's tutelage is also impressive, even allowing for the 45-year time span involved and for the fact that 70 of these dissertations treated philosophical topics not directly related to experimental psychology.[29] Despite the evident productivity of Wundt's laboratory, and his strenuous efforts to combine experimental psychology and systematic philosophy, the institute did not receive a regular budget and its courses were not listed in the university catalogue until 1883, four years after its founding.[30] The response of Wundt's colleagues and the educated public was not immediately favorable, either. After traveling to Leipzig and other German universities, Emile Durkheim reported in 1887 that "Wundt's example is hardly followed at all; in his own country he even encounters stiff resistance."[31] In the same report, he noted that the majority of the students in Wundt's laboratory at the time were not taking their degrees in philosophy, but in mathematics and natural science. Wundt made the same observation in 1893, adding that most of the philosophy students were preparing to become Gymnasium teachers.[32] Thus, the student public for Wundt's experimental psychology was quite similar to that of physics, described earlier. These future *Bildungsbürger* were participating in discovering nothing less than the causal laws of their own subjectivity. Their carefully recorded observations bear comparison with the fine distinctions and judgments of taste that created a cultural code for bourgeois French citizens in the same period.[33]

As Wundt's international reputation increased and a stronger economy helped make more money available for science in general, the government of Saxony finally provided spacious quarters and a substantial budget for the institute in 1897.[34] But the expansionist strategy that had been so successful for physiology in the medical faculty yielded less impressive results in the case of Wundt's laboratory. By his own count, Wundt had had seventeen assistants by 1909. Thirteen scholars wrote their *Habilitationsschrift*, or second thesis, under Wundt, not all of whom served as assistants. By 1914, only seven of these students or associates had obtained full professorships in German-speaking universities outside Leipzig; eleven ended up teaching either in Leipzig or in foreign countries. The others apparently did not go into academic psychology.[35] Of course, Wundt could not have been expected to train all of Germany's experimenting psychologists himself, and he did not.

The other people who institutionalized experimental psychology in Germany were not philosophical autodidacts like Wundt, but were themselves trained philosophers. Philosophy participated in the trend to specialized scholarship in nineteenth-century Germany. The experimenting psychologists' aim was to reform philosophy from within, and thus to gain the advantage their competence in a distinct scientific specialty gave them in what was becoming a highly competitive discipline.[36] For support in this effort they aligned themselves with important people in other fields, especially physiology. The list of names on the masthead of the first issue of the *Zeitschrift für Psychologie und Physiologie der Sinnesorgane* (Journal for Psychology and Physiology of the Sense Organs), founded by Hermann Ebbinghaus and Arthur König in 1890, included five of the most distinguished physiologists of the day: Hermann Aubert, Sigmund Exner, Ewald Hering, Johannes von Kries, and William Preyer – and also the world's most prominent former physiologist, Hermann von Helmholtz. The philosophers, in addition to Ebbinghaus, were Theodor Lipps, G. E. Müller, and Carl Stumpf. "The tasks and goals of the journal," the editors announced, "are indicated by just these names. The journal strives for a unification of persons and views in the scientific service of a great, unified cause."[37] This was an oblique reference to Wundt's journal, which had included contributions from none of these authors.

The connection to physiology involved more than names on an editorial board. In their statement the editors noted the increasing use of natural scientific methods and precision apparatus in biology and physiology, and espoused the cultural values embodied in that apparatus by proclaiming their intention to use measurement to help psychology become "an exact science as far as possible."[38] The experimenting psychologists also adopted the institutional innovations Wundt had imported from physiology, particularly the research laboratory and its corresponding assistantships. Four of these philosopher-scientists – Ebbinghaus, Müller, Stumpf, and Oswald Külpe, a student of Wundt's who took a decidedly independent line – established or significantly expanded seven of the thirteen psychological institutes that existed in Germany by 1914 (see Appendix 1, Table 1).

These professors, along with Wundt's student Ernst Meumann, were in many respects the real founders of experimental psychology in Germany as a network of institutions strictly bound neither to the traditional rubrics of the university nor to the confines of a single university laboratory. The *Zeitschrift für Psychologie* and the *Archiv für die gesamte Psychologie* (Archive for All of Psychology), founded by Meumann in 1903, became the dominant journals of experimental and general psychology, respectively, open to contributions from the entire German-speaking world and beyond. The Society for Experimental Psychology was organized in 1904 by Müller, Giessen psychiatrist Robert Sommer, and Friedrich Schumann, then assistant to Stumpf in Berlin. Its 104 original members included professors of philosophy and physiology, physicians, and Gymnasium teachers, representatives in fact of "all lines of thought, insofar as they based themselves on experimental psychology," as Sommer later stated.[39] It was in the pages of these journals and especially at the Society's biannual meetings that younger experimenters reported their results and demonstrated their abilities.

The establishment of both the *Zeitschrift für Psychologie* and the Society for Experimental Psychology was notable for the conspicuous absence or at most the ritual invocation of Wundt's name. Laden with significant ambiguity was the unanimous vote of the Society at its first meeting, "at the request of Herr Külpe . . . to send a telegram of greeting to Herr Geheimrat Wundt as the Nestor of experimental psychology."[40] The term "Nestor" was conventionally given to esteemed older colleagues. Classically educated as they were, however, these German academics surely remembered that although Nestor, the revered sage of the Greek hosts in the *Iliad,* was always heard with respect, his advice and dark warnings were rarely heeded.

Specifically rejected here were Wundt's definition of the psychological subject as the "psychical individual" and his limitations on the experimental method. Ebbinghaus, Müller, Külpe, Meumann, and their students proclaimed the "corporeal individual," or more elegantly, the organism to be their proper subject, and developed techniques for the experimental study of memory, aesthetic judgment, abstraction, even thought itself. In his 1885 study of memory, for example, Ebbinghaus applied what he called "the method of natural science" to phenomena that Wundt had specifically excluded from experimental investigation. His innovation was the use of statistically manipulable units, series of syllables, consisting of two consonants with a vowel between. He measured the time or the number of trials required to learn series of these selected at random, then computed the saving in later relearning as a percentage of the original figure. The resulting curves of forgetting became standard in experimental studies of learning and memory, and his methods entered the canon of laboratory instruction in Germany.[41]

In even more flagrant disregard of Wundt's views, some of these psychologists, particularly Ebbinghaus, Meumann, William Stern, and Hugo Münsterberg, searched for ways of applying their methods to social issues. Common to all these efforts was an emphasis on measurement, performance, and efficiency in keeping with Germany's rapid, technology-led industrial growth. Examples included Ebbinghaus's sentence-completion test, developed in 1895 to determine whether students performed better earlier or later in the school day and thus address the issue of fatigue, Meumann's "experimental pedagogy" based on the efficient distribution of what he called "mental work," and Hugo Münsterberg's "psychotechnics," skills testing intended to help assign apprentices in industry to jobs appropriate to their abilities. Wundt, who originally supported the founding of Meumann's *Archiv*, resigned from the editorial board in protest after only one year, suggesting pointedly that the name of the journal ought to be changed from "Archive for All of Psychology" to "Archive for Education and Psychology."[42]

Despite their organizing zeal, their evident dedication to experimental methods, and their efforts to turn those methods to practical account, these scientist-philosophers did not press at first for independent chairs and institutes for experimental psychology. Instead, they competed with other philosophers, sometimes with one another, for appointment to already existing philosophy professorships, and then made funding and facilities for their experimental work a condition of their acceptance. This was a fruitful strategy during the generation from 1890 to 1910. While the number of full professorships of philosophy in Germany increased only 10 percent in those years, from forty-four to forty-eight, the number of those positions held by experimenting psychologists more than tripled in the same period, from three to ten. It was in this period, and not earlier, that German replaced French as the leading language of publication in psychology.[43]

For the longer term, however, precisely this success posed difficult academic-political problems. In principle, there were enough philosophy chairs to accommodate some experimenting psychologists, and the money was available to provide the laboratory space and equipment they said they required. In practice, there were other philosophical specialties besides psychology, and even at the best-financed universities professors of philosophy were still expected to teach more than one of them, a requirement that younger psychologists would find increasingly difficult to meet as experimental methods became more demanding and research more time consuming. Nor was it by any means clear that expertise in psychology as a specialty within philosophy necessarily required mastery of experimental technique. Despite the vast social and economic changes that had occurred in Wilhelmian Germany, the concept of humanistic *Bildung* and the idealistic conception of "pure" science in which philosophers took such pride had not lost their hold on either the educated public or the academic community, experimenting psychologists included.

2

Carl Stumpf and the training of scientists in Berlin

Stumpf's early career and appointment in Berlin

The career of Carl Stumpf, teacher of the Gestalt theorists and many of their leading associates in Berlin, exemplifies the complex situation of experimental psychology as a philosophical specialty in Germany.[1] Born in 1848 in the village of Wiesenthied in Franconia, Stumpf came from an academically trained family. His father was a provincial court physician. After completing the Gymnasium in Aschaffenburg, he began his university studies in Würzburg in 1865. There he met and became a disciple of Franz Brentano, who inspired him with ambitious plans for a revival of Christian philosophy based on rigorous study of classical and Scholastic thought as well as empirical psychology.

Brentano propounded an empirical rationalism based on the "evidence," or evident truth, of both the logical axioms and the facts of introspection, which he called "inner perception." In his view, evident truth inheres in the object of knowledge, not in our grasp of it. As he explained in a lecture Stumpf heard in 1869, "Even God cannot make it evident to us that red is a sound or $2 + 1 = 4$. His will (by which he made the world as it is and not otherwise) would thereby contradict itself." By the same token, we also know with evident certainty *that* we have this or that sensation, wish, or emotion, whether these objects of consciousness are "actually" illusions or not. In his major work, *Psychology from an Empirical Standpoint* (1874), Brentano derived this claim from the doctrine of intentionality, for which he is best known. This is the idea that all experience involves directedness toward an object. He also called intentionality "*immanent objectivity*," because for him "Every mental phenomenon includes something as object *within itself.*" That is, an object of some kind is in our "inner" perception, and it is these objects about which we have certain and not merely probable knowledge.[2]

But we still need to know how we are constituted so that we might have access to such knowledge. This, for Brentano, is the task of empirical psychology – to

28

discover and classify the facts of "inner perception" and the basic types of mental states that make it possible for us to apprehend and understand the world as God created it. He had this combination of metaphysical goals and logical and psychological analysis in mind when he boldly proclaimed in his inaugural lecture in 1866 that "the true method of philosophy is none other than that of the natural sciences." As Stumpf recalled, he and his fellow students found this to be "a new, incomparably deeper and more serious way of understanding philosophy."[3] Brentano and his devoted followers constituted a network equal in strength and intensity within German-speaking philosophy to that of the varied schools of Neo-Kantianism.

After a year of work together, Brentano sent Stumpf to Göttingen, where he received a thorough grounding in epistemology from Hermann Lotze and also studied physiology and physics, the latter with Wilhelm Weber. Stumpf completed his doctoral thesis on the relationship of Plato's God to his idea of the Good in 1868, then returned to Würzburg for more study with Brentano, and even considered following him into the priesthood. Encouraged by Lotze, he finally decided for the university, and submitted his second thesis on the mathematical axioms in Göttingen in 1870. Though he had begun his career in pure philosophy, he had learned from both of his teachers to respect psychological research.

In his first major book, on the psychological origins of the idea of space (1873), Stumpf applied the methods he had learned from Brentano to refute a central claim of Lotze's theory of perception. One of the problems in understanding vision is why we see the world in three dimensions when the pattern of excitations on the retina has only two. To explain this, Lotze had hypothesized additional retinal factors called "local signs" as cues for depth. Against this, Stumpf claimed that the quality of being extended in space is an immediately given datum of the same order as the sensory qualities themselves. According to Stumpf, simple judgments such as "the color is extended," or "the extension is colored" both refer to the same psychologically real "state of affairs" (*Sachverhalt*), of which extension is a "partial content." He thus presented spatiality as an attribute of what Brentano and Lotze both called the unity of consciousness, a psychological fact that needs no explanation in terms of Kantian categories of apperception or of learned interpretations of physiological "local signs."[4]

Stumpf's skillful defense of "psychological nativism" and the endorsements of both Brentano and Lotze won him one of the two chairs of philosophy in Würzburg in 1873. This came three months after Brentano left the Catholic church and resigned his associate professorship at the same university because he refused to accept the doctrine of Papal infallibility. Brentano's appointment to a full professorship in Vienna the next year put him in an excellent position to propose Stumpf for a professorship in Prague, which was offered and accepted in 1879. Stumpf later said he was glad to get away from Würzburg, where a "heretical Catholic" such as he had become could not feel at home.[5]

Though he lacked laboratory facilities in Prague, Stumpf nonetheless completed the first volume of his most important scientific work, *Tonpsychologie,* in 1883. The bulk of this book consisted of careful observations made on himself, which showed that psychological factors are immediately influential in hearing and thus implied that the strictly physical approach taken by Helmholtz twenty years earlier was insufficient. He emphasized both his affiliation with philosophy and his identification with Brentano's conception of it by saying that psychophysical research of this kind would eventually lead to a "measuring theory of judgment" (*messende Urteilslehre*).[6] For this research he was named to positions in Halle in 1884 and Munich in 1889. Stumpf was commited to an empirical, but not empiricist, philosophy – the belief that philosophy could respect the facts of human experience without reducing all knowledge to experience. His dedication to that ideal had brought him appointments to four full professorships before he was forty-five.

Such rewards showed that there was indeed room in the philosophers' guild for experimenting psychologists. But Stumpf was well aware of Neo-Kantians' disdain for merely empirical knowledge and their insistence that the primary aim of philosophy was to make and sustain transcendental claims that would be valid for knowledge or values as such. In his 1891 essay, "Psychology and Epistemology," he claimed that Kant's "greatest error" had been his "neglect" of psychology, and he sharply reminded Neo-Kantians that no discussion of a priori judgments or forms of understanding can avoid presupposing some notion of what judgments are. Still, the two words describe different tasks, and it would be inappropriate to reduce one to the other. Epistemology's role is to seek out and determine the logical grounding of "the most general, immediately evident truths," such as the geometrical axioms. Psychology's purpose is to investigate "the origin of concepts" empirically, by determining "the most exact characterization of the aspects and modes of alteration of ideas" as they appear to experiencing subjects.[7]

Stumpf's use of the term "origin" (*Ursprung*) had little to do with historical or ontogenetic causation. Rather, he had in mind Lotze's distinction between the genesis (*Genese*) or emergence and the validity (*Geltung*) of ideas, originally intended to differentiate between epistemology and logic.[8] This was an ancestor of Hans Reichenbach's now-common distinction between the (psychological or sociological) contexts of scientific discovery and the (mainly logical) contexts of justification. For logical empiricists like Reichenbach, and for many other philosophers, that distinction was a disciplinary demarcation line. Despite his immersion in empirical research, Stumpf's project remained a philosophical one – to reconcile realism and rationalism by accounting for both the origins and the validity of knowledge claims. Different as the tasks of psychology and epistemology might be, Stumpf insisted that they do not belong to different disciplines; for neither can be done without the other. Theorists of knowledge "cannot ignore the issue of the origin of the concepts" whose general meaning they analyze, but "must be fully engaged in the depths and difficulties of this problem" as experts. Psychologists, for their part, must also be theorists of knowledge, "not only because judgments of knowledge are a special class of judgment phenomenon," but primarily because they "must have clarity about the fundamental basis of all knowledge, as anyone must for whom science is more than a trade [*ein Hand-werk*].[9] Stumpf thus stressed philosophy's need for empirical expertise about the process of knowing while simultaneously depicting experimentation for its own sake as a respectable but definitely lower-status activity.

Wilhelm Dilthey, the gray eminence of Berlin philosophy in those years, evidently found Stumpf's conception of empirical psychology and the idea of psychological research as a propaedeutic to higher philosophical concerns attractive. In his important essay, "Ideas Concerning Descriptive and Analytical Psychology," published in 1894 (discussed further in Chapter 5), he cited Stumpf's work on tonal fusion as evidence for the inadequacy of the "dominant" psychology, which would "explain the constitution of the mental world" according to hypotheses about its components, forces, and laws, "in the same way as physics and chemistry explain the physical world." Dilthey was not opposed to experimentation, but called for a psychology that valued such causal explanations "only secondarily, with an awareness of their limits."[10] As early as 1884, Dilthey had written to Friedrich Althoff, the Prussian official in charge of university affairs, supporting Stumpf's appointment in Halle. He had expressed interest in bringing Stumpf to Berlin as early as 1892; Stumpf was offered a professorship there the next year. In a letter to his friend Count Paul Yorck von Wartenburg in 1895, Dilthey stated that he had arranged the appointment, and claimed that "my intervention prevented the complete radicalization of philosophy here by the natural sciences."[11]

Behind Stumpf's appointment was a complex mixture of traditional academic intrigue and Friedrich Althoff's semimodern approach to science policy.[12] Philosophy was not central to Althoff's agenda of making Prussia, particularly the University of Berlin, a world leader in natural science and humanistic scholarship.[13] But when Berlin's Philosophical Faculty showed support for experimental psychology by promoting Hermann Ebbinghaus, pioneer of quantitative research on memory, to an associate professorship in 1886, Althoff approved small grants totaling 1,600 Marks to support his experimental exercises and authorized two rooms for a teaching and research laboratory in 1892. The opportunity to expand this base arose when Dilthey and his much older colleague Karl Zeller proposed a temporary third professorship to lighten their teaching loads. The position became a replacement for Zeller on his death.

The Philosophical Faculty recommended Stumpf ahead of G. E. Müller and Benno Erdmann in July 1893, because, they said, he was qualified to lecture in both psychology and history of philosophy, and was also the person best equipped to build a psychological institute at Berlin that could compete with Wundt's in

Leipzig. But Dilthey, who wrote the Faculty's recommendation, also emphasized that Stumpf "will keep away from trespassing on physiological territory" and avoid "wasting students' time in unproductive series of experiments." This may well have been a cutting reference to Hermann Ebbinghaus, who had been promoted to associate professor in Berlin in 1886, but whose concentration on psychophysics and his own highly quantitative approach to the study of higher mental processes appear to have antagonized Dilthey (see Chapter 5).[14]

When Althoff followed his already famous "system" and solicited advice from senior and junior scholars on the state of the field and the qualifications both personal and scholarly of the candidates, he got a similar message from a Theodor Lipps, then professor in Breslau. Lipps opposed Müller as "a physiologist" whose appointment would be a false signal of official support for that sort of psychology.

Nonetheless, in August Althoff developed and submitted to Prussian Finance Minister Miquel a proposed budget for the new institute, including an initial outlay of 30,500 Marks and an annual budget of 5,090 Marks, more than double Leipzig's yearly figure at the time. Cannily, if implausibly, he represented this expenditure as a net saving, since Berlin would be getting a philosopher and a psychologist in one appointment. Such an offer might well have been attractive to Stumpf. The closest he had ever come to having a laboratory of his own was the space he had had in Munich, which "consisted of the attic floor of a high tower and a cabinet in the hall, where he kept tuning forks that he could use in the lecture room on Sunday."[15]

Surprisingly, Stumpf reacted negatively to Althoff's grandiose plans. First he asked for time to think the matter over, and to consider any counteroffer the Bavarian government might make to keep him in Munich. Then he refused the appointment. A flurry of letters followed, as both Dilthey and Althoff tried to persuade Stumpf to change his mind.[16] When Althoff finally wrote that he was firmly resolved "to win you for Berlin" and asked him to name his conditions, Stumpf replied that his main motive for hesitating was "the worry that I would not find the peace and concentration necessary for the completion of scientific projects on which I am now at work." Instead of a full-scale institute that would drain his time and energy, he proposed that "only a psychological seminar be established, with the task of supporting and supplementing the lectures with laboratory exercises and demonstrations. Carrying out scientific work for publication would naturally not be excluded but would not be among the essential purposes of the seminar." Moreover, he added, "I am in any case of the opinion that large-scale research in experimental psychology has objective difficulties as well . . . I could not decide, now or later, to follow the example of Wundt and the Americans in this direction."[17]

Dilthey and Althoff quickly alleviated Stumpf's worries. Althoff sweetened the offer with a generous increase over his Munich salary and promised adequate space for lectures, demonstrations, and his own research. Acceptance followed quickly. The new seminar, founded with an initial outlay of 6,000 Marks, a single assistant, and an annual budget of only 1,000 Marks, began operating when Stumpf arrived in Berlin in the spring of 1894. Dilthey had achieved his aim of installing a version of experimental psychology in Berlin that was congenial to him. At the same time, in an 1895 report to his Ministry, Althoff wrote proudly that with this appointment "psychophysics [sic], the most modern branch of philosophy, is given the possibility of proving itself by the fruit of its labors in Prussia also." In his official letter recommending Stumpf to the royal household, Althoff took care to note that "in confessional matters Stumpf is a man of very mild views," as shown by the fact "that he married a Protestant and let his children be raised in the evangelical [Lutheran] faith." This remark was significant in view of the royal family's Protestantism, and also because Stumpf had studied with Brentano, who was strongly opposed in some quarters for his alleged Scholasticism despite his dissident views on papal infallibility.[18]

The growth of the Berlin institute and Stumpf's conception of psychology

The dynamics of specialization, however, proved difficult to resist. As time went on, enrollment in the seminar's laboratory courses increased dramatically from 25 in 1894 to more than 50 in 1907, while attendance at Stumpf's psychology lecture course quintupled from 50 to 250 in the same period.[19] In response, Stumpf petitioned the Ministry repeatedly for additional funds, space, equipment, and staff, making pointed comparisons each time to the other leading German institutes in Leipzig, Göttingen, and Würzburg.[20] In 1900, the seminar moved into new and larger quarters and officially became a Psychological Institute; by 1912 it could boast a budget of 4,400 Marks, more than four times that of 1894. As Table 1 in Appendix 1 shows, this made it physically the second largest psychological laboratory in Germany and the financially best supported. Unfortunately, Stumpf never received a second full-time assistant. He attempted to fill the gap with part-time and voluntary help, but still took over many of the beginning courses and exercises himself.

Other activities also made demands on Stumpf's time, especially his increasing interest in ethnomusicology. He had begun to study the music of non-Western peoples in the 1880s. His move to Berlin, then a world center of both music and musicology and an important departure point for ethnographic expeditions linked with Germany's new colonial empire, gave him new opportunities to expand this research. In 1900 he set aside a room in the new institute to house the *Phonogramm-Archiv,* a collection of Edison cylinders containing samples of music recorded by traveling scholars and amateurs. At first Stumpf supported the collection himself. From 1904 to 1909 came regular grants from the Virchow Foundation of the Prussian Academy of Sciences and additional private gifts. The

Ministry provided single appropriations of 3,600 Marks in 1910 and 5,000 Marks in 1914, and the newly established Albert Samson Foundation of the Prussian Academy guaranteed an annual grant of 5,000 Marks in 1912, which was later raised to 7,000 Marks. The activity required to raise these sums and the amounts – more than the institute's budget – clearly show how committed Stumpf was to this enterprise.[21]

Despite the pressures of this and other commitments, and the lack of sufficient paid staff to help with teaching, Stumpf did not limit student membership in his institute, as Wundt had done. Rather, he continued to see it as more a pedagogical than a research tool. As he wrote in 1910:

> In such a young research tendency [*nota bene:* not 'science' or 'discipline'] with so little developed methodology, so many sources of error, with great difficulties in the exact setting up and carrying through of experiments, it could not be the main goal [of the institute] to produce as many dissertations as possible. Instead, the leading aims must be these two: first, support of the lectures by means of demonstrations and exercises; second, providing the necessary aids for the experimental work of the director, the assistants, and a few especially advanced workers.[22]

He maintained that conservative stance when asked for an opinion on the status of experimental psychology in connection with the founding of the Kaiser Wilhelm Society in the same year. Rather than proposing an independent, nonuniversity research facility, he favored increased funding for existing institutes – including his own.[23]

Among the "few especially advanced workers" Stumpf mentioned were nearly all of the founders or leading co-workers of Gestalt theory: Max Wertheimer, Kurt Koffka, Wolfgang Köhler, Adhemar Gelb, Johannes von Allesch, and Kurt Lewin. All but Wertheimer received the doctorate for work done in Berlin from 1906 to 1913, Koffka, Köhler, and Lewin for experimental research under Stumpf's direction. Wertheimer worked for two years in the Berlin institute before completing his dissertation under Oswald Külpe in Würzburg in 1904, and then returned often to Berlin for discussion and research, especially with his close friend Erich von Hornbostel, Stumpf's assistant at the *Phonogramm-Archiv.* All of these scholar-scientists also studied at other universities. But Stumpf was the master under whom the Gestalt theorists learned their trade as experimenters. For this reason alone it is well to take a closer look at both the theoretical and the methodological training offered in Berlin between 1900 and 1910, especially at Stumpf's opinions on the purpose and meaning of experimental psychology.

In his acceptance speech on joining the Prussian Academy of Sciences in 1895, Stumpf defended himself against the charge that he had often "left the circle of philosophy" in his research. He had no intention "of replacing philosophy with specialized investigations or positivistic worship of the facts." Rather, his aim was "to investigate issues of basic importance with the concrete material of specific phenomena and in closest connection with the specialized sciences," that is, "to grasp the general in the particular."[24] In his first year at Berlin, he wrote to Althoff to complain that his seminar had been incorrectly designated the "Seminar for Experimental Psychology" in the university catalogue. He had deliberately suggested the broader name "Psychological Seminar," he wrote, "to avoid giving the impression that only experimental work is planned, when I am also planning to link such work to theoretical exercises in philosophy." The narrower designation, he feared, would keep talented students away and "instead attract a certain sort of American, whose whole aim is to become Dr. phil. in the shortest possible time with the most mechanical work possible."[25]

This combination of empirical training with broader philosophical intent and disdain for "American-style" narrowness and pragmatism was a constant motif at Stumpf's institute throughout his tenure. In his inaugural address as rector of the university in 1907, Stumpf contrasted the bitter battles of self-righteous, idealistic system builders with the division of labor and cooperative spirit of modern research science, and contended that only this spirit in both the humanities and the natural sciences could lead to a "rebirth of philosophy." He recommended that all philosophy students "lay hands" on some field of concrete research during their education, and specifically suggested work in the natural sciences and physiology for psychologists.[26]

Though Stumpf thought that hands-on science could be useful to young philosophers, he plainly shared Wundt's opposition to applied psychology. By this time, classroom teachers were becoming important consumers of educational psychological research in Germany; and teachers' associations served as local pressure groups for the new psychology in Leipzig, Berlin, and Hamburg. The teachers, allied with psychologists like Eduard Meumann and William Stern, hoped to put pedagogy on a scientific footing and to persuade government officials of the need for school reform.[27] In 1900, Stumpf helped to organize a society for child psychology to encourage teachers to make and report observations in the schools. In his address to the group, however, he criticized the overuse of statistics by the "half educated" (*halbgebildet*) instead of the "properly educated" (*vorgebildet*) and quoted Wilhelm Scherer on the brothers Grimm: "Art and science are not goods that can be acquired by the association and organization of the masses." This elitist stand may have pleased some teachers; but as Stumpf later recorded, the society later declined because applied psychology and the school reform movement were so prominent that "there was little room left for a society with pronounced theoretical aims."[28]

What conception of psychology was best suited to philosophy's progress? Like nearly all of the German philosophers of his time, Stumpf based his answer on a general theory of science, which rested in turn on firmly held epistemological convictions. Interestingly enough, he only voiced these commitments after be-

Photo 1. Carl Stumpf in rectoral robes. Courtesy of Dr. Siegfried Jaeger.

coming director of a major laboratory forced him to consider the implications of his double identity more closely – "after one of the philosophical disciplines, psychology, has itself apparently or really entered the special sciences." For him "the immediately given" was the basis of all science, in physics as well as psychology. The difference lay in the way the sciences proceed from this common beginning. For the natural sciences, phenomena are "only the starting point" for research about "a world of things, existing independently of consciousness but connected within itself according to causal laws." For psychology, the given is the stuff of science; its laws are not inferred from, but observed in, the appearances.[29]

Stumpf's emphasis on the lawfulness of the given came directly from the Brentano tradition. His phenomenological perspective was quite different from the phenomenalism of David Hume or Ernst Mach, in which both psychological and physical laws are derived by conclusions from or connections among simple sensations (see Chapter 4). Moreover, Stumpf's list of immediately given "appearances" included far more than simple sensations of color or tone. To these he added impressions of spatial and temporal extension and distribution, memory images, and also the relations among these appearances, such as similarity, fusion, or gradation. The laws of these relations are neither causal nor functional but "immanent structural laws." Their discovery thus belongs to phenomenology; "we have only to describe and recognize them."[30]

In advocating the reality of relations, Stumpf followed his long-time friend William James and his student Edmund Husserl, who dedicated his *Logical Investigations* (1900) to Stumpf (see Chapter 5). But Stumpf expanded the given still further to include "psychical functions." These included the noticing, grasping, and judging of appearances and their relations, the construction of concepts, and the emotions, desires, and the will. For most psychologists, such functions are the inferred causes for the manifestation of appearances. For Stumpf they are immediately given, though they are conceptually independent of appearances. Intensity, for example, cannot be predicated of an emotion and of a color sensation in the same way. Stumpf acknowledged that functions like attention could sometimes change the structure of appearances, but he thought that "the structural laws of the appearances are more likely to be attributable to physiological causes."[31]

There were two ambiguities in this schema. One was Stumpf's use of the term "appearances" in two different senses, both for sensations and relations as opposed to psychical functions and for all of the immediately given, including psychical functions, as opposed to inferences from the given. The second was Stumpf's conception of reality. For physics reality is inferred from the given; for psychology it is immanent in the given. Stumpf clearly thought the two realities were combinable in principle, but did not say how to proceed from one level to the other. In 1896 he had taken an interactionist position on the relation of mind and body. Ten years later he maintained that the psychology of sensation may be strictly deterministic, but the psychical functions are not necessarily subject to mathematical law. Thus he established the reality of the psychical realm only by implicitly denying it the same kind of lawfulness as that of the external world.[32]

Stumpf was familiar with the functionalistic psychology advocated in the United States by William James, John Dewey, and James Rowland Angell; both views emphasized the active aspects of consciousness. But he gave little emphasis to the other defining feature of American functionalism, the organism's adaptation to its environment. In America, that feature made functionalism an important step toward the conceptual differentation of psychology from philosophy and its simultaneous redefinition as a socially applicable science. Precisely this was absent from Stumpf's conception. The purpose of his psychology was the same as it was for Brentano – to develop a phenomenologically accurate philosophy of mind. The term "function" referred to the operations of that mind, whether they were in relation to the external world or not.[33]

Training young scientists in Berlin

Despite its limitations, the richness of Stumpf's conception of consciousness, the idea that psychical functions are observable at all and the admission that their laws remained to be discovered challenged the young and eager. Stumpf's introductory lecture awakened such hopes; the course was indispensable for students at the Berlin institute, because Stumpf never published a systematic text. According to Herbert Langfeld, who attended the course together with Kurt Koffka in the winter semester of 1906–1907, Stumpf defined psychology in his initial lecture as "the science of the elementary psychical functions," and asserted that "psychology is engaged in the observation of daily life, in order to raise such observations to a science through methodical treatment."[34] After this hopeful beginning, however, Stumpf presented not a fully developed account of psychical functions, but a highly technical survey of issues, literature, and results, in visual and aural sensation, space perception, and psychophysics. The first part of the course, "Individual Psychology and Psychophysics," which treated factual issues, contained seventeen chapters; the second, "General Psychology and Psychophysics," dealing with universal principles, had only three. Despite his claim that psychology was a necessary propaedeutic to philosophy, Stumpf had allowed his dedication to empirical research to take him far away from his original philosophical goal.

Stumpf's emphasis on facts over theory might have been exciting to the initiated, but any listener might well have wondered whether all the results he presented really were advances toward the stimulating goals he had set at the start. Some philosophy students looking for an intellectually and emotionally exciting worldview were repelled. One such student gave this account of a visit to Stumpf's lecture course:

> I entered his lecture hall and left it just as quickly; for a larger-than-life-sized picture of an ear labyrinth hung on the blackboard. Obviously I had wandered into a medical course. I finally discovered that the psychology of Professor Carl Stumpf was not at all what I understood under that name . . . we sang: Philosophy here is no big deal / It's being destroyed by Stumpf and Riehl.[35]

Not only such students had their doubts. Kurt Koffka later recounted this episode from his student days:

> A colleague of mine with whom I was going home asked me the question: "Have you any idea where the psychology we are learning is leading us?" I had no answer to that question, and my colleague, after taking his doctor's

degree, gave up psychology as a profession and is today a well-known author. But I was less honest and less capable, and so I stuck to my job. But . . . his question never ceased to trouble me.[36]

The colleague was probably Robert Musil, who completed his degree in the same year as Koffka, 1908. Originally trained as an engineer but already a published writer by 1904, Musil turned to philosophy and experimental psychology in 1905 in the hope of resolving his inner conflict between intellect and feeling, science and art. He worked for a year in Stumpf's institute, designed an instrument for controlled presentation of color stimuli in 1906, the year he published his first novel, *Young Törless,* and wrote a dissertation critical of Ernst Mach's philosophy of science. Though he continued to respect experimental psychology, he found its results unsuited to his purposes:

> But what fame awaits the writer who penetrates the depths! And precisely there is where I seek the extraordinary! And introspection is obviously such an unsatisfactory tool! It is senseless to invest one's ambitions here![37]

Those who persisted, like Koffka, were introduced to the working life of the institute in theoretical seminars and practical exercises. These were not survey courses like those offered at American universities then and now, but intensive introductions to selected issues, with emphasis on the methods for investigating them. Students were expected to obtain general orientation from Stumpf's lecture course and the literature he recommended. The seminar topics ranged widely, from "The Influence of Tragedy" (summer 1900) to "Legal Psychology" (1903 and 1904) and "The Mind–Body Problem and the Law of the Conservation of Energy" (summer 1908).[38] Students learned what Stumpf meant by experimental phenomenology in the exercises for beginners, introduced in 1902. These were conducted by the institute's assistant – Friedrich Schumann until 1905, then Erich von Hornbostel and Narziss Ach for one semester each, and Hans Rupp from 1907 on (see Appendix 1, Table 3). The assistant employed exemplary problems in psychophysics, memory, or space perception to demonstrate research techniques and the use of apparatus, drawing upon the institute's instrument collection, which was, of course, especially strong in acoustics. It included, for example, a set of tuning forks donated by the director of the Berlin conservatory, the violinist Joseph Joachim, a "pipe organ" of glass tubing designed by Stumpf, and a "tone variator" invented by William Stern to regulate electronically the presentation of ascending and descending tonal series. Equipment for vision studies included a radial tachistoscope, which looked like a bicycle wheel with slits, designed by Schumann for brief exposure of visual stimuli such as groups of letters at controlled intervals. As student interest in the exercises increased, Stumpf expanded the instrument collection, incurring budget overruns in the process.[39]

Obviously, Stumpf was not opposed to using precision apparatus, but he did not believe that laboratory technology could replace controlled introspection, or "self-observation," as he called it. In the 1890s he had had a bitter dispute with Wundt about research by one of Wundt's students in a problem in acoustics. The quarrel indicated that the nature of scientific expertise in psychology remained contested. Stumpf's position was completely consistent with his philosophical and social standpoints. He insisted that a single judgment from a musically informed observer in an appropriate setting such as a concert hall or a church "has more weight than a thousand from unmusical and unpracticed observers." Against Wundt's argument that his student had assembled more than 110,000 judgments, Stumpf contemptuously replied that one might as well decide the matter by vote.[40] As he put it later, instruments could be "useful and necessary" to "fix the conditions under which self-observation occurs as exactly and objectively as possible"; but this is "only an introduction and aid to subjective self-observation, which remains decisive as before." Schumann called self-observation not a science, but "an art, which can be acquired only by conscientious practice."[41]

Although experimental dissertations from Berlin contained tables of data, it was evidently not the primary aim to produce them. The largest amount of space by far was devoted to verbal reports of what subjects had just experienced under given conditions. Using such material presupposed the accuracy of short-term introspection, which Stumpf, following John Stuart Mill, called "primary memory." On this basis experimenters could be subjects for others and their own subjects as well, as Stumpf continuously showed in his acoustical research.[42] Thus there were differences between Berlin and Leipzig procedure, but there were also similarities. For both Wundt and Stumpf, the purpose of the experiment in general was not to predict and control the behavior of naive subjects, or to determine individual differences in their responses, but to characterize the phenomena under discussion precisely. The all-important goal of scientific training in both of Germany's most prestigious institutes was to create an elite of suitable experimenters, who could accurately observe and report their own experiences. This was a natural-scientific version of the German universities' primary purpose, the training of elite civil servants.

Given this aim, it is understandable that one of Stumpf's students later described him as "fatherly, friendly, but rather critical" in judging students' work, and that only a few of those who enrolled in seminars and exercises in Berlin began or completed experimental dissertations.[43] Between 1900 and 1915, four theoretical and ten experimental theses on psychological topics were completed under Stumpf's direction (see Appendix 2). The total for Leipzig in the same period was fifty-nine.

The Berlin institute was a very select institution indeed. The implication was that those who showed persistence and skill would have Stumpf's support in their later careers. Stumpf said that his negative attitude toward the production of dissertations was justified by the exacting demands of experimental research. He was not interested in forming a school to promote his approach to psychology, nor even in dominating a single research area by assigning pieces of it to doctoral candidates to be worked up, as he could well have done for the psychology of hearing. Instead, he pursued his research largely alone, and recruited others to help him with his other activities, such as the *Phonogramm-Archiv*. Still, at least two guiding principles were there for students to follow: allegiance to "the immediately given" as the source of psychological knowledge, with measurement used only to specify the given more precisely; and opposition to Neo-Kantian critiques of knowledge, with the intention of establishing a realist worldview based on empirical research.

Max Wertheimer chose to emphasize just these aspects of Stumpf's teaching – his attitude toward method and his focus on larger philosophical principles – in a speech on the master's seventieth birthday in 1918. It is revealing of both master and student that Wertheimer expressed his praise in aesthetic and moral terms:

> There are researchers who attack nature as though it were an enemy, set traps for her, take the offensive, try to bring her down like an opponent, like cool technicians or sportspeople who want to feel their own strength, achieve successes. Others are like travelers who write amusing essays; others have their card boxes, write thick books in which everything swims confusedly together; still others work busily on a certain small piece, have real C.N.V.F. [concentric narrowing of the visual field] . . . , are blind to the right and to the left.
>
> How different you are! For you the facts are not objects of attack, nothing that should bring flashing results. For you the facts are as though they were in a father's hands. In Africa there is a custom in one tribe: When one wishes to show trust to a guest, the mother lays her nursling in his arms and says, hold the child. So do you hold the facts in your hands, and so you have taught us: devotion to the real. . . .
>
> And second: so round, so clean, so complete in itself is everything for you – nothing of C.N.V.F.! We always feel: This is not just any individual fact, but everywhere principles are at work. Here is no splitting up of any large-curved [*grosskurvige*] results, but rather always building stones, that is, always work with the most important issues of specialized science in view. . . .
>
> And third: In yet another sense there is no C.N.V.F. for you. As much as you love and support work in specialized science, you have nonetheless taught us to keep our gaze directed to larger questions of principle, to work toward the fruitful cooperation of psychology and the theory of knowledge, with the highest problems of philosophy in view. *None of us wishes to be locked up in the workroom of specialized science.* . . .[44]

3

The philosophers' protest

By the time Wertheimer had spoken his words of praise, the potential danger of "being locked up in the workroom of specialized science" had become clearer than ever before. The young scientists who emerged from the Berlin institute and other psychological laboratories in Germany were experimenting psychologists who genuinely deserved the name, since the largest part of their training consisted of laboratory work. They may have taken up this specialty in hopes of discovering empirical solutions to weighty issues like the mind–body problem.[1] But their investment of time and energy in acquiring experimental techniques – their immersion in laboratory life – led to a changed attitude toward the parent discipline. Some of their teachers, like Stumpf and G. E. Müller, had already been caught in the spider's web of methodological precision. Even so, a philosophical worldview remained at least a distant hope for them. For some members of the new generation, philosophical reflection had become only a source of theories to be tested by experiment.

An incident at the congress of the Society for Experimental Psychology in Innsbruck in 1910, where Köhler and Koffka gave their first papers and Wertheimer was a spectator, reveals how dogmatic such attitudes had become. At that meeting philosopher Moritz Geiger presented a major address entitled "On the Essence and Meaning of Empathy." In the discussion period, Karl Marbe, who had just become director of the Würzburg institute, remarked that the significance of such theorizing lay only in stimulating experimental investigations: "The method employed by the proponents of empathy theory is in many ways related to the method of experimental psychology the way the method of the pre-Socratics is to that of modern natural science."[2]

Philosophers responded in kind to the challenge of such scientistic rhetoric, and to the more concrete threat posed by the experimenters' increasing numbers and productivity. Freiburg professor Heinrich Rickert, for example, a leader of the Southwest German school of Neo-Kantianism, asserted that the goals of psychology and natural science were the same – to explain mental processes by subsuming them under general natural laws. "Logically viewed, psychology is therefore a natural science." In contrast, history deals with unique events and individuals, while philosophy seeks normative principles and not empirical laws. The implication was clear: neither history nor philosophy has any use for experimental psychology.[3]

The tone of Rickert's mentor, Wilhelm Windelband, was considerably uglier. In a 1909 portrait of nineteenth-century German philosophy, for example, he asserted that:

> For a time in Germany it was almost so, that one had already proven himself capable of ascending a philosophical pulpit [*Katheder*] when he had learned to type methodically on electrical keys and could show statistically in long experimental series carefully ordered in tables that something occurs to some people more slowly than it does to others. That was a none too satisfying page in the history of German philosophy.[4]

He hoped that this "temporary dominance of psychologism" would give way to a renewed Hegelianism better equipped to provide truly philosophical answers to "the great problems of life, the political, religious, and social questions."[5]

In view of the statistics cited in Chapter 1 showing the relatively small number of psychological institutes in Germany, Windelband's claim that experimental "psychologism" had been or was becoming "dominant" in philosophy was plainly a gross overstatement. Historians have long taken such comments as at least indirect evidence of growing resistance by traditional German elites to the changes in social values associated with rapid industrialization and urbanization. University professors expressed this resistance most vividly in their long, ultimately unsuccessful struggle against granting equal rank to engineering and technical disciplines.[6] Philosophers who were so disposed apparently found experimental psychology, this intruder in their midst, to be an eminently suitable object on which to project such resentments. Windelband's contemptuous reference to key-punching marks a traditional distinction between mental and hand work. It also refers to activities associated with the emerging new middle class of employees, many of whom worked as recordkeepers in insurance companies and banks. His remark manifestly expresses fear of cheapening an elite discipline by making entry too easy to achieve. But his use of the traditional term *Katheder* for the professor's chair signaled another fear – that of losing the spiritual potency and moral influence believed to flow from reflection on general truths. Just this antiutilitarian commitment had become increasingly problematic as the practical demands of industrial society for specialized expertise grew.[7]

But the conflict was not only one of elite *Bildungsbürger* (educated middle-class) values against mere technology. The leading sectors of German industrialization, chemicals, and electricity, were precisely those in which scientifically trained management was at a premium, and university disciplines such as chemistry and physics had received enormous institutional and intellectual impetus from this fact. Experimental training in the natural sciences could be and was accommodated in the universities, so long as scientists proclaimed the continued primacy of high theory and basic research over applied work, and emphasized the pedagogical over the technological value of laboratory training.

In Stumpf's institute, too, experimentation had a pedagogical purpose, giving philosophy students hands-on contact with science, while the director and his associates employed it in the service of high theory. But other experimenting psychologists tried to respond to societal demand for usable science by converting laboratory techniques into practical tools. The most prominent examples were Hermann Ebbinghaus's word or sentence completion test (aimed at saving mental effort), William Stern's differential psychology, Ernst Meumann's concept of "mental work," and Hugo Münsterberg's psychotechnics (see Chapter 1). Stern tried to go in both directions by helping to found Germany's first journal and laboratory for applied psychology in 1908, while developing his scientific concern with individual differences into a personalistic philosophy.[8] For many philosophers, however, neither this level of effort nor the conscious self-limitation of people like Wundt and Stumpf to high theory and basic research was sufficient. They attacked not one or another approach to or use of experimental psychology, but the experiment itself.

In his 1911 essay, "Philosophy as Rigorous Science," for example, Edmund Husserl accused the "experimental fanatics," as he called them, of confusing their "cult of the facts" with a genuine analysis of consciousness. The essential qualities of consciousness, for example its intentional or directed character, are "in principle different from the realities of nature." They must therefore be studied with methods and described in terms different from those of the natural sciences. According to Husserl, asking how often subjects make, for example, a judgment that object *a* is the same as object *b* under given conditions cannot tell us anything about the act of judgment as such. Only an "essential" or phenomenological analysis of consciousness can make philosophical sense of "the gigantic experimental work of our times, the plenitude of empirical facts and in some cases very interesting laws that have been gathered. . . . Then we will again be able to admit that psychology stands in close, even the closest relation to philosophy – which we can in no way admit with regard to present-day psychology."[9]

Perhaps Carl Stumpf could have agreed with this argument, had it been put more moderately. In fact, Husserl named Stumpf and Munich professor Theodor Lipps as exceptions to the trend he criticized. Most important in this context is that Husserl also pointed to the academic politics at work. After attacking "that sort of specious philosophical literature that flowers so luxuriantly today," offering "theories of knowledge, logical theories, ethics, philosophies of nature, pedagogies, all based on the natural sciences, above all on experimental psychology," he added this remark in a footnote:

> This literature receives support not least because the opinion that psychology – and 'exact' psychology of course – is the foundation of scientific philosophy has become a firm axiom, at least among the natural scientists in the philosophical faculties. These faculties, giving in to the pressure of the natural scientists, are very zealously giving one chair of philosophy after another to scholars who in their own fields are perhaps outstanding but who have no more inner sympathy for philosophy than chemists or physicists.[10]

Clearly, the institutional situation of experimental psychology in German-speaking universities was no longer satisfactory by 1911, if it had ever been. The original strategy of advancement within philosophy accorded with both the commitments of the experimentalists themselves and the structure of the university system, which accommodated innovation to the preservation of professorial privilege. That strategy had succeeded for twenty years, but its very success was bound to lead to a Malthusian conflict sooner or later. The limits of philosophers' willingness to tolerate this state of affairs had been reached. A new disciplinary boundary line was being drawn with the help of the attack on "psychologism," and the experimenting psychologists were being placed firmly outside the boundary.

The experimenters were well aware of this. During the opening ceremonies of the fifth congress of the Society for Experimental Psychology in Berlin in 1912, several speakers impressed upon the ministerial officials and other politicians present the need for more government support, especially new professorships, for experimental psychology. The Prussian minister for culture and higher education responded with a vague reference to the society's small size, compared with medical groups. The mayor of Berlin stated bluntly that he hoped to see concretely applicable results from psychology, especially in such areas as the examination of witnesses in court and the determination of moral responsibility in cases of insanity.[11] Nothing came of this exchange, but the terms of discussion had clearly been set. If psychology were to get more support, its representatives would have to prove its usefulness to society, and redefine its discourse accordingly. If this was not possible for psychology as a philosophical specialty, then alternative locations would have to be found.

Oswald Külpe proposed such a relocation the same year. Külpe had published extensively in both general philosophy and experimental psychology, and had made the establishment of well-equipped laboratories a condition of all three of his professorial appointments.[12] Even so, he recognized that the next generation of experimenting psychologists no longer had either the time or the patience to deal with the demands of a double identity: "for the newly rising breed it is becoming practically impossible to serve both masters, to do one thing and not let the other lapse, if they do not wish to sink into dilettantism and superficial busy work." Külpe's solution was clear, and on the surface quite practical – to establish "a psychological institute with corresponding work rooms, budget, and staff" in all university medical faculties, as part of the training in basic science already required of physicians. He justified taking such a step by pointing out the need for better empirical research in psychopathology. Though Külpe conceded that psychiatrists and neurologists had already begun to develop a range of empirical methods as diagnostic aids, he contended that these relied on "vulgar experience alone" and ignored "the available specialized research in psychology."[13]

Some reviewers agreed with Külpe's critique of psychiatry but did not discuss his proposed remedy. Others wondered where the money could be found to finance so many professorships. Most incisive was the response of physician and psychologist Willy Hellpach in the *Zeitschrift für Psychologie*. Though he endorsed the reform plan in principle, Hellpach pointed out that not only psychiatrists, but practicing physicians of all kinds would have to be persuaded that experimental psychology could help them in diagnosis, therapy, and providing expert testimony. "A lot of water will flow into the sea before psychology officially moves into the medical propaedeutic," he predicted. Fortunately or not, he was correct. Psychology did not officially enter medical education in Germany until the 1970s.[14]

Külpe's essay also found two more sympathetic readers – Kurt Koffka and Wolfgang Köhler. In a review for a nonspecialist journal, Koffka, who had worked briefly with Külpe in Würzburg and had just begun his teaching career in Giessen, praised the proposal unstintingly:

> As things are, intervention is only to be expected when the authorities are convinced of the usefulness, yes of the indispensability of experimental psychology. . . . We younger psychologists have every cause to thank Külpe for having pointed out this route, not least because of his understanding and sympathy for our scientific difficulties.[15]

Köhler, who had just begun teaching as an instructor in Frankfurt, also became involved in the controversy, but he avoided taking a clear position. Responding to a negative review of Külpe by Georg Anschütz of Leipzig, Köhler took the author to task for merely repeating Wundt's attacks on Külpe's approach to psychology and offering only "petty little criticisms" besides. It does not matter what opinion one has on the relationship of psychology and philosophy, he declared, so long as the opponent in a dispute is treated fairly.[16]

During this discussion, in 1912, the event occurred that transformed the issue into a crisis. To fill the chair left vacant by the retiring doyen of Marburg Neo-Kantianism, Hermann Cohen, the philosophers there recommended Ernst Cassirer, who held related views. The natural scientists had been pressing for years for the appointment of an experimenting psychologist. According to

Cohen's colleague and friend Paul Natorp, state officials, faced with the difficult choice between settling the dispute expansively by establishing a new chair for experimental psychology, or satisfying the philosophers by naming Cassirer, a Jew of known liberal politics, did neither. Instead they appointed Erich Rudolf Jaensch, a student of G. E. Müller and a philosophically trained experimenter, to the existing chair. In answer to this, six of the leading philosophers in Germany – Natorp, Rudolf Eucken, Edmund Husserl, Alois Riehl, Heinrich Rickert, and Wilhelm Windelband – circulated a petition against appointing any more experimentalists to philosophy chairs.[17]

The petitioners recognized "the highly gratifying advance of this discipline" in recent years, but found that naming its representatives to philosophy chairs deprived students "of the opportunity to obtain systematic direction from their professors about general questions of worldview and philosophy of life, especially in these philosophically troubled times." A total of 107 university teachers signed the document, including 27 (40.9 percent) of the 66 full professors of philosophy in Germany and Austria. Excluding the experimentalist full professors, who of course did not sign, the percentage was exactly half. The philosophers sent their statement not only to the philosophical faculties of the German-speaking universities, but also to the ministries of culture and education of all the German states. This act of open lobbying was an unprecedented step in the history of a discipline whose members thought of themselves as being above politics.[18]

In response to this action and to Külpe's proposal, Wilhelm Wundt published a widely cited polemic, "Psychology in the Struggle for Existence," in 1913. Wundt castigated the philosophers for thinking only of their "property rights," and argued for preserving the status quo on both theoretical and practical grounds. To separate one type of psychological research from the others by excluding the experimenters from philosophy would be dangerous and unjustified; for psychology's subject matter, mental life, is in principle a unified whole, only parts of which are accessible to experiment. In any case, he reminded the psychologists, "nearly half of the entire psychological literature is made up of books and articles that extend into the fields of metaphysics or the theory of knowledge." The "liberation" of experimental psychology would not mean that psychologists would no longer concern themselves with such issues, but only that they would do so without proper philosophical training. "Then the time would truly have arrived when psychologists become tradesmen (*Handwerker*)," Wundt warned, "but not exactly tradesmen of the most useful sort."[19] By using this class-based metaphor, he showed that he shared the elite values of the philosophers, such as Windelband, who had used similar terms to condemn the work done in laboratories like his own.

In any case, Wundt pragmatically argued, since psychology still lacked obvious applications and would therefore attract few students on its own, the only way it could justify its presence in the university system was to remain part of the general philosophical education of Gymnasium and university teachers. It was highly improbable that government authorities would provide funds and facilities for an independent psychology that they had not given to psychology as a part of philosophy. It was therefore in the interests of philosophers and psychologists alike that no one be allowed to teach in psychology "who is a mere experimenter and not at the same time a psychologically and philosophically educated man, filled with philosophical interests."[20]

The intense and at times bitter debate on the status of psychology that followed lasted for more than a year. Everyone agreed on the rejection of one-sided specialists, but the battle raged on, with ugly words on both sides.[21] Leipzig historian Karl Lamprecht, defending his friend and colleague Wundt in Maximilian Harden's journal *Die Zukunft* and thus bringing the dispute to the attention of a wider public, presented the issue bluntly as a political contest. Self-styled "pure" philosophy, which he disparagingly called "conceptual poetry" (*Begriffs-dichtung*), was enjoying a revival in the universities, he noted. This was no problem in itself, but now the "pure" philosophers were trying to reserve more positions for themselves by pushing the psychologists into the natural sciences. This was nothing more than an importation of "the idea of power politics" into the university.[22]

Judging by the intense reaction to it, Lamprecht's reference to power politics touched a raw nerve. In general, however, his opponents confirmed his analysis without realizing it. In his reply to Lamprecht in *Die Zukunft,* for example, the philosopher and social theorist Georg Simmel argued, correctly, that the historian was wrong to see the philosophers' petition as the work of only one school of thought. Rather, he contended, the discipline as a whole was defending itself against being starved out by a field that had yielded nothing of direct significance for philosophy "excepting perhaps Fechner's [psychophysical] law."[23]

The thoughtful youth of today, Simmel claimed, want "something more general or, if you will, more personal" from philosophy than this:

> One may call this a mere secondary product of science, or philosophy as science [*Wissenschaft*]; but where it is no longer offered to the young, they turn to other sources that promise to fulfill their deepest needs: to mysticism or to that which they call "life," to social democracy or literature, a falsely understood Nietzsche or a skeptically colored materialism. Let us not delude ourselves: the German universities have given up the leadership of youth to forces of this kind. Certainly the movement away from philosophy in the older sense to experimental psychology is not the only reason for this turn . . . but the substitution of chairs of experimental psychology for chairs of philosophy proper puts the seal on this tendency and gives it increasing support.[24]

The dominant motifs of the academic thought of the day were all present in this virtuoso passage, especially the fear of social democracy and materialism and the nostalgic idea that German philosophers had once had "the leadership of youth." All that was missing was the slogan-word *Weltanschauung*. Better proof could not be had of the deeper political and cultural significance of this dispute, or of the underlying motives of the philosophers' campaign. It was not only a matter of disciplinary demarcation lines, but a question of discursive regulation – of fervent hopes, and dreams, about how "real" philosophers should talk.

A consensus did gradually develop. If experimenters were to be kept out of philosophy professorships, cooler heads thought, their place was still in the philosophical (that is, arts and sciences), not the medical faculty. As the Munich pedagogue Aloys Fischer wrote, "it is time now to establish positions for [psychology] in the philosophical faculty. Philosophers and psychologists should support this demand together, instead of fighting one another." Oswald Külpe accepted this position in 1914.[25]

Yet it was not to be. Experimenters without philosophical pretensions, like Friedrich Schumann, could seize opportunities for expedient solutions as they arose. Schumann had been appointed head of the psychological institute of the commercial academy in Frankfurt in 1909. When the academy became a full-fledged university with separate philosophical and natural science faculties in 1914, he chose to be assigned to the natural sciences for the simple reason that "of the humanities, psychology is the most expensive; of the natural sciences, it is the cheapest."[26] Karl Marbe's solution in Würzburg was to secure psychology's right to count as a major field alongside philosophy and pedagogy in state teachers' and doctoral examinations.[27] In general, however, experimental psychology's institutional base remained limited to single chairs of philosophy in the German universities until World War II.

This, then, was the social situation that young investigators like Wertheimer, Köhler, and Koffka faced as they sought a theoretical and practical orientation for their lives in science. The psychological laboratories that had been established in Germany by this time and the multidisciplinary community gathered at the meetings of the Society for Experimental Psychology gave them an institutional context within which they pursued their research interests. As far as permanent positions were concerned, however, that network was not rich in locations, and the philosophers' protest showed that the limits to growth had been reached. Experimental psychology had achieved a foothold in the German scientific establishment, but only on condition that its practitioners demonstrate that they were "filled with philosophical interests" – prepared, that is, to create and sustain a mode of discourse consistent with the "higher" intellectual and moral values allegedly upheld by the German philosophical tradition.

This was not the only way things could have gone. By 1914 the relevance of experimental psychology to a wide variety of other disciplines and its potential applications in law, psychiatric diagnostics, education, and industry had become abundantly clear. In the same period, American psychologists, blessed with a much less restrictive university structure, traded effectively on similar claims of technocratic potential to separate their discipline from philosophy and to redefine its aims. When John B. Watson proclaimed in 1912 that the purpose of psychology was "the prediction and control of behavior," he gave a name to this development.[28]

Wilhelm Wundt pointed to one of the reasons why this did not happen in Germany when he took note of the American situation and warned that psychologists separated from philosophy would become "tradesmen." Carl Stumpf's sentiments in this matter were little different. Such attitudes were not limited to the older generation. Even militant advocates of applied psychology, such as Karl Marbe, or other younger researchers, such as Franz Hillebrand, thought that psychology's place was in philosophy. Hillebrand was sure that experimental research could make valuable contributions to metaphysics and the theory of knowledge: "How many useless controversies could be avoided in those fields, if only the investigations of the essence and origin of our perception of space and time . . . were deemed worthy of notice by the 'pure' philosophers! Instead, the fiction of their a priori character still does its mischief today."[29]

This was precisely the sort of scientistic discourse that the philosophers wanted no part of. The social roots of their dissatisfaction lay in a combination of antimodern *ressentiment* and territoriality. But the most prominent reason they gave for it was the gap they claimed to discern between experimental psychology's philosophical ambitions and its actual results. We have seen in the case of Carl Stumpf that such criticism was based on something more than an aversion to doing philosophy with instruments. There was indeed a gap between the sober, pragmatic world of psychological facts and the "higher" values that German philosophers wished to preach. As they worked in Stumpf's institute, Koffka, Köhler, and Wertheimer directly confronted both the promise and the problems of experimental psychology as a philosophical discipline. As Koffka's response to Oswald Külpe's proposed reform shows, they were well aware of the debate about psychology's status, and understood that it had clear implications for their own future.

It has often been said that in Germany, intellectual revolutions tend to become confused with the genuine article – "revolutions in the head," Hegel called them. Here was a situation in which a radical reform, if not a revolution in psychological thinking could have practical effects. Such a radical change was what the Gestalt theorists tried to bring about by reconciling holism and natural science. In order to understand their attempt more completely, it is necessary to examine more closely the intellectual situation in which it was rooted.

4

Making a science of mind: Styles of reasoning in sensory physiology and experimental psychology

What is science; what is mind; how can one be brought to bear upon the other? Contrary to linear accounts of experimental science replacing armchair speculation, a common context for psychological discussion endured in Germany throughout the nineteenth century, extending from physiology and philosophy to pedagogy and literature.[1] Experimenting psychologists entered this discursive field rather late, and never achieved scientific or cultural control over more than a small portion of it. As they struggled to create a science of subjectivity, they organized their efforts around two problematics. One, derived primarily from Kant, centered around the problems of knowledge and aesthetic judgment; the other, rooted in *Naturphilosophie,* was the mind–body interface, which came to be called the psychophysical problem.

Theirs was a complex challenge: to mobilize the instrumental and conceptual resources of the natural sciences, in particular sensory physiology, and bring them to bear on these central concerns of nineteenth-century German philosophy; and at the same time to assure the autonomy and reality of the subjective phenomena they wished to explain. The natural-scientific side of that balancing act will be considered in this chapter, the philosophical one in the next. In a sense this divide is artificial; for the conceptual foundations physiologists and experimenting psychologists proposed for a natural science of mind were themselves "implicit philosophies," with quite distinct cultural resonances.[2] The frame of reference in each case is not the entire cultural or scientific scene, but the portion of it with which the Gestalt theorists became acquainted during their initiation as scientists. In these debates they encountered during their training both the mode of thinking against which they rebelled and the cognitive and instrumental resources from which they created their alternative approach.

51

Helmholtz versus Hering: Mechanical and organic explanatory styles in sensory physiology

Gestalt psychology began as a radical revision of three overlapping dualisms that dominated perceptual theory and research in the last third of the nineteenth century: sensation versus intellect, peripheral versus central processes, and physiological versus psychological categories. For Hermann Helmholtz, unquestionably the leading sensory physiologist of his time, these three dualisms were equivalent. In their efforts to reform this schema, the Gestalt theorists mobilized alternative assumptions and theoretical models proposed by Helmholtz's leading opponent, Leipzig physiologist Ewald Hering. The controversy between Helmholtz and Hering has traditionally been presented as a contest between "nativist" and "empiricist" theories of vision (discussed in the Introduction). The terminology conforms to contemporary usage – actually Helmholtz invented it, to Hering's consternation.[3] But the issues were far broader and deeper than this. Theirs was a dispute about how to construct a science of sensation, based on deeply conflicting philosophical commitments, styles of theorizing, and investigative strategies.

In his classic paper "On the Conservation of Force" (1847), Helmholtz stated clearly that his scientific language, like that of Isaac Newton, depended on a conception of "the universe as consisting of elements with inalterable qualities." Only in such a world would it be possible to explain natural phenomena mechanically – by referring only to the motions of objects treated as point-masses having "inalterable motive forces that are dependent only on spatial relations."[4] His theories of vision and hearing were, in essence, attempts to employ this mechanistic and deterministic language in the study of the senses. Helmholtz described visual sensation, for example, as a causal chain of local motions. Excitations are transmitted by a multitude of nerve fibers from the cones on the retinal surface. These are so distributed that each cone receives stimulation from corresponding bright points in the visual field. Each fiber proceeds "through the trunk of the optic nerve to the brain, without touching its neighbors, and there produces its special impression, so that the excitation of each individual cone produces a distinct and separate effect on the sense."[5]

The so-called Young–Helmholtz color theory, which specified retinal receptors and corresponding neural processes for the primary colors red, green, and violet, was simply an application of this idea. In the case of tone sensation, Helmholtz postulated a collection of nerve fibers leading from the arches of Corti in the inner ear – later it was the more than 4,000 transverse fibers of the basilar membrane – each of which, in his view, resonated best in response to a different frequency.[6] He emphasized the strictly mechanical nature of sensory processes, the absence of input from other sources during their course, the separation of the nerve fibers from one another – and the intimate linkage of his vocabulary with then-current technology – by comparing the whole with a network of telegraph wires.[7]

As Helmholtz acknowledged, these mechanical operations did not produce an exactly accurate image of the physical world. Sensations are not true copies but "signs" of that world. Helmholtz assured the correspondence of the two worlds by two very different means, which together point to a fundamental tension in his thinking. On the one hand, he took an empiricist and pragmatic position emphasizing humans' interaction with the physical world; as Timothy Lenoir puts it, he "treated the eye as a measuring device which the brain uses in constructing a practically efficient map of the external world."[8] On the other hand, he held a Kantian belief in the objectivity of natural knowledge; sensory signs are thus parallel to the physical world in the sense that they "produce a lawful order by a lawful order."[9] For the disciplinary project of making a science of sensation, the primary difficulty in this approach was that the information provided by Helmholtz's sensory signs is insufficient to account for a number of important perceptual facts. The most significant of these is that we see objects in three dimensions, even though the pattern of excitations on the retina is two-dimensional. It was in such cases that Helmholtz invoked psychological language.

To deal with the problem of depth perception, Rudolph Hermann Lotze had posited "local signs" in 1854. These were additional excitations, independent of those for color and assigned by the succession of feelings associated with eye movements, which convey the location of stimulated retinal points with respect to the point of clearest vision (see Chapter 2). Helmholtz accepted these and added tactile sensations. Frequent association of the latter with the former lead to "unconscious conclusions" of spatiality, which he portrayed as nonverbal versions of the inductive conclusions described in John Stuart Mill's *Logic*. The major premise is an inductive generalization from previous experience; the minor premise are the current sense data; the conclusion is a set of perceptions of the size, shape, and distance of objects. Thus, for Helmholtz the psychology of perception is scientific inference writ small.[10] Instead of accepting Mill's view of mind as a passive recipient of sensations, however, Helmholtz called his inductions "an unconscious and involuntary activity of the memory" and likened the learning involved to a form of "experimentation." The perceptions produced by this process are so irresistible that they are treated "as if" they were themselves sensations. Nonetheless, Helmholtz thought that the component sensations, in this case the flat retinal images, are recoverable, given the appropriate direction of attention and practice.[11]

Commentators have noted a certain tension between Helmholtz's emphasis on active inference in his general account of perception and on passive association in specific accounts of vision and hearing. His recourse to unconscious conclusions from and judgments of sensation evoked thoughts of a Kantian active intellect. He later indicated some affinity for the views of the idealist philosopher Fichte, calling our experiences of the lawful regularity of phenomena "a force acting in opposition to us in a manner analogous to the workings of our own will."[12] Helmholtz did not think, like his colleague Emil Du Bois-Reymond, that psychical phenomena are "outside the realm of causal law," that is, beyond the reach of mechanical explanation.[13] Yet when he spoke of will in this way he implied that a mechanistic science of sensation was possible only at the cost of adopting a nonmechanical philosophy of mind.

To Helmholtz and those who followed his example, the assumption that the nerves are indifferent carriers of excitations which played no role in the ordering of sensation meant that the sense organs could be studied as mechanical instruments, as Helmholtz did with the eye as a camera oscura. The pattern of measurable physical stimuli could be compared with the corresponding pattern of elementary sensations, and searches undertaken for anatomical arrangements in the organ itself capable of adequately transforming the former into the latter. This was more than sufficient as a research program for sensory physiology.[14] For experimenting psychologists, Helmholtz's assumption of the physicist's world instead of the experiencing subject as his point of departure had led him to relegate some of our most impressive introspective experiences, such as color contrast, to the realm of illusion or unconscious judgment. In a memorial essay for Helmholtz published in 1895, Carl Stumpf identified such issues as the fundamental dividing line between the physiology and psychology of the senses.[15] It was also the place at which the alternative, organic style of explanation presented by Ewald Hering as well as the psychophysical parallelism of Hering and Ernst Mach made their marks.

Hering presented easily comprehensible demonstrations to bring home the insufficiency of perceptual theories based on physical assumptions alone. When we look at a piece of white cardboard from which a zigzag piece has been cut, for example, we might see either a hole in the cardboard and a dark space behind it, or a black patch in the plane of the cardboard; yet the retinal image is the same for each impression.[16] To attribute such phenomena to judgments or other psychological acts, Hering argued, was comparable to earlier "explanations" in terms of vital forces. More fundamentally, he asserted that Helmholtz, because he started out from the outside – the physical stimulus – and not from sensation per se – what observers actually see – could not explain the subtleties of the eye's adaptation to light, color contrast, after-images, our perception of four, rather than three, primary colors, or the nature of black and gray. The point, he insisted, is not whether experience plays an important role in perception – it surely does – but to develop a truly physiological explanation of the facts.[17]

Hering's real aim was to avoid Helmholtz's dualistic recourse to hypothetical psychological processes operating on sensory material. To achieve that purpose, however, he redefined the object of discourse in the theory of vision by giving sensation more of what Helmholtz had thought of as perception, and thus making at least part of the psychical not an underlying cause of phenomena, but the thing in need of explanation. The core term in Hering's alternative language was not a strictly physiological one, but the distinction between real and "seen objects" (*Sehdinge*), which he made in 1861 and retained throughout his career. The example just given shows what he had in mind – the dark space behind the zigzag hole in the cardboard and the black patch on its plane are both "seen objects." For him this distinction was not a question for epistemological debate; whether these phenomena are "objective" or "subjective," whether they are "really" experienced directly or concluded from "unnoticed sensations," was beside the point. Accepting the psychological reality of seen objects was a methodological necessity, "an indispensable prerequisite for understanding the visual function and its laws."[18]

Hering's seen objects were plainly not equivalent to Helmholtz's signs. For him, the sensory order is not a map of the external world, but the only reality we have. As he wrote in 1862: "The retinal image is the final effect known to us of real objects on our bodies. What transpires beyond the retina, we do not know."[19] Within that reality, relationships are not something concluded, unconsciously or otherwise, but something immediately apprehended. For example, the fact that we do not call the brighter part of a gray paper white but only brighter when light is reflected onto it "*corresponds to a difference in perception*" from that of a white paper lying next to it. This does not mean that the sensation is really the same in both cases, but is only interpreted differently: "*I believe rather that the 'sensation' is fundamentally different in the two cases.*" This is possible because not only peripheral sense organs, but also the parts of the brain to which they are connected are actively involved in organizing sensation from the beginning: "light sensation is not simply a function of the stimulus and the momentary state of the first affected neural structures, but also depends on the state of the part of the brain related to the visual activity, in which the optical experiences of one's whole life are contained and in some way organized."[20] Hering later admitted that he did not know what the physiological correlates of relations like "darker" or "lighter" might be. But it was clear that this broadened concept of sensory processes required a different style of explanation from that offered by Helmholtz, one that employed organic rather than machine metaphors and placed more weight on functional interaction within the organ of sense, and between peripheral and central organs, than on mechanical signal or energy transfer.

Encouragement to develop such a model came from research Ernst Mach conducted in the 1860s on contrast phenomena. Among these were the gray, ringlike areas that appear to observers between the black and white bands of a rotating disk at certain speeds of rotation. These have since been called "Mach bands" in his honor. Mach showed that such phenomena occur whenever corresponding brightness relations exist; these gradients, he asserted, appear in the retinal image itself, and are "of the utmost significance for the perception of external objects." He attributed them to interaction among excited retinal areas, such that "the light on a single point is thus evaluated in terms of the light on all other points."[21] Hering accepted Mach's claims, but criticized him for not going far enough. As he later argued, retinal interaction must be significant for many other phenomena besides simultaneous brightness contrast; it was necessary, if not sufficient, for the correct perception of color in general.[22]

To support this claim, Hering extended the principle of mutual interaction from the retina to the entire "visual space" (*Sehraum*), including the retina, the optic nerve, and related parts of the brain. This he called the "inner eye" in contrast to the lens, pupil, and retina, which he called the "outer eye." Borrowing a term from metabolism – an organic, not a mechanical process – he postulated four self-regulating processes of "assimilation" and "dissimilation" in the "inner eye," one for each of the primary colors and one for black and white. External stimulation activates the corresponding "assimilation" or "dissimilation" process, which reaches a given midpoint between the relative levels of stimulation and of autonomous nervous activity, then returns to a state of equilibrium. This leaves a residuum or "trace," which remains in the "inner eye" for activation when suitable stimuli arrive from without. This, Hering thought, was the explanation of relative color constancy – the fact that we do not perceive the full range of ambient illumination, but only a narrow range of relatively stable colors. Activated traces of previous color experiences compensate for changes in illumination and thus produce what Hering called "memory colors." He asserted that not only the organic structures involved but the traces themselves were heritable.[23]

Hering's organic model offered a genuine alternative to Helmholtz's mechanistic conception of sensory processes. Two fundamental differences in these styles of theorizing remain to be considered – in the meanings conveyed by the term "functional," and the cultural resonances of those meanings. For Helmholtz, the mind's task is wholly pragmatic – to localize and identify objects for the purposes of life. Perception is thus explained when its function is identified in this sense. This was the reason for Helmholtz's emphasis on sensations as signs for external objects. Given this functional orientation, it follows that positing organic processes in the visual part of the brain does not explain our experience of space. For Hering, a functional theory referred to the operation of the organism itself rather than to its relation to the external world. Hence, starting from physical objects in geometrical space cannot explain the experiences we have of "seen objects" in phenomenal space.

Helmholtz's empiricism was fundamentally linked with his commitment to science as "the intellectual mastery of nature." He assigned the central role in his physics and physics-based physiology to the concept of *Kraft,* i.e., force, only later renamed energy, which had been derived in part, as he himself acknowledged, from human labor power. Physiologists and others soon applied the idea of energy conservation recursively to human labor in attempts to create a science of work intended to make the "human motor" run more efficiently, an effort extended to "mental work" by Emil Kraepelin and others by the end of the century.[24] For Helmholtz, using this metaphor allowed him to consider sense organs as machines and to employ technological analogies like that of the telegraph for nerve transmission. At times he even subordinated his Kantian commitment to the objectivity of knowledge to this discourse of functional control. "The law of sufficient reason," he wrote, "is really nothing more than the urge of our intellect to bring all of our perceptions under its own control." Causality is hence no more than a tendency of intellect to project this urge onto nature.[25]

Hering is often falsely labeled as an idealist, but he in turn denounced Helmholtz as a "spiritualist" for invoking psychological processes to overcome difficulties in his theory of sensation.[26] His scientific credo, published in 1870, was entitled "memory as a function of organized matter"; as will be clear shortly, he firmly believed in assigning mental events to material causes. Still, his emphasis on the primacy of experienced realities over those of the physicist was bound to be attractive to the aesthetically inclined. Further, his preference for an organic chemical over a mechanical version of functionalist discourse distanced him from metaphors of domination and control.

The reception of Hering's ideas in sensory physiology was ambivalent. His four-process model of color sensation was widely adopted, or incorporated into others, but his overall approach was not so widely accepted.[27] For experimental psychologists, Hering's perspective opened up two exciting prospects. They could adapt his phenomenological approach and use his investigative techniques to open up a whole realm of psychological phenomena that could be studied empirically and ordered mathematically, without invoking vital forces. And they could listen to the siren song of Hering's and Mach's psychophysical parallelism.

Hering once likened the relationship between physiological and psychological research to that between teams of tunnel-borers working from opposite sides of a mountain. If the two teams would only cooperate and follow the same rules, the odds favored their meeting in the middle: "The goal is knowledge of the causal relation of all physical events on the one side, of all psychological events on the other; *the presupposition is the lawlike dependence of both types of events on each other.*"[28] Ernst Mach had proposed such a "heuristic principle of research" as early as 1865:

> Every psychical event corresponds to a physical event and vice-versa. Equal psychical processes correspond to equal physical processes, unequal to unequal ones. When a psychical process is analysed in a purely psychical way into a number of qualities a, b, c, then there corresponds to them just as great a number of physical processes, α, β, γ. To all the details of psychical events correspond details of the physical events.[29]

Kurt Koffka later acknowledged that this version of psychophysical parallelism "looks identical" with the one formulated by Wolfgang Köhler in 1920, but argued that Mach's philosophy prevented him from making effective use of it.[30] In an 1896 essay, G. E. Müller summarized Mach's and Hering's formulations in a set of five "psychophysical axioms," adding the principle that every qualitative change in sensation corresponds to a qualitative change in the "psychophysical process" connected with its occurrence. Stumpf later pointed to the exciting prospect that such rules for theory construction held out. If Müller's axioms proved correct, he wrote, psychology could become "the giver, not the taker discipline" with respect to physiology.[31] The Gestalt theorists took up that offer, while radically reformulating Müller's axioms (see Chapter 11).

Hering's student and co-worker Franz Hillebrand showed that it was possible to do effective research with seen objects, without speculating about brain events. In an 1893 paper, he found that objects viewed with both eyes are not seen at the average midpoint of their lines of direction, as predicted by Helmholtz. Hering had said that objects appear instead along the horizontal horopter, the line at which objects are seen single. Hillebrand modified this by showing that nearer objects are seen in a curve concave, further objects in a corresponding curve convex to the horopter. This came to be known as the Hering–Hillebrand horopter deviation. He drew the sweeping conclusion that "the location of seen objects is generally not in agreement with that of the corresponding real objects."[32] For psychologists, the message was that such subjective phenomena could be studied for themselves, and the deviations involved measured, with or without physiological theorizing.

Hillebrand's paper was concerned with the problem of apparent distance of objects from observers. Wundt's student Götz Martius had already measured phenomena of apparent size in 1889. Hillebrand and others extended this work and developed mathematical models to regularize the results. In 1909, Walter Poppelreuter then added an additional dimension, that of form (*Gestalt*). Apparent form, he pointed out, only corresponds to objective form in frontal parallel orientation. The wall cabinet next to the door of a furnished room, for example, is objectively just as much a rectangle as the door; but if we look directly at the door, we perceive the wall cabinet as a rhombus. Since rhombi can be measured as well as rectangles can, Poppelreuter rewrote the mathematical models already developed in order to take the new factor into account.[33]

Such progress did not come without problems. Between 1908 and 1911, two of G. E. Müller's students, Erich Jaensch and David Katz, used sophisticated versions of Hering's methods in space and color perception to achieve results that cast doubt upon the adequacy of his theoretical framework. Katz's work will be discussed here. Using reduction screens of the kind that Hering had employed to demonstrate color constancy, Katz distinguished six fundamentally different ways in which colors appear: surface colors (*Oberflächenfarben*), the two-dimensional colors we normally see on the surfaces of objects; film colors (*Flächenfarben*), which have no localization or precise spatial characteristics, like an expanse of blue sky in the mountains on a cloudless day; volume colors, the three-dimensional colors of transparent media such as colored liquids; and finally luster, luminousness, and glow.[34]

Katz's demonstration that the same color could appear in so many different ways enriched psychological description considerably but posed thorny theoretical problems. The same was true of his experimental results. In one experiment, observers were asked to compare shades of gray that had been mixed on a color wheel – an instrument which, when rotated, gives a uniform mixture of two colored strips – placed near a window with a light gray paper presented in a dark corner of the room. When he placed a reduction screen with two holes between the observer and the two matched grays, so that one hole was filled by light coming from the paper and the other by light coming from the disk, the hole filled by the color wheel appeared considerably lighter. When he varied the shadings coming from each direction and asked observers to tell him when the colors were the same, he found that perfect matching was impossible; some difference, difficult to describe in words, always remained. He concluded that color constancy is more complicated than Hering had thought. Not only the quality, but the brightness (which Katz called the "illumination") and the saturation (*Ausgeprägtheit*) of a color must be immediately perceived.[35]

These results were fully compatible with neither Helmholtz's nor Hering's theory of perception. Observers' ability to deal with all of these stimulus dimensions could not be attributed to retinal interaction alone. Katz postulated "central transformations" occurring along with retinal interactions, attributing color constancy to the former and contrast to the latter processes. He regarded film colors as the most purely "physiological," and the other phenomena, including surface colors, as products of experience. These were the outlines of what he called a "psychological" theory of color. He thus retained the dichotomy of "physiological" and "psychological" processes; yet he insisted that his "central transformations" were not unconscious conclusions, but part of seeing itself.[36]

Experimenting psychologists had used Hering's seen objects as an entry route for systematic research on perception – with empirically exciting but otherwise ambiguous results. For Helmholtz, the lines of disciplinary demarcation were clear. Sensation is a physiological process that occurs in a peripheral organ, providing the data on which perception, a psychological process, operates. For Hering and those who followed him, perception still operates on sense data, but these have been preprocessed, as it were, in the retina and the inner eye, and thus have more about them of what Helmholtz had thought of as perception. Freiburg physiologist Johannes von Kries invoked dualistic epistemology to reinforce strict boundaries between physiology and psychology, as Helmholtz had done.[37] Followers of Hering, both physiologists and psychologists, tended to take up monistic or parallelist positions. Viewed from this perspective, it is clear that the central issue was not the choice of a nativist or empiricist account of vision or depth perception, but of competing styles of explanation in an epistemic space that was still unified theoretically but increasingly fragmented institutionally.[38]

Mechanism, elementism, and positivism in experimental psychology

Gestalt psychology is traditionally presented as a reaction against the so-called "structural" psychology of Wundt and Cornell professor Edward Bradford Titchener, who came from England to study with Wundt in the 1890s. Recent research has undermined this presentation by showing that Wundt's and Titchener's psychologies were very different from one another.[39] The point that is relevant in this context is that the Gestalt theorists did not oppose the views of any single individual, but rather the elementistic and mechanistic assumptions about consciousness shared explicitly or implicitly by all attempts to present psychology as a natural science in the nineteenth century. The breadth and vehemence of their attack lend credence to the suggestion that theirs was part of a widespread revolt against positivism in European thought at the turn of the century. And yet, as they drew upon Hering's theory of vision, the Gestalt theorists also mobilized and reinterpreted observations and conceptual resources from their positivist predecessors.

Other than a general adherence to mechanical physics and associationist psychology as norms for scientific discourse, the deepest source of elementistic assumptions in psychology was methodological – the belief that it is necessary to assume the existence of some basic unit of consciousness in order to measure its effects, and thus to create any science of mind at all. Gustav Theodor Fechner, the Leipzig physicist and philosopher who founded psychophysics, held that belief, though he was certainly not a mechanist. In the 1850s, his Leipzig colleague Ernst Heinrich Weber had posited a proportional relationship between stimulus values – the distances between two pin pricks on the skin, for example – and those at which observers noticed differences in their sensations – the so-called two-point limen or threshold. Fechner used similar methods in vision and other sense modalities to establish the psychophysical law in 1860. This modified Weber's formula to the logarithmic relation, $S = k \log R$ (k = constant). Fechner, an early advocate of atomism in physics, thought he was measuring sensations, which he assumed to constitute psychological atoms or elements that could be added in series if they were of the same kind. He also thought he had found a mathematical expression for the unity of mind and body that was the central ideal of German *Naturphilosophie*. Yet he was as much a physicist as a metaphysician. The formula had come to him in a dream in 1850, but he took care to present it as the result of empirical research, valid independently of any supposition about the essence of mind and firmly based upon "the great energy principle." Helmholtz found the Weber–Fechner law, as it came to be called, quite congenial to his view of sensation.[40]

Wundt's elementism was also heuristic, but it was an artifact of his conception of causation, not measurement. In his mature psychology he broke with Helmholtz's mechanistic approach to sensation, and opposed Johann Friedrich Herbart's "mechanics of mind" (*Vorstellungsmechanik*) as well as the association-ism of John Stuart Mill. Instead he presented consciousness as an *unitas multiplex* of will, feeling, and cognition held together by a dynamic, volitional process he called "apperception."[41] His concept of "psychical causality" stated, in essence, that the products of psychical processes have values that can increase or decrease independently of the physical inputs or outputs involved. Thus the "subjective value" of our desire for something can increase over time, while the accompany-ing physical acts, such as muscular movements, obey the energy principle.[42]

Yet despite all these moves away from mechanistic discourse, Wundt did not give up elementism. He admitted that the tables of reaction times and judgment thresholds obtained in the Leipzig laboratory were sufficient only to determine the coexistence, succession, and duration of psychical processes. To yield genuine psychological laws comparable to Newton's laws in physics, Wundt believed, further assumptions were needed. Most important of these was the idea that psychical processes and objects have components, the most primitive of which are simple sensations and feelings. Here Wundt left himself open to the charge of inventing nonexistent entities. When accused of this, Wundt pleaded guilty as charged in a little essay entitled "Invented Sensations." Simple sensations, he wrote, are "never given to us in immediate internal perception but are the results of a psychological abstraction." This abstraction is necessary because science does not merely accept the existence of phenomena but tries to explain them causally as the results of underlying processes.[43]

Thus, despite his invocation of "psychical causality" and his conception of consciousness as process, Wundt had not abandoned Helmholtz's concept of "the universe as consisting of elements with inalterable qualities." What, then, were the "motions" of the psychical elements, and how did they combine with one another? In some places Wundt used a chemical analogy similar to the one Mill used to describe the fusion of associations. "Psychical entities" (*psychische Gebilde*) such as spatial presentations, rhythms or affects, are "in a way somewhat analogous to a chemical compound, the characteristics of which also cannot be determined by listing the qualities of the chemical elements of which it con-sists."[44] When he described the general laws of psychical causality, however, the chemical analogy disappeared in favor of the much broader principle of "creative synthesis":

Every perception is divisible [*zerlegbar*] into elementary sensations. But it is never merely the sum of these sensations, rather something new with

specific qualities arising from their connection. Thus we combine the presentation of a spatial form [*Gestalt*] from a number of light impressions.
Perception remains constantly creative over against the sum of sensations
which is its substrate.[45]

This view that perceived wholes emerge from the associative connection or combination of elements has often been cited as a forerunner of Gestalt theory.[46] True,
in this view elements combine in nonsummative ways, but Wundt did not specify
whether the elements change or stay the same, disappear, or merely remain unnoticed but potentially conscious after they have been creatively synthesized. In
any case, his language still presupposes elements that combine, and thus in a sense
cause the "psychical entities" we apprehend. Attractive as the words "creative
synthesis" sounded, the principle was in fact little more than a synthetic creation
of Wundt's all-encompassing system. The case of "creative synthesis" was symptomatic of the difficulties Wundt encountered in his attempt to reconcile his
recognition of the specific character and complexity of psychical entities and
processes and his commitment to a style of natural-scientific explanation that
required real, causal powers underlying manifest phenomena. Younger scientists,
including some of Wundt's own students, turned instead to the positivism of Ernst
Mach and Richard Avenarius. In doing so, they articulated the specific version of
elementism against which the Gestalt theorists revolted.

When Mach clearly set out his philosophy of science for the first time, in 1872,
his foils were Wundt and Helmholtz. Their shared assumption that physics could
only refer to matter and motion, he argued, is fundamentally flawed: "Matter is
possible appearance [*mögliche Erscheinung*], an appropriate word for a gap in our
thinking." The laws of physics are no more than economical summaries of tables
of appearances. They have "only the value of a memory aid or formula, the shape
of which, being arbitrary and indifferent, changes very easily with our cultural
standpoint."[47] Thus, Mach agreed with Wundt about the primacy of immediate
experience and the virtues of gaining a mathematical hold upon it, but denied
anything more than instrumental value to such attempts. He articulated the profound implications of his concept of science for psychology most fully and
forcefully in *The Analysis of Sensations* (1886).[48] There is no substance, mental or
physical, behind the appearances: "thing, body, matter are nothing apart from the
combinations of the elements – the colors, sounds, and so forth – nothing apart
from their so-called attributes." The things we see are only collections of sensations, their names only convenient designations for these collections; and the
emotions, too, are constituted of "sensations of pleasure and pain." The self that
allegedly perceives these things and feels pleasure or pain is only a designation for
a more strongly coherent portion of the "mass of sensations." When we die, "only
an ideal, mental-economical unity, not a real unity, has ceased to exist."[49] This is
not Hering's heuristic phenomenalism of seen objects, behind which real objects

safely, if uncertainly, rest, but a far more radical phenomenalism, similar to that of arch-empiricist David Hume.

Mach's use of the word "elements" in one sentence and "sensations" in another was not an isolated mistake, but his constant practice. At one point, "elements" are simple sensory qualia, "the colors, sounds, and so forth"; at another, "the table, the tree, and so forth are my sensations." There are two difficulties here. If all we perceive is a uniform "mass of sensations," then we should see a collection of browns and greens, not a tree. Mach could have avoided this problem by saying that the sensory tree is somehow composed of brown and green "elements," or that the perceptual tree is composed of sensations of brown and green; but that is evidently not what he wished to do. The second difficulty is implied in the words "my sensations." If not only patches of color but also houses, trees, and sky belong to me as sensations, then "the ego can be so extended as ultimately to embrace the entire world." Mach was accused time and again of solipsism or "subjective idealism."[50]

One of the many serious implications of solipsism is that there can be no hope of obtaining shared knowledge either of nature or the mind. When he addressed this issue directly, Mach therefore placed the "elements" in an all-inclusive third realm, where the word "sensations" can be applied to them in their functional dependence upon a perceiver, whereas in another functional connection they are treated as physical objects. Thus the proper method of science remains every-where the same – the determination of functional relationships among observable qualities, or "connecting the appearances." The best science is the simplest, most "economical" set of such connections. Economy of thought serves the same practical purpose in science as the ego does for sensations – "to provide the fully developed human individual with as perfect a means of orientation as possible."[51] This doctrine later came to be called "neutral monism." The destructive implications of all this for any conception of causality, including "psychical causality," were obvious. For Mach "there is no rift between the physical and the psychical, no inside and outside"; such dualistic distinctions are "solely practical and conventional."[52]

Wundt's successor as professor of "inductive philosophy" in Zurich, Richard Avenarius, employed a somewhat different line of argument that enabled him to advocate a monistic viewpoint without having to accept a conventionalist concep-tion of science. As an alternative to conceptions of experience that employ con-cepts like "soul," "consciousness," or "inner experience," which ultimately rest on metaphysical dualism, he proposed that all science be based on a "natural world-concept" drawn from what is found in "pure experience."[53] Within this realm of pure experience Avenarius discerned two regions, ego and environment, which are always given together and in relation. Within the environment he distinguished things and other selves, which are both experienced as "not I," but in different ways. The important point is that he was not talking about absolute distinctions between different realities, or between "mediate" and "immediate" experience (as Wundt did), but about immediate experience as a unity with different aspects. To summarize his argument he quoted some lines from Goethe, which Wolfgang Köhler would cite as well decades later:

> Always in your view of nature
> One thing above all remark;
> Nothing's inside, nothing's outside;
> What's within, that is without.
> So then grasp without delaying
> This so sacred open secret.[54]

From this basis Avenarius developed a clear conception of the task of psychology. While philosophy's purpose is to establish the concept of experience as such, and physics is concerned only with experience as independent of the individual, psychology treats experience as dependent upon the body, or part of the body, which does or has it. Avenarius spoke of "system C," a "part of the nervous system" which he tended to identify with the brain. It is possible, at least in principle, to compare perceptions, thoughts, and feelings with changes in "system C," and thus to arrive at an empirical instead of a metaphysical form of psychophysical parallelism.[55]

As shown earlier, Mach had proposed a similar program as a "heuristic principle of research" in 1865. He restated that program even more explicitly in *The Analysis of Sensations:*

> If I see figures which are the same in size and shape but differently colored,
> I seek, in connection with the different color-sensations, certain identical
> space-sensations and corresponding identical nerve-processes. If two fig-
> ures are similar (that is, if they yield partly identical space-sensations) then
> the corresponding nerve-processes also contain partly identical compo-
> nents.[56]

In later editions of the book, Mach indicated that he meant the word "seek" quite literally. Given the appropriate physical and chemical apparatus, he imagined that he or someone else would be able to confirm psychophysical parallelism directly by observing the brain processes that occur during sensation. In his *Elements of Psychophysics,* Fechner had distinguished two versions of that enterprise. "Outer" psychophysics established functional relations of stimuli and sensations; "inner" psychophysics would eventually do the same for sensations and brain events. Mach was deeply influenced by Fechner in the 1860s. Though he took care to distance himself from Fechner's panpsychism – attributing "soul" to plants, for example – the romantic ideal of mind–body unity lived on vibrantly in his psychophysical program.[57]

Such views justified the combination of physical, physiological, and psychological research that Mach had pursued for decades. Observations drawn from this work filled the pages of *The Analysis of Sensations.* One of the most famous of these was that of a square which, when presented with one corner pointing downward, ceased to be a square at all and became, psychologically, a diamond, although its geometrical dimensions remained the same. Mach used this and similar figures to bring home a point Hering had also made – that the geometrical space of classical mechanics is not the same as the "space of sensation." By this he meant that the visual field is not a simple grid in which any location is in principle like any other, but has three unequal directions – up is more important than down, the right side usually more interesting than the left, the portion in front of perceivers more significant than the one behind them. Thus, in Mach's view, in order for two visual forms to be perceived as similar, not only the sensations associated with their geometry but also the associated "sensations of direction" (*Richtungsempfindungen*) had to agree. The Gestalt theorists made a great deal of the anisotropy of visual space, though they firmly denied Mach's sensationist interpretation of it.

Carl Stumpf wrote a highly critical review of Mach's book the year it appeared. Given his own view of space perception, which stated that color and extension are given together and immediately apprehended as such (see Chapter 2), Stumpf was bound to take offence at Mach's theory that visual space is built up from sensations of eye movements, and his constant allusion to new kinds of sensations. In Stumpf's opinion, naming the phenomena and then adding the suffix "sensation" was an extraordinarily heavy-handed way of dealing with complex psychological facts.[58] Such criticism helped to give Mach the reputation of being an amateur in philosophical circles; a second German edition of *The Analysis of Sensations* did not appear until 1900. Nonetheless, his influence grew. Some of the central features of his and Avenarius's philosophy, particularly the elimination of a separate psychical causality, the emphasis on sensations as immediately accessible data, and the hope of discovering the functional dependence of these sensations upon organic processes, became central assumptions for the post-Wundtian generation of experimenting psychologists. Kurt Danziger goes so far as to speak of a "positivist repudiation of Wundt."[59]

The most wholehearted participant in the uprising, at first, was Oswald Külpe. In the early 1890s he was *Privatdozent* and Wundt's assistant in Leipzig, and the master apparently regarded him as his right-hand man. At the very beginning of the book he asked Külpe to write for the institute's lecture and laboratory courses; however, Külpe rejected the notion of the "psychical individual" as the subject of psychology, substituting the "corporeal individual." He devoted nearly half the work to sensations and counted them with extraordinary elan, discovering, for example, 696 brightnesses. He added space and time to the sensory attributes, as Mach did, and called ideal presentations (*Vorstellungen*) not products of active psychical processes but "centrally excited sensations."[60]

The functional relations Hermann Ebbinghaus discovered in his pioneering study of memory (discussed in Chapter 1) were not precisely the kind that Mach had in mind. There was no mention of sensations or physiological correlates, only what Ebbinghaus called "statistical constants," such as the curve of forgetting, which plotted the number of items retained or lost over time. He explicitly stated that such functional orderings of data are all that can be had in the study of memory, and compared them to social statistics.[61] In this sense his, like Mach's, were functional, not causal explanations. He associated himself with the positivist trend in the first volume of his textbook of psychology, published in 1897. There he substituted the term "organism" for Külpe's "corporeal individual." In an address to the third International Congress of Psychology in 1900, he alluded to an increasing shift from physical to biological terminology encouraged by the writings of Darwin and Spencer as the main sign of psychology's scientific progress, along with the growth of measurement. He modified the strictness of Machian sensationalism along these lines, admitting, for example, that pleasure and pain could not be treated as simple sensations. But the tendency to construct the organism from separate units and to avoid central guiding processes was evident throughout Ebbinghaus's systematic psychology.[62]

In 1908 Ebbinghaus brought the particular version of elementism behind that procedure into the open. It is false, he asserted, to say that the "atomistic splintering" characteristic of scientific method makes it impossible to regain the "living totality" with which one began:

> From the beginning the body is for the biologist nothing else but a unified whole, no less unitary than the soul, originally simple, later much more rigidly structured; and this whole has brought forth the parts, not the other way around. But in order to come to know it in detail and show this to others, as is after all desired, one must necessarily proceed *as if* the opposite were true; one must begin with the observation of the parts and separate these from the whole in which alone they exist by analysis and by abstraction, or try to differentiate them within it. Precisely this is the intention and procedure of the psychologist.[63]

The group that came together at the biannual meetings of the Society for Experimental Psychology was unified by little more than a definite idea of what psychology should not be – philosophical speculation – and a vague idea of what it should become – an empirical science. In a much-cited overview published in 1900, William Stern wrote that, "other than the strongly empirical tendency and the use of experimental results there are hardly any common features" to the new psychology. To portray the situation he evoked a resonant political image: "Indeed, gentlemen, the psychological map of today is as colorful and checkered as the political map of our fatherland in the dearly departed period of *Kleinstaaterei*."[64] The implication was that unification on some basis, even at the risk of excluding some part of the field, was preferable to the present disunity, as Prussia had achieved German unity by excluding Austria.

Nonetheless, both the psychophysical methods developed by Fechner, Wundt, G. E. Müller, and others into a fine art and Ebbinghaus's memory experiments proved to be durable "exemplars," as Thomas Kuhn would call them – theory-laden models of procedure. The fact that these procedures were taught in the laboratory courses at Stumpf's Berlin institute shows that their use did not require full acceptance of Mach's or any other positivist epistemology; but it did involve fundamental presuppositions about the conduct of science. In the natural sciences, the reliance on precision measurement and instrumentation instantiated the belief that the manifestations of physical – or biophysical – forces could be given numerical values and the forces themselves thus placed under human control.[65] By participating in this culture of precision with their heavy brass instruments for measuring reaction times and their own ways of deriving graphable data, experimenting psychologists did more than acquire scientific respectability. Rather, by manipulating the resources of objectivity-discourse, they reconstituted the object to which that discourse referred. What had been mental and moral capacities became psychical functions; and the sensing, perceiving, conscious mind became an instrument that functions, or fails to function, in a measurably "normal" way.

But the elementistic presuppositions – more precisely, metaphors – that appeared to justify that enterprise had never gone unquestioned. Common to Wundt's earlier statement about the use of "invented sensations" and the declaration from Ebbinghaus just cited was their tone of militant defensiveness. Accusations of the kind Ebbinghaus referred to had been common currency in German – and not only German – philosophical circles for nearly twenty years by this time. These critiques went to the heart of the justifications offered for experimental psychology's procedures, and proposed alternative dynamic and holistic vocabularies aimed at overcoming the alleged separation of science and experience.

5

Challenging positivism: Revised philosophies of mind and science

The struggle within German academic philosophy was only part of a far broader cultural and ideological conflict that was not limited to one country. One contemporary writer spoke of an "idealistic reaction against science" in European thought at the turn of the century. Intellectual historians have generalized this to a "revolt against positivism."[1] Taken too far, such accounts can lead to easy dichotomies between science and antiscience, or modern and antimodern positions. In philosophy of mind, however, this was also a time of attempts to discover a middle path between idealism and positivism, represented most prominently in America by William James and John Dewey, on the Continent by Henri Bergson, Dilthey, and Edmund Husserl's *Logical Investigations*. All of these writers opposed the views of Mach and Avenarius, described in the previous chapter. Bergson, Dilthey, and James worried about the challenge to traditional values posed by technology and doubted that science could offer sufficient certainty to serve as a cultural compass. And yet Bergson and James used the term "empiricism" to describe their methods, and all three thinkers were fascinated enough by natural science to present their alternatives as research programs.

There was no linear ordering of antipositivism to cultural conservatism, either. Dilthey's attack on the "separation of science from life" expressed such a standpoint. But James's and Dewey's "via media" in philosophy ultimately grounded the political theories of social democracy and progressivism. Because they paid so much attention to empirical psychology, David Lindenfeld describes these thinkers' work as a "transformation of positivism"; James Kloppenberg speaks of a more expansive version of naturalism, a "radical theory of knowledge" based on the immediacy of lived experience, that "made possible new ways of thinking."[2] When neovitalist Hans Driesch brought his challenge to mechanistic metaphors from developmental biology into philosophy, he joined that chorus.

The challenges to scientistic discourse on mind came both from within and from outside the community of experimenting psychologists. In Germany, that community was still tied to philosophy both institutionally and intellectually; the struggle was in part one for position in a common epistemic space. These debates, like those in sensory physiology, provided both intellectual challenges and conceptual resources to the founders of Gestalt theory. Because of their disciplinary location between philosophy and natural science, they were aware of the philosophers' attacks. They accepted their criticisms, and ultimately even employed some of their terminology, but did not conclude that a natural science of mind was therefore impossible.

Philosophies of mind as process: Bergson and James

From a different starting point and by different arguments from those of Mach, French philosopher Henri Bergson arrived at similar conclusions about the conventional character of scientific practice and its incompatibility with lived experience. Bergson's major works were all translated into German between 1908 and 1912. Their publication constituted part of the revival of metaphysics propagated there by Rudolf Eucken and Southwest German Neo-Kantians, like Wilhelm Windelband.[3] At the same time, Köhler and Koffka taught Bergson in their earliest seminars. They took note of his critique of associationistic psychology and his call for psychological categories better suited to lived experience, but pointed to his proposed alternative as an example of what they wished to avoid.

In his "Essay on the Immediate Data of Consciousness" (1889), Bergson argued that experience viewed as a succession of separate, thinglike states is no less an abstraction from lived consciousness than time as measured by the hands of a clock. Both are fundamentally spatial. Lived consciousness, on the other hand, is a spatiotemporal continuum, "like a mutual penetration, a solidarity, an intimate organization of elements, each of which is representative of the others and neither distinguished from nor isolated by abstracting thought."[4] Associationist psychology clearly exemplifies the difficulties of applying "spatial" science to mind, according to Bergson. We can describe the movement of an object in space, for example, by postulating an infinite number of reference points, through which the object may be said to move. But "they are not parts of the movement; they are so many views taken to it; they are, we say, only supposed stopping points. Never is the mobile reality *in* any of these points; the most we can say is that it passes through them."[5] But instead of rejecting metaphysics, as Mach did, Bergson proposed an intellectual division of labor. Scientists could retain their analytical methods, while metaphysicians would strive for a "true empiricism" that would seek by intuition to keep "as close to the original itself as possible."[6]

Bergson's reduction of metaphysics to a question of method can be seen as an important concession to positivism. Still, he had equated point-to-point analysis with natural science in general, thus making it seem inherently incommensurable with his dynamic, holistic model of experience. The consequences of this logic became clear in Bergson's most popular work, *Creative Evolution* (1907). There he asserted that intuition was rooted in an all-encompassing life-principle called the élan vital. Though he protested that this vital principle was only a heuristic construct, he had clearly left his original basis in direct experience. There was no way of experiencing the élan vital even by intuition, since it was what made intuition possible.[7]

In Stumpf's lectures Köhler and Koffka also learned of William James's *Principles of Psychology* (1890). There James challenged associationism in terms similar to those Bergson used, condemning the conception of consciousness as a collection or succession of constant, retrievable "ideas." In James's words, that view sacrifices "the continuous flow of the mental stream" and substitutes for it "an atomism, a brickbat plan of construction," for which there was no introspective support.[8] For him this atomism was one instance of "the psychologist's fallacy," the confusion of thought as experienced with a properly objective psychologist's view of it. James singled out the doctrines of unconscious inference and unnoticed sensations as prize examples of where such confusions could lead. Helmholtz, for example, attributed visual form perception to the fusion of retinal images from the two eyes, and contended that we can discover the original sensations by first opening and closing one eye, then the other. To James this involved the assumption, unsupported by introspective evidence, that sensations present in monocular vision remained present but unnoticed in binocular vision.[9]

The dynamic concept of consciousness as a continuously flowing stream for which James thought there was good introspective evidence was obviously different from Mach's "mass of sensations." Within that stream James placed an "Object" consisting of "things" cognized intellectually, which we know "about," and an immediately "felt fringe" of expectation and relation which we know only by active "acquaintance." The "fringes" he attributed to "the influence of a faint brain-process upon our thought, as it makes it aware of relations and objects but dimly perceived," for example when we hear a thunderclap and immediately have the vague expectation that something – a lightning bolt – is about to happen.[10] James, like Bergson, resorted to nonintellectual factors to account for the unity of the mental stream. The two writers also agreed that we grasp only part of the flow at any given moment, and that practical interests act as a principle of selection. In emphasizing the role of attention directed by interest directed in turn by need or utility, James accepted the Darwinian metaphor of selection as the governing principle of order in experience. Darwin called natural selection the mechanism of evolution, but James opposed machine metaphors and substituted aesthetic images like that of the stream, or that of mind as an artist, working on the "mere matter" of sensation to produce objects, "as a sculptor works on his block of stone."[11]

Despite his criticism of associationism and his use of dynamic and aesthetic rather than mechanical metaphors for consciousness, James never relinquished the positivist hope of causally relating all psychical phenomena to organic events. His "empirical parallelism" was similar in logical structure to the doctrines of Mach and Avenarius, but not in its content. Mach demanded for two similar figures corresponding nerve-processes with "identical components," a point-to-point isomorphism. For James, perception and thinking are integral processes. The "organic conditions" corresponding to them must therefore share this characteristic. Research on the loss and recovery of function in brain-damaged patients showed "that the whole brain must act together if certain thoughts are to occur. The consciousness, which is itself *an integral thing not made of parts*, 'corresponds' to the entire activity of the brain, whatever that may be, at the moment."[12] In an 1884 paper, he proposed a more specific analogy from electrodynamics to describe brain action that Wolfgang Köhler later developed in detail (see Chapter 11):

> The whole drift of recent brain-inquiry sets towards the notion that the brain always acts as a whole, and that no part of it can be discharging without altering the tensions of all the other parts. The best symbol for it seems to be an electric conductor, the amount of whose charge at any one point is a function of the total charge elsewhere.[13]

This – by 1890 quite widespread – theoretical commitment to locating the causes of mental events in the brain rather than the mind underlies many of the conceptions in James's psychology, not least that of "the stream of thought." Even so, he found it difficult to reconcile his empirical parallelism with his ethically derived allegiance to the doctrine of free will, which presupposed an active mind. In 1890 he declared that such issues could only be decided on philosophical, not empirical grounds. Yet even after he took the step to metaphysics more than a decade later, problems remained. James's treatment of so-called mental compounds reveals the limits of his attempt to reform empiricism. The awareness of the alphabet, for example, is for James indeed "something new" compared with twenty-six awarenesses, each of a separate letter. But he did not say that it is something fundamentally different, presumably because it is as immediately present in consciousness as the others. He preferred to treat that additional awareness "as a twenty-seventh fact, the substance and not the sum of the twenty-six simpler consciousnesses," and to attribute the different sorts of awareness to simpler or more complex "physiological conditions."[14] James did not guess what these "physiological conditions" might be; nor did he explain how postulating a twenty-seventh "awareness" would solve the problem of its intrinsic relation to the others. James's pluralistic universe remained, in essence, a universe of pluralities. His radical empiricism revised, but did not abandon the traditional version.[15]

Carl Stumpf regarded James as a both a colleague and a friend. He drew support from *The Principles of Psychology* for his own cautious extension of empiricism, in particular his claim that relations are as directly experienced as sensations (see Chapter 2). He recommended the book to his students for this reason, but advised them to treat its theoretical conclusions with caution.[16]

Starting from the whole, and from "the things themselves": Dilthey and Husserl

In the meantime, two other philosophical reformers, Wilhelm Dilthey and Edmund Husserl, developed alternative conceptions of consciousness comparable to James's, but drew more negative conclusions about the possibility of a natural science of mind. Max Wertheimer took a seminar with Dilthey while a student in Berlin, and read Husserl's work shortly afterward; he drew upon their ideas in his own, natural-scientific refashioning of holistic thinking (see Chapters 7 & 8).

Dilthey had good reason to cite James in support of the dynamic and holistic conception of consciousness he presented in his 1894 essay, "Ideas on Descriptive and Analytical Psychology." For him, as for James, conscious experience is not a collection of simple sensations and their corresponding "ideas," but "a structured whole" combining intellect, feeling, and will. This whole is not static but dynamic, a "living, unitary activity within us."[17] The "dominant psychology" cannot grasp this reality, Dilthey argued, because its representatives persist in reducing "all phenomena of consciousness to elements imagined to be atom-like" and constructing the psyche from these strictly hypothetical entities. Dilthey did not question the achievements of hypothetical thinking in the natural sciences, and he granted that experimental and quantitative research had proven useful for the understanding of phenomena at the mind–body interface. He nonetheless insisted that such methods would never help us to grasp the central aspects of mental life, especially the role of will.[18]

When he spoke of "the dominant psychology," Dilthey meant the ideas that had been dominant in his youth, particularly the "intellectual mechanics" (*Vorstellungsmechanik*) of Johann Friedrich Herbart, the associationism of Mill, and the empiricism of Helmholtz, all of which drew heavily on analogies from physics and chemistry. He congratulated Wundt somewhat sardonically for recognizing the dynamic character of mental life and the importance of will in his principle of "creative synthesis," but quite rightly refused to exempt the Leipzig master from his general indictment. He joined Wundt in angrily dismissing the extreme sensationalism of Mach and the younger generation of experimentalists, which he called "a declaration of intellectual bankruptcy." Such thinking could only lead to "increasing skepticism, a cult of superficial, unfruitful fact-gathering, and thus *the increasing separation of science from life.*"[19]

Dilthey's alternative was an injunction that became the common wellspring of Gestalt theory and the so-called Leipzig school of "holistic psychology" (*Ganzheitspsychologie*) – the injunction to proceed "from the whole to the parts."

He maintained that in inner life the "experienced connection [*erlebter Zusammenhang*]" of thought, feeling, and will is primary, while "the distinguishing of its individual members [*Glieder*] comes afterward." For him this meant beginning with an intuitive description of inner experience more systematic than that carried out by poets but at the same level of sympathetic understanding. In his 1894 essay, such description yielded continuous cycles of thinking, feeling, and willing in interaction with a changing milieu. Aspects of the environment are taken up as ideas, affectively treated, then converted into willed action, which in turn alters the milieu, and so on. This dynamic flow of interrelationship between the "totality of human nature" and the world Dilthey called, simply, "life" or "life itself."[20] The cultural coding in this terminology is evident. The "life" Dilthey meant was that of an educated elite of cultivated individuals, capable of comprehending the thought and feeling of past cultures by actively "co-experiencing" them.

In his *Introduction to the Human Studies* (1883), Dilthey had spoken of the person as a "psychophysical whole," and presented this as the appropriate unit for the human studies, rather than the physical "atoms" of the natural sciences. In 1894, he emphasized in addition that individuals are formed (*gestaltet*) by their own history, and by that of their society and culture. The product of such interactions between individuals and their cultural circumstances he called "the form of the psyche" (*Gestalt der Seele*). The teleological end of Dilthey's psychology, then, was not a monadic individual but a socially and culturally formed or organized personality, which he called "character," or, in one place, *Gestalt*.[21] The word *Gestalt* is often used in German to designate a historical personage. In this sense it refers both to persons and to their significance, to the figure they cut on the historical stage. Such figures were the center of Dilthey's research interest throughout his career; he called biography "the most philosophical form of history." He did not consider each individual unique, however, but referred to "typical people," or to "forms of individuality." Though all kinds of personalities could offer material for such a psychology, he thought it was better to begin with "the developed person of culture" (*entwickelten Kulturmenschen*). Only when such personalities had been studied and classified with all the available methods would psychology "become a tool for the historian, the economist, the politician and theologian."[22]

In principle, Dilthey had outlined an alternative research program to that of the experimentalists, one that was in better agreement with the values of Germany's traditionally educated elites than the technological vocabulary of the experimentalists. But he did not elaborate the practical side of that program in any detail. Eduard Spranger, Karl Jaspers, and others later tried to create such usable tools — systems of personality "types" based in part on Dilthey's thinking (see Chapter 17). In the 1890s, experimenting psychologists reacted to such views, predictably, with the disdain of practitioners for the armchair expert. In his vehement rebuttal in the *Zeitschrift für Psychologie*, Hermann Ebbinghaus asserted that Dilthey's "descriptive" psychology was nothing new; experimenting psychologists, too, begin with "the given" and analyze it into its experiential elements, in order to discover the laws of their connection. From this perspective Dilthey himself was guilty of placing hypotheses before facts. Only a vague sense of "connectedness," not the structure of consciousness, and certainly not that of personality, are accessible to introspection. Thus Dilthey's psychology, too, required acts of intellectual "construction."[23]

This point was well taken. In his reply, Dilthey acknowledged that our sense of totality is given only in "partial experiences" of connectedness. These must be supplemented by elementary logical operations, such as abstraction and generalization, before they can become genuine functional connections. He nonetheless denied that experienced structure is a hypothesis; for these operations, too, are parts of experience. Thus we do have an immediate, though vague awareness of a partial experience's place in and value for the whole.[24] In 1894 Dilthey did not explicate in detail how this awareness operates; he attempted to do so in his later analyses of poetic experience. Both his reply to Ebbinghaus and the essay on the forms of individuality that followed it were rather brief, and he soon laid aside his psychological system.

In the 1890s, then, Dilthey did not offer a method that could do justice to lived experience as he conceived it and still compete with the apparent precision of the experimentalists. Edmund Husserl thought he had done so. He presented his alternative in his *Logical Investigations* (1900–1901), according to one contemporary "perhaps the most influential single work published in philosophy" in the first twenty years of this century.[25] Like his friend and colleague Stumpf, to whom he dedicated the book, Husserl began his second, philosophical career under the influence of Franz Brentano. He hoped that careful descriptions of the act of counting and of the role of symbols in arithmetical thinking would help him to solve fundamental problems in the philosophy of his original field, mathematics. However, criticism from Paul Natorp and Gottlob Frege, his own discovery of Bernard Bolzano's logic and his realization of intrinsic difficulties in his original scheme ended these hopes. Frege in particular pointed out a problem Husserl had also seen – that our experience of counting does not include concrete ideas (*eigentliche Vorstellungen*) of large numbers. The validity of the sum 999,999 + 1 surely cannot be traced to the experience of counting apples; nor can the mathematical significance of concepts like infinity be reduced to ordinary counting.[26]

In the first volume of the *Logical Investigations,* entitled "Prolegomena to Pure Logic," Husserl attacked in broadest terms the doctrines he called "psychologism" and "anthropologism," the attribution of the validity of logical propositions to inductions from experience or to supposed "natural laws of thinking." All such ideas lead inevitably to "naturalism" or relativism, he claimed. Though he named Mill, Alexander Bain, Wundt, and others as his most prominent opponents, it was the "modern Humeans" Mach and Avenarius who brought the issue most clearly into the open. The fundamental question is whether or not "ideal objects of thought," such as the propositions of arithmetic, are really no more than efficient signs for "mere individual experiences, collections of presentations, and judgments about isolated matters of fact."[27]

Given the term "pure logic" in the title, there was little doubt that Husserl intended to side with idealism. What comes forward most strongly in the *Prolegomena* is a Platonistic account of logic as a science of the ideal laws governing concepts and propositions conceived on the model of Bolzano's realm of entities "as such" (*an sich*). As the studies in the second volume of the *Investigations* show, however, Husserl did not separate such concepts and propositions entirely from the realm of mental acts. Instead, he held, with Lotze, Brentano, and Stumpf, that "logical concepts as valid units of thought must have their origin in concrete intuition; they must have grown up by abstraction on the basis of certain experiences." The only way to overcome psychologism was thus not to ignore psychology, as Frege and the Neo-Kantians did, but to examine more closely than even Kant had done the experiences we have when we think. This is what Husserl meant when he said that "phenomenology is descriptive psychology."[28]

The prototypes of these "logical experiences" were acts of perception. "Experienced sensation," Husserl asserted, "is animated [*beseelt*] by a certain act character, a certain grasping [*Auffassen*] or referring [*Meinung*]." Sensations are "an analogical building material" for the objects presented through them; but we see the objects, not their components.[29] Husserl meant this meaning-giving act and its reference to an object when he spoke of the intentionality of consciousness. In this respect, there is little difference between Husserl's psychology and that of Brentano. His innovation was to assert the psychological primacy of such experiences, of perception over sensation. Failure to recognize this was, in his view, the fundamental error in Hume's account of experience, and thus of the "modern Humeans" as well.

In the second logical investigation, devoted primarily to the problem of abstraction, Husserl accused Hume of ignoring the fundamental difference between an object as it appears to us, for example a smooth, white globe with its uniform coloring, and the sensations supposedly contained in that object. The object as it appears is something essentially different from a complex of sensations, and Husserl argued that we must accept its existance as it appears in intuition, rather than saying it is really something else. An object can appear in various ways, show various "sides" of itself according to the interests of the observer. Nonetheless we know *that* we perceive it with nearly self-evident certainty: "I can delude myself about the existence of an object of perception, but not about whether I perceive it as determined in such and such a way, and that it is in the referring [*Meinung*] of this perception not a totally different object, e.g. a pine tree and not a June bug."[30] This was a version of psychological realism similar to that advocated by James. Husserl credited the American philosopher with helping him to liberate himself from psychologism. In fact, his position harkens back to Brentaro's "immanent objectivity" (discussed in Chapter 2); and it soon became one of the foundations of Gestalt theory.

Husserl then applied this version of the intentional model of consciousness to what he called "the experience of truth." By this he meant "the agreement . . . between the experienced sense of a statement and the experienced state of affairs [*Sachverhalt*]." Insofar as a proposition "makes sense" to us, it has, or can have, that sort of (self-) evidence which amounts to our experience of its truth. Thus, propositions about mathematical, fictitious, or nonexistent entities can also acquire evidence, even though we do not have direct experience of the objects to which they refer. The truth of the statement "Pegasus does not exist" is as self-evident to us, on this account, as that of the statement "I see a round, smooth globe," which can be true even if the globe is an illusion, because it refers to the person's own seeing. The validity of such statements is therefore guaranteed, even though there is no presentational content, no "real" object involved.[31]

With such arguments Brentano's students used the techniques of thinking they had learned from him to couple the structure of cognition with that of propositions. Instead of relating the structure of experience to that of the person, as Dilthey did, the key question in logic and the theory of knowledge became for them not "How can we know?" but "How can we describe the act of knowing and formalize the results?" This shift to what has been called logical realism was a step on the way to the reform of analytical philosophy later in the century. However, the implicit, and soon quite explicit formalism of this approach ironically led to the very disciplinary split between logic and psychology that Husserl, Meinong, and Stumpf wished to avoid.

A different but related tension, between formal ontological order and psychological fact, emerged in Husserl's third logical investigation, on the theory of whole and part. In this discussion Husserl employed two key concepts. The first was that of "moment," or aspect. Stumpf, as we saw in Chapter 2, spoke of color and extension as "partial contents" of a "state of affairs" (*Sachverhalt*) that is experienced is a unity. He was not talking about separable parts of an object, but about interrelated aspects of experience. Husserl had already used the concept of "figural moment" in his *Philosophy of Arithmetic* to describe the feature of intuitions in which the apprehension of like sensory contents and their subsumption under a collective concept seem to occur in a single act, when we speak, for example, of a heap of apples or a swarm of birds. In the second logical investigation, Husserl extended this usage to perception in general, speaking of the color or form (*Gestalt*) "moments" of the experience of objects. In the third investigation, he generalized the concept still further, employing the term "moment of unity" taken over from Neo-Kantian philosopher Alois Riehl for this class of experiences. He then made an additional, crucial distinction between "phenomenological" and "objective" moments of unity. The former "give unity to the psychological experiences and experience-parts," while the latter "belong to the intentional and nonpsychological objects and object-parts." A contemporary example of an objective moment of unity might be the current flowing through a computer system. Without it, the system would only be a collection of equipment pieces, but its unifying status is not dependent on the successive mental acts or states of any given perceiver.[32]

From this distinction Husserl developed an a priori classification of part–whole concepts according to what he called their "unity of foundation." Here, too, he could easily have meant a psychological act, but he did not. In the second logical investigation, he described perception as an act of "grasping" based or "founded" on sensory material. Here, his use of the term is more like the idea of a building that rests upon a foundation, or is built up on a "foundation" of bricks and mortar. Thus, to use one of his examples, when we see a group of stars drawn on a piece of paper, we are presented with a "hierarchy of foundations" in which points found the lines, the lines found the individual figures, and finally the figures found the collection or group. This hierarchy exemplifies what one commentator calls a relation of "one-sided" foundation: *a* founds *b,* but not vice-versa. Stumpf's claim that color and extension necessarily require or presuppose one another is an example of mutual or two-sided foundation.[33]

Now, these are obviously not psychological statements about how wholes are experienced, but claims about their "objective," that is, formal ontological structure. Had he been making an empirical statement, Husserl could not have implied that we see a collection of lines and a group of stars in the same way. According to his own theory of perception, we see the group of stars whether we sense the lines or not. Yet for all his emphasis on conscious activity and the primacy of perception, Husserl did not give up elementary sensations. In fact, he required them as one sort of analogical building block for perceptions. Thus, though he carefully distinguished them, his objective and psychological realms had a parallel, hierarchical structure.

Experimenting psychologists drew a number of important lessons from Husserl's work. His arguments against "psychologism" gave them a clear idea of the limitations of older psychologies, without denying the legitimacy of psychological investigation as such. When he said that "phenomenology is descriptive psychology," he hastened to add that such a psychology could not be the same as that of the experimentalists, but in 1900 he left the door wide open for the use of his observational procedure as an alternative data base in psychology.[34] This was the method that Carl Stumpf had used in his psychological research from the beginning (see Chapter 2). Its first rule of procedure – that investigators should begin by examining the conscious contents they actually have, not what empiricist assumptions say they ought to have – meshed comfortably with Ewald Hering's injunction to begin with "seen objects" rather than those of physics. In other words, experimenters working with Hering's methods could obtain philosophical support for what they were already doing. In his 1911 work on color (discussed in Chapter 4), David Katz acknowledged that he had been encouraged to adopt a "general phenomenological attitude" by lectures and seminars he had heard with Husserl, who had been teaching in Göttingen since 1901. Katz later said that to him, "phenomenology, as advocated at that time by Edmund Husserl, seemed to be the best connection between philosophy and psychology."[35]

If this was a misunderstanding, then Husserl himself was partly responsible for it. In his Göttingen lectures of 1905 on the phenomenology of time consciousness, he drew in part on William Stern's experimental research on the "psychical present" to support a far more sophisticated view than Bergson's undifferentiated flow of *durée*. Dissertations written under Husserl's direction in Göttingen, for example by Wilhelm Schapp on color and Alfred Brunswig on comparison, could easily be taken as contributions to descriptive psychology, though they lacked experimental support. Schapp's work, in particular, is notable for his claim that qualities such as color, form (*Gestalt*), and movement are immediately given in perception, as means of presenting "the thing itself."[36]

The most important use of Husserl's method and vocabulary in psychology came in the work of former positivist Oswald Külpe and his students in Würzburg on the psychology of thinking. The Würzburg experimenters discovered numerous psychical contents in acts of thought different from the sensations, images, and feelings called for in classical associationism; and they made observations on the purposive, directed character of thinking that bore a close resemblance to Husserl's ideas. Narziss Ach, Henry Watt, and Külpe himself found, for example, that the experimenter's instruction produced a "mental set" (*Einstellung*) that played a dynamic role in observers' thinking. Külpe presented a series of four syllables, each one printed in a different color and position, and asked observers to focus on one specific aspect, for example, to name the letters, then to tell the color or the shape. As expected, responses were more accurate for the aspects observers were told to focus on, regardless of how often a given object had appeared before. Ach showed that the process could be unconscious by achieving similar results with subjects under hypnosis, and attributed the "ordered and goal-directed course of mental events in general" to hypothetical "determining tendencies."[37]

In 1906 and 1907, August Messer and Karl Bühler introduced Husserl's vocabulary and his model of consciousness into the Würzburg vocabulary. In his systematic treatise on sensation and thinking, Messer described Husserl's act of "intending" as the "imageless element" that "gives meaning [*Sinn*] to the word in consciousness." Bühler distinguished three classes of thought contents: simple "thoughts," or content correlates of Ach's determining tendencies, "thought-memories" distinguished by consciousness of a rule, and "intentions," in which "the act of meaning comes to the fore and not what is meant."[38] Finally, Külpe himself made the directedness of thinking a cornerstone of his epistemology: "For us the fundamental characteristic of thinking is referring [*meinen*], meaning aiming at something."[39]

Considerable controversy swirled about these claims, primarily on methodological grounds. Wundt called the Würzburg procedures "pseudoexperiments" and alleged that they were open invitations to suggestion.[40] More important for the emergence of Gestalt theory was Ernst Cassirer's substantive criticism that the Würzburg experimenters tended to treat the contents and processes they discovered as new units alongside the conventional ones, rather than treating thought processes as wholes.[41] Nonetheless, Külpe, the former positivist, did not hesitate to use his school's results to proclaim the death of sensationalism. Mach's epistemology fails, he wrote, because sensations are not all there are in consciousness: "Modern psychology teaches that sensations are products of scientific analysis. . . . We do not discover elementary contents, such as simple colors or brightnesses, tones or noises, elements of any kind in our investigation of what is given in consciousness."[42] Külpe was equally explicit about the significance of this work for the status of psychology as a philosophical discipline:

> Associationist psychology, as founded by Hume, has ended its solitary reign . . . There are still psychologists who have not risen above this standpoint. Their psychology can rightly be accused of unreality, of moving in an abstract region where it neither seeks nor finds entry into full experience. These are the psychologists who offer stones instead of bread to those representatives of the human studies who are asking for psychological support.[43]

Thus, Külpe and his followers relied on Husserl's methods and his model of cognition for support in their effort to establish a fruitful middle position in philosophy, located between traditional, sensation-based empiricism and idealism, without forsaking experimental method.

Husserl himself retracted his designation of phenomenology as "descriptive psychology" as early as 1904. He and his students insisted that their efforts yielded evidence about the nature of consciousness as such, while experimental procedures only provided provisional laws about the range of variation of certain conscious experiences under given conditions. Between 1906 and 1907, he decided that only a "transcendental phenomenology" would overcome the danger of relativism once and for all. His 1911 essay "Philosophy as Rigorous Science" (discussed in Chapter 3) helped prepare the way for this step.[44] Experimenting psychologists, Stumpf included, did not follow him along this road.

From neovitalism to psychovitalism: Hans Driesch

By this time, Hans Driesch had joined the challenge to mechanistic discourse in philosophy of mind through a side entrance, coming from a biologists' debate about the nature of organic development. In doing so, he underscored the ontological dimension of both arguments; for both revolved around the issue of whether and how discursive norms derived from mechanical physics could be extended to all forms of reality. At the same time, talk of wholes determining parts, and of inherent organizing dynamics coming from within rather than without, was central to his position.

Most nineteenth-century German biologists were trained in physiology, and the commitment to mechanistic categories in that discipline remained strong. However, both the study of organic form and the theoretical tradition created by Hans Blumenbach and Immanuel Kant, which Timothy Lenoir calls teleomechanism or "vital materialism," persisted in Germany throughout the century. That tradition posed a challenge even the most committed mechanists continued to face – how to deal with the apparently inherent purposiveness of ordered growth and development without invoking occult vital forces.[45] Wilhelm Roux proposed to provide such an account in his "developmental mechanics": a "causal morphology of organisms" that would reduce the origination and maintenance of these forms to "movements of parts."[46] His self-styled "mosaic theory" was compatible with the teleomechanist tradition, in that it described development as the "self-differentiation" of "hereditary potentialities," with irreversible functional differentiation among the cells. This hypothesis was supported in part by Roux's own experiments at the marine biological station in Naples. When he killed one of the first two cleavage cells in a frog's egg, the surviving cell, as expected, gave rise to only half of a normal embryo.

In 1891, working at the Naples station with a different organism, the sea urchin *Pluteus,* Driesch separated the paired, multicellular preembryonic structures in their eggs, called blastomeres, from one another without killing any by shaking them apart in sea water. Each developed into a whole, normal, though somewhat smaller than average adult. Even crushing a section under a glass plate, thus confusing the structures completely, did not prevent the development of anomalous but functioning individuals. Jacques Loeb obtained comparable results, also with sea urchin eggs, in 1893. Loeb opposed ideas of unfolding inherent natural order and relied heavily on machine metaphors that equated knowledge with control; for example, he thought of the tendency of certain plants to grow toward light as a kind of reflex response. Apparently he saw no threat to his mechanistic position in these results, but Driesch clearly thought differently. In his "proof of vitalistic events" in 1896, he subsumed all organisms under a single teleological schema, in which the goal is the whole organism, and used an aesthetic metaphor from music to oppose Loeb's reliance on a machine metaphor. Developmental potential could be realized in two ways. In "determined equipotential systems," potential was divided among the parts of the system; in "harmonious equipotential systems," like his sea urchins, it was not. The existence of "harmonious equipotential systems," he declared, was proof of "dynamic teleology" in nature.[47]

Little of this would have been significant for the emergence of Gestalt theory if Driesch's own development had ended there, but it did not. First he applied the concept of "harmonious equipotential systems" to phenomena of organic self-regulation, such as the regeneration of organs and restoration of function. He then searched for a more general teleological principle to justify the expansion of his ideas beyond biology. He found it in a central concept of German Romanticism, *Seele,* which he took from Eduard von Hartmann's critique of recent psychology, published in 1901. In his book, *The Soul as an Elementary Factor of Nature* (1903), Driesch argued that the nature of human activity cannot be explained mechanically. In a physical-chemical system there is a one-to-one ordering of stimuli and reactions; within the constraints of given boundary conditions, systems can react only to present stimuli. In human action, however, the effects of past experience are "present for me"; this fact, he contended, makes for a "free divisibility and combinability of elements" in a lawlike, but not mechanical act.[48]

Driesch illustrated this claim with cases in which the alteration of a single stimulus element has a "total effect" (*Ganzwirkung*). For example, when someone calls across the street to a friend, *"Mein Vater ist schwer erkrankt"* (My father is seriously ill), the effect is completely different from that obtained when he says, *"Dein Vater is schwer erkrankt"* (Your father is seriously ill); yet only one letter of the message has changed in German. Taking the opposite case, the message "your father is dead" produces in a single hearer the same reaction in different languages, even though all the stimulus elements – i.e., the letters – may be different. This is what Driesch meant by the claim that past events – in this case the presence or absence of a loved parent – can be "present for me." Today, such examples are material for research in information processing. For Driesch they were evidence that the brain is a "harmonious equipotential system with respect to its possible performance"; he was sure that no inorganic system could function in this way, and concluded that psychophysical parallelism must be rejected.[49]

This transfer of neovitalism to psychology touched off a controversy in the *Zeitschrift für Psychologie.* The debate was discussed in Stumpf's seminar in the summer of 1908, when Wolfgang Köhler and Kurt Koffka were both students in Berlin.[50] Erich Becher, a student of Halle psychologist and philosopher Benno Erdmann, praised Driesch for pointing to the link between vitalism and mind–body interactionism, but criticized his justification for it. Aside from the psychological "naiveté" of the formulation that past experiences can be "present for me," he said, Driesch had needlessly confused the general principles of physical and chemical events with those of man-made machines. Becher recognized the importance of "total effects" of the kind Driesch had described, but pointed out that these can also be found in machines; a keywound toy is an example. On the other hand, the unlimited connectability of stimulus and reaction is characteristic of any telephone system. On this basis, Becher argued, all we can conclude is the existence of "preestablished harmony" between the mental and the physical worlds, which is little more than parallelism in disguise. Instead, Becher espoused the version of interactionism Stumpf had offered in 1896. Although "a certain nervous process in a certain part of the cortex is the regular precondition for the occurrence of a certain sensation," he argued, the sensation itself "absorbs no physical energy, and its relationship to the [physical] conditions cannot be expressed in mathematical concepts and laws." Though he depicted psychical phenomena as emerging from physical causes in the brain, he also claimed that they were capable of altering physical effects, as in hearing or in acts of will.[51]

The year after Stumpf's seminar focused on this debate, Driesch elevated his neovitalism to the status of a full-blown philosophy in his book *The Science and Philosophy of the Organism* (1909). He retained the arguments in his earlier work, but tried to avoid the criticism that he had reinvented a mental substance called *Seele* by making more extensive use of the term "entelechy," which he had already introduced in 1896. "Entelechy" for Driesch is the formative power of the organism, the end for which physical and chemical factors are only the means. This power is itself neither spatial nor temporal, but manifests itself in its effects. The logical sign of such a manifestation is the fact that in organisms wholeness is retained despite division into parts: "Is it possible to imagine that a complex machine, unsymmetrical in the three planes of space, could be divided hundreds and hundreds of times and still remain intact?" For Driesch, such observable "individuality" meant that we can "see the world, the world of entelechy, as it is in its immediacy." Entelechy itself can only be inferred, but its effects are most evident to our own self-observation, particularly in our awareness of acts of will. With this Driesch completed the step to psychovitalism he had begun in 1903.[52]

In some respects Driesch's critique of mechanism was similar to Bergson's critique of associationism. However, though his entelechy was like Bergson's élan vital in being only indirectly knowable, he did not speak of it as a higher reality grounded on intuition, as Bergson did. He insisted that, despite his use of terms like *Seele,* he was not a Romantic but a realist. His entelechy was another aspect of the single reality of nature, qualitatively different from but in constant interaction with the aspects studied in physics and chemistry. The idea that purposive action in nature could be reconciled with at least a loose form of mechanism was widespread by this time. Thus some biologists, like Herbert Spencer Jennings, were prepared to see the positive side of Driesch's critique of mechanistic thinking. Whether it was necessary to postulate an unknowable entelechy was another matter. Here it seemed as though Driesch had substituted metaphysics for investigation.[53]

Philosophers were not so allergic to metaphysics. Driesch's ideas were similar to current philosophical reflections about the mind, for example Husserl's account of perception as the "besouling" of sensation. Driesch himself was aware of the step he had taken toward philosophy, and drew the consequences. In 1909 he obtained the right to teach natural philosophy in the natural science faculty of the

University of Heidelberg, where he had been living as a private scholar. Two years later he shifted to the philosophical faculty, with the express approval of the chairholder, Wilhelm Windelband. The pragmatic Oswald Külpe had pointed out to him that with the few existing natural science faculties in German-speaking universities, he would have better chances of advancement in the philosophical faculty.[54] Driesch took up Husserl's ideas and the research of Külpe's Würzburg school when he worked out the logical and epistemological implications of his philosophy in greater detail.

Thus, just when Driesch's ideas had ceased to be interesting to most biologists, they became so for philosophers and psychologists. In many respects his thinking paralleled that of the philosophers described earlier. None of these challenges to positivism was as antiscientific or irrationalistic as is often claimed. James's commitment to empiricism was hardly in doubt; but even Bergson, Dilthey, and Husserl presented their ideas as alternative conceptual and research programs for a philosophical science, though it was clear that this would be a human and not a natural science. These critiques gave further weight to the argument that it is impossible to reconcile the phenomena of mind with the claims of natural science on the basis of mechanistic categories. At the time, however, it was difficult to see what other basis there could be.

6

The Gestalt debate: From Goethe to Ehrenfels and beyond

In the generation from 1890 to 1910, the heuristic assumption that consciousness consists of elements continued to unite experimenting psychologists in the way that the dualism of peripheral and central processes organized physiological and psychological research in perception. From 1890 on, however, both heuristic elementism and conventional dualist frameworks faced challenges, not only from philosophers, but also from experimenting psychologists who pointed out anomalies in the given theoretical or descriptive order. The result at first was a proliferation of research problems and partial theories designed to plug the leaks. This led to an impression of hopeless plurality, not of linear progress. Nowhere was that impression stronger than in work on the related problems of recognition and form perception.

In 1889, the Danish philosopher and psychologist Harald Höffding distinguished a quality not mentioned in the traditional schema of associationist psychology, "in which a direct differentiation of several elements is not possible for us." This was the "quality of familiarity" characteristic, for example, of situations in which we vaguely recognize that we know a name but cannot place it. Höffding's countryman Alfred Lehmann and Wundt treated this quality as an anticipatory expectation prior to the appearance of a memory image. Höffding insisted that the phenomenon appears with the immediacy of a sensation, often without additional reproduction or recall, yet does the work of an idea. We can look at a house, for example, and retain a memory only of an ornament about a certain window; later we see the ornament and immediately say, "I know this house."[1]

Experimental studies soon revealed the theoretical and practical significance of such phenomena. In research on reading that had potentially important implications for education, Benno Erdmann and Raymond Dodge found that observers could retain up to four or five times more letters when they were presented as parts of a word or sentence than when they appeared as unrelated jumbles. As the logician Erdmann pointed out, both letters and words are perceived as such "not only on the basis of the optical components" into which they can be analyzed, "but rather as a result of the configuration of these components" that is characteristic for them. For example, the word "feeling" written vertically with the last letter at the top is read one letter at a time, but the same word printed normally is not. However, Erdmann and Dodge saw no need to revise associationism on this account; instead they attributed word recognition to frequent repetitions of "gradually more strongly associated complexes of sensations."[2]

Subsequent research challenged this interpretation. Friedrich Schumann, for example, found that observers could see words of up to twenty-five letters presented with his tachistoscope clearly and distinctly in all their parts, even though the word they saw was not exactly the same as the stimulus word. It seemed difficult to imagine that practice alone could account for this result. Schumann espoused a version of Wundt's assimilation theory: "In the act of recognition the images of former perceptions of the same object are reexcited, fuse with the sensations, and give to the perceptual process its 'quality of familiarity.'"[3] But this fusion of sensations and images, aside from being unobserved, did not explain Höffding's finding that recognition could occur without reproducing any specific contents from memory. Evidently there was some relationship between the organization of stimulus elements and their recognition and retention, but there was little agreement on the nature of that relationship. One study reported a total of fourteen theories on the subject.[4] The same situation obtained in form perception.

Reframing Gestalt discourse: From art to epistemology

Both the problem of form and its connection with the problem of whole and part are as old as philosophy. When Aristotle said that "the whole is prior to the parts," he meant to make a statement about the relationship of "form" – the essence, or substance of things – and its "matter," or attributes, not a statement about the intelligibility of things.[5] When the philosophical poet and romantic scientist Johann Wolfgang Goethe introduced the Gestalt concept to nineteenth-century German thought, he combined these agendas in a manner all his own.

In Goethe's morphology, the term "Gestalt" referred to the self-actualizing wholeness of organic forms. Goethe considered all advanced structures of a plant or an animal to be transformations from a single fundamental organ. He accounted for similarities among the members of a species by formal laws of (self-) organization, ultimately derived from an ideal type he called an *Urbild,* and attributed the differences to environmental effects. This was similar to the view of Kant and Blumenbach (discussed in Chapter 5), for whom the functional role of an organism's parts is determined by a law inherent in the whole – a conception of cause and effect different from that of mechanics.[6] It was only a short step from positing self-actualizing morphotypes in the organic world to conceiving the human person also as the result of an ongoing process of self-creation, or *Bildung*. Like many of his contemporaries, Goethe enjoyed the then-popular pursuit called "physiognomics," originated by Johann Caspar Lavater. This involved interpreting the outlines of human facial features exhibited in silhouettes called "shadow pictures" as expressions of the subjects' character. For some this was surely no more than an amusing parlor game; for Goethe and others there was more to it than that. Romantic artist, physician, and philosopher Carl Gustav Carus elaborated the notion that there were morphotypes of personality into a full-blown theory in the 1840s.[7] But Goethe had stated the metaphysical commitment that underlay that enterprise long before.

Though he referred to the morphotypes he postulated as "pure ideas" of nature in Platonic fashion, Goethe insisted that they were both real and nonmaterial. This fervent belief in the fundamental unity of material and nonmaterial reality, and an equally fervent belief in the unity of science and art, also grounded his opposition to Newton's theory of color. In the historical part of the *Farbenlehre,* Goethe wrote that "we must necessarily think of science as an art if we expect any kind of wholeness from it.[8] He said that he "came to color theory from the direction of painting, from the perspective of the aesthetic coloring of surfaces." He was therefore concerned to properly delineate, organize, and account for the color phenomena, not reduce them to something else. Newton wanted to improve his dioptric telescope, and was therefore concerned with sharp images and the dependency of color on the refrangibility of light. Newton's concept of colors as corpuscular, and of the prism as a "dissector" of white into fragments, seemed fundamentally mistaken to Goethe, because it explained something self-evident – the perceived simplicity and uniformity of white sunlight – by something obscure and hypothetical. We are not dealing with two sciences here, with Goethe's being that of perception, Newton's that of optics. For Goethe, rather, the two were poles of a single, dynamic unity. As he put it to Eckermann, "you see, there is nothing outside us, that is not at the same time in us, and as the external world has its colors, so does the eye as well."[9]

Goethe articulated this belief most clearly in the poem "Epyrrhema," already quoted (see Chapter 4), and in the novel *Elective Affinities,* which tells a story of human passion in the language of magnetism and chemical attraction. The most famous line of "Epyrrhema," "For what's within, that is without" (*Denn was innen, das ist aussen*), expresses the polarity of essence and appearance, not of body and mind. For Goethe nature's images are clues to the workings of the organs and the mind that comprehend them – or, more appropriately, to nature's law in us. In his view, that law is dynamic, not static, filled with comings and goings, not technomorphic mechanical operations. The ideal end results of these dynamic interactions are classically proportioned forms, signs of balance, lawfulness, and order realizing itself in nature, not imposed upon it by an ordering mind. Hermann Helmholtz put it well when he wrote in 1875 that "what Goethe sought was the lawlike (*das Gesetzliche*) *in the phenomena.*"[10] The Gestalt theorists later cited both *Elective Affinities* and "Epyrrhema" to indicate their commitment to just this ideal.

In stark contrast to all this, late nineteenth-century discussion of the Gestalt problem among academic philosophers and experimenting psychologists had shifted from being to experience. The issue was what it had always been for the British empiricists, and for Kant – whether to ascribe our experience of forms, or wholes, to sensation or to intellect. Ernst Mach continued in this vein, but reached a dead end. When Christian von Ehrenfels took up where Mach left off, he transformed the discussion by marrying Goethe's terminology and aesthetic priorities with long-standing problems in epistemology and logic.

For Kant, the unification of impressions is achieved first by "a survey of the manifold and then its summation," which he called "the synthesis of apprehension." For J. F. Herbart, connections of the sensory manifold are "an immediate achievement of the unity of the mind," based on the reproduction of fused presentations.[11]

In opposition to Herbart, Mach, in a lecture of 1865, attributed perception of both space and form to "muscular feelings" similar to the kinesthetic sensations Wundt invoked to explain depth. According to Herbart's theory, Mach argued, it is impossible to understand how similar but differently colored shapes, for example, are thought to be the same; for in Herbart's psychology, "only simultaneous and similar" ideas can call one another into consciousness. This demands "qualitatively equal presentations [*Vorstellungen*] in both series. The colors are different. Therefore there must be equal presentations connected with the colors but independent of them." These are "muscular feelings" of the eye, which are the same for both shapes. To make this idea plausible Mach gave examples of the effect of position on the perception of objects. The letters d, b, p, and q, for example, represent the same figure in different positions, but they are nonetheless seen differently, because the muscular apparatus of the eye is asymmetrical. With these observations he had tried to convince a scientific audience in 1861 that the space of sensation and that of geometry are fundamentally different. As he later recalled, the question "was accounted not only superfluous, but even ludicrous."[12] The theory of "muscular feelings" fared little better.

A more general remark Mach made at the end of his 1865 lecture had a different reception. The situation just described for letters also obtains for melodies: "We can choose the melodies in such a way that not even two partial tones in them are the same. And yet we recognize the melodies as the same." In fact we recognize similar melodies more easily than the keys, and similar rhythms more easily than the tempi in which they are played. Thus, Mach argued, not only similarity of visual form, but "all abstractions" must be based on presentations (*Vorstellungen*) of special quality. "But where is psychology supposed to find all these qualities?" he asked. "No problem! They will all be found, just like the muscular sensations for the theory of space. The organism is rich enough for the time being to fulfill the needs of psychology in this regard, and it is time to take seriously the 'bodily resonance' of which psychology so readily speaks."[13]

In *The Analysis of Sensations,* Mach eliminated the word *Vorstellung* and replaced it with the word "sensation," in keeping with the work's epistemology. He was more cautious about "muscular feelings" or "bodily resonance" than before, but retained the strategy of positing additional qualia, giving the same example as in 1865:

> In examining two figures which are alike but differently colored (for example, two letters of the same size and shape, but of different colors), we recognize their sameness of form at the first glance, in spite of the difference of color-sensation. The sight-perception, therefore, must contain some identical sensation-components. These are the space-sensations – which are the same in the two cases.

In the case of figures like the square and the diamond, which are geometrically similar but are seen differently, Mach invoked "sensations of direction" and postulated the requirement of homologous position. When this is not given, intellectual effort is required to make "the affinity of form" apparent. In his opinion, such phenomena therefore marked the boundary between sensation and intellect. But when he again extended the observation to melodies, he failed to find a suitable "sensation" to cover the case.[14]

This is where Christian von Ehrenfels, a young Austrian nobleman who studied with Brentano in Vienna and Brentano's student Alexius Meinong in Graz, entered the discussion. His essay, "On Gestalt Qualities," published in Richard Avenarius's *Quarterly for Scientific Philosophy* in 1890, became the founding document of Gestalt theory. Ehrenfels accepted Mach's observations but rejected his interpretation of them. As he pointed out, a simple colored extension such as a red patch is immediately recognized as similar to other such surfaces, even though they are continuous and therefore have no "dividing walls" from which space sensations might emanate. In the case of melody, he noted, Mach had used the term "sensation" in an unusually wide sense; ordinarily it is absurd to think of sensations as extended in time. But instead of accepting Mach's notion, Ehrenfels restructured the discussion by taking melody as his paradigm case for deciding what such forms "are in themselves" (*an sich seien*). Noting, as had Mach, that we can recognize two melodies as identical even when no two notes in them are the same, he argued that "these forms must therefore be something *different* from the sum of the elements." They must have, that is, what he called "Gestalt quality."[15]

Ehrenfels had identified a fundamental weakness in Mach's philosophy that made it impossible in principle for him to account for our actual experience of both spatial and temporal complexes as wholes. In Mach's "mass of sensations," all elements are related to or dependent on all other elements. Only practical interest decides, for example, whether a particular collection of elements is regarded from a "physical" or "psychical" point of view. Ehrenfels maintained that Gestalt quality is not something projected onto sense data but itself "a positive content of presentation" given along with the "elementary presentations" that serve as its "fundament" (*Grundlage*). He clearly thought that Gestalt qualities showed how ideation (*Vorstellung*) "brings more" to sensation, but he accepted Mach's claim that they are immediately given, not results of abstraction or comparison. As one recent study puts it, Ehrenfels's Gestalt qualities, in contradistinction to Machian elements, are in a relation of "one-sided," not all-sided dependence to sense data.[16]

With the aid of this definition, Ehrenfels discovered Gestalt qualities everywhere in experience. The idea embraced, for example, complex objects founded on the perception of inner states, such as moods. Relations, such as the judgment that red is not green, are also based on but not reducible to their elements. From here it was a logical step to claim that the same relationship holds for what he called "Gestalt qualities of a higher order," for example terms like service, theft, marriage, and war – concepts that retain their identity even though the examples that instantiate them change, and are therefore something other than mere names for "aggregates of ideas." Thus, he concluded, "Gestalt qualities comprise the greater part of the concepts with which we operate," and "the greater part of our associations" take place through their mediation.[17] Ehrenfels had imported the term "Gestalt" from aesthetics into epistemology and psychology, and he emphasized that "higher-order" Gestalt qualities are particularly evident in literary forms, as well as in music. Examples would be our feeling for a musical or literary style, or even the analogy of a crescendo to the sunrise or a rising expectation. Understood in this way, he thought, Gestalt qualities would do much to explain and justify "the shaping power of fantasy" in consciousness.[18]

That was a suitable conclusion for someone who was a musician, a poet, a playwright, and a devoted patron of Wagnerian opera as well as a philosopher. Ehrenfels wrote that the tone painting in Wagner's music dramas "provides an inestimable wealth of material for the comparison of Gestalt qualities of all kinds." A man of powerful emotions and great hopes, he had made a pilgrimage on foot from his family seat in Lower Austria to Bayreuth for the premiere of *Parsifal* in 1882. For a time he hoped that the emotional uplift of Wagner's music could substitute for his lost religious belief.[19] His Jewish students later testified that he was free of anti-Semitism himself (see Chapter 8), but he was an associate of fellow Wagnerian Houston Stewart Chamberlain, author of the racialist Bible, *Foundations of the Nineteenth Century*. Chamberlain posited a rather different higher order Gestalt – the racial Gestalt, a unity of physical, mental, and spiritual attributes allegedly corresponding to the intellectual and moral qualities of a people – and claimed he could perceive this intuitively: "Descartes pointed out that all the wise men in the world could not define the color 'white'; but I need only to open my eyes to see it and it is the same with race."[20]

For philosophers and psychologists in the 1890s, the more strictly philosophical implications of the notion of Gestalt qualities were more worrisome than its potential ideological resonances. Ehrenfels's essay linked the problem of form with the perception of wholes and parts in general. This brought out the issue's connection with fundamental problems in ontology and the theory of knowledge. Ontologically speaking, the objects that have Gestalt qualities are not mere collections of properties; they are structures, not sets.[21] But Ehrenfels himself did not clarify the nature of the complexes of which Gestalt qualities are qualities. Psychologically speaking, Gestalt qualities were neither sensations nor judgments. According to then-accepted categories, they were thus neither physical nor psychical. Whichever they were, the relation of such contents to the acts that generated them was unclear. These problems were so obviously relevant to both philosophical and broader cultural debates that nearly every philosopher and psychologist of the day had something to say about them. A few of the more significant positions will be discussed here, in order to create a feeling for the stress and strain with which writers tried to squeeze Gestalt qualities into existing frameworks of discourse.[22]

In essence two kinds of positions can be distinguished. One group of writers posited two levels, foundational contents and psychological processes, either intellectual or volitional, superadded to them. The other group insisted on the givenness and sensory immediacy of Gestalt qualities, but drew different conclusions from this fact. Meinong, himself a musician and composer, shared his student and friend's aesthetic inclinations, and he concluded that Ehrenfels and Mach had uncovered an undeniable fact. But he rejected the term "Gestalt quality" as an unwarranted intimation that "extrapsychological realities" were involved. Instead he cited Ehrenfels's use of the term "fundament" and called Gestalt qualities "founded contents" (*fundierte Inhalte*). These he attributed not to relations, but only to "complexions," as he called collections of elements. This allowed him to account for the fact that both Gestalt qualities and relations are logically dependent on their members without having to equate them psychologically.[23]

Meinong did not claim that intellectual acts could alter the contents on which they are founded. Munich philosopher and psychologist Hans Cornelius soon drew that conclusion from facts about tonal fusion described by Stumpf. Like patches of color, overtones, glissandos, and the like exhibit no phenomenal discontinuities, but are experienced as continuous wholes. Stumpf called fusion "that relation of two contents, especially sensory contents, in which they make not a mere sum, but a whole." He also stated, however, that in a musical chord, for example, the individual tones remain unchanged in consciousness but are not noticed as such, though we can analyze them out when we redirect our attention to them. Cornelius, following James, denied the latter claim and maintained that attention could indeed transform sensory contents. This implied that Meinong's term "founded content" was inappropriate in such cases, for its use assumed the constancy of the "fundaments."[24]

In his systematic psychology text of 1897, Cornelius generalized his opposition to atomistic psychology and repeated his critique of "unnoticed sensations." Gestalt qualities he called "attributes" (*Merkmale*) of complexes that do not belong to the elements, but are produced by "relations of similarity" recognized in perceptual judgments. But he also imputed Gestalt quality to feelings, opposing the traditional view that feelings were elements that were associated in the same way as ideas. In 1900, he went so far as to propose that psychology ought to begin with such qualities, and not with analytically derived elements. His student Felix Krueger took up this view and applied it in extensive studies of hearing after the turn of the century, accusing Stumpf of "hypostasizing elements," "mistaking physical for psychological concepts," and holding a "rigid concept of sensation."[25]

Cornelius's senior colleague in Munich, Theodor Lipps, adopted Meinong's two-level framework, but changed its content. For him Gestalt qualities were products of unconscious psychical processes of "apperception," ultimately traceable to an ego projecting itself into the world. Against Cornelius, he argued that feelings are the symptomatic effects of these processes, not characteristic features of complexes per se.[26] Friedrich Schumann rejected such elaborate apperceptive processes and also Meinong's "founded contents." The only intellectual work needed to perceive form, he argued, is attention. In the case of Mach's square and diamond, for example, subjects need only to focus long enough on the lower corner of the square or one side of the diamond to recognize the similarity of the two figures.[27] It takes little intellectual effort to recognize the similarity of Schumann's position to that of Mach.

Attention was the favorite source for higher mental processes among psychologists, and Schumann's view is traditionally portrayed as a skeptical retreat to "the classic, preholistic position."[28] Yet he recognized that there were phenomena which did not fit that view. Most interesting were cases of grouping. Although the lines presented in Figure 1a are evenly spaced, subjects spontaneously see a series of "fence posts." When we alter the distances only slightly, as in Figure 1b, effort is needed to separate a single line from the resulting "fence posts." Even more striking are cases of subjective contour, in which boundaries appear although they are not in the figure as drawn. In Figure 1c, for example, we immediately see a square. Schumann acknowledged that neither successive summation nor focused attention suffices to account for such observations: "We probably have to do here with a sensory aspect which is not further definable, a final fact of consciousness, which can only be made clear by examples."[29]

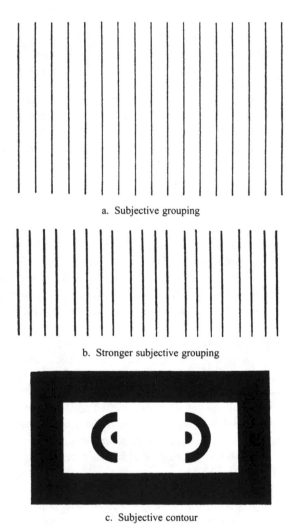

a. Subjective grouping

b. Stronger subjective grouping

c. Subjective contour

Figure 1. Subjective grouping and contour (Schumann 1900). Friedrich Schumann, "Beigräge zur Psychologie der Gesichtswahrnehmungen," *Zeitschrift für Psychologie,* 23 (1900), pp. 7, 10, 13. Reprinted by permission of Verlag Johann Ambrosius Barth.

Schumann admitted that similarity of spatial forms is not the same as that of elements, or "fictive parts," as he called them. Still, he claimed, Gestalt qualities are not new contents or qualities, but features of the elements as given. That we recognize a square in different positions, for example, implies that we have a presentation of the four sides' connection and of the figure's "right-angledness" (*Rechtwinkligkeit*). "At most" one could say that such impressions had a "special capability" of summating the others, but it is simpler to ascribe this capability to the lines themselves.[30] In 1904, G. E. Müller expanded this argument by stating that certain factors increased the "degree of coherence" (*Kohärenzgrad*) of stimuli and thus made their "collective apprehension" as a "unified complex" easier. Among these factors he named nearness, symmetrical position, "inclusion in common contours," habit, and expectation. But Müller did not pretend to know how these factors worked, nor did he say how he could reconcile their existence with even a modified associationism.[31]

Research on Gestalten: The Graz school

By this time Meinong had worked out a model of cognition that became the basis for systematic research on Gestalt perception in the so-called "Graz school." The model described thoughts as intentional judgments in which "higher-order" objects are built up on the basis of "lower-order" contents that may be given in sensation. Taking up an idea from Brentano's student Kasimir Twardowski, he argued that higher-order objects may exist in perception, but need not do so. The words of a proposition about round squares, for example, exist as mental contents, but the objects to which they refer do not. In the case of relations, the similarity between the copy of a picture and the original does not exist as picture and copy do; in Meinong's terminology it "obtains" (*besteht*). Relations and the complexes into which they enter, such as melodies and forms, were "objects of a higher order" in this sense.[32]

Meinong's co-workers in Graz, Stefan Witasek and Vittorio Benussi, tried to validate this model experimentally. Ambiguous figures and illusions provided excellent material for this purpose, because subjects make different judgments about them on the basis of the same sense data. In the case of ambiguous figures, there seemed to be no difficulty for the model. In Meinong's terms, a single set of contents – the perceptual representations of the lines in the figure – is directed toward two different objects, according to whether the perceiver judges them to be one or the other. In the case of illusions, however, there was a problem. In Figure 2, for example, a version of the Müller–Lyer illusion, the lines *ab* and *bc* are equal, but are usually judged to be unequal. Witasek reasoned that if we have an "adequate" idea or representation of the line *abc* in which the presentations *ab* and *bc* are equal, then the judgments we make of them must also be equal. Instead, the illusion persists even after we are convinced it is an illusion. Witasek therefore concluded that such cases were "illusions of sensation," not of judgment.[33]

Benussi preferred to postulate an intervening process between sensation and judgment. In doing this he drew an analogy to the notion of "suppositions" based on inadequate or incomplete introspective evidence, which Meinong was developing at the time. Since melodies and figures do not exist but "subsist," Benussi asserted, "they cannot affect our senses." If we nonetheless have presentations of them, then this must be due to prejudgmental "processing" (*Bear*

Figure 2. Müller–Lyer figure, as employed by Benussi (1904).

beitung) of sense data, which he called "production" at Meinong's suggestion. If merely seeing points or hearing tones is insufficient to grasp Gestalten, then such production processes must supplement these acts and thus "lead to the construction of a Gestalt presentation [*Gestaltvorstellung*]."[34]

To test this hypothesis Benussi conducted a thorough study of the Müller–Lyer illusion, which had rapidly become a major battleground for conflicting physiological and psychological interpretations. He asked subjects to concentrate alternately on the whole figure and on one or another of its parts, and found that the "analytical" attitude reduced, but did not eliminate the illusion, while the "Gestalt" attitude increased it. Other experimenters, like Schumann, had noted these effects of attitude, and had said that concentration, careful comparison of the lines and practice could counteract it. Benussi found that even in such cases the illusion did not always disappear, which indicated that physiological processes must be involved. But current physiological hypotheses were not sufficient. If the illusion were due to eye movements – a peripheral process – for example, then differences of illumination should make no difference, since the eyes adjust to them. Yet the illusion varies with changes in illumination. Thus, for Benussi the results showed that our tendency to produce ideas of higher-order objects from "inadequate" sense data was the basis of the illusion. He found further evidence for the production theory in a variety of other phenomena, including reversible and perspective drawings, illusions of hearing and touch, and the perception of time.[35]

Benussi's work with attitudes and cognitive styles may seem perfectly reasonable to psychologists today; but it was nearly totally ignored at the time, for two reasons. Processing models of perception and cognition invoking psychical functions like attention or summation were already common currency, but no one could give introspective evidence or any other direct account of the "production" process. Benussi later dropped the term "production" for the rather strange formulation, "ideas of extrasensory provenance."[36] Still more problematic was his dependence on Meinong's model of consciousness. The distinction between contents of psychical acts and their existent or nonexistent objects had the great advantage for philosophy of making it possible to analyze the logic of sentences with fictional objects, such as round squares, and sentences with real objects in the same terms. On this view, however, Gestalten need not exist as psychical contents;

Benussi relegated them to the realm of "irreal," potentially fictive objects. This line of thinking was certain of rejection by experimenting psychologists, who had an obvious interest in establishing the reality and experimental manipulability of such phenomena.

In 1913, Karl Bühler became the first scientist to determine psychophysical difference thresholds for Gestalt perception, specifically for perceptions of proportion (see Chapter 9). But he called the production theory "untenable" and mentioned so little of Benussi's research that Benussi wrote a long article to establish his priority in work on the topic. Perhaps the sharpest criticism came from Neo-Kantian philosopher Ernst Cassirer. If such complications were required to save the idea of fundaments, he wrote, perhaps it would be better to discard it, along with the claim that elements "subsist" at all outside of connections and relations: "The question here can never be how we go from the parts to the whole, but how we go from the whole to the parts." Brentano never accepted Gestalt qualities as anything more than "a particular kind and sum of relations," as his student Anton Marty put it.[37]

Perhaps this was sufficient for logical analysis. As Schumann's observations had shown, however, in psychology some relations had different effects from others. As to why this was so there was little more clarity than before. By 1910, the Gestalt problem had become "one of the most current issues in psychology," as Stumpf's student Adhemar Gelb wrote. But psychologists seemed unable to accommodate the sensory immediacy of forms and other wholes within their various categorical frameworks. Consistent associationists like Ebbinghaus preferred simply to accept relations and Gestalt qualities as ultimate data alongside sensations, as Schumann did, without trying to explain them systematically. Such "honest poverty," he said, was preferable to "the appearance of wealth."[38]

Carl Stumpf recognized that there were cases of "immanent structure," such as tone color and tonal fusion, but he refused to include Gestalt qualities among them. Melodies, rather, were special cases of "aggregates" (*Inbegriffe*), in which similar tone and rhythm relations and their associated images and feelings are summarized in consciousness. For these he preferred the name "form" from ordinary language. Form is not in itself an appearance, but "the correlate of the summarizing function" responsible for the appearance of complex collections. Gelb adopted this position and stated the argument behind it: Since relations are contents of consciousness like any others, no new content was needed. But he admitted that this view did little justice to "the unity, the self-enclosedness of entities like melodies, shapes and so forth."[39]

The Gestalt debate in neurophysiology

The problem of form also exposed weaknesses in the prevailing framework of neurophysiological theory. The Berlin neuroanatomist Wilhelm Waldeyer neatly expressed the firm belief on which that framework rested when he reportedly said to Charles Sherrington that the brain is "a thinking machine [*Denkmaschine*]."[40] This view rested on the assumption that nerve impulses travel by conduction along fixed paths. We have already seen (in Chapter 4) that this idea was fully compatible with associationist psychology. The discovery of the synapse showed that nerves were not exactly like telephone wires, as Helmholtz had thought, but this did not alter the principle of linear transmission.

The growing success of the neuron theory developed by Waldeyer, Ramón y Cajal, and others, and increasingly detailed anatomical atlases of the nervous system seemed to support the idea that the tracing of pathways would lead, sooner or later, to understanding all functions, from the reflexes to the higher mental processes. Sherrington's attempt to carry out this program showed that the task would not be so easy, since complex interactions of nerve and muscle groups were the rule even for simpler reflex reactions. But these results were only beginning to have an impact in Germany by 1910. In that year one physiologist stated bluntly that a truly natural-scientific psychology "would overcome all of the wordy explanations of the old psychology" by explaining "all mental events as expressions of conduction or pathway processes."[41] We are back to the old struggle for position within a shared epistemic space. Instead of letting mental processes operate on mechanically transmitted sensations according to their own (associationist) laws, some imperialists were apparently intent on extending the machine model to mind.

Incisive warnings against such a strategy came from the Freiburg physiologist and philosopher Johannes von Kries. Neurophysiologists, he suggested, placed too much confidence in concepts drawn from peripheral processes – that is, in the sense organs and the nerves emanating from them – because of their superior knowledge of these. Central processes, for example in the cortex, are "completely different." Specifically, the conduction theory could explain the strengthening of an already existing association, but not "the associative influence of complexes." The excitations produced by the sight of a single object in different positions, for example a horse seen from the front or the side, have "absolutely nothing in common," not even their spatial relation; yet we nonetheless recognize the object as the same. This is also true of temporally extended phenomena, such as melodies and rhythms. One way out of such dilemmas is to say the terms like "horse" are only names we give to associated experiences. But this is of little use, in von Kries's view, for the fact that we designate things in this way presupposes the physiological facts that need explaining.[42]

The only current attempt to solve this problem was the hypothesis propounded by Prague physiologist Sigmund Exner and by Mach, which invoked additional "accompanying phenomena," such as kinesthetic sensations. To von Kries this was untenable – how can the six muscles responsible for eye movements account for all the rich variety of form phenomena? Moreover, the hypothesis was theoretically unsatisfying. Mach in particular violated his own canon of simplest description when he demanded "another special quality, another point of agreement each time" for melodies, visual forms, and so forth. There was no need for such ideas "if we limit ourselves to the proposition that the physically similar corresponds to the psychically similar." As to how the "physically similar" might be constituted, von Kries suggested that the differentiation required to deal with the horse example would be achieved if each of a series of functionally similar brain cells retains a trace of the excitation caused by an object, but in a slightly different way. "New thoughts" about the plasticity of neurons might also offer interesting perspectives.[43]

The philosopher-psychologist Erich Becher took up von Kries's criticism in 1911 and offered experimental evidence for the inadequacy of the conduction model. He presented meaningless, asymmetrical line figures with a tachistoscope in such a way that they projected only to one specific portion of the retina, and had observers associate them with letters and numbers in given sequences. He then changed the exposure position so that the drawings projected to a different portion of the retina, hence, according to then-current assumptions, to different parts of the visual cortex. The results were clear. Observers reproduced the associated items even when the figures were presented in different colors and positions, although the cells stimulated in the second series should have contained no trace of the previously learned associations. After an exhaustive review, Becher concluded that no current neurophysiological theory could account for "the unique reproductive effect" of forms "independent of their size, position in sensory space, and quality."[44]

Von Kries had suggested that any part of a figure evoked associative effects, so that there could be at least punctual overlap in the two sets of presentations. Becher showed that this assumption did not always hold. Removing a point from the lower portion of a cross made of dots, for example, does not appreciably change the effect; but removing either the uppermost point or the ones to the right and left does. It is not immediately clear how any physiological hypothesis could explain the special significance of one point in a figure for its perception as a whole. His conclusion was categorical: if physiological hypotheses cannot explain these relatively simple form phenomena, then they cannot be retained for simple sensory qualia either. The fact that they had been demanded at all was probably due more to "historical than objective reasons," particularly the higher status of the natural sciences.[45]

This should have been enough to put physiology in its place and to justify an independent scientific position for psychology. Instead Becher went much further, repeating his earlier argument for an independent "psychical causality" alongside the physical variety (see Chapter 5), and asserting that mind–body interaction did not necessarily contradict the energy principle.[46] Becher clearly thought that he had discovered in Gestalt perception the definitive proof he needed to establish the irreducible reality of psychical phenomena.

The challenge of all this to experimenting psychologists at the time was clear. The inability of the discourse constituted by the unification of sensation-based associationism and mechanical physics, and its concomitant assumptions about science, to deal with the sensory immediacy of form perceptions seemed established; but no equally comprehensive categorical framework had come along to replace it. To accept the new facts about form as ultimate data – as Ebbinghaus, Schumann, and others did – and to resolve the problem by quietly reforming associationism might have been sufficient if experimental psychology had already been established as an institutionally autonomous discipline. In Germany, however, this was not the case. For the philosophical critics of experimental psychology, these difficulties were only further evidence for their denial that such a science could ever provide reliable or philosophically relevant evidence about mental life, much less address the burning cultural questions of the day.

Ernst Cassirer described the situation clearly in 1911. Psychologists, he said, had attempted to avoid the skepticism to which a phenomenalist stance would normally lead by declaring the "elements" of consciousness to be psychological reality. By doing this they had committed the error of confusing that reality with its representation:

> The ultimate parts which we can conceptually discriminate become the absolute atoms out of which the being of the psychical is constituted. But this being remains ambiguous in spite of everything. Properties and characteristics constantly appear in it that cannot be explained and deduced from the mere summation of the particular parts.[47]

The lived experience of meaningful relationships so important to cultivated elites – and in ordinary life – refused to submit to the ordering discipline of scientistic discourse. Experimentalists like Ebbinghaus retreated to heuristic elementism – the position that atomistic or sensationist assumptions were necessary to maintain the appearance of scientific procedure. But his admission of "honest poverty" in the face of Gestalten showed that even this "as if" stance was difficult to maintain.

Some chose to leave the experimentalists to their own devices and seek a solution elsewhere. James's answer was pragmatism. Bergson chose the metaphysics of the élan vital, while Driesch and Becher offered neovitalism and psychovitalism. The Southwest German Neo-Kantians spoke of transcendental norms or values, and Dilthey approached them when he elaborated the concept of "objective spirit" in his later philosophy. Husserl, too, later invented a transcendental phenomenology. The parlous situation in experimental psychology aided and abetted this exodus to supposedly higher ground. Others sought the depths instead – Nietzsche in his aphoristic meditations on the Dyonisian wellsprings of desire beyond good and evil, Freud by combining Darwinian and hydraulic metaphors to create a dynamic theory of unconscious conflict. In an age of specialization and fragmentation in economy and society as well as in academic life, calls for a unified worldview acquired increasing intensity. Experimenting psychologists may have shared these hopes, but they remained committed both to natural-scientific method, and to a definition of their subject matter centered around the problem of knowledge and the mind–body interface. The Gestalt theorists responded to this complex challenge by accepting both commitments while proposing radical changes in their conceptual and methodological content. To their story we now turn.

Part II

The emergence of Gestalt theory, 1910–1920

7

Max Wertheimer, Kurt Koffka, and Wolfgang Köhler

The generation of experimenting psychologists to which the creators of Gestalt theory belonged faced a complex orientation problem. Long-accepted discursive norms derived from mechanical physics and empiricist epistemology appeared inadequate either to deal with important facts about mind, particularly facts of aesthetic experience, or to sustain claims to cultural authority that were important to Germany's cultivated elites. Yet, in Germany at least, psychologists' self-assigned task remained the development of an empirically based philosophical worldview. Max Wertheimer, Kurt Koffka, and Wolfgang Köhler responded to this challenge with a radical reconstruction of psychological thinking intended to satisfy the requirements of both science and philosophy, of method and mind, in a way that expressed the values of their culture. Theirs was a common effort; but each of them made distinct contributions to it, because each was initiated into experimental psychology's social world in his own way. The following pages document those encounters until 1910, the year the three met in Frankfurt and began their common enterprise.

Max Wertheimer

In his jubilee tribute to Carl Stumpf, Max Wertheimer credited his teacher with inculcating in his students the ideal of psychology in the service of an empirical, yet not empiricist, philosophy (see Chapter 2). He was prepared in many ways to accept such a view of science and to make it his own. Born in 1880 into a German-speaking Jewish family in Prague, his intellectual and cultural efforts were supported from the beginning. His mother, Rosa Zwicker, daughter of the supervisor of nurses at Prague's Jewish hospital, was well-versed in German literature and music. Young Max learned to play the piano, took violin lessons, and spent many evenings making music with her. His father, Wilhelm, had only a grammar school education, but became a wealthy and respected businessman and founder of a

Photo 2. Max Wertheimer circa 1912. Historical Museum, City of Frankfurt am Main.

successful commercial academy. The system of instruction he devised for his school relied on personal tutorials rather than on rigid drill. Max Wertheimer advocated such individualized methods in his last work, *Productive Thinking*.[1]

The Wertheimers' family life and Max Wertheimer's education was like that of many German-speaking Jewish families in Prague at that time – a combination of successful acculturation to the larger society with attempts to maintain Jewish identity in the home.[2] Max was sent to a Catholic elementary school run by the Piarist fathers, then prepared for university at the famous Neustadt or Kleinseite Gymnasium; but he and his older brother Walter also studied Hebrew and learned the Torah. When his maternal grandfather, Jacob Zwicker, gave him a volume of Spinoza's works for his tenth birthday, the roots of his later worldview began to sprout. As one biographer reports, "the boy's complete absorption in the book led his parents to restrict his reading, but he continued to read Spinoza secretly with the connivance of the maid, who concealed the book in her trunk." Five years later, during a family argument about the weekly Torah portion, Max showed that he had taken the rational humanist's vision to heart: "'It's completely unnecessary for religion, or for knowledge of the Bible, to know what the *Sidra* [Torah portion] happens to be this week. Demanding that would be nonsense.'"[3]

When he entered Prague's Charles University in 1898, Wertheimer registered in the legal faculty. However, courses in philosophy, physiology, music, and art history soon began to dominate, and after five semesters he shifted to the philosophical faculty. Among his teachers were physiologist Johannes Gad, neurologists Arnold Pick and Sigmund Exner, criminologist Hans Gross, educator Otto Willmann, and philosophers Emil Arleth, Anton Marty, and Joseph Schultz. But the name which appears most frequently on his record is that of Christian von Ehrenfels, who had become a full professor in Prague in 1896. Ehrenfels's impressive personality, his combination of philosophical, literary, and musical gifts, and his intellectual independence exerted a powerful attraction on his students, including those who later created the so-called Prague school – among them Max Brod, Felix Weltsch, and Franz Kafka. From "Prague's great philosopher," as Brod later called him, Wertheimer heard lectures and seminars in psychology, theory of value, and other topics.[4] His hold on Wertheimer seems to have endured even beyond 1907, when he published a series of essays on sexual ethics in which he applied his concept of higher-order Gestalten to society and advocated, among other things, the eugenic selection of superior males who would be permitted to engage in polygamy, while others would be forbidden to breed. Wertheimer read and carefully underlined the first of these essays, accompanying the key points of the argument with exclamation points in the margin, but he did not do the same with the others.[5] As will be shown, however, the two men remained in contact (see Chapter 8).

From 1902 to 1904 Wertheimer worked with Stumpf and Friedrich Schumann in Berlin, becoming so involved in the new experimental methods that he read through all twenty-five volumes of the *Zeitschrift für Psychologie* that had appeared up to that time. There, too, he met and became friends with Erich von Hornbostel, Stumpf's assistant at the *Phonogramm-Archiv,* and deepened his knowledge of music by attending four courses with musicologist Max Friedlaender. Other lectures he heard were by Wilhelm Dilthey in history of philosophy, Friedrich Paulsen in pedagogy, and Adolf Wagner and Gustav Schmoller in political economy.[6] He also took an active part in Berlin cultural life through his friendship with industrialist Georg Stern, Stern's wife Lisbeth, and her sister, Käthe Kollwitz. Evenings were spent at the theater, concerts, and lectures, or at home with the Sterns, where they played chamber music almost every night and had long discussions about art, music, education, politics, and philosophy.[7]

Wertheimer seemed to have found his niche in Berlin. Yet he did not complete his dissertation there, but went instead to Oswald Külpe's Würzburg institute, where he received his degree summa cum laude in the autumn of 1904 after working with Külpe and Karl Marbe. Why he made this move is not clear. Kurt

Koffka later referred to a "difference of personalities" between Wertheimer and Stumpf, but gave no details.[8] Surely there was no objection to Wertheimer's topic, the psychology of testimony, as Stumpf offered seminars in legal psychology at just this time. More probable is that Wertheimer had chosen not to apply the methods then in use at the Berlin institute, but to continue research he had already begun in Prague with his law school friend Julius Klein.

A programmatic article the two men published in a journal edited by their law professor, Hans Gross, one of the founders of criminology, clearly indicates the high hopes they had for their work. In it they criticized current methods of assessing the truth of statements made in court, which concentrated too much on testing the credibility of witnesses or suspects and not enough on the issue of "whether a specific event ever really happened or not." They proposed to develop "not a psychology of statements," such as William Stern advocated, but a method for "the psychological diagnosis of states of affairs or facts."[9] The major problem in the way of such an effort was lack of cooperation from the suspects themselves. To counteract this, Wertheimer used a word-association technique as an elaborate ruse.

The supposition was that a criminal act, like any other, leaves behind a characteristic "complex" of associatively linked perceptions, judgments, and feelings, all with a strong "feeling tone." In a case of theft from a luxurious home, words that were likely to be part of such a complex, such as "silver," "garden," or "doorstep" could be presented along with irrelevant words, and suspects' reactions tested. Determined suspects could easily succeed in avoiding any outward sign of recognition, but could not prevent words or ideas from the complex from occurring first in reaction to the relevant words in the series. They would then either reveal themselves by responding with these words, or seek others and thus take longer to react. Though the principle was simple enough, the methodological problems were not. Wertheimer and Klein therefore proposed a number of variations, including physiological techniques of the kind employed in modern lie detection.[10]

In his dissertation, Wertheimer presented to each observer the diagram of a house in which a theft had occurred, but told them nothing of his purpose. He then presented them with word series, recorded their reactions and reaction times, and asked them to describe their experiences in the order that they had occurred. The last procedure was the method then being developed in Würzburg for the study of thought. Then observers went through similar test series again, but with the explicit instruction to deceive the investigator about their knowledge of the emotionally charged "complex" words. As a control experiment, individual words or word pairs were offered instead of longer series. The results confirmed expectations both objectively, in that reaction times for the "complex" words were appreciably longer and distributed differently, and qualitatively, in that "artificial" behavior patterns – embarrassed pauses, for example – emerged with the intent to deceive, in which a "conscious disposition" (*Bewusstseinslage*) appeared, but no word or image.[11]

Wertheimer and Klein had proposed to determine by psychological means whether events had actually occurred. In the dissertation Wertheimer stated instead that he wished "to draw conclusions about the existence and character of psychological facts from . . . forms of external behavior."[12] This was a more correct description, since his subjects had committed no crime. The language Wertheimer employed, including terms like "task" and "conscious disposition," came from the Würzburg school; and three of his five observers, Karl Marbe, Henry Watt, and Ernst Dürr, were leading members of that group. However, the Würzburg method of "re-presenting" experiences (*Kundgabe*) appeared in only one section of the thesis, and did not reappear in any of his later work. Furthermore, the theory of "complexes" was based squarely on classical associationism. Wertheimer stated that "whole complexes" should be presented where possible, but he meant only that series of words are preferable to single words or syllables. The only statement in the thesis that contained even a hint of concern with the Gestalt problem was the claim that, although the objective "unity" of a complex could influence the results, it was not the determining factor.

The importance and potential applicability of Wertheimer's technique were recognized quickly. The dissertation was frequently cited, and Wertheimer was invited to write a handbook article on the subject as late as the 1930s.[13] A priority dispute soon arose between Wertheimer and Carl Gustav Jung, who developed a similar technique on a related theoretical basis while working with patients at Eugen Bleuler's Burghölzli Clinic in Zürich. In the end, Jung acknowledged Wertheimer's priority in a letter. But he had lost the battle and won the war, as both the term "complex" and the association technique entered the armamentarium of psychoanalysis. Wertheimer pursued this line of investigation for two more years, but then stopped for reasons still unexplained.[14]

The years from 1905 to 1910 were a time of wandering and searching. Supported by his father, Wertheimer traveled throughout central Europe, doing research on alexia, aphasia, and cortical blindness – deficiencies of reading, speech, and vision due to brain damage – at Otto Pötzl's neuropsychiatric clinic and Julius Wagner von Jauregg's neurological clinic in Vienna, in Gad's institute in Prague, and later in Frankfurt. Apparently this work was intended to lead to his habilitation, but only a brief research report was ever published.[15] Wertheimer returned to Berlin frequently in this period, staying with the Stern family and working with his friend von Hornbostel at the *Phonogramm-Archiv.* The first published results of this work, a paper on the music of the Vedda, a tribe in Ceylon, show that something had changed in Wertheimer's thinking by 1910. Wertheimer called this music "among the most primitive known," since it is produced without instruments and consists only of brief fragments of no more than two or at the most three notes. Yet analysis yielded not constant repetition of melodic or structural formulas, but a variety of "definite forms with definite structural laws" which remain constant through all variations of the same 'song' or fragment. "We may say," Wertheimer concluded, "that a melody is not given by specific, individually determined intervals and rhythms, but is a Gestalt, the individual parts of which are freely variable within characteristic limits."[16]

Evidently Wertheimer had begun to investigate the problem of form. During this period, or perhaps still earlier, he read Husserl's *Logical Investigations* on the problem of whole and part (see Chapter 5) and noted on a scrap of paper that Husserl had not sufficiently considered "the ontological aspect" of the problem. Perhaps it was also at this time that he wrote a letter to a friend and colleague identified only as "dear K.," in which he detailed ambitious plans to test the laws of formal logic empirically and to study melody as a Gestalt.[17] The 1910 paper gave an indication of the alternative direction Wertheimer wanted to take. In the same year he had the idea that would lead to his habilitation, and to the first experimental research on Gestalt theory (discussed in Chapter 8).

Kurt Koffka

Kurt Koffka's acquaintance with both Berlin culture and Berlin psychology was more immediate and longer than Wertheimer's.[18] Born there in 1886, Koffka came from a family of lawyers, including his father, Emil Koffka. His younger brother, Friedrich, became a judge. His mother, born Luise Levy, was of Jewish descent, but listed herself as Protestant – an important distinction in the upper-bourgeois society of the imperial capital. Koffka's upbringing was thoroughly cosmopolitan. He learned English from an English governess, and went to the Wilhelmsgymnasium, one of the best-known schools in the city. Later he perfected his English by spending the academic year 1904–1905 at the University of Edinburgh. He remained a lifelong Anglophile. Influenced largely by his mother's brother, a biologist, Koffka became interested in philosophy at an early age and enrolled in that subject when he entered the University of Berlin in 1903. Though he began his studies under Neo-Kantian Alois Riehl, he changed to Stumpf and psychology after his return from Edinburgh because, he later stated, of his "resistance to idealistic philosophy; I was too realistically minded to be satisfied with pure abstractions."[19]

After hearing Stumpf's psychology lecture in the winter semester of 1906–1907, Koffka immediately put into practice the master's conception of psychology as "the study of daily life raised to a science" by studying his own color weakness in the physiological laboratory of Willibald Nagel, thus also beginning the physiological side of his training.[20] The psychological side was his dissertation research on rhythm, an outgrowth of Stumpf's seminar on that topic held in the summer of 1906. The work was completed in 1908 and published the next year. Koffka's task, assigned to him by Stumpf, was to determine whether rhythm

Photo 3. Kurt Koffka. Archives of the History of American Psychology.

could be elicited visually as well as aurally. Koffka achieved this by projecting figures, such as circles or lines, onto a screen and systematically varying the interval between projections with a clever arrangement of electrical circuits. He recorded reaction times and carefully monitored observers' accompanying movements, such as foot tapping, but the primary data were detailed verbal reports of self-observations. Among the practiced observers were Hans Rupp, Erich von Hornbostel, and Adhemar Gelb. He got "particularly extensive data" from Mira Klein, daughter of the publisher of poet Stefan George, whom he subsequently married.[21]

From his results he drew two conclusions. The first was that "in general, association had no perceptible influence on the course of the experiments," although very strong associations, such as melodies, did occasionally occur. The second was that "grouping," determined or structured by an "accent," was of fundamental importance for the experience of rhythm, no matter how it was elicited. Koffka argued that the then current theory of rhythm, which stressed kinesthetic sensations, did not explain the role of grouping, but only shifted the problem to another level of explanation. In any case, kinesthetic responses such as foot tapping and involuntary movements sometimes occurred before and sometimes after verbal responses.[22] Both Koffka and Herbert Langfeld later recalled that Stumpf's students were discussing Ehrenfels's concept of Gestalt qualities at the time. In fact, it was the seminar topic for the winter of 1906–1907, and one of Koffka's subjects used the term in a response. However, Stumpf opposed this terminology. Koffka adopted the term "unitary form" (*Einheitsform*), a compromise between Husserl's "moment of unity" and Stumpf's "form." This he supposed to be "the psychical entity that corresponds to the function of summation [*Zusammenfassung*] when objective relations exist among the summated parts" – a formulation perfectly consistent with Stumpf's views. Koffka had obviously recognized the significance of the Gestalt problem and cast about for ways of grappling with it in terms acceptable to his elders. As he later remembered, he "tried to put as much of these ideas into my dissertation as the master would permit."[23]

Not this work, however, but his research on color-blindness brought Koffka his first assistantship, with physiologist Johannes von Kries in Freiburg. There was also a modest psychological department in the philosophical seminar, run in 1909 by instructor Jonas Cohn with the formal support of the chair-holder, Neo-Kantian Heinrich Rickert; but Koffka remained only one semester in Freiburg before moving to Oswald Külpe's Würzburg institute in the summer of 1909.[24] While in Freiburg, he became acquainted with independent philosopher Fritz Mauthner, author of a critique of language strongly influenced by Mach and Nietzsche. Mauthner maintained that language is not an abstract logical system, but only "an infinitely complex activity" developed in response to practical human needs; "a people knows no other reason, no other logic than that of its language."[25] Such thoughts later impressed Ludwig Wittgenstein. Koffka apparently saw in their author a suitable confidant for his worries about the future of experimental psychology, and about his own future as well. In birthday greetings he sent to Mauthner from Würzburg in May of 1909, he wrote: "Unfortunately the situation of psychology is quite sad at the moment. As Külpe informed me, the Prussian government is behaving scandalously in its new appointments. It is said that a psychologist will not even be named to Ebbinghaus' chair. So the complaints of Rickert and company are ridiculous!"[26]

Obviously the research atmosphere was different in Würzburg, and Koffka joined the work of Külpe and his students with a series of experiments on association by similarity, which grew into a full-length monograph on imagery. Koffka wrote to Mauthner that he got along "famously" with Külpe. When he finally published his results in 1912, he dedicated the book to Külpe "in grateful remembrance of summer semester 1909," thanking him both for "scientific stimulation" and for "valuable personal influences which made working with him unforgettable." He recognized Stumpf as his first teacher, but noted that he now preferred "to solve many a problem in a way other than that which I learned from him."[27]

Koffka's engagement in his work and the affective bonds that long hours shared in the laboratory could create came through in his thanks to his observers for "the trust" they showed in the experimenter, leading to the formation of "close personal relationships."[28] In addition to Külpe's assistant Karl Bühler and Mira Koffka, his nine observers included Wilhelm Stählin, a Protestant minister with whom he later founded a journal for the psychology of religion (see Chapter 14), and Robert M. Ogden, an American who was later instrumental in bringing Koffka to the United States and publicizing Gestalt theory there.

Koffka, like his Würzburg co-workers, plainly regarded the statements of philosophers as propositions to be tested experimentally. His chief target was Hume's conception of images as pale copies of sensations, which he derided as "popular psychology." In his view, the total of 944 observations he obtained from his observers' verbal reports of their imagery supported Husserl's contention that images can appear just as vividly in consciousness as sensations. Although he retained Külpe's designation of images as "centrally excited sensations," Koffka argued that perception delivers the material from which images arise. This was consistent with Husserl's general model of ideation as the intentional ordering of images from earlier experience rather than their mechanical reproduction. He acknowledged, however, that the problem of how this ordering comes into being had not yet been solved.[29]

Unfortunately, there was no place for Koffka in Würzburg. In the middle of his year there, Külpe left to take up a professorship in Bonn, taking Bühler with him as his assistant. Külpe arranged to have Koffka kept on for one more semester, but his successor Karl Marbe had promised the position to someone of his own choice for the following semester. An obvious alternative was the commercial academy in Frankfurt, where Stumpf's former assistant Friedrich Schumann had succeeded Marbe. Koffka remained in Frankfurt three semesters, but in 1911 he earned the rank of *Privatdozent* in experimental psychology and pedagogy in nearby Giessen under August Messer, also a member of the Würzburg school. The stage seemed to be set for a routine academic career at a provincial university, with a professorship there or elsewhere sometime in the future. But in the meantime, during his first semester in Frankfurt, Koffka had had the meeting with Max Wertheimer that he later called "one of the crucial moments of my life."[30]

Wolfgang Köhler

An account of Wolfgang Köhler's ancestry reads like that of a German professor of the early nineteenth century.[31] He was born in 1887 in Reval (today Talinn), Estonia. His father, Franz Eduard Köhler, then director of the German-language Gymnasium in that city, was a minister's son who came originally from Thuringia and studied at the universities in Jena and Göttingen, receiving the doctorate in philology from the latter in 1865. Franz Eduard's brother, Ulrich Köhler, was professor of ancient history in Berlin. Köhler's mother, Wilhelmine Girgensohn, a minister's daughter, came from a long line of "Baltic Germans," who had settled

Photo 4. Wolfgang Köhler. Courtesy of Dr. Siegfried Jaeger.

in the region in the late seventeenth and early eighteenth centuries but had never given up their strong cultural patriotism. In 1893 the family moved to Wolfenbüttel, near Braunschweig, where Franz Köhler again became Gymnasium director. Wolfgang grew up and attended school there, graduating in 1905.

Köhler's inspiration to become a natural scientist came in large part from an unusual teacher – Hans Friedrich K. Geitel, a physicist and mathematician of international reputation who published on the conduction of electricity in gases, radioactivity, and other topics, and corresponded with such luminaries as Planck, Becquerel, and Rutherford, but refused to leave Wolfenbüttel for personal reasons, despite repeated calls to university posts. Köhler acknowledged Geitel's impact on his thinking in a letter written for his teacher's sixtieth birthday in 1915. "When I expressed the intention of studying philosophy," he wrote, "you said to me that in your opinion only a reasonably thorough study of mathematics and natural science would give hope for any achievements in that field. I have endeavored to follow your advice, and I must say today that I shudder at the thought that I should have neglected it."[32]

Follow it he certainly did. Of the courses listed on Köhler's university transcripts, more were in mathematics and natural science (twenty-five) than in psychology and philosophy (twenty-three). After studying philosophy, history, and physical science in Tübingen for a year, Köhler went to Bonn, where he made his first acquaintance with experimental psychology under Benno Erdmann. The decisive stage in his academic career began when he moved to Berlin in the fall of 1907, with the specific goal of writing a dissertation with Stumpf. In addition to working in Stumpf's institute, he continued to hear lectures in philosophy and natural science. In the former his teacher was Alois Riehl, a specialist in the philosophy of mathematics and natural science. In the latter his teachers included Walther Nernst, cofounder of physical chemistry, and physicist Max Planck, who was then in the midst of intense debates on quantum theory. Further exercises came in Emil Fischer's organic chemistry laboratory, as well as in physiology. In addition, he attended the colloquium of experimental physicist Heinrich Rubens, in which nearly all of the important physicists in Berlin took part. For a time he financed his studies with support from his family; he later supplemented this by tutoring in history of philosophy and with a small stipend arranged by Stumpf. Little is known about his free-time activities in this period, but one detail stands out – like Wertheimer, he was a musician. He played violin and piano with skill and enjoyed making chamber music with Geitel's brother, a government official, on Sundays.[33]

At first Köhler was frustrated by a lack of direction from Stumpf, who appeared preoccupied with his duties as university rector. In November 1907, he wrote Geitel that he was considering going to Munich and get a physics degree there with Wilhelm Wien. Erich von Hornbostel took an interest in his work, however, and encouraged him to apply his expertise in physics to psychological issues. Originally Stumpf had asked him to follow up on work by physicist Lord Raleigh on the localization of acoustical phase differences, but Köhler gave up this topic when it became clear that Raleigh's phenomena were the result of physical intensity differences, not psychological overtones. Von Hornbostel then encouraged him to try to develop a method for measuring the variability in the tone color of vowel sounds and musical instruments against partial tones of constant frequency.[34] After considerable effort and many setbacks, he hit upon a method that produced exciting results.

The development of new technologies for communicating sound, such as the telephone and the phonograph, had made acoustics an exciting multidisciplinary research field. The challenge here was analogous to that of psychophysical research in general – to find ways of gaining instrumental access to and "objective" measuring control of a dimension of human subjectivity. Many theories had been presented to explain the role of the tympanic membrane in hearing, especially in the fixation of tones. Ernst Mach had attempted unsuccessfully in the early 1870s to determine the effects of sound vibrations on the living ear by introducing sound through a tube and making observations with the help of a microscopic mirror. In *The Analysis of Sensations* he suggested that an indirect procedure using reflected light might achieve the desired results, and Köhler took up this idea. With the help of Stumpf's son, a medical student, and of a cooperating physician from the medical faculty, Köhler first had a tiny mirror placed onto the eardrum of an unemployed tailor. Unfortunately, the mirror became lodged in the canal in front of the man's eardrum, and had to be removed in the university's ear clinic. After that, Köhler continued the experiments with the mirror in his own ear. According to a later account, he demonstrated that the mirror was in place, and also that he had kept his sense of humor through it all, by appearing in the office of the cooperating physician with the thread that had been used to secure the mirror hanging from the other ear.[35]

Köhler had a room in the Berlin institute set up so that a steady stream of light from a stationary source could be directed at the inserted mirror, then reflected to yet another mirror and from there to a recording apparatus. The results deeply impressed him. As soon as a strong tone began to sound near his ear, a thin line appeared on the screen in front of him, then jumped to a new location and remained there as long as the tone remained steady. As the tone grew weaker, the image gradually returned to its starting point, but with no noticeable broadening. "The only explanation," Köhler wrote to Geitel, was that the tensor tympani is "an accommodation muscle which reflexively tenses the eardrum more strongly the greater the sound intensity, just as the corresponding muscle regulates pupil width in the eye." He hoped to magnify the images on the screen and thus measure the shifts corresponding to intensity differences. Such results would obviously be useful to physiologists and otologists. In addition, they might also provide a physiological explanation for Fechner's psychophysical law, which specified a logarithmic relationship between stimulus increments and reported sensations (see Chapter 4). Köhler hoped to show that the tensor tympani inhibits the vibration of the eardrum with increasing intensity "in just such a way that a logarithmic function appears." Happy as he was with these results, he worried about his future: "I cannot very well get a doctorate with just so much physiology."[36]

One month later, in April 1909, that problem disappeared. When Köhler tried to magnify the images by using a stronger lens and a smaller distance to the screen, he found that the images broadened from circa .75 millimeters to .75 centimeters, "that is, enough to get the most beautiful curves." Now he could show for the first time that the eardrum vibrates in direct response to acoustical stimulation, and also solve the question of vowel and instrument tone color that von Hornbostel had originally posed. In Stumpf's assessment of Köhler's dissertation, he emphasized that obtaining visual images of the eardrum's vibration alone was "a very important step forward and an extraordinary achievement."[37]

The curves that Köhler thus obtained revealed an unexpected pattern. Apparently the ear's response to particular frequencies corresponded quite closely to its response to spoken vowels. Köhler tested this further, presenting some preliminary findings in the second portion of his dissertation. He continued the work in Berlin, serving as an unpaid assistant to take advantage of the institute's superb collection of turning forks and other acoustical apparatus. Selecting thirty forks covering a range of four octaves, Köhler sounded them each fifteen times in random order for three observers, who were not informed of the purpose of the experiment, and asked them to judge the tones for their similarity to vowel sounds. Other observers with native languages other than German were brought in for less extensive testing, and some of the results were rechecked using a Stern tone variator (described in Chapter 2). The correspondence was confirmed. Most surprising of all was that the pitches judged most often to be pure or unmixed vowels in German formed a series of ascending octaves, from *u* (English: *oo*) to *i* (English: *ee*).[38]

Köhler made the radical implications of his findings quite clear. Helmholtz had discovered relationships similar to these years before, with some differences in the higher frequencies. Yet he continued to maintain that since physical tones possess only frequency and amplitude, the tones we hear should be described only in terms of the corresponding characteristics, pitch and loudness. Köhler agreed with the physical side of Helmholtz's theory, but denied its simple extension to psychology. The vowel qualities, he claimed, are "not qualities the tonal region has alongside others"; they are "the only qualities that it has at all." In fact, knowledge of a tone's vowel character is necessary to produce a judgment of pitch.[39]

Köhler believed he had shown for hearing what Hering had shown long before for vision — that the world of psychical qualities is not a mirror image of the physical world. The laws of each realm can be determined exactly, but "precisely the laws which determine a phenomenal system as system" deviate so much from those of the physical stimuli "that only a complex function can represent the relations" between the two systems. The point requires emphasis, Köhler said, because it is "ignored by a theory of knowledge that is still widely held."[40] He did not say which theory of knowledge he meant, but cited Mach, Stumpf, and Brentano together in support of his attempt to replace "the meager concept of pitch" and do justice to "the actual content" of perceived tones.[41]

The concept Stumpf introduced to supplement pitch was "tone color," the effect of the combination of fundamental and overtones. Stumpf treated "tone color" and the "clang color" exemplified by the different sounds we hear when the same pitch is sounded on different musical instruments as complementary aspects of the same phenomenon. In both cases it was simply a matter of the partial tones noticed or not noticed at particular pitches. The data from the measurements on Köhler's eardrum confirmed the presence and impressiveness of tone color. Low bass tones, for example, were not heard as such, but seemed to consist entirely of their color. According to Stumpf's view, clang colors should vary in the same way as tone colors. Instead, Köhler found, "these colors retain a recognizable similarity, in some cases despite a substantial shift of all the components." The sound of a tenor horn, for example, remains more or less constant up and down the scale, despite significant differences in the relative intensity of the partial tones. Köhler found himself forced to disagree with "my respected teacher, without whose instruction I would not know how to work in acoustics at all."[42]

Clang colors were similar to the so-called "interval colors" on which Mach based much of his discussion of tone sensation. Köhler acknowledged the similarity by using the terms interchangeably. Clearly both were complex qualities, as Hans Cornelius and Felix Krueger had defined them, since they depended not on individual tones or partials but on their relations (see Chapter 6). Köhler saw this, too, but carefully distinguished these phenomena from the so-called "difference tones" to which Krueger attributed complex qualities in his polemics against Stumpf. "Much has been written somewhat hastily about complex qualities," he wrote, "and in such a way that one must look about for things which do not fall under this category."[43] Köhler's findings forced Stumpf to defend his theory, and then to do further research of his own in order "to clear up this important problem of phenomenology." Subsequent measurements by others yielded values somewhat different from those obtained by Köhler, but confirmed the existence of vowel qualities. Decades later, the theoretical issues were still not satisfactorily resolved.[44]

Important here is what this episode indicates about Köhler as a young scientist. Though he had demonstrated a remarkable degree of intellectual independence, he had carried it off with a suitable show of respect for his teacher. He had made initial contact with, but had diplomatically avoided getting involved in, the Gestalt debate. But the central point was that he had combined physical and psychological observation in the service of a psychological goal. He explained his rationale for doing this at the beginning of his second acoustical article:

> The objective processes which evoke sensations have no special interest for modern physics, which strives as far as possible to lay aside problems carried over from anthropomorphism. . . . In the meantime, the psychologist's need for knowledge of the physical conditions which are involved in his experiments has only increased. For the reason just given, he must often help himself; and he can all the more readily become a physicist for the moment the more he is convinced that nothing hinders his science more than the treatment of the objects of consciousness as physical objects, or as photographs of them.[45]

Max Planck had used the word "anthropomorphism" as part of a massive, widely publicized attack on Mach's philosophy of science in a 1909 address to the Assembly of German Natural Scientists and Physicians in Leiden. Of course it is not possible to eliminate the idiosyncrasies of individual scientists from physics, Planck argued, but to recognize this is not to say that physical laws can only be convenient summaries of their sensations. This "anthropomorphism" fails to recognize the goal for which all scientists strive – to discover precisely that which is "invariant" in nature. "This constant element, independent of every human (and indeed every intellectual) individuality, is what we call 'the Real'."[46] Planck's hope for a unified physical "world-picture" to be achieved by hypothetical-deductive reasoning was rooted in deeply held religious and ethical convictions, as well as in the faith that the laws of nature correspond to the laws of thought. Commentators have given his views the name "rational realism" to distinguish it from Mach's inductivist, pragmatist relativism.[47]

Köhler read Planck's lecture as soon as it appeared, and developed his own position on the issues raised there.[48] He insisted on the separation of physical and psychological laws. Yet he used the term "invariants" for the vowel qualities he had measured, thus making them seem like phenomenal analogies to the abstract theoretical principles that Planck had presented as the highest goals of physical theory.[49] Köhler remained committed to discovering such invariant phenomena in the psychical realm, and to establishing the "complex function" relating physical and psychological processes, for the rest of his scientific life.

Max Wertheimer, Kurt Koffka, and Wolfgang Köhler had each responded positively, in his own manner, to Carl Stumpf's call to "lay hands on" psychological reality as a way of approaching philosophy. The setting of their response was not limited exclusively to the rooms of the Berlin psychological institute, or to the corresponding apartments in Würzburg and Frankfurt. Wertheimer, Koffka, and Köhler all came from the cultured upper middle class milieu that nurtured most of Germany's academic scientists. Each of them had been prepared to begin a successful scientific career in the German universities without the problems and pressures of upward mobility. During their university studies, each had acquired broad background in both philosophy and natural science.

Most important, however, was that all of them had mastered at least two of the available research methods in psychology, and had confronted both the conceptual presuppositions and the experimenter-subject relationships involved in doing so. Each of the three made a significant gesture revealing the intensely personal nature of his commitment, from Wertheimer's reading of the entire *Zeitschrift für Psychologie* to Koffka's and Köhler's experimenting on themselves. In addition, each had learned that psychological science was a cooperative effort by an elite of trained observers, a collaboration that could lead to lifelong relationships. Thus, they had been initiated in a variety of ways into the social world they had chosen.

But Wertheimer, Koffka, and Köhler also experienced quite directly the inability of prevailing psychological discourse, including that of their common teacher Stumpf, to deal with the facts that phenomenological observation yielded. Köhler chose the route of respectful opposition from within; Koffka learned what he could from Stumpf, then carried on elsewhere. Wertheimer did both of these things, and then began to think further independently. When the three men came together in Frankfurt, they were ready for something new, and they created it together.

8

Laying the conceptual and research foundations

The radical reconstruction of psychological theory that Wertheimer, Koffka, and Köhler proposed proceeded in four stages. First, Wertheimer presented the theoretical foundations in the form of a new epistemology and linked them with experimental research in his famous paper on motion perception. Next, Köhler and Koffka further developed Wertheimer's theoretical perspective and applied it first to perception and human behavior, then to animal problem solving. Köhler then extended the Gestalt principle to the external world and the psychophysical problem. Finally, Wertheimer presented studies of productive thinking as the prolegomena to a new logic. This and the following chapters consider each step in turn.

While on the way from Vienna to the Rhineland for a vacation in the autumn of 1910, the story goes, Wertheimer had an idea for an experiment on apparent movement when he saw alternating lights on a railway signal. He got off the train in Frankfurt, bought a toy stroboscope and began constructing figures to test the idea in his hotel room. He then telephoned the psychological institute at the commercial academy in the city and spoke about the plan with Köhler, who had just begun working as an assistant there. Köhler obtained space for him in the laboratory and the use of Schumann's tachistoscope, then served as the primary subject. Koffka and his wife Mira joined them later, after which Schumann himself invited them to work at the institute.[1] With these experiments, Gestalt theory entered research. But as will soon be clear, conceptual developments came first.

The experiments took place in the Academy of Social and Commercial Sciences in Frankfurt, founded in 1900 and funded by private interests in collaboration with the city government (Photo 5). This was a new form of support for higher education in Germany, and a correspondingly progressive spirit was evident when the experimenting psychologist Karl Marbe assumed the first philosophy professorship in 1904. Rooms for a psychological institute were funded by the Carl Christian Jügel Foundation, which also endowed the professorship. The

Photo 5. Frankfurt Commercial Academy (later University) ca. 1910. Courtesy of Psychological Institute, University of Frankfurt am Main.

institute officially opened in 1906, and laboratory courses began in 1908. With initial outlays of 10,000 Marks for books and apparatus and an annual budget of 1,500 Marks, the institute was well financed for the time. Later accounts mention its superior apparatus, and its personnel included two assistants, one budgetary (*planmässig*) and the other extrabudgetary (*ausserplanmässig*) but continuously renewed.[2]

In 1910, when Friedrich Schumann came from Zürich to replace Marbe, who had succeeded Külpe in Würzburg, a second chair for systematic philosophy and history of philosophy was funded by the Speyer Foundation. Its first occupant was Hans Cornelius, the critic of Stumpf's doctrine of unnoticed sensations. Although he did no experiments, Cornelius frequently had lively discussions with the experimenters, and later recalled benefiting particularly from his talks with Wertheimer. In 1911, Adhemar Gelb came to Frankfurt to replace Koffka as assistant, but Koffka returned regularly from nearby Giessen to talk with his friends and to hear Wertheimer lecture. In the beginning, however, there were only Wertheimer, Köhler, Koffka and Mira Koffka. As Koffka later recalled: "It began with Wertheimer and Köhler in Frankfurt with me as a third; we liked each other personally, had the same kind of enthusiasms, same kind of backgrounds, and saw each other daily discussing everything under the sun." It was in this close-knit atmosphere that Gestalt theory was born.[3]

Expanding the Gestalt category: Number concepts in "naturally thinking" people

Though the experiments they conducted together were completed by the spring of 1911, Wertheimer did not submit the results for publication until January 29, 1912. By that time he had already presented important aspects of the emerging Gestalt outlook in a paper on concepts of number in primitive peoples, based on research he had done in Berlin and completed by 1909. Wertheimer's methodological stance in that paper was an extension of phenomenology to cultural anthropology: "It is insufficient to ask what numbers and operations of our mathematics the peoples of other cultures have," he asserted. "The question must be: what units of thought do they have in this field? What tasks for thinking? How does their thinking approach them? What achievements, what capabilities are required?" Considering the thought of so-called primitive peoples from a Western standpoint prevents us from even asking these questions, and thus "blocks the way to a true knowledge of the actually given." Given this approach, it is not surprising that Wertheimer soon discovered a vast array of examples showing that Western arithmetic, epitomized in John Locke's assertion that all of mathematics is reducible to two concepts, one and plus-one, was inadequate to account for the rich numerical life of "naturally thinking people."[4]

Examples of the difference could be very simple: one horse plus one horse equals two horses; one person plus one person equals two people; but one horse plus one person equals a rider. Such changes of designation indicated to Wertheimer — justifiably or not — that a fundamentally different unit of thought was involved. Such a claim could hardly be persuasive by itself. However, the distinction was not always a matter of using different names. A builder goes in search of pieces of wood for a house: "One can count them. Or, one can go with an image of the house in one's head and get the pieces of wood that are needed. One has a group image [*Gruppengebilde*] of the posts, which is quite concretely related to the form of the house." In such cases, Wertheimer maintained, we are not dealing with counting at all in the Western sense, but with "number forms," entities that are "less abstract than our numbers but fulfill analogous functions" and thus are "logically a middle thing" between Gestalt qualities and concepts. Similarly, "the boat builder does not think of the number, but the arrangement, form, and direction – there the end, there that piece – and has a total image [*Gesamtvorstellung*] of the whole."[5] The term *Gesamtvorstellung* came from Wundt's ethnological psychology (*Völkerpsychologie*), but Wertheimer clearly did not think it appropriate to speak of such wholes, or whole-images, as emerging or being synthesized from component sensations.

In this kind of thinking, characterized by "the preponderance of form," people see and count objects according to their place and role in functional units. It is not always possible to divide such forms arbitrarily into countable pieces, as ought to be the case if the algebraic concept of number were applicable. Consider the following example:

> I break a stick in two. One approach says, I now have "two." (Two what? That's immaterial; I have two – new – units.) The arithmetic makes a jump: first there is one, a stick, then there are two fragments, and between the first and second state of affairs there is a gap which is ignored. This becomes particularly clear if it is not a stick but a spear: the result is not two *x*, but perhaps a piece of a spear (with the tip) and a bit of wood (the rest of the shaft). This other orientation says "two" only if there are two genuine pieces, two parts, not two "units." (I go not from one to two in this case, but from a perfectly functional spear to a useless, broken one, or no spear at all.)[6]

Psychologically speaking, Wertheimer concluded, it is "a fiction that everywhere any arbitrary division is completely equal to every other; the things (and our Gestalt grasp of them) make certain divisions more likely."

Wertheimer had expanded the Gestalt category beyond Ehrenfels's original definition. Ehrenfels had posited Gestalt qualities given along with sense data – qualities that, by virtue of their "one-sided dependence" on these data, make wholes different from sums of their parts (see Chapter 6). Wertheimer claimed instead that specifiable functional relations (*Funktionsbezüge*) exist which decide what will appear to be or function as a whole and as parts. He called the operation involved "centering the category," or, more generally, "predetermination." It is important to note that he stressed two aspects of this operation: "predetermination on the Gestalt of the whole," as in a pastry whose shape "decides" how it is to be sliced; and "the (not necessarily conscious) tendency to make natural, unified wholes [*Gestalten*] the result of the division."[7]

The implications of such an idea for the philosophy of mathematics were clear, and Wertheimer did not hesitate to draw them. He cited numerous cases of "primitive" thinking in which the commutative law does not hold. In the language of the Ewe, a West African people, for example, two things in each of three places are not the same as three things in each of two places. Actually, this is the case for Western people also, since intuitive perceptual similarity is not the same as mathematical equality, an abstract construction. But Wertheimer's point was that the construction of equations in general is "quite a strange thing for natural thinking." Even Giuseppe Peano's "metaarithmetic," known today as the base system, in which reckoning changes according to whether the base is 1, 2, 6, 10, and so forth, can be seen to have "real foundations" in something so prosaic as telling the time, an example of calculation in base 12.[8] Wertheimer had no intention of questioning the validity of mathematics, or the technical value of abstract, arithmetical thinking in civilized society. Instead, he wished to show that neither traditional Aristotelian nor modern mathematical logic were suitable guides to psychological reality. Husserl and Meinong had attempted to connect the psychological genesis and the rational validity of concepts by combining the logic of intentional acts of cognition with that of propositions (see Chapter 5). Wertheimer undermined that purpose by pointing to the different psychological objects that consciousness can have under different cultural conditions.

Wertheimer's perspective might have led to some form of cultural relativism. But he argued instead that the fundamental form-connectedness of his examples showed that "centering the category" is a naturalistic logical operation that occurs in all cultures. If there are general laws of thought, Wertheimer implied, then their explication ought to begin with such operations. Immediately after the case of the builder seeking wood for his house, given as an example of "the preponderance of form," Wertheimer added this one: "A somewhat blunted triangle, not a rectangle or a hexagon, as it would have to be called from a merely numerical point of view."[9] There is nothing functional about a figure drawn on a piece of paper, and Wertheimer did not claim that we perceive the triangle in this way because we have learned to do so, or because it is in our biological interest. The blunted triangle *is* a triangle for us – its being a triangle is an objective characteristic of its existence in our perception. In such observations Wertheimer went beyond descriptive psychology and pointed toward a new epistemology that he did not yet name, based on the claim that not sensations, but structured wholes, or Gestalten, are the primary units of mental life.

Gestalt theory is born: Frankfurt, 1913

In the next two years Wertheimer consolidated his right to make such claims by offering courses in both psychology and philosophy, including "Ethnological

Psychology" (*Völkerpsychologie*), "Theory of Knowledge," "Psychology of Memory," and "The Origins of Philosophy."[10] In the lectures on epistemology in the summer semester of 1913, he presented in explicit and general form the doctrine that he had only implied in the paper on number concepts. One of his listeners was Gabriele, Countess von Wartensleben, a teacher who received her doctorate under Schumann that year for experiments on the reading of letters, in which Wertheimer and Köhler were subjects.[11] In 1914, she published the following summary of Wertheimer's views in a footnote at the beginning of a book with the extraordinary title, *An Ideal Portrait of the Christian Personality: A Description 'sub specie psychologica'*. To make clear the fundamental change in perspective that Wertheimer proposed, the passage is given *in extenso:*

M. Wertheimer's Gestalt theory (which has not yet appeared in print, but about which I learned in a lecture on epistemological problems which he gave at the Frankfurt Academy in the summer semester of 1913, and in many private conversations) contains the following basic thoughts:

1. Aside from chaotic, therefore not, or not properly, apprehensible impressions, the contents of our consciousness are mostly not summative, but constitute a particular *characteristic* "*togetherness*," that is, a segregated structure, often "comprehended" from an inner center, which can be different according to the nature of the ideational content [*Vorstellungsinhalt*]; e.g., an optical or acoustical, or also a dynamic or intensity center. To this the other parts of the structure are related in a hierarchical system. Such structures are to be called "Gestalten" in a precise sense.

2. Almost all impressions are grasped either as chaotic masses – a relatively seldom, extreme case – or as chaotic masses on the way to sharper formation, or as Gestalten. What is finally grasped are "impressions of structure" [*Gebildefassungen*]. To these belong the objects in a broad sense of the word, as well as "relational contexts" [*Beziehungszusammenhänge*]. They are something *specifically different from* and more than the summative totality of the individual components. Often the "whole" is grasped even before the individual parts enter consciousness.

3. The process of knowing – knowledge in a precise sense of the word [*im prägnanten Sinne*] – is very often a process of "centering," of "structuring," or of grasping that particular aspect which provides the key to an orderly whole, a unification of the particular individual parts that happen to be present; what results is that a structured unit emerges as a whole due to, and through, this centering. The result of just this knowledge process is a springing forth [*Herausspringen*] of the Gestalt from the "not yet formed" [*noch nicht gestaltet*]. Certain appearances [*So-Färbungen*] of the parts result from the specific total conception; parts and specific states now become "understandable" on this basis.

The entity that results from the knowledge process depends in many respects not only on the object, but also on the observer. Thus there are several ways of grasping many phenomena, but generally only one can be correct: that which makes all states understandable and derivable from the central "idea" and thus gives meaning [*Sinn*] to the entire given.

The same statements made about different entities can have completely different directions, according to the way in which they "sit" in the entity [*drinsitzen*], e.g., whether they are nearer or further from the center. Thus, e.g., in the case of "the wall is red," "red" "sits" differently than in the case of "blood is red" (though the logical situation becomes more complicated here). Thus something completely different is meant by a complex connection such as "drinker philosopher" [*Trinkerphilosoph*], according to whether the drinker is thought to be in the philosopher or the philosopher in the drinker.[12]

This passage contains nearly all of the fundamental principles of Gestalt theory, with their implications for logic and the theory of knowledge duly noted. Wertheimer had proposed something like an ultimate transformation of positivism. While retaining the injunction, central to positivist thinking since Hume, that knowledge can be derived only from phenomenally given sense data, he radically altered the meaning of the term "sense data" by turning the primacy of sensations on its head. The most important features of the doctrine at this stage were: Wertheimer's use of the term *Gestalt* to refer to both individual objects and the organization of objects in the psychological field; his differentiated conception of consciousness, which does not exclude the existence of "elements" but takes them to be unusual, boundary cases; and his use of the term *Gestalt* not only for the objects of consciousness and the system of their relations, but also for the "knowledge process" thought to underlie their appearance in consciousness. Wertheimer later applied the concept of restructuring, or "recentering," to human problem solving and thinking. The distinction between "natural" and abstract "arithmetical" thinking thus became a dichotomy between "reproductive" applications of traditional logic and "productive," or genuinely original, dynamic thinking (see Chapter 12).

Both the epistemological core and the methodological standpoint of Gestalt theory were thus already developed by 1913. Given their obscure location, however, it is not likely that many psychologists or philosophers noticed these words when they first appeared. Koffka and Köhler referred to the paper and lectures just discussed only infrequently, and cited the paper on apparent motion as the beginning of the Gestalt movement. They were correct for several reasons. First, as Wertheimer himself said, much was only "fragmentarily suggested" in the paper on number concepts. More important was that, despite Wertheimer's expressed intention of setting up a systematic research program in ethnological psychology, it was not immediately clear whether or how the viewpoint expressed there and in his 1913 lectures could be sustained by laboratory, as opposed to field, research. Finally, though Koffka and Köhler doubtless discussed the findings of the first essay with Wertheimer, they experienced the phenomena of apparent movement themselves. Wertheimer's paper on number concepts marked the beginning of Gestalt theory as a new mode of conceiving psychological problems, one with philosophical implications. The paper on motion launched Gestalt theory as a transmittable model of scientific thought and experimental practice.

Letting Gestalten live in the laboratory: The phi phenomenon

In keeping with the common tendency of research in perception and cognition, then and now, to follow technological innovation, the literature on apparent move-ment, particularly the so-called stroboscopic effect, had grown rapidly with the rise of motion pictures after the turn of the century. The phenomenon itself – motion seen between two stationary light sources flashing at given intervals – had been observed as early as 1850 by the physicist Plateau. Sigmund Exner, one of Wertheimer's teachers, had obtained it with two electric sparks in 1875. However, no one had satisfactorily explained the phenomenon, because various theories assigned it a different status. Exner, for example, called it a "sensation" to empha-size the immediacy of the impression, and cited it as evidence for a sensationalist and physiological theory of perception, as opposed to the psychological construc-tion or conclusion of motion from elementary sensations. Yet he also showed that apparent motion produces negative after-images in the same way as real motion, and therefore proposed an explanation in terms of central, not peripheral phys-iological processes. Mach, too, spoke of "movement sensations," attributing them to eye movements, but maintained that stroboscopic motion is anomalous, "a peculiar process, which has nothing to do with the other sensations of movement."[13]

Others insisted on a psychological explanation. Johannes von Kries, for exam-ple, objected to calling the phenomenon a sensation and asserted that it was a judgment. Schumann referred to "illusions of judgment." Karl Marbe attributed seen motion to fused after-images of the stationary stimuli. Finally, Paul Ferdi-nand Linke, a student of Hans Cornelius and Felix Krueger, characterized the impression as a "complex quality" and said that its appearance depends on the subjective impression of a moving object to which sensations could be related. He speculated that processes like "assimilation" as described by Wundt are in-volved.[14] Thus, Wertheimer's paper addressed a decidedly current issue that raised anew the questions of central and peripheral processes, sensation and perception, and their relationship to one another. His careful experimental work

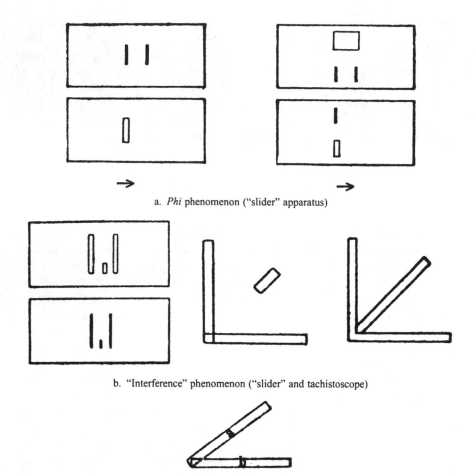

a. *Phi* phenomenon ("slider" apparatus)

b. "Interference" phenomenon ("slider" and tachistoscope)

c. Dual part motion (tachistoscope)

Figure 3. Setups for Wertheimer's motion experiments (1912). Max Wertheimer, "Experimentelle Studien über das Sehen von Bewegung," *Zeitschrift für Psychologie*, 61 (1912), pp. 262, 264. Reprinted by permission of Verlag Johann Ambrosius Barth.

successfully refuted the theories of Mach, Marbe, and Schumann while supporting Exner's.

Wertheimer did not discuss these theories in detail at first, but immediately recast the issue in phenomenological terms. "One sees movement," he asserted, not an object that is first in one place and then in another, but movement per se. The first question must therefore be, "What is the psychic reality, what is the *essence* of this impression?"[15] His experimental design made it possible to answer that question in a new way.

Wertheimer used two instruments. One was a simple "slider" (*Schieber*) apparatus of his own design. This consisted of two metal (sometimes cardboard) sheets with openings cut in them as shown (Figure 3a), one a fixed frame and the other with a handle to slide it under the frame. When images of the sheets were projected on a wall and the "slider" sheet was pushed in and out, observers saw a single moving line below and two lines in succession above. After repeated exposures, Wertheimer reversed the openings above and below without telling the observers and asked if they saw any difference. Even with this instrument it was possible to demonstrate that observers experienced no difference between apparent and real motion.

For exact experimentation Wertheimer used a tachistoscope, an instrument specially constructed by Schumann to study successive expositions (see Photo 6). In normal viewing, observers looked through a small telescope at a spot aligned with the outer ring of the tachistoscope wheel. The experimenter then placed the object to be viewed in an adjustable, slitlike opening on the wheel and rotated the wheel at varying rates. Wertheimer put a prism near the telescope objective so that it covered the lower half; light thus entered from straight ahead in the upper part and from the side in the lower part of the lens. One exposition slit on the wheel left the upper half of the objective free, another the lower half. Thus, when the wheel rotated, observers saw first one exposition field, then the other. Wertheimer took care to place the prism close enough to the objective so that "with each exposure the whole circular surface of the exposure field is seen."[16]

This arrangement permitted what is now called a parametric procedure – systematic isolation and controlled quantitative variation of single factors while holding other conditions constant. Wertheimer varied (1) the time interval between stimulus expositions (usually by regulating wheel rotation speed); (2) stimulus distance; (3) form; (4) color; (5) intensity; and (6) the relations of the two stimulus objects to each other. In addition, he introduced additional objects into the exposition field and considered effects of "variations in subjective behavior," including prolonged observation, set or attitude, and attention to a fixation point.[17]

In the standard experiment, a white stripe was placed on a dark background in each slit while the rotation speed of the tachistoscope wheel was regulated so that the time required for the light to pass from one slit to the next varied. Above a specific threshold value, approximately 200 milliseconds, observers saw the two lines in succession. With much faster rotation, about 30 milliseconds, the two lines flashed simultaneously. At the so-called "optimal" stage, approximately 60 milliseconds, observers saw "a definite motion: a line moves clearly and distinctly from an upper position into a lower one," when the slits were aligned horizontally,

Photo 6. Max Wertheimer with modified Schumann tachistoscope. Historical Museum, City of Frankfurt am Main.

or from one side to the other, when they were aligned vertically (see Figure 3b). Most important both epistemologically and psychologically was that in most cases apparent motion could not be distinguished from real motion. In fact, when observers said one was "better," that is, clearer and more impressive, it was most often the illusion.[18]

More important still for the development of Gestalt theory was what happened when the time interval was increased slightly beyond 60 milliseconds. After repeated exposures at this interval, observers saw motion without a moving object. Wertheimer gave this phenomenon the neutral name *phi* to distinguish it from other motion phenomena. The *phi* phenomenon, he claimed, was "simply a process, a transition," or "an across in itself," which appears in exactly the same way that color or form do. Contrary to other psychical data, however, such events "are dynamic, not static in nature; they have their psychological flesh and blood in the specific character 'across,' *and this cannot be composed from the usual optical contents.*"[19] Although he did not cite Bergson here, Wertheimer had provided empirical evidence for the claims the Frenchman had made in his critique of associationism (see Chapter 5).

According to the conventional view of motion, he argued, we see an object take on several positions successively and something is "added" subjectively. If this were correct, then an object would have to be seen moving, and at least two "positions," the starting and the end points, would be required to produce seen motion. Neither of these conditions always held. Indeed, the phenomenon of objectless motion was most convincing when the stimuli were so far apart that they could no longer be seen through the viewing device. Then there was "nothing of intermediate positions; the stripe . . . has not passed through the field, the ground remained quite black – but the motion goes across," as one subject said. In another case, a third object was placed between the two stripes, so that if something were seen to be in motion between them the observer would notice the interference. Instead, observers saw motion going behind the third object. In still other cases, a single object was seen to move between two positions, although there was no second stripe there at all. After repeated exposures of the two stripes shown in Figure 3c, one observer saw "a small but distinct movement towards the vertical line from the right, a rotation of about 40°. The horizontal? They must have taken it away." In fact it was still there; but when Wertheimer quickly removed it, there was no change in the phenomenon.[20]

To test current theories of motion perception, Wertheimer gradually varied the exposure intervals and the arrangement, size, and color of the stimulus objects, and also examined the role of attitude and attention. He dealt easily with Schumann's "illusions of judgment" by showing that the phenomenon does not disappear, but rather improves with increased attention or prolonged observation. Against Mach's attribution of seen motion to eye movements, he argued that even ordinary motion pictures were too complex for such an explanation; there were too many motions in too many directions at once. He dealt with Marbe's theory just as easily. Afterimages often remained in the locations corresponding to those of the stimulus objects, yet observers nonetheless observed movement between them.[21]

With Linke's "identification" theory Wertheimer was more cautious. An object was in fact seen most of the time in the optimal stage, and the impression of identity, when it occurred, could be very strong. But the *phi* phenomenon showed, he argued, that such impressions are not in fact "constitutive" for motion; for here there is no object to identify as being in motion. In addition, study of the transitions between the three stages yielded phenomena of "dual part motion" – motion seen proceeding a short distance from one stimulus object, then "reemerging" toward the other – and "fusion" phenomena, described schematically as $a \phi \quad b$ or $a \quad \phi b$, instead of the paradigmatic $a \phi b$. The identification theory explained none of these variants easily. When there was uncertainty, it was about identity, not motion.[22]

Implications

Given the nature of the *phi* phenomenon and his confirmation of Exner's observation that apparent motion produces negative after-images in the same way as real motion does, it was to be expected that Wertheimer would propose a physiological model to explain his results. According to recent neurological theory, he noted, the excitation of a central neural point causes a disturbance in the point itself and in the area around it. Increased attention heightens this disposition. If two neighboring points are stimulated within a given time interval, "a kind of physiological short circuit, a specific passage of excitation from *a* to *b*" should occur. If process *a* is at its peak when process *b* enters, "then a crossing of excitation occurs" and the *phi* phenomenon appears. The negative after-images are after-effects of prolonged stimulation, a "flooding back" of the current flow. With a succession of exposures under optimal conditions, "a unitary continuous whole-process" would be produced. This would account for cases in which *phi* is seen with only one stimulus object, and for the fusion phenomena.

Wertheimer made it clear that he did not mean to equate "nearness" in the perceptual world literally with "nearness" in the brain. Rather, "the neighboring points of the retina must be thought of as being in a special, especially strong and especially 'near' reciprocal functional connection with corresponding afferent central points."[23] Going still further, he suggested that there are transverse and total processes that "result as specific wholes from the excitation of individual cells over a large area." Two connected lines, for example, would then "appear as a *duo in uno,* as a compelling total Gestalt [*Gesamtgestalt*]. Not two lines coming from a single point, but an angle is there . . . a unified whole process resulting from the individual physiological excitations as a whole – a simultaneous *phi-*function." The existence of such functions could pose "specific experimental questions," which could lead to important results. To explain the recognition and reproduction of such objects, for example, "the appearance of the previously existing physiological total form [*Gesamtform*] of the unified process would be essential, not the reproduction of specific individual excitations."[24]

For Kurt Koffka, this was the most important part of the paper for the emergence of Gestalt theory. Here is how he put it in a 1931 lecture:

> To have proved that movement as experience is different from the experience of successive intervening phases meant a good deal at that time. But it might not have meant more than the introduction of a new kind of sensation. . . . Wertheimer did very much more: he joined the movement experience, the movement *phi,* to the psychology of pure simultaneity and of pure succession, the first corresponding to form or shape, the second to rhythm, melody, etc. This was the decisive step: just as movement as a psychologi-

cal event has its specific properties as a *shift* of process over an area or through space, thus has specific properties qua distribution of process, the same point of view is applicable to rhythm and melody, where the rhythmic group as a whole, the melodic phrase, has these particular psychophysical characteristics and is not resolved into a number of individual part-contents.

I remember the actual moment perfectly well, when I learned of this new view. It was in Wertheimer's room in Frankfurt, when he told me, who had been his perfectly submitting [sic] subject for several months, of the result of his work and of his conclusion. I can still feel the thrill of the experience when it dawned on me what all this really meant. Of course, at that time I had the merest inkling of it, none of us saw as yet very far, but I saw that much, that now at last form had become a subject that could be handled; it [had] made its final entry into the system of psychology.[25]

Wertheimer's paper and the experiments that led to it served as an alternative exemplar of theory-laden scientific practice for these younger scientists, offered by an independent brother figure whose only power over them was that of his ideas and personality. Methodologically, his entire procedure assumed Hering's heuristic phenomenalism, the accessibility and scientific primacy of "the psychically specific, observable given," discovered by "observation without postulating." Although such access often required precision instruments and carefully controlled conditions, the chief instrument was the observer. Hence Wertheimer's confidence in the validity of his results despite the use of only three observers: "the characteristic phenomena appeared in every case unequivocally, spontaneously and compellingly."[26]

This was common methodological ground for many experimenting psychologists in Germany by this time. The difference was that Wertheimer took the epistemological implications of the method seriously, accepting the objectivity even of phenomena that ought not to exist according to accepted theoretical perspectives. Instead of naming or renaming the phenomena according to the dualistic schemata provided by both empiricist and rationalist epistemology, Wertheimer proposed to restructure the schemata. This meant revising not only the terms of psychological description, but those of physiological explanation as well.

Wertheimer was quite clear about the hypothetical-deductive conception of scientific theory that underlay his physiological speculations:

[A physiological theory has two functions in connection with experimental investigation. On the one hand it must encompass the diverse individual results and general rules in a unified manner and make them deducible [from the theory]. On the other hand – and this seems to me an essential point – a theory must further *advance research* leading to the posing of concrete *experimental questions,* which first test the theory itself and then lead to deeper understanding of the regularities of the phenomena.[27]

Wertheimer did not employ psychological concepts to supplement physiological assumptions, as Helmholtz did. Nor did he use physiological language to supplement experimental phenomenology, as Stumpf did. Instead he proposed a dynamic interaction of both, in which detailed phenomenology comes first, physiological models are created to fit the results, and the models then used to guide further inquiry. Koffka later said he was "enthralled" to hear "that psychological and physiological events had to be pulled together *under the lead of psychological facts rather than of physiological hypotheses.*"[28] Mach, G. E. Müller, and others had enunciated this goal as a general principle, but it was quite another thing to hear such an idea expressed as an implication of phenomena one has seen oneself, and still more exciting to hear about a detailed, predictive physiology of "continuous whole processes" rather than associated combinations of elementary excitations.

It must have seemed quite strange for a cautious investigator like Friedrich Schumann to witness the radical changes going on in his institute. At the 1912 congress of the Society for Experimental Psychology, he announced the publication of Wertheimer's work at the end of his own paper on visual perception. In response to a discussant, he emphasized that the "sensory something" (*sinnliches Etwas*) of which Wertheimer had written was "not merely postulated," but "directly observed under various conditions." Still, he claimed that attempts to explain *phi* and other phenomena as illusions of judgment "cannot be regarded as hopeless."[29] As time went on he would concede more ground to his young co-workers (see Chapter 14).

Some of those co-workers became converts as well. Adhemar Gelb, Koffka's successor as Schumann's assistant, had taken Stumpf's position on the status of Gestalt qualities in his 1911 dissertation (see Chapter 6), but he changed his outlook after coming to Frankfurt. At the 1914 congress of the Society for Experimental Psychology, in Göttingen, he reported experiments on the perception of space and time with unusual results. He switched on three series of three lamps each at equal time intervals, but varied the amount of separation between them. He asked observers to fixate the middle point and judge whether the time interval between the first and second points on the left was larger or smaller than that between the second and third. To his surprise, in many cases observers spontaneously reported changes in the configuration of the points such that their arrangement was symmetrical to the judged time differences. If, for example, observers judged the time interval to be smaller, they said that the point on the right was closer to the middle point than when they judged the interval to be larger. Gelb got similar results in other sense modalities, and suggested that the explanation "lies essentially in the direction of the 'Gestalt laws'" discovered by Wertheimer, but not yet published.[30]

At the same meeting Wertheimer announced the most important of these "Gestalt laws" – fundamental principles of perceptual organization – during the discussion of a paper on tactile motion perception by Graz psychologist Vittorio Benussi. Specifically, he said that he had discovered "among several Gestalt laws of a general kind a law of the tendency toward simple formation [*Gestaltung*] (law of the *Prägnanz* of the Gestalt), according to which visible connection of the position, size, brightness, and other qualities of components appears as a result of altered subjective Gestalt apprehensions."[31] His meaning could not possibly have been clear from this statement. Unfortunately, the promised publication did not appear until 1923.[32]

But Wertheimer already began paying the price for his intellectual independence, and his association with the equally independent Christian von Ehrenfels, in 1914. While the Society met in Göttingen, he was being considered for a position in Prague, with the support of both Stumpf and Ehrenfels. He sent reprints of his work to Ehrenfels directly from the meeting. In his accompanying letter, he pointed out that his empirical work had been well received thus far, and that his epistemological lectures also discussed work in physics, such as the ideas of Poincaré and Einstein. In later letters he informed Ehrenfels of his Gestalt investigations and his work in progress on language disturbances, said he would be glad to lecture in pedagogy, and suggested that Frankfurt neurologist Ludwig Edinger would write on his behalf to his colleague Arnold Pick, as "he has quite a high opinion of my research." In addition, he responded to criticism of certain comments he had made against the Brentano school in the past:

> A certain blockage of the horizon versus other psychological and philosophical efforts and research was sad to me then; I have never denied the contributions (of the Brentano school), especially in relation to logical exactitude and clarity, and I hope myself to be essentially indebted also to my Prague schooling in logical exactitude.[33]

The criticisms may have come from Prague philosophers Anton Marty, Oskar Kraus, or Alfred Kastil. All were devoted disciples of Brentano, who had opposed candidates suggested by Ehrenfels in the past because they had appeared to deviate too far from the master.[34]

All this was to no avail. Wertheimer appeared less upset about being rejected than about the transparent absurdity of the reasons given. In particular, he wrote, "it is in no sense the case that the basic thinking of my motion paper came from Schumann . . . the theory I developed has nothing to do with Schumann's ideas." Equally unjustified was the claim that he lacked laboratory experience, given the independent research he had done in the institutes of Stumpf, Külpe, and Exner.

Schumann, he added, was surprised to hear of this and offered to write immediately to correct the false impression. It is not clear whether other factors influenced the outcome. Max Brod later wrote that Ehrenfels, at least, "always sharply
opposed anti-Semitism,"[35] but his colleagues may not have felt the same way. An
alternative explanation is that Wertheimer paid a price for not presenting himself
as the devoted follower of an established school. He would not achieve advancement according to his merits until 1929.

9

Reconstructing perception and behavior

Meanwhile, Wolfgang Köhler had developed his own critique of his elders' psychology. While continuing his research on hearing in Frankfurt with a variety of subjects, including schoolchildren, he found, among other things, that the act of "hearing out," or listening for the partial tones in a chord or a vowel sound could produce tones that did not exist before. Under normal listening conditions, he asserted, such partial tones are not only unnoticed; they are not there at all: "What is heard out are not 'the' partial tones, but leftover remnants of them that seem to be relatively indifferent for the total character" of the chord or vowel sound.[1] This supported a claim made by Hans Cornelius against Stumpf twenty years earlier. In his essay, "On Unnoticed Sensations and Errors of Judgment," published in 1913, Köhler worked out the implications of that finding, which merged with those of Wertheimer's research.

Challenging the constancy hypothesis

The focus of Köhler's critique was what he called the "constancy hypothesis" (*Konstanzannahme*). The designation did not refer to the relative constancies of color, size, and form under changing ambient conditions described and researched by Hering and others (see Chapter 4). Rather, these "constancies" were evidence against the view Köhler now challenged, the tendency shared by Helmholtz and Stumpf "to regard perceptions and sensations as much as possible as unambiguously determined by peripheral stimulation."[2] This assumption had been under attack for over twenty years. Even Helmholtz had said that sensations do not present an absolutely exact, point-for-point image of the physical environment. Stumpf had gone still further in this direction, especially in his treatment of tonal fusion. The criticisms of Cornelius, Hillebrand, and others, all cited by Köhler, and particularly the long debate with Felix Krueger had forced Stumpf to admit that this " 'atomism with regard to sensory phenomena' " – the claim that

135

there were elementary components of such complexes that persisted, but remained unnoticed – was a hypothesis, not a proven fact. Its value must thus be determined by the coherence it gives to research results, or by its usefulness in experimental practice. Köhler attacked the notion on both grounds.[3]

The constancy hypothesis, he argued, was untenable without assuming entities and acts that could neither be verified nor falsified, the "unnoticed sensations" and "errors of judgment" in his title. There were and could be no independent criteria to decide when these auxiliary hypotheses could or should be applied in particular cases. The idea that attention and practice can bring previously unnoticed sensations to consciousness, for example, is circular on its face, according to Köhler. If sensations were unnoticed before, or are not being noticed now, we can know this only by inference, not by observation; but the premise from which the inference follows is the constancy hypothesis.[4]

A second reason usually offered for the existence of "errors of judgment" is that such errors were correctable under proper conditions. To this Köhler replied that no one denies that even the reports of experienced subjects can be unreliable. He also recognized what psychologists now call the demand character of the experimental situation, the tendency of subjects to produce the phenomena called for. But in most cases, he maintained, "errors of judgment" are really only slight deviations from the expected observations. "Where even naive subjects make definite statements with the greatest confidence . . . that do not fit the stimulus determination of sensations," there was little reason not to accept such reports. The third criterion, situations in which judgments of this kind are impossible, actually decides against the constancy hypothesis. Children, too, possess size and color constancy; this cannot be due to errors of judgment, since these are supposed to depend on experience.[5]

Thus, Köhler concluded, the attempt to salvage the constancy hypothesis by invoking "unnoticed sensations" resulted in an attitude that stood in the way of "research in the true sense," for "the interests of a conservative system can be overwhelming in the absence of independent criteria." Here Köhler's tone became more passionate: "The mere term ('unnoticed sensations' or 'error of judgment') sometimes carries more weight than the most careful observations – just the phrase alone!" He even accused two – unnamed – laboratories of suppressing contrary evidence. The assumption of a one-to-one correspondence between stimuli and sensations might once have been a useful way to bring order to "the young science" of psychology, he conceded. Now, however, "the actual effect of these expedients which guarantee the system is often enough mainly to discredit our one way of moving forward – observations and the pleasure in them – and thus to paralyze the will to advance. Fortunate are those who consider these words exaggerated!"[6]

In essence, Köhler had challenged his teacher, Stumpf, his current employer, Schumann, and other experimenting psychologists to do what Wertheimer was doing and Ewald Hering, too, had advocated – to try to explain, rather than explain away, what trained and naive observers actually report, letting the theoretical chips fall where they may. He recommended "tentatively" that the constancy hypothesis "be given up entirely." As a result, "a simplicity of sensory psychology which I believe is premature and artificial will thereby be sacrificed"; for now "we assume" that central factors – processes in the cortex – also play "an essential role" in perception, and that simpler relations dominated by peripheral factors are "limiting cases." Giving up the constancy hypothesis will make the theoretical situation more complicated at first, he admitted; but the result in the end may be "a deeper understanding of the whole field." Until now many properties have been neglected that "are often much more important for a psychology of perception than the usual sensory attributes . . . This applies particularly to the psychological correlates of stimulus complexity, and specifically to the everyday perception of *things*."[7]

The real basis of experimenting psychologists' commitment to the constancy hypothesis was their belief that making such an assumption was in accord with proper scientific discourse. Köhler therefore took care to state that he, too, believed that sensations and perceptions conform to natural law, though he did not specify what concept of natural law he meant. He also assured his readers that he did not intend to deny the correspondence of physiological and psychological processes: "I regard the other variables also as physiological in nature." But he did not specify the variables he had in mind. His only reference to Wertheimer's ideas about the brain processes corresponding to motion perception was a vague allusion to "plausible physiological hypotheses" in a footnote.[8]

Another basis for commitment to the constancy hypothesis was the belief that it is biologically functional. As Köhler noted, however, size and color constancy, which do not conform to the constancy hypothesis, have "enormous teleological value." In any case, even biologically neutral illusions require explanation. The case in point Köhler cited was Vittorio Benussi's work on the Müller–Lyer illusion (see Chapter 6). Benussi had found that it is difficult to compare just those lines of the figure in isolation that the task required subjects to compare. This "can only be a matter of the characteristics of the perception itself. Ultimately the whole illusion must be rooted in these properties. Present-day psychology of perception does not have a satisfactory answer" to this problem. The facts might be more easily explained, Köhler thought, "if we regard as the 'immediately given,'" and certainly as the biologically primary reality, "not 'sensations,' but (for the most part) *things*."[9]

Whether he meant by "things" real objects or Hering's "seen objects," it was clear that Köhler had proposed nothing less than restructuring the entire categorical system in sensory psychology by doing away with the limiting distinction between "physiological" and "psychological" processes in its standard form – as a distinction between sensations and judgments built up or founded upon sensation.

Though he abstained from epistemological arguments, he had implicitly also questioned the validity of a principle that had been a cornerstone of both empiricist and rationalist theories of knowledge. However, he had given only the briefest indication of the sort of conception that he wished to put in its place. As Koffka later recalled, "Köhler was originally very cautious in regard to" the new Gestalt thinking.[10]

Swimming with a changing tide

Wertheimer, Köhler, and Koffka were not alone in their efforts to recast psychology's vocabulary. At the 1912 meeting of the Society for Experimental Psychology, where Schumann announced the results of Wertheimer's motion paper, Kiel professor Götz Martius directly opposed his teacher Wundt's principle of "creative synthesis" from hypothetical conscious "elements," and proposed to begin instead with phenomena taken as wholes in order to establish lawlike relations among them. Otto Selz, presenting work begun under Külpe two years earlier, pointed to the importance of the "total task" (*Gesamtaufgabe*) in reproductive thinking. When he asked subjects to designate the "whole" or a "part" of an object named in a stimulus word, or something "superordinate" to or "coordinate with" it, they often performed other operations first that also had "task" character, such as calling up a memory. In Selz's view, this showed, contrary to associationist views, that thinking is not a matter of mechanical information retrieval, but a dynamic "actualization" of "dispositional" knowledge by means of reflexlike "schemata."[11]

At the same meeting, Karl Bühler presented research on immediate "impressions of proportion," which, like Selz's work, came from Külpe's institute in Bonn. Given, for example, two rectangles alongside one another, with bases the same length, equal angles, and parallel sides but different heights, he found that subjects judged one "slimmer" or "plumper" than the other. By obtaining threshold values for such judgments with standard psychophysical methods, Bühler claimed to be the first investigator to give measurements for what he called "the Gestalt quality of the compared rectangles." He attributed these impressions of proportion to hypothetical operations taking place between sensation and judgment.[12] Clearly, a consensus was building for integrating some kind of holistic vocabulary, or at least holistic-sounding descriptions, into psychological discourse.

In a series of review essays written for nonpsychologists and in his introduction to a projected series of research papers from Giessen, Kurt Koffka joined this chorus. In doing so, he staked Gestalt theory's broader theoretical claims for the first time and prepared the ground for recasting all of perceptual theory on Gestalt lines. As he put it in 1914, a complete transformation had occurred in this field, for which Ewald Hering had opened the way more than thirty years before. Formerly, "one approached the study of perception not without prejudice, but *sub specie sensation.*" Now scientists begin with "the immediately presented perceptual facts," so that in extreme cases "sensation is understood from the point of view of perception, instead of the other way around." The work of Bühler and others was part of this transformation, because it established the independent existence of Gestalt perceptions. Yet Bühler accepted only "simple" Gestalten and wished to build up more complicated ones from them. However, Koffka claimed, Bühler's own results showed that an object's "total form" plays a role in the perception of proportion, even when subjects do not attend to it.[13]

The real turning point for Koffka was, of course, Wertheimer's "fully new Gestalt theory," the attribution of both motion perception and stationary Gestalten to physiological "whole-processes." If the implications of this thought were fully carried through, he claimed, then *"psychological analysis (in the traditional sense) is dispensed with,"* because sensation is "no longer the simplest elementary experience."[14] Koffka made clearer what he meant by the last statement in a 1915 polemic against Vittorio Benussi that became the first full statement of Gestalt theory as a psychological system. Because the controversy began over differing interpretations of similar research results, these will be sketched briefly here before turning to Koffka's position.

Confronting the Graz school

Koffka began to apply Wertheimer's experimental design as early as 1912, using laboratory space at the psychiatric clinic in Giessen provided by Robert Sommer, who purchased a Schumann tachistoscope for Koffka's use. Though he was formally not permitted to oversee dissertations, Koffka assembled a small group of co-workers. He presented their initial results in 1913 as the first installment of a series he called "Contributions to the Psychology of Gestalt and Motion Experiences," later retitled "Contributions to Gestalt Psychology." In his introduction to the series, he claimed, in opposition to both the Graz school and to Bühler, that the successively presented stripes in Wertheimer's experiments were "merely stimuli for the movement experience, not contents on which it is founded."[15] He proposed to take this idea further by systematically altering the conditions of Wertheimer's original experiment.

Wertheimer had supposed that the two stimuli he presented were different neither in extent nor in intensity, and that the positions of the seen objects in visual space corresponded to positions in real space. In the first paper of the series, officially authored by student Friedrich Kenkel, Koffka eliminated the last of these assumptions by using Müller–Lyer figures as stimuli, and asked whether an illusory difference in size can produce movement in the same way that differences in real size can. In the figures shown, for example (Figure 4), the lines AB, CD, and EF are objectively equal, but subjects see CD < EF < AB. When Kenkel

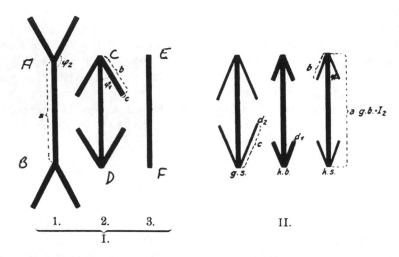

Figure 4. Setups for Koffka and Kenkel's motion experiments (1913). Adapted from Friedrich Kenkel, "Untersuchung über den Zusammenhang zwischen Erscheinungsgrösse und Erscheinungsbewegung bei einigen sogenannten optischen Täuschungen," *Zeitschrift für Psychologie,* 67 (1913), p. 448.

exposed the first two of the three figures one after the other in the tachistoscope, subjects saw the arrows moving back and forth in a normal case of apparent motion. With successive projection of first the middle line EF and then of the extending arrows, the figure seemed to "grow" arrows, then shrink back into a line. In series like the one shown, the first figure seemed to expand and contract as the arrows flipped up and down. The difference in the apparent size of the middle lines had clearly induced an additional motion phenomenon. Subjects called the effect "colossal" and "fantastic." Kenkel named the new motion "alpha" and normally observed motion "beta."[16]

While Koffka and Kenkel were at work, Benussi published demonstrations of the same phenomena, which he labeled *S* and *s* movement. Kenkel said he had no quarrel with Benussi's observations, only with his interpretation. According to Benussi, the difference in the relative positions of successive "phase images" becomes "the basis [*Grundlage*] of an image of motion founded on the component parts of the figure." The object cannot be positively identified in successive phases because of the conditions of presentation. We therefore have an image "produced" from inadequate sensory material. Kenkel replied that, as Henri Bergson had said in "unsurpassed" fashion and as Wertheimer had shown empirically, seen motion is not a series of phases, but "something completely unitary and not composed of individual parts; indeed the experience as such is destroyed by analysis."[17]

Reviewing Kenkel's paper, Benussi emphasized his priority of discovery as well as the general agreement between Kenkel's findings and his own, and suggested that the differences in their views were either purely terminological or a misunderstanding. A succession of phases is a condition for the appearance of an image of motion, but such a succession need not be in the phenomenon itself, nor had he ever held such a position. However, he wrote, if Koffka and Kenkel mean that alpha and beta or *S* and *s* motion arise in the same way, then this is "epistemologically not justified"; for, as Kenkel had also noted, alpha movement can be eliminated by more critical observation, while beta movement cannot. Thus, these phenomena are "not unambiguously determined by the stimuli," but are products of different ways of apprehending stimuli.[18]

Of course, Koffka, too, did not think that motion phenomena are "unambiguously determined" by stimuli; the difference was the process thought to intervene between stimulus and percept. As he argued in 1914, if differences in apparent size produce the same motion effects as differences in real size, this would be the same kind of proof for the central origins of alpha motion that Exner and Wertheimer had given for apparent motion in general. All such phenomena would therefore be results of process shifts in the brain, and need not be attributed to intervening "psychological" processes.[19]

Since his 1904 study of the Müller–Lyer illusion, Benussi had extended his work on modes of apprehension or cognitive styles to a wide variety of other illusions, and had expanded both his experimentation and his theorizing to ambiguous figures in general.[20] But he had retained the fundamental distinction between "apprehension" and objective states of affairs that he had derived from Alexius Meinong's distinction between content and object. For him, unified acts of production were still the primary condition for the appearance of Gestalten in consciousness. As a prize example he cited the role of grouping in the experience of rhythm, confirming results from Koffka's dissertation with generous acknowledgment. By 1914 he had developed a full-scale typology, in which subjects' modes of apprehension conditions the time required to perceive Gestalten. He argued, further, that such cognitive styles are linked with character traits; subjects who are more "critical" would exhibit an "analytical" rather than a *Gestalt* attitude.[21] However, he also acknowledged that cognitive style is a condition, not an explanation, of Gestalt perception. In the case of the Müller–Lyer figure, the outward or inward direction of the arrows has the effect of increasing or decreasing the "noticeability" (*Auffälligkeit*), or salience, of the center line with respect to its surroundings for both kinds of subjects.[22] This meant that the organization of the figure itself, not cognitive style alone, must be involved.

Max Wertheimer had "long Gestalt talks" with Benussi at the 1914 congress of the Society for Experimental Psychology, where he announced his own Gestalt laws. As he wrote to Ehrenfels, "we understand each other outstandingly well theoretically, which we were pleased to note to our mutual astonishment again and again."[23] Wolfgang Köhler praised Benussi's work in 1913, and suggested that other psychologists had ignored it, despite its wealth of assembled facts, because of its connection with Meinong's theory of objects. Benussi acknowledged that his work had been largely ignored, but remained loyal to Meinong's theory, since his experimental results had been obtained in order "to determine the truth value" of that theory.[24] Any dispute with Benussi, it seemed, would have to be an argument about epistemology and philosophy of mind as well as psychology.

Koffka's polemic avoided these issues. Instead, he said he wished to juxtapose the two theories to clarify their relative success "in formulating experimental problems." Koffka's tone was friendly but uncompromising. He did not wish to overlook "that which is common to both" approaches; yet he undermined the basis for cooperation by excluding Meinong's theory of objects at the start in order to keep this a "specifically psychological discussion."[25] Koffka was, indeed, more a psychologist than a philosopher. But confining the argument to psychology was also a clever debater's gambit; instead of engaging in difficult epistemological discussions, Koffka could fight on ground selected by himself.

Since Benussi could produce no introspective evidence for the process of "production" – a fact that he himself admitted – it had only heuristic value. But it could have that value only if it helped to make clear descriptive distinctions among phenomena, or led to clear experimental decisions about their theoretical treatment. Koffka tried to show that neither Benussi's criteria for "inadequate Gestalt apprehension" nor the hypothesized "production" process fulfilled these requirements.

Koffka concentrated on three of Benussi's criteria for "inadequate Gestalt apprehension": the discontinuity between stimulus and phenomenon, stimulus ambiguity, and stimulus inadequacy. Useful though these criteria may be as starting points for description, he argued, they cannot distinguish exactly between sensations and Gestalten. Take, for example, the case of ambiguity. On the one hand, sensory presentations sometimes display ambiguities that can be quite significant. It is possible to select a series of gray papers such that they will be judged a = b, b = c, but a ≠ c, so that the difference in stimulus intensity remains below the difference threshold in the first two cases, but not in the third. If one retains the idea that sensations are unambiguous, this result must be explained by resorting to the "unnoticed sensations" exposed as undemonstrable by Köhler. On the other hand, there are Gestalten that are quite unambiguous, such as an expanse of blue sky on a cloudless morning. The same holds for the components of the ambiguous figures Benussi so often used – lines, angles, and the like – which are themselves Gestalten.[26]

More serious still was Koffka's claim that, despite Benussi's denials, the conception of sensation on which Benussi's criteria were based presupposes the constancy hypothesis. Benussi himself referred to "sense impressions which remain constant" and "images [*Vorstellungen*] of figures which can be different from one another." Speaking in this way assumes not only that sensations exist but that they have a fixed relationship to physical stimuli. Benussi can only mean this when he refers to "produced" images as being "without stimuli" or as "ideal" or "inexistent objects." If any facts of direct observation could be offered in support of such "objects," Koffka argued, then it would mean that both Gestalten and sensations would be simultaneously accessible to introspection, which Benussi denies. If constant sensory contents cannot be directly observed, yet are objectively necessary, they must be "unnoticed" – another example of the way of thinking excoriated by Köhler. The sensory contents upon which Benussi's Gestalt images are allegedly built, Koffka asserted, are "merely hypostasized." Koffka acknowledged that Benussi expressly agreed with Köhler's opposition to the constancy hypothesis, and declared that "the great value" of Benussi's work is that it "frees" many experiences from their "bondage to the stimulus." But with the "production" theory and its presuppositions, Benussi took away what he had just given.[27]

In essence, Koffka had argued that Benussi's theory did not conform to the rules of theory construction he and Köhler had set up. In his introduction to the "Contributions" series, Koffka presented those rules as follows:

> We attempt on the one hand to describe the phenomena as exactly as possible, and on the other to find lawlike dependencies between them and the objective existing processes (between experience and stimulus); we do without every kind of postulated consciousness, unnoticed sensations, unnoticed activities. . . . Finally, we hope to be able to use the results of our investigations for theoretical decisions.[28]

Required was a model that incorporated Benussi's results while conforming to those rules.

Revised conceptions of experience and behavior

Koffka's version of that model was based on a revised conception of conscious experience. The fundamental break with the Graz school was a radical revision in the meaning of the word "stimulus." In this conception, this word no longer referred to a pattern of excitations on a sense organ, as it had throughout the nineteenth century, but to real objects outside of and in functional relation to a perceiving and acting organism. Thus, "The same object can be for the same organism at one time a 'sensory stimulus' and at another time a 'Gestalt stimulus,' depending on the state of the organism." A hungry fish, for example, snaps at a worm; a satiated one leaves it alone. Koffka was referring here to the concept of "biologically adequate" stimulation, which he had probably encountered in the writings of Frankfurt neurologist Ludwig Edinger.[29]

In this view, wholes are not constructed in the mind by hypothetical psychological processes, but are "direct experience-correlates of the stimuli." The relation between a perceived whole and a stimulus pattern thus has the character of the relation between sensation and stimulus in traditional psychology. Therefore, "one cannot predict merely from knowledge of the stimulus object what the experience will be." Since the state of the entire nervous system is involved in every act of perception, "*a properly type of psychological, that is, descriptive analysis is thus ruled out,*" because "there is no proof . . . that the 'sensations' found by analysis were present in the original perception." In a footnote, Koffka put this point still more strongly: "The unambiguous sensation exists only for the psychologist; it is a product of the laboratory."[30]

Ironically, Koffka cited a monograph from arch-sensationist E. B. Titchener's laboratory in support of this claim. Oswald Külpe had drawn the same conclusion from the Würzburg school's research. In 1893, Külpe, following the positivist Avenarius, had substituted the organism, or the "corporeal individual," for the "psychical individual" as the subject of psychology. Later, he returned with Husserl's help to the primacy of perception over sensation in experience, but without giving up the idea that perception was dependent on states in a perceiving organism. Koffka took over and reworked Külpe's thinking on the basis of Wertheimer's Gestalt concept. In his enthusiasm, he did not note that the *phi* phenomenon, too, was "an artifact of the laboratory"; for it could be experienced in pure form only under carefully controlled conditions, and even then only after prolonged stimulation. Though Wertheimer claimed to have observed it in ordinary life, he acknowledged that this took practice.[31]

At the descriptive level, Koffka drew the most radical conclusion possible from Wertheimer's 1913 lectures: "A pure description of one's experiences cannot be oriented to the concept of sensation; its point of departure is, rather, that of the Gestalt and its properties." By "Gestalt," he clearly meant, as did Wertheimer, not only individual objects (Köhler's "things") and Hering's "seen objects," but also the changing relations of objects to one another in the psychological environment. Taken seriously as a guide to research, such a conception would inevitably expand and complicate the task of description. But Koffka went still further, extending the Gestalt concept from perception to action. Given his relocation of the stimulus from the peripheral sense organ to the external world, he claimed, "We may in fact place the experiencing of Gestalten squarely beside that of creating Gestalten; to sing or play a melody, dash off a sketch, write, and so forth, are not cases where one sings or plays *tones,* or draws or writes *strokes. The motor act is an organized whole process.*" By this he apparently meant that, just as the products of activity are meaningful wholes that are fundamentally different from a sum of their components, the actions, too, have a structure that cannot be reduced to a bundle of reflexes. After this, the final step to physiology was a foregone conclusion. As Koffka boldly asserted, "there are real Gestalten." For him, the physiological correlates of both perception and action were "not the individual excitation of one brain area plus association," but unified events with "whole properties," which can be "significantly different according to whether we experience Gestalten or sensations."[32] This was a bold claim indeed, given that neither Koffka nor anyone else at the time had knowledge of brain physiology adequate to cash it out.

The opposition to the Graz school could hardly have been more complete. Stefan Witasek, for example, had said that only the processes involved in retinal stimulation, which certainly do not have Gestalt character, are physically real. For him, "produced" images had a qualitatively different, psychological reality. Koffka denied the need for such dualities. By equating the process of *Gestaltung* – forming or shaping – with its resulting Gestalten, Koffka had assimilated the intentional model of consciousness into a more comprehensive, functionalist framework. The object-directed mental acts that were the basis of psychology for Brentano and his students now took on the same structure as goal-directed acts of behavior, and both were correlated with "organized whole processes" in the brain. Thus, in Koffka's view, "this division of the psychical into constant and variable sections disappears," and "one avoids all notions of Gestalten as bizarre and unique phenomena."[33]

This conception of experience, particularly the attribution of structure to both perception and behavior and the emphasis on the interaction of organism and environment, had obvious affinities with American functionalist psychology. Indeed, Koffka cited John Dewey's *Essays in Logical Theory* (later reissued as *Essays in Experimental Logic*) directly. As early as 1896, Dewey had criticized the simple "reflex arc" as a linear, mechanical link of stimulus, sensation, and response, and called for a model based on the interaction of organism and environment. In the work Koffka cited, he maintained that logic and psychology could be reconciled on the basis of an "experimental logic," in which "the various types of conceiving, judgment, and inference are treated, not as qualifications of thought per se or at large," but as "*adaptations for control of stimuli*." Moreover, Dewey had added that the structure of these "adaptations" is dictated by the situation: "We keep our footing as we move from one attitude to another, from one characteristic quality to another, because of the position occupied in the whole movement by the particular functions in which we are engaged."[34] Thus, Dewey's position and the hopes he had for it appeared quite similar to Wertheimer's.

There was, however, one important difference between this viewpoint and Gestalt theory. Dewey's position was explicitly evolutionary, developmental, and instrumentalist. For him, the purpose of psychology was to discover the origin of each structure of thought and behavior, and to trace the "successive modifications through which, in its response to changing media, it has reached its present conformation."[35] For Koffka, as for Wertheimer, the primary aim was ontological. It was evidently more important to demonstrate the structured character of experience in general than to discover the evolutionary roots of specific structures. His use of the snapping fish as an example of a Gestalt stimulus underscored this point; statements about the relevance of "the state of the organism" to its behavior were applicable in principle to any species. The central idea for Koffka was that "elementary" sensations are neither psychologically, logically, nor historically primary.

Koffka made it clear that his recentering of perceptual theory had radical implications for conventional psychological categories. The awareness of a task (*Aufgabe*), for example, became "nothing other than a subjective condition of experience" correlated with "an alteration of the total state of the central organ." Further, he discarded the conventional attribution of some phenomena, such as contrast, to "peripheral" and others, including most illusions, to "central" factors. He cited an experiment from one of Wertheimer's lectures to show that contrast could also be centrally conditioned. Seen naively, a gray ring exhibited on a background that is half green and half red appears to be a homogeneous gray. But when a narrow strip of paper or a knitting needle is placed on the ring at the line dividing the two halves of the figure, the half on the "red" side suddenly appears greenish, while the half on the "green" side appears reddish. Put simply, under these conditions, two half-rings look different from one whole ring, even though the sensations are the same. To Koffka, this showed that contrast phenomena are not different in principle from the perceptual styles exhibited in Benussi's work with the Müller-Lyer illusion (discussed in Chapter 6). Both require explanation in terms of "whole processes" in perceiving organisms.[36]

From psychological Gestalten to "real Gestalten"

As the example of the snapping fish showed, however, Koffka clearly realized that the implications of his approach went beyond psychology. Zoologist Siegfried Becher had recently written that form (*Gestalt*) is characteristic of "the activity of living substance" in general. For philosopher Erich Becher, Siegfried Becher's brother, such statements supported psychovitalism (see Chapter 5). Koffka claimed that his redefinition of the stimulus concept and his equation of formed action with its formed objects "have built a bridge" from psychology to theoretical biology. But crossing this bridge meant accepting the existence of "real" Gestalten in the brain, which were hypothetical, to say the least. He was therefore quite cautious about that claim, stating that "even if there are no physical Gestalten, there might nevertheless be stimuli for Gestalt presentations." In any case, for systematic psychology, the new view of the relation of stimulus and experience was the important point; that could be held even if physical Gestalten do not exist.[37]

Looking to the future, Koffka claimed that his theory offered new opportunities for research, which "could also be important for the theory of knowledge." As shown in Chapter 7, Koffka was aware of the problematic situation of experimenting psychologists in German universities. Here was a redefinition of experience that promised to give psychologists important research to do while assuring them that the outcome had potential relevance to philosophy. But Koffka did not specify the epistemological implications of his theory. Benussi's teacher Meinong had sought a theory of knowledge that would do justice to both the facts of perception and the demands of valid reason. This was also Wertheimer's goal. Koffka's system stressed the former at the expense of the latter; he did not even address the troublesome question of the relation of "seen objects" to real stimulus objects. Moreover, the claim that all psychical objects are in principle real for the experiencing subject overlooked the fictitious entities that had stimulated the development of Meinong's theory. Though these are important for logic and epistemology, they are of no interest to the psychology of perception, so Koffka avoided the issue. But he clearly hoped for more. His use of the term "Gestalt stimulus," his conflation of form and function and his suggestion that there are "real Gestalten" lent his rhetoric an ontological dimension it had not had before.

10

Insights and confirmations in animals: Köhler on Tenerife

Even before Koffka published his first systematic statement of Gestalt theory, Köhler had carried out the first series of his now-famous experiments on anthropoid apes on the island of Tenerife. These provided evidence for Koffka's extension of Gestalt language from perception to action, and also for his claim that Gestalt perception is not limited to human cognition. Köhler fleshed out these claims in a way that clearly expressed his own commitment to aesthetic rather than technological values in science.

The anthropoid research station on Tenerife

Köhler went to Tenerife as the second director of the anthropoid research station of the Prussian Academy of Sciences in December 1913. The idea for the station originally came from Max Rothmann, a Berlin physician, in 1912. The hope was to compare the gestures, language comprehension, color perception, and other behavior of various anthropoid species, in order to determine their respective places on the developmental scale on the way to humans. Tenerife, one of the Canary Islands off the northwest coast of Africa, was chosen for such studies because the environment there was less foreign to the anthropoids' original habitat than European zoos, while the place was more accessible and the climate more tolerable to Europeans than the jungles of Cameroon.[1]

Like many scientific enterprises of the period, the station was established with a combination of private and state support. Financing came initially from the Selenka and Plaut foundations in Munich, and a consignment of chimpanzees from the German colonial government in Cameroon. The newly founded Albert Samson Foundation of the Prussian Academy provided long-term financial security. Samson, a banker interested in science, had heard the anatomy lectures of Wilhelm Waldeyer in Berlin. When he decided to leave the largest portion of his estate to science, he asked Waldeyer to oversee the new foundation's affairs in

Prussia. Waldeyer assembled a distinguished advisory board, including Max Planck and Carl Stumpf. The foundation's purpose, according to its official statute, was to support inquiry into "the natural, biological bases of individual and social morality" in the fields of ethnography, anatomy, physiology, and experimental psychology. Waldeyer applied this to the anthropoid station by saying that its goal was to discover whether animals exhibit living conditions and behavior "that go beyond the simple life of instinct and approach the level of ethical and moral expression in human life."[2] Stumpf's *Phonogramm-Archiv* was another foundation beneficiary (see Chapter 2). Excluding the director's salary, the operating expenses of the anthropoid station in its second year were 5,000 Marks, slightly more than the budget of the Berlin psychological institute at the time.[3]

The fact that the station was the first of its kind evoked a certain amount of national pride. When the American animal psychologist Robert M. Yerkes inquired about the facility and mentioned his plan to establish one in the United States, Eugen Teuber, the station's first director, emphasized that it was "no longer planned" but in full operation, so that "there can be no doubt about our priority in this area and the Americans must reckon with our results when they begin."[4] Teuber, a doctoral candidate who had studied with both Wundt and Stumpf, came to Tenerife in December 1912, with architect's plans to build a station in the style of a Bavarian chalet. Fortunately, land prices made that scheme impractical, so he rented a small country house and banana plantation from Don Melchor Luz, a former mayor of Orotava on the island's northern coast. Because it was protected from the winds, the site was good for the chimpanzees. Teuber and his wife, who acted as his assistant, settled with their family into the house. Next to it a large area was fenced in and covered with wire netting supported by a pole five meters high. This enclosure kept the apes confined while giving them room to roam. On one side of it were the animals' sleeping quarters and a laboratory with facilities and equipment for photography, filming, and sound recording for the *Phonogramm-Archiv.*[5]

In addition to studying the apes' physical condition and determining that five of them were between five and six years old, Teuber made general observations of their social and other behavior. He took particular note of their upright gait, games, and "dances," gestures, facial expressions, and sounds. A repeated short "o, o, o," for example, was a sign of greeting and happiness; a short "ae, ae" was a warning; a high "e, e, e" showed fear; and "u" sounds indicated sorrow or rage. Rothmann and Teuber reported successful communication with visual signs for banana, orange, or sugar, and suggested that the signing of the deaf might be more suitable for the apes than vocal language – a point not taken up again for nearly sixty years. Most significant for Teuber's successor, however, were "intelligent achievements" Teuber claimed to observe in a young male named Sultan, who quickly learned to open the doors of the compound by putting a key in the lock, and used sticks as tools to reach distant food. Teuber reported that "more exact scientific investigation" of the sensory functions and intelligent achievements of individual animals was needed; but he returned to Berlin at the conclusion of his one-year contract to complete his doctorate. Wolfgang Köhler was chosen to succeed him, on Stumpf's recommendation.[6]

Köhler, who was only twenty-six at the time, negotiated advantageous start-up conditions with true panache. Pointing out to his far senior colleague Waldeyer that he was "without independent means," and that leaving Germany was a career risk, he requested and received an annual salary of 6,500 Marks, double that of his predecessor and more than six times his assistant's salary in Frankfurt. In addition, the foundation paid the rent and furnishings of the house attached to the station, travel expenses for Köhler and his wife, and even the salary of his substitute in Frankfurt, to ensure that his assistantship would be held for him.[7]

Köhler immediately showed his mettle as both an investigator and an administrator. Shortly after his arrival, he wrote an extensive report to justify the purchase of an additional chimpanzee in his absence, offering to pay for it himself. When the animal died shortly thereafter, he wrote an equally detailed account of its symptoms and treatment, blaming himself for not having examined the animal more closely before purchase. Waldeyer quickly reassured him, and urged that he "spare no expense" on the construction of better facilities. He also offered to extend Köhler's appointment until April of 1915 and "perhaps still further."[8] Waldeyer had reason to be pleased with his young appointee. By the time he wrote these letters, in the spring of 1914, he had already received word of remarkable results Köhler had obtained with simple "intelligence tests" on the apes, which have since become classics in the history of psychology. In order to understand their significance at the time, it is necessary to place them briefly in context.

Background: On clever and stupid animals

Darwin's assertion that evolution extends continuously from other animals to humans raised the issues of the nature, origin, and limits of animal learning and intelligence, and the relation of these to their human counterparts. In the English-language literature there were three major lines of opinion on these issues. One extended the properties normally associated with intelligent behavior or thought to a wide range of other species. A second attributed apparently intelligent behavior in animals to the imitation of human beings. The third applied the principle enunciated by Conwy Lloyd Morgan: "In no case may we interpret an action as the outcome of the exercise of a higher psychological faculty, if it can be interpreted as the outcome of the exercise of one which stands lower on the psychological scale." To Morgan, referring to such "higher" faculties other than behavior required by environmental conditions meant invoking a level of psychological complexity that natural selection could not explain. Thus, his "canon" was a regulative device that decided both what was to be observed and how observations should be interpreted to give evolutionary theory a more secure scientific basis.[9]

In his 1898 dissertation, entitled "Animal Intelligence," Edward Thorndike cast his lot unreservedly with Morgan and aimed "to give the *coup de grace* to the despised theory that animals reason." He enclosed animals, mainly cats and dogs, in so-called puzzle or problem boxes, and measured the time required for them to pull a string that opened the door and permitted them to reach food they could smell outside. Thorndike thought that what was learned was a tendency to react. Combining an analogy from Darwinian natural selection with assumptions from Lockean empiricism and associationism, he argued that animals try random behaviors on the basis of an instinctive motor impulse until the correct solution occurs by chance. They then learn the necessary acts gradually in sequence, as the pleasure that comes with success "stamps in" an associative connection between sensory impressions and motor impulses.[10]

Thorndike's methods were attacked on publication because of their unnaturalness for the animals. Nonetheless, the "trial and error method" became one of the fundamental procedures in animal learning studies, since it provided quantitative data and thus legitimated the field as natural science. However, when Thorndike applied his methods to primates, particularly gibbons and macaques, in 1901, he discovered that a smaller number of trials was needed, that the time required for solution dropped abruptly rather than gradually, and that the solution remained fixed longer than in cats and dogs. He thus took care to exempt primates from many of his strictures about animal stupidity. But he retained his basic schema, adding only that primates, because of their larger brains, could have more "free ideas" and thus associate more readily. Thorndike also hoped to use his method with human subjects, and thus to help solve social problems, particularly in education. This program and his research style exemplified the technocratic hopes of many psychologists in the United States at the time.[11]

German-language discussion of these issues focused primarily on the spectacular feats claimed for animals like the horse "Clever Hans." Such investigations were a way to learn about the capabilities of animals, but were also a means of reinforcing the gap between expert scientists and credulous amateurs. When he investigated Hans in 1904, Oskar Pfungst, then a co-worker in Stumpf's institute, showed that the horse's alleged ability to multiply and divide was actually due to its perception of signals unconsciously transmitted by humans. Once he excluded these by using blinkers and had others pose problems to which he himself did not know the answer, the animal failed completely. Pfungst credited Hans with high perceptual ability and an adequate memory for images, but certainly not with the ability to think.[12] Pfungst offered a similar interpretation in a study of more than two hundred apes of different species reported at the 1912 congress of the Society for Experimental Psychology. Arguing against enthusiastic reports by Alexander Sokolowsky based on observations in Hagenbeck's Zoo in Hamburg, Pfungst, like

Thorndike, argued that anthropoids could be considered intelligent only in the sense that they learn more rapidly and form and retain more complicated associations than other animals. However, both Pfungst and Sokolowsky admitted that observations in zoos were insufficient to decide the issue, so it was opportune that Max Rothmann announced plans for an anthropoid research station at the same session.[13]

It was there, too, that Köhler first publicly showed interest in comparative psychology. During a visit he had made to the recently famous Elberfeld horses, which were supposed to be able to do fourth roots, he said, they "could not even achieve what Clever Hans easily managed, according to all reports." He suggested that "all enthusiastic reports are attributable to insufficient attention to sources of experimental error and the calculation of probabilities."[14] Köhler continued the discussion from Tenerife, offering to challenge the owner of the Elberfeld horses to try some of the new "intelligence tests" on "his stallions," and thus end the publicity once and for all. However, though he was obviously a partisan of science over amateurism, he was certainly no negativist about primate intelligence. As he wrote to Waldeyer, his results clearly spoke for Sokolowsky and against Pfungst in their debate.[15]

Köhler's observations: "Genuine achievements" and "imitations of chance"

All the performances Köhler observed were variations on a single theme – overcoming an obstacle to reach a goal object, usually food. He presented the variations in order of difficulty; but as the dates he provided show, the tests were not originally given in that sequence. Evidently Köhler ordered his results systematically after the fact. The case Köhler presented at the beginning showed that his true aim was to study intelligent behavior in general, not only that of anthropoids, for no ape was involved. He constructed an enclave in which he placed in turn a female dog, several chickens, and his infant daughter, who had just learned to walk. Each time a fence separated an attractive object from the animal in such a way that it had to make a detour to reach it. For the dog and the child, the solution came quickly and showed many of the same features – first a slight turn of the head, then "a kind of jerk" or sudden movement, followed immediately by a single, smooth motion around the corner of the barrier and toward the goal (see Figure 5a). The real difficulty of such apparently simple solutions became clear in the case of the dog, when food was placed so close to the fence that she did not move, but remained "fixed" to the spot by the smell.[16] For the chickens, the solution came more slowly and in a different way. They continually ran up against the barrier, and only came to a solution when they more or less accidentally landed at the right spot in the course of their zigzag wanderings (see Figure 5b). In general, Köhler concluded, there was "*a very obvious difference in form between*

a.   b.

Figure 5. "Genuine achievements" (a) and "imitations of chance" (b) Wolfgang Köhler, *Intelligenzprüfungen an Menschenaffen,* 4th ed. (Berlin, Heidelberg, New York, Tokyo: Springer-Verlag, 1974), p. 11. First published 1917. Reprinted by permission of Springer-Verlag.

genuine achievements and imitations of chance."[17] Köhler constructed the entire argument of the book around this distinction, but the theoretical problems it raised are best discussed after a more complete account of the results.

The apes dealt with such detours rather easily. When Köhler tried to learn more by making the tests more difficult, they responded with new kinds of "genuine achievements" involving the use and making of tools. In the simplest examples, the apes employed already available objects as extensions of their bodies, such as using a stick to pull in a piece of fruit lying outside their enclosure, climbing up an open door and "riding" it toward a piece of fruit hanging from the ceiling of the compound, using "jumping sticks" like vaulting poles, or even using people as "ladders" or other chimps as "footstools" to reach the same goal when it hung too far away from the door.[18] In other tests, the apes constructed box towers to reach particularly high pieces of fruit (see Photo 7a and b). Here they used "tools" that they were not likely to have used before, and certainly not in this way. Here, too, both individual differences among the apes and fundamental differences between chimpanzee and human perception became noticeable. Of the six apes, only three, including Sultan, managed constructions of more than two boxes. Köhler remarked that the apes seemed to possess nothing of the "naive statics" of human beings. They showed no appreciation of the arrangement required to build a stable tower, nor of the notion that placing one box on top of another is a "mere repetition" of putting the first box on the ground.[19]

The most spectacular instance of tool making was again the work of Sultan. He was given two hollow bamboo sticks with openings of different diameters, neither of which was long enough to reach a banana lying outside the bars. More than an hour's trying yielded no solution, though he did make the "good error," as Köhler called it, of using one stick to push the other toward the fruit. The experiment was

Photo 7a. Apes stacking boxes. Berlin-Brandenburg Akademie der Wissenschaften, Archive.

Photo 7b. Apes stacking boxes. Berlin-Brandenburg Akademie der Wissenschaften, Archive.

then interrupted, but the rods left with the animal. While playing with them, he brought them into a straight line. Suddenly he placed one inside the other, jumped up, ran to the bars and pulled in the banana with his double stick. "The proceedings seem to please him immensely," Köhler wrote, for he kept on pulling in pieces of fruit and other objects without stopping to eat or handle them (see Photo 8 of a later repetition). In other instances, the apes demonstrated both the extent and the limits of their ability to "see" what Köhler called the "situational value" of objects. Needing a stick but finding none, the apes would catch sight of a tree, tear off a branch, and use it, thus "seeing" the branch as a tool. On the other hand, Köhler reported, they were quite incapable of "seeing" a box as a tool for climbing when another ape was lying on it, or when it was in a corner, and thus "merged" with the adjoining wall.[20]

Such a dry listing does little justice to the emotional range of the apes' behavior or to the liveliness of Köhler's account. He points again and again to the role of affect, especially the animals' eloquent, but generally unsuccessful, pleas for help in solving a problem, or the fits of rage into which they fell when they failed, when they went tearing and screaming about the enclosure, smashing the offending instrument against the walls. Often sudden calm, and a successful solution, followed such rages. Köhler also emphasizes that these primates were thoroughly social animals, but notes that this sociability – shown, for example, by the "skin treatment" they gave one another with evident pleasure – should not be confused with altruism. Once, when one of the apes tried to push a box to a spot under the hanging objective and found it too heavy, others quickly joined in and pushed along also. But as soon as they were near enough to the goal, one of them quickly jumped up on the box, snatched down the fruit and ran off, making no move to share its booty with the others.[21]

After recording such observations, Köhler knew that he would have to defend himself against charges of anthropomorphism. He therefore made numerous photographs and films of the apes' behavior "to convince the doubters," as he wrote to Waldeyer.[22] As he insisted, the way the apes went about solving problems was directly observable, and there was no need to read any notions from human behavior into it. In his accounts, he emphasized, "nothing is said about the 'consciousness' of the animal, but only about its 'behavior.'" Nor was he trying "to prove that the chimpanzee is a marvel of intelligence"; his results indicated the animals' limitations as clearly as their admittedly great abilities. They possessed "no inclination or gift, for instance, for the study of fourth roots or elliptic functions," something Köhler obviously regretted having to point out in "a serious book."[23] Köhler did not rule out imitation as one possible source of these performances. However, he pointed out, the chimpanzees generally learned by imitating one another, not by imitating human beings. This was shown by their frequent "fashions," in which the use of jumping sticks, for example, spread rapidly from its inventor to the other animals. Yet even in such cases, Köhler contended, "it is

Photo 8. Sultan with double stick. Berlin-Brandenburg Akademie der Wissenschaften, Archive.

most difficult for chimpanzees to imitate anything unless they themselves under-stand it."[24] Words like these showed that, despite his denial of anthropomorphism, Köhler felt justified in concluding that chimpanzees showed "a type of behavior which counts as specifically human," the ability to act with "insight."

Insight

This choice of term proved to be unfortunate, because both scientists and non-scientists continued to read mentalism into it. Köhler's best defense was the apparent simplicity and testability of his criterion for insight: "*the appearance of a complete solution with respect to the structure of the field,*" which is "the product of a complete survey of the whole situation," and is usually characterized by "a smooth, continuous curve, sharply divided by an abrupt break from the preceding behavior."[25] Köhler thought that two kinds of observations offered convincing support for this criterion. The first involved "good errors," one of which has already been noted. These were not undirected or uncoordinated acts, but clear-cut attempts at a solution that were consistent with the structure of the situation but nonetheless false. The second was the pause that often occurred after a few unsuccessful attempts, in which the animal would look back and forth from tool to objective, scratch its head, and appear quite literally to be surveying the situation. This alone was enough to convince at least one visiting expert, Köhler reported.[26]

Simple as this criterion seemed, it actually rested on a number of assumptions. The first of these was that a complete survey of the situation is necessary. Köhler provided a negative example of this himself, when he threw food over a wall or into an adjoining room and the animal ran after it.[27] The second was that such surveys give rise to behavior. Actually, they may be succeeded by behavior, but cannot be said to cause it directly, without the intervention of a process in the animal. Köhler realized this, but said little about such processes. Third, the statement that the animal responds with "the behavior required for the solution" sounded as though there could be only one "required" behavior. Yet the apes often surprised Köhler by finding solutions different from the one he thought they would discover, such as the use of jumping sticks. Most problematic of all was the idea that the apes' performance gives "the appearance of a complete solution." Köhler claimed that this was merely a description, but the notion that phrases like "the structure of the field" are descriptions at all presupposes Wertheimer's claim that such structures are immediately given in perception. The idea that "curves of solution" occur is an extension of Wertheimer's conception to action, similar to Koffka's attribution of Gestalt characteristics to seen objects and behavior.

Apparently Köhler had come to such views without the support of American functionalism. He had, however, read *Mind in Evolution* by the British social philosopher Leonard Trelawney Hobhouse, and was "very astonished," as he reported to Waldeyer in May of 1914; for he "had never heard" of the "excellent experiments" in it before. As he later acknowledged, many of his own experiments were borrowed or modified from Hobhouse's work with a rhesus monkey, a chimpanzee, and other animals. Hobhouse, too, noted cases in which the animals seemed to come upon a solution "in a flash." He also spoke of their "critical successes" or "decisive trials" and even, in a sense, of their "good errors."[28] There were also broad areas of theoretical contact between the two. Hobhouse's aim was to find an adequate place for mind in nature without reductionism. Opposing Thorndike's reliance on the associative connection of instinctive impulses, he argued that the behavior he observed should be called "perceptual learning." By this he meant habit formation controlled by the concrete perception of results. The animals appeared to him to have knowledge of objects "as the centers of many relations." They must therefore be capable of "drawing inferences from one object to another similar only in the relation of its parts." Borrowing a term from G. F. Stout, Hobhouse called such operations "practical judgment," a reproductive "synthesis of perceptual elements" qualitatively different from association or assimilation.[29]

Instead of using such terms, however, Köhler employed descriptive and interpretive language based on Wertheimer's epistemology. That language emphasized the organism's immediate grasp of and functional adaptation to the logic of its present situation, not a reproductive synthesis of present and past experience. Though he had evidence that the apes' memories could be very good, he nonetheless asserted that chimpanzees live in a limited time frame: "One never saw them deliberately concentrate on the successful choice with an eye to the future."[30] Given this emphasis on the immediate present, Köhler could only explain the animals' ability to reproduce successful solutions months, even years later by referring to "structured whole processes" in the cortex, as Wertheimer had done in the case of the *phi* phenomenon. But he did not discuss such processes here. Even at the level of description, applying Wertheimer's Gestalt concepts to primate behavior posed a methodological problem that went deeper than the issue of anthropomorphic language, and Köhler faced it in a later essay. How do we know that our observations of the apes yield correct descriptions of what they see and do? Is it possible, he asked, that *"there are realities in the animals investigated which are perceptible to us only in these total impressions"*; and if so, "in what manner do the total processes in and on the body of the ape produce total impressions in our perception?" He rejected the doctrine of inference by analogy, but recognized that he had no replacement for it, only "new questions."[31]

Köhler versus Thorndike: Aesthetics versus technocracy

Nonetheless, Köhler claimed to have made "crucial tests" that were decisive against Thorndike's or any other explanation of these observations in terms of chance. In one such experiment, he tied a rope to a heavy stone, then wound it around a piece of fruit and laid it obliquely to the bars with the free end extending between them. According to Köhler, Thorndike's theory demands, first, that the original solution must occur by chance, and, second, that the movements required for the solution must then be learned individually, and the habits gradually associated with one another to form a whole act. After first pulling in the direction of the string, however, four of the animals solved the problem by passing the rope hand over hand along the bars until the fruit was in reach with no hesitation, as soon as they saw the rope. There was no collection of random motions.[32] Köhler had made his demonstration appear crucial by construing Thorndike's theory as literally as possible. Still, he had touched that theory's sore point, the problem of explaining successful learning with only one trial. "After this," he wrote, "I did not think it necessary to make the same experiment with the other animals." In response to one likely criticism, Köhler admitted that he could not know whether the apes had already developed the motions involved before their captivity and were only applying them in this case. He insisted, however, that this was unlikely, and that even if it were so, the transfer itself must still be explained.[33]

Köhler's primary methodological criticism of Thorndike was similar to that made in America: the puzzle box situation so limited the animals' perception of the situation that the first successful solution could only occur by chance. In his own work, Köhler declared, "everything depends upon the situation being open to the subject." Indeed, he often moved a stick or changed other aspects of the arrangement – usually when the apes were not looking – to make a "genuine" solution easier.[34] Köhler's primary purpose was the same as Wertheimer's had been in his experiments on motion perception – to construct situations in which "good" phenomena would happen, so that their "essence" could be revealed.

Such an approach reflected an aesthetic rather than a technocratic attitude toward science. Although he repeated his tests often during his stay on Tenerife, Köhler included only a few later observations in his report, because he found the mechanization that resulted from later repetition and practice "ugly," "constrained," and "indifferent to the essence of what has been demanded." Indeed, he admitted that he liked the chimpanzees' behavior during their tenth or eleventh repetitions of a solution less than that in the first or second; he thought that "something is spoilt" in the interval. Essential for Köhler was not the achievement of grasping the fruit, but the beauty and order of the structured process directed toward that goal. Consistent with this was his attitude toward the application of his results. He proposed using his methods with children of different ages, arguing that they are "certainly as scientifically valuable as the intelligence tests usually employed"; but he also said that "it does not matter so much if they do not become immediately practicable for school or other uses."[35]

This aesthetic orientation to science was intimately bound up with Köhler's theoretical commitments. In Thorndike's model, the experimenter stands outside the situation and measures. The result is a learning curve, which is not itself visible in the animal's behavior but is abstracted from it. Köhler's procedure presupposes the existence of the situation as a structured unit "extending from the animal to the fruit," as Koffka later put it.[36] The essence of the phenomenon is not a statistical curve, but the "continuous curve of solution," which is experienced, not constructed by the experimenter. Thorndike's long-range goals were technocratic, and the relation of experimenter and subject in his work was correspondingly manipulative. In Köhler's scientific training, psychological observation was a means to philosophical ends, and the experimenter–subject relation was a partnership for the production of "good" phenomena. Applying such an approach to primates presented obvious problems, the main one being that only one of the "partners" can report or interpret his or her own experiences.

Science in wartime: Köhler and Yerkes

Köhler's observations were substantially complete by June of 1914, but publication was delayed until 1917 by the outbreak of war. As a reservist, he was called up in the general mobilization of July, but was forced to remain on Tenerife, because German, Italian, and Spanish vessels all refused to take any of the sixty German men of military age on the islands for fear of "difficulties" from Allied ships. Köhler was frustrated by the situation, above all by the lack of German news: "We hear of the war almost entirely from Paris, and such obvious lies that one cannot read the newspaper." Still, he hoped for a rapid end to hostilities "for the sake of all culture." Two months later his patriotism was undiminished: "Every time we have any news from Germany there is a feast day, and we read the newspaper – normally so quickly leafed through – like a work of literature." The contrast between the "spiteful, low" style of the English and Spanish papers and the "outraged conviction" of the German was obvious to Köhler. But "I cannot think of a trip to the homeland, hard as it is to remain here. A steamer with five hundred reservists on board was captured recently, and only five got through." Under such conditions, there was no choice but "to try to continue with science."[37]

By January 1915, mail delivery was assured with the help of Waldeyer's colleagues in neutral countries, C. U. Ariens Kappers in Holland and Richard Herbertz in Switzerland. Köhler could thus order and receive books and other materials.[38] Among the books ordered was a midwifery manual; Köhler and his wife had three children during their stay on the island. With relatively secure contact to the outside world and comparative calm on the islands themselves, thanks to Spain's neutrality, he produced in the next three years three long research monographs and a book, a total of nearly six hundred printed pages. However, his wartime situation was not without difficulty. Letters took four to five weeks in each direction, and were occasionally censored or intercepted. Köhler complained often about delays in the receipt of page proofs, manuscripts, and especially of necessary money transfers.[39]

Still more seriously, he was rumored on two different occasions, once before the outbreak of war and once a year later, to be spying on British shipping and communicating with or supplying German submarines from the station. The first time, Köhler wrote to his former physics teacher that "an Englishman has contrived to put it about that the apes are only a pretext for us to engage in espionage, where perhaps a Zeppelin could land! But no one takes that seriously and we even enjoy special protection of the authorities thanks to occasional recommendations from the Foreign Office in Berlin." The second time the British consul went so far as to lodge an official protest. In a more customary use of his key word, Köhler reported that "the insight of the Spaniards" saved the situation, as Spain had become more friendly to Germany in the interval. Despite insinuations to the contrary in a recent popular book on the subject, no direct evidence has been found that Köhler was a spy.[40]

One of Köhler's helpful contacts was Robert Yerkes. After his correspondence with Rothmann and Teuber, Yerkes tried to arrange a visit to the station to conduct research. When he wrote to Köhler in early 1914, Köhler supplied full details of the living conditions and research facilities at the station, and immediately offered to exchange reprints. Later he wrote enthusiastically about the idea of founding a research station in the United States. It would be important to study more animals, due to the individual differences among them, and in any case, he wrote, "you in America investigate these issues somewhat differently and would choose different problems for research than we, and so both institutes would supplement one another beautifully."[41]

The two scientists developed an active correspondence by way of Havana, which lasted as long as America remained neutral. Köhler reported extensively to Yerkes about his research, and Yerkes helped him to obtain books and had the films he had made of the apes' exploits developed.[42] Unfortunately, the literature exchange did not succeed as hoped. A long delay in receiving reprints of Rothmann and Teuber's description of the station and of Köhler's first research paper prevented Köhler from sending copies to Yerkes. Yerkes sent the requested literature from America, but it was diverted, turning up a year later in a Madrid library. In the meantime, Yerkes had published a monograph in which he described behavior in an orangutan named "Julius," observed from February to August, 1915, that was quite similar to that of Köhler's chimpanzees. Though he used the word "insight" once, Yerkes generally preferred to call this behavior "ideational learning."[43] Köhler received a copy of this work while writing up his own results, and wrote to Yerkes that he regretted not hearing of it earlier. The kind of learning exhibited, and the description of behavior with boxes, sticks, and so forth "could just as well be that of the chimpanzees in my experiments," though "I have, of course, been able to observe very much more for a longer period with more animals." Though Köhler expressed himself courteously, the undertone of competition is unmistakable. Somewhat less courteous was Köhler's response to other remarks in Yerkes's monograph. Yerkes had called the Academy's station "modest" and suggested that an American station could be established "without regard to this initial attempt of the Germans." Kohler stiffly asked whether "the future American station needs to begin its activities with pronouncements which appear to indicate a low opinion of the older enterprise." Judgment of a scientific project should depend on its leadership, "about which you certainly do not know," not upon the expense incurred.[44]

Yerkes hastened to deny any anti-German feeling, offering to translate and publish an abstract of Köhler's work in the *Journal of Animal Behavior,* and to introduce it with an apology for not giving Köhler full credit in his monograph, "as soon as I have the necessary data in hand." Actually, Köhler had conducted his trials a full year earlier than Yerkes, and had described this and other work in his letters. Yerkes had mentioned, for example, that Köhler had also employed box-stacking problems, but in a manner that appeared to diminish both the work's significance and Köhler's priority.[45] America's entry into the war shortly after this incident ended communication between the two researchers. The wounds of war, and in Köhler's pride, took some time to heal. Asked his opinion of a possible American takeover of the station in the fall of 1919, Köhler replied that "even before the States broke relations with us, Mr. Yerkes wrote so arrogantly and woundingly about German research in general and our station in particular that I must refuse to submit to further incivilities from this man in possible negotiations." The tone of Köhler's response also reflects the general atmosphere of hostility among scientists on both sides, which persisted for several years after the end of the war. However, when Yerkes offered to resume correspondence in 1921, Köhler helped reactivate contacts between German and American scientists by sending him a list of his German colleagues and their addresses.[46]

"Structural functions" in animal perception

In the meantime, Köhler had completed two experimental studies of animal perception that clarified the theoretical significance of the earlier work. In the first study, carried out from late 1914 to early 1915 with assistance from Martin Uibe, a German physics student, and his wife Thekla, Köhler tried to determine how the visual perception of chimpanzees differs from that of humans.[47] To do this he developed a number of ingenious variations on standard experiments on space perception and size and color constancy, thus making the apes, in a sense, into experimental subjects in the classic mold. He even used the German term *Versuchsperson* in one place, without quotation marks. In the case of size constancy, the results were quite new. Köhler trained two of the animals with food rewards to choose the larger of two wooden boxes painted white and placed equidistant from them. He then changed the position of the boxes so that in certain trials the distant, larger surface produced a smaller retinal image than the nearer, smaller one. In all but one of 120 trials, and in all of the 60 "critical" trials, the apes chose the objectively larger surface. He obtained similar, though not so clear-cut results in tests of color constancy, using black and white papers on the box fronts.[48]

The most surprising result of all came in the final series of trials, using chickens instead of apes. Kohler trained the chickens to peck at seeds lying above a glass plate, below which were two papers of different shades of gray. He trained two animals to choose the lighter, and two the darker nuance, with the rough but very effective technique of snatching away the entire board, seeds, glass, and all, when the chickens made a "wrong" choice. As Thorndike would have predicted, the learning curves were flatter than those of the chimpanzees. Between 400 and 600 trials were required to achieve a basis for controlled tests. In these, the chickens were allowed first to peck whether the choice was "right" or "wrong." There followed 100 "critical" trials, in which the formerly lightest paper was presented with a still lighter one, or the previously darkest with a still darker paper; thus the "correct" choice would be a shade of gray the animals had not been presented before. Of these trials there were only 4 "errors," and in 72 "easy" tests only 5. Also impressive was the "phenomenology" of choice, as Köhler called it. If the chicken's head was first placed between the two papers, it seemed to be pulled in the "right" direction. Apparently, he wrote, the transition to (achromatic) color constancy occurred far lower on the evolutionary scale than previously thought.[49]

Köhler concluded that such results challenged all theories that attribute the seeing of surface colors to "transformations from more retinal vision, and ascribe primary influence in this process to experience." He was clearly referring to David Katz's theory of color vision, presented four years before (see Chapter 4). Katz, in a review in the *Zeitschrift für Psychologie,* suggested that the animals had really learned to react to the difference in brightness between the grain and the paper, which could remain constant under different illuminations. Köhler carried out additional experiments to exclude this possibility, which he published the next year.[50]

His theoretical account of these results came in a comprehensive monograph completed early in 1917, but not published until 1918. In essence, Köhler proposed to determine whether the constancy hypothesis he had attacked in 1913 held for animals of different levels of development.[51] If the principle were correct, he reasoned, then chickens trained to peck seeds from the lighter of two gray papers would choose the paper they had been conditioned to choose, even when it was paired with another paper that was much lighter or much darker than its original counterpart. Two choice situations might be used to test this prediction. Animals trained to choose one paper 2 over the darker paper 3 could be offered a choice between a lighter paper 1 and 2, so that the originally lighter paper appears as the darker; or they could be offered a choice between papers 3 and 4, so that the originally darker paper appears as the lighter. In the first situation, the choice of paper 2 would have to occur despite both the change in the relational character of paper 2 and the animals' tendency to prefer known over unknown objects. If the animals nonetheless chose paper 1 over paper 2 in such a test, Köhler argued, this would be conclusive proof that the relation of the papers to one another had been decisive, and not the absolute quality of one paper or the other.

After more than 1,000 preliminary trials, Köhler offered the chickens a mixture of "critical" trials, with the new pairing, and training trials. The results were not as clear-cut as they had been in the earlier experiments on color constancy, but their tendency was undeniable. Of a total of 100 critical trials for one animal, 77 choices were according to the criterion of "structure;" in the 120 training trials, the ratio was 113 to 7. Köhler then experimented with apes, using the same box apparatus as before, but with slightly larger papers. He gave results for only one animal, Chica; of 40 critical choices, 37 were "structural." After this, Köhler deemed it unnecessary to make further tests, for which Yerkes later criticized him. Köhler then added nearly perfect results from similar experiments with his three-year-old son. Most notable here, he thought, was the matter-of-fact character of the child's choosing. For the child, he concluded, "the structure of the color pair seems to be determinative to an extent no longer true for the adult."[52] Finally, Köhler applied a similar procedure to tests with colors other than black and white, using a variant of the color wheel that was the standard apparatus for such tests in humans. Though it took much larger intervals between colors to achieve discrimination thresholds than with human subjects, the principle and the results were the same. Each time the apes chose the lighter mixture, even when it was objectively as dark as, or darker than, the originally darker mixture.[53]

David Katz and Géza Révèsz had already found in 1907 that chickens could discriminate triangles and circles, and had trained them to peck the second or third item in a series of kernels. In 1912, Karl Lashley found that albino rats could be trained to discriminate the larger of two circles in similar fashion. Lashley's work started a debate among the small group of American animal psychologists as to whether animals could actually discriminate forms, or whether they possessed "only a more or less crude pattern vision," as Walter Hunter maintained.[54] In essence, Köhler had shown that the relational principle behind all experimentation with difference thresholds also held for nonhuman animals. No one was prepared to attribute judgments to chickens; Köhler preferred the more neutral term, "structural functions."

But the way he interpreted that term belied his apparent caution. He approvingly cited Otto Selz's research on "total tasks" (see Chapter 9), but he did not accept Selz's explanation of the way parts and wholes are experienced. Nor did he accept the claim offered by Hans Volkelt that animal consciousness is a "totality" (*Ganzheit*), in which complexes take effect "as wholes" in some unspecified way. Instead he argued that "the whole reproduces on the basis of its specific structure," meaning that the relation "lighter than," once learned, has an independent psychological reality of its own.[55] Though Köhler attributed this claim to Christian von Ehrenfels, it was actually an application of the views Wertheimer had expounded in 1913, with one important change. Instead of referring to a hierarchy of functional relations organized around a "center," as Wertheimer did, Köhler spoke of "systems." This language hinted at much more general implications to come (see Chapter 11).

More explicit were the implications for comparative psychology. Köhler acknowledged that the time and effort the animals required to deal with these relationships shows the difference between them and humans. But his findings questioned the exclusive reservation of such functions to animals gifted with language. His research, along with biologists' observations cited by Koffka that even frogs and lizards react to structured stimuli, showed, in Köhler's opinion, that "only a portion, and hardly the essential portion, of the reactions of even the lowest organisms can be understood as mere juxtapositions and successions of absolute stimulus influences in isolation." It followed that models of evolutionary history based on the primacy of sensation are "worthless."[56] It also followed that methodological injunctions such as Lloyd Morgan's "canon" are not only worth-less but positive hindrances to research, like the constancy hypothesis in percep-tion. Köhler called the injunction to explain animal behavior on the basis of the simplest and lowest functions possible "a dangerous demand": "Who says in the present state of animal psychology with what kind of function we have supposed too much of an animal – which is the simpler and lower function in a given case? If one wishes to economize, then one must first know the value of the different kinds of money."[57]

Here again Köhler argued on practical scientific grounds against research rules he thought arbitrary. Yet, as in his 1913 reference to "the joy of observing," or his 1917 claim that there is "something ugly" about mechanized, habitual behavior, aesthetic values lay close to the pragmatic surface. His results, he said, represent "a gain in status" for the lower forms of life: "Where up to now, in order to satisfy the unfortunate principle of economy, a reaction could not be explained woodenly enough, we now have the right and the duty to view nature as somewhat richer and more colorful. One can have the greatest interest in exact procedure in research and still be pleased with a result of this kind."[58]

The end of the station

Clearly, Köhler had managed to carry on with science under difficult conditions. But as the war neared its end, the difficulties began to mount. The station's operating expenses increased steadily, from 10,149 Spanish Pesetas in 1914 to 16,322 in 1918 and 19,202 in 1919. Inflation on the islands was exacerbated by inflation in Germany, which put increasing strain on the Samson Foundation's finances. By 1918 the cost in 1914 Marks of running the station had tripled, while the foundation's cash income had increased roughly 1.8 times. The station's budget, including salaries, now totaled 31,200 Marks, approximately three-fifths of the foundation's income. In addition to this and to the communications prob-lems already described, Köhler and his wife became ill shortly before the end of 1917, and moved to another part of the island for several months to recover. Then, in November of 1918, just as the war ended, the land on which the station stood was sold to an English firm, and the station had to be moved. Amazingly, the foundation was able to cover all these expenses. Köhler, now fully recovered, wrote to Stumpf in May of 1919 full of plans for the future.[59]

But the rapid decline of the Mark in the summer and fall of 1919 made con-tinuation impossible. A bank loan was arranged while various alternatives were considered.[60] Despite the express wish of Prussian officials to keep the station alive, all attempts proved fruitless. Renewed illness forced Köhler to return to Germany at the end of May 1920, before the final decision to close. After his return he arranged to have the apes sold to the Berlin zoo. They arrived in October, one month after Waldeyer resigned the directorship of the Samson Foundation because of failing health. He died soon after. The apes survived a while longer, providing material for more observations, including an extensive description of a chimpanzee birth.[61] But within a few years they, too, had succumbed, probably to the change in climate.

It was a sad end to an important chapter in the history of science. Köhler did no more primate research. Already in 1915 he had confided his increasing boredom to Geitel: "Two years of apes every day; one becomes chimpanzoid oneself, and the scientifically uncomfortable [aspect is that] one no longer notices something about the animals as easily." Four years later, he wrote that he stayed on after the war ended only to feed his family, which he could not do on an assistant's salary: "Both objectively and as far as my health is concerned I have long since ceased to belong in this position."[62] In any case, he was less interested in primate research for its own sake than in the light it shed on more general theoretical issues.

11

The step to natural philosophy: *Die physischen Gestalten*

Before he returned from Tenerife, Köhler published a book with a provocative title: "Physical Gestalten at Rest and in a Stationary State." This was his philosophical masterwork, an extension of the Gestalt concept from perception and behavior to the physical world, and thus an attempt to unify holism and natural science. The Albert Samson Foundation paid for its publication because it was written on Tenerife, though it was not directly concerned with anthropoids; but the book had had a long gestation period.

A long and difficult birth

In the winter of 1913–1914, the semester after Wertheimer presented his Gestalt theory in lectures, Köhler offered a course in Frankfurt called "The Physical Basis of Consciousness." After only seven months on Tenerife he already saw the significance of the apes' behavior for that topic. In July 1914, he posed the issue in a letter: "Can a conclusion be drawn from the kind of solution as well as the whole behavior of the animal about the processes that occur in the animal (brain-physiological or if one will also psychological)?"[1] Apparently he did not consider writing a book on the question until one year later. In a letter to Stumpf in May 1919, he told more of the story:

> The first thoughts about it came to me in the spring of 1915. However, after I saw that little could be accomplished in a brief period or without extensive preliminary work in physics, I decided that it was my duty to remain true to my anthropoid assignment, and therefore worked through the "intelligence tests" and the paper on "structural functions." But after that (in the summer of 1917), I believed I could return to the larger and more general task for the sake of the issue and for my own sake, even if that meant paying less attention to the apes for a while. Unfortunately, first my own fatal [sic!] illness, then the illnesses of my wife and children, then simultaneously

departing from the old and founding the new station, and beyond this the difficulty of the task itself, have delayed its completion for so long.[2]

Köhler submitted the book for publication in one of the series of the Prussian Academy of Sciences, as his contract required; but he doubted that it would be accepted, since its treatment of physical and mathematical issues was "too broad and elementary." In 1917 he had requested that it be given to Max Planck or Walther Nernst for an opinion, since "the content falls outside the competence of the psychologist but not within that of the neurophysiologist and physiologist." Actually, the book's terrain lay primarily between philosophy and physics; in the subtitle he called it "an investigation in natural philosophy." He asked Stumpf, to whom he dedicated the book, to defend its "philosophical content in its close interconnection with physics," if necessary.[3] It will soon be clear that such territorial issues created considerable difficulties for the book's acceptance.

Köhler also remarked to Stumpf that he had kept the presentation "largely independent of the anthropoid research," because "the consequences go far beyond questions of animal psychology." He had not kept the account of his anthropoid research free of metaphors from physics. He noted there, for example, that "many animals run terrified directly in front of an automobile as though they were following lines of force, when a slight swerve to the right or the left would save them.' In his critique of Thorndike, he argued that to satisfy a "genuine" — that is, a physically and mathematically precise – concept of chance, there should be "no essential difference" whether we speak of the Brownian motion of gas molecules in an enclosed chamber or "the so-called chance impulses of a chimpanzee."[4] Köhler conceded that no one actually thought of advocating such an idea. Instead, Darwinists preferred to let the undefined concept of instinct supply the otherwise missing direction and coherence to behavior. Philosophers like Eduard von Hartmann and Henri Bergson expressed dissatisfaction with Darwinian chance, invoking instead entities like "the unconscious" or élan vital. Köhler, too, rejected explanations based on chance. But "the alternative," he maintained, "is not at all between chance and factors outside of experience. Great parts of physics have nothing to do with chance. . . . It seems particularly surprising from the standpoint of physics that one should continually insist on speaking here of 'either-or,' when *after all there are quite other possibilities.*"[5]

These "possibilities" were actually reducible in Köhler's opinion to one – that there are "physical Gestalten," the laws of which correspond to those of behavior and psychological experience. The prepublication tribulations of his text made it obvious from the outset that it would be difficult to sustain or even express an argument for such a proposition in a way that would be clear and convincing to both physicists and psychologists. He sent the manuscript to Stumpf and his old teacher Geitel, but asked Geitel to submit it to Vieweg, a respected natural science publisher located in Braunschweig, near Wolfenbüttel. Stumpf "has no good connections with publishers," he explained, and also no understanding of the higher mathematics in the book; "Planck already had to speak to that before the Academy." But despite his choice of publisher – and the separate introductions he later wrote specifically for philosophers and natural scientists – he acknowledged to Geitel that the book was "written for psychologists." He therefore did not describe in detail "what very concrete meaning the whole 'Gestalt question' has for contemporary psychology (and biology in general)."[6]

Precisely that omission made it difficult to make Köhler's intentions clear to certain physicists. Geitel recommended the manuscript enthusiastically to Vieweg as "a deep and independent creation," and wrote to Köhler that he was "convinced that the Gestalt concept will also be introduced into physics and show itself to be fruitful." But the editor of the monograph series in which the book was to appear, Eilhard Wiedemann of Erlangen, sharply criticized Köhler's foreword and introduction. In his first sentence, Köhler had written that his observations on animals had shown that their behavior was "immediately determined by simple Gestalt factors." Such language, Wiedemann wrote, "must frighten any reader. It is strange how even physicists, once they have been thoroughly infected by philosophy, cannot express themselves so that any reasonable person knows what they want."[7] Geitel saw immediately what was at work in this reaction – "opposition to the natural-philosophical viewpoint in modern physics" by measurement-oriented experimental physicists. He had already suggested, more gently, that Köhler make his introduction easier for less theoretically oriented physicists to understand. Köhler was outraged at Wiedemann's "philistinism," and successfully pressed to have the book published outside the monograph series. But the publisher continued to ask for changes in the introduction; apparently it took a favorable recommendation by Albert Einstein to assure its publication in the original form. The dispute did, however, convince Köhler to drop the offending sentence from the foreword, to add some clarifying remarks about Gestalt qualities, and to write a second introduction for physicists.[8]

In the first introduction, now directed not only to psychologists but to philosophers and biologists, Köhler stated clearly the issue that concerned him – the unity of experience. Psychologists have come to speak more and more of Gestalten, he wrote, but felt "a slight inhibition in this practice." Seeking theoretical as well as empirical legitimacy for this usage, Köhler proposed to turn to physical science, "where the ground has been secure for a long time, to find a control or at least analogical confirmation." Should such help not be forthcoming, then psychologists would have to accept "the unsatisfying idea . . . that experience is not in itself a closed system."[9] In his 1913 and 1918 papers, Köhler had employed the motif of psychology as a young science to support his call for trust in observation against constraints imposed by an ideal of science taken from physics. Here the insecurity of youth became a reason for turning to just that science for aid.

In the introduction for physicists, Köhler took more an insider's tack. It has become a matter of course in natural science, he noted, to treat states that remain relatively stable over time as the results of continuous processes tending toward minimum energy displacement and maximum entropy. Psychologists, however, still believe that physical (and psychical) systems are discontinuous "mosaics" of elements. The correctness of this view for a limited range of psychophysical events gave it authority for a long time, but as soon as psychologists made greater demands on it its weaknesses became apparent. Nonetheless, they had retained this thinking because they believed that natural scientific exactitude requires it, and because they were not familiar with the more modern style of physical thinking. Given such familiarity, he asserted, it would become clear that the behavior of optical Gestalten at rest "requires no more for its explanation than ascribing characteristics of a physical system to the optic sector of the nervous system." The physiological hypothesis presented by Wertheimer and Koffka would thus represent "only the transition to a properly physical view of nervous events."[10]

Later in the book it became clear that Köhler was proposing to treat the neurophysiological processes underlying Gestalt phenomena in terms of the physics of field continua rather than that of particles or point-masses. His license to do this came, he thought, from the creator of field physics himself, James Clerk Maxwell. As he wrote in one of his last lectures, "Under Planck's influence I had dimly felt that between Wertheimer's new thinking in psychology and the physicists's thinking in field physics there was some hidden connection. What was it?'[11] He found his answer in the preface to the *Treatise of Electricity and Magnetism* (1873). There Maxwell compared the methods of then-current mathematical physics, which proceeded from the motions of individual particles, to those of Faraday, which began with the "whole," that is, the field of forces acting upon the particles, and clearly stated his preference for Faraday's method. Actually, Maxwell qualified Faraday's holism by depicting field action as occurring only at contiguous points of the field and describing it in series of partial differential equations. By doing this he hoped to preserve the consistency of his equations with Newtonian mechanics; but to later generations it was clear that the idea that fields of force have an independent existence was a first step away from the assumption that the world consists of pieces of matter from which forces emanate. There remained, however, a certain tension between regarding fields as "wholes" and their mathematical treatment, which required positing a finite number of discrete entitites. As Planck later said, field theory threatened to divide physics into two realms, one of "corpuscles" and one of "continua."[12]

The "Ehrenfels criteria"

In fact, however, the categories with which Köhler proposed "to look for Gestalten in physics" were not physical at all, but psychological, or rather, psycho - logical. Köhler called them the two "Ehrenfels criteria." The first of these was "suprasummativity" (*Übersummativität*). According to it, physical states and processes could be called Gestalten if they have qualities or produce effects "that cannot be derived from the similarly constituted [*artgleiche*] qualities and effects of their so-called 'parts'." This criterion, Köhler maintained, is "necessary but not sufficient," for it seems to require only the addition of "Gestalt qualities" to the sensations fixed by individual stimulus components. It should be supplemented by "functional nearness," the requirement that stimuli or component parts be so situated as to be able to influence one another reciprocally. Needed also was a second criterion, "transposability" (*Transponierbarkeit*), the retention of relations in the same order despite shifts in the parts. This condition is "sufficient, but not necessary" for demonstrating the independence of Gestalten from their parts, as such, for there are Gestalten to which it does not apply.[13]

Actually, these were not strictly Ehrenfels's criteria. Ehrenfels had used the second criterion, the transposability of melodies, for example, to prove the existence of "suprasummative" Gestalt qualities. In this sense, then, one criterion was inferior to the other. Alois Höfler, an Austrian colleague of Ehrenfels, was the first to introduce transposability as an independent Gestalt criterion; Köhler, too, gave the two criteria equal standing.[14] Far more important, he dropped the idea of "Gestalt qualities" as something additional given on and with each of the elements or parts of an entity, while accepting Ehrenfels's wide application of the concept to melodies and their tones, to relations in general and their members, and even to the sense (*Sinn*) of a sentence and its component words. This shift presupposed the epistemological perspective that Wertheimer had presented in 1913. Accordingly, Köhler did not speak of entities that have Gestalt qualities, but of objects or processes that *are* Gestalten. Doing this put Gestalt theory clearly at odds with the still-predominant tendency in modern analytical philosophy and natural science toward methodological and ontological individualism – the denial, for example, that human groups have any reality other than that of their members and their relations with one another.

This transformation became clearer in Köhler's treatment of the category "sum." He used the term primarily to refer to collections of "similarly constituted" (*artgleiche*) characteristics, entities or components joined by what he called "and-connections" (*Und-Verbindungen*), borrowing a term from Wertheimer. This meant that a grouping or collection may be called "summative" only when its parts or pieces can be removed or changed with no alteration in either the whole or the other parts or pieces. For example, three stones, one in Australia, one in Africa, and one in the United States, can be said to be a group in the formal sense; but displacing one stone leads to no noticeable displacement of the others, nor to any change in their relation to one another. The group of stones is therefore a sum; this would also be true, Köhler claimed, if the stones were one meter apart. Against the objection that he had constructed this criterion in such a way in order to make it easy to discover "suprasummative" entities in physics, Köhler replied that physics contains numerous "pure sums" of just this kind. The concept applies not only to arithmetical sums, but also to the scalar and vector sums commonly used in mechanics. In this sense, he acknowledged, sums play "a dominant role in physics."[15]

From a logical point of view, these criteria raised numerous questions. The term "part," for example, remained ambiguous. Köhler recognized this, but said that a comprehensive logical treatment of these issues would be "far too much effort," and was in any case unnecessary for the task at hand. This rough-and-ready attitude toward logical analysis corresponded to Köhler's open disdain for formalistic approaches to philosophy in general. In his opinion, most systematic philosophers preferred to clarify fundamental concepts "from the outside and far removed" from experience. He proposed instead to apply his criteria not to "the physical" as such, but to physical systems as they appear to the physicist "from close up."[16] Such applications fill half the book. Since the form of proof was nearly the same in every case, a few examples will suffice.

"Strong" and "weak" Gestalten and their significance

Köhler found what he was searching for in the field of electrostatics. In a well-insulated ellipsoidal conductor, for example, the density of charge is greatest at the points of greatest curvature and smallest at the points of least curvature. The distribution of charge in such a conductor thus has a definite pattern of organization that depends on the shape of the conductor, which Köhler called the system's "topography," but is independent of the materials used or the total quantity of charge involved. Köhler later added that the electrical fields which accompany such distributions of charged particles also belong to their "topography," and demonstrated that these, too, fulfill the Ehrenfels criteria. It is impossible to build up such structures piecemeal, for example, by feeding charged particles first into one part of the conductor and then another. In such cases the charge immediately redistributes itself in its characteristic form over the entire surface. As Köhler put it, "The 'natural structure' [*Eigenstruktur*] assumed by the total charge is not described if one says: 'at this point the charge-density is this much,' and 'at that point the charge-density is that much' . . . the occurrence of a certain density at one point determines the densities at all other points."[17]

Such physical systems Köhler called "strong" Gestalten. In them, the mutual dependence among the parts is so great that no displacement or change of state can occur without influencing all the other parts of the system. Indeed, Köhler wrote, "strong" Gestalten actually do not have "parts" at all, but only interacting "moments of structure" that "carry" one another. Here he took up a concept Husserl had used in the discussion of wholes and parts in his *Logical Investigations* (see Chapter 5). Köhler, however, was referring to real physical processes, not to subjective "moments of intuition." He specifically contrasted this usage with the "foundation" concept also used by Husserl and Meinong. It makes no sense, he maintained, to speak of such wholes as being "built up" of static parts, as we can with geometric figures. "The moments of a structure, unlike the items of a geometrical grouping, are *not logically prior* to the total structure," and "*a physical structure upon a given topography is not logically secondary relative to its moments.*"[18]

Köhler showed that stationary electric currents, heat currents, and all phenomena of flow are "strong" Gestalten in this sense. These he distinguished from what he called "weak" Gestalten, which are not immediately dependent on the system's topography. An example is a group of isolated conductors connected by fine wires. As in "strong" Gestalten, the "natural structures" for each of the conductors are in principle dependent on conditions in the system as a whole; but their specific articulation is not influenced by events in remote parts of the system. The mathematics for such systems are summative, according to Köhler's own definition. Because there is a finite number of "regions" in the system, it is possible to leave the connecting wires out of account and compute the linear algebraic functions for the electrical potential in each of them. Nonetheless, the system is a Gestalt, because a shift in the current input produces a change in the whole system.[19]

The most important problem in dealing empirically with "strong" Gestalten, Köhler noted, is that they are destroyed or radically altered on contact with measuring instruments. Such structures have therefore "all but disappeared from experimental physics." This makes their pervasiveness in theoretical physics all the more impressive. "Weak" Gestalten are satisfactorily treated with simultaneous linear algebraic functions, but "strong" Gestalten can be described either with integrals or with series of partial differential equations. The latter method, as Köhler pointed out, was the one preferred by field theorist Maxwell. Köhler had thus demonstrated to his own satisfaction that the systems he had defined as Gestalten are as accessible to measurement as anything physical. Since the mathematics of such systems are applicable to a wide variety of other phenomena, he concluded that "temporally constant, continuously extended total entities are present, it seems, in nearly the whole of physics."[20]

Ludwig Boltzmann had argued that the use of differential equations meant postulating the existence of atoms. In his *Eight Lectures in Theoretical Physics,* Max Planck had attempted to reconcile this view with the idea that "there exist no countable elements in completely continuous entities" such as electromagnetic fields. The result was a two-level view of physical reality, dominated at the micro level by corpuscles and at the macro level by continua.[21] Köhler did not deny the existence of atoms, nor did he find recent developments like the electron theory at all incompatible with the concept of continuously structured fields. In fact, it was Hendryk Lorentz's view of charge and field as two aspects of a single entity that had encouraged him to think of current and conductor as two "sides" of a total physical "topography." However, he emphasized the continuity aspect and gave it an ontological gloss. He noted that the Laplace equation, which summates series of partial differential equations in the form $\nabla V = 0$ (in current American notation, $\nabla^2 V = 0$), is commonplace in many areas of physics, and argued that "one must suppose that this single, constantly returning equation determines something common to all these phenomena."[22] Unfortunately, he did not explain why one "must suppose" any necessary connection between such mathematical regularities and ontological similarity.

Though Köhler rejected the assumption that nature, or mind, is composed only of discrete elements, he also took care to deny that this claim had anything to do with "the kind of romantic-philosophical inspiration" behind the view that everything is related to everything else. Such statements may be true in a formal sense, he admitted, but to take them seriously would mean to deny the existence of physical Gestalten "in a fruitful and scientifically very real sense of the word." A system's degree of independence from or dependence upon its surroundings varies with the boundary conditions. It is the task of both physics and psychology to determine these in specific cases. Köhler had just as little use for the idea that only consciousness as a whole is given: "With this kind of reality one really cannot do much." Such "totality" theories miss the important point, namely, "the existence of self-enclosed, finitely extended Gestalten with scientifically determinable, natural laws."[23]

Ernst Mach had said much the same thing in his own brief discussion of physical systems in *The Analysis of Sensations*. Precisely because everything is related to everything else, he had claimed that it is more effective to concentrate upon contiguous areas, as Maxwell had done, and insisted on the assumption of maximum continuity within these. Thus, for Mach, the selection of a particular portion of the whole universe for discussion was dictated by the practical needs of research. Köhler clearly accepted this claim, but he also insisted that the idea of physical Gestalten as topographically determined systems is a "much richer and above all [more] objectively determined concept" than that of a sum of elements. In his view, we cannot always choose to think arbitrarily of one arrangement of points instead of another. Rather, "the Gestalt law observed by such material and the specific structure spontaneously and objectively assumed by it prescribe for us what we are to recognize" as a Gestalt.[24]

Here, for the first time, the philosophical implications of Köhler's 1913 critique of the "constancy hypothesis" became clear. Köhler wanted to oppose not only Helmholtz, but also Descartes, both the mechanist and the geometer. To Köhler, Descartes's error was to assume that both nature and the psyche actually function the way his mathematics did, "to reduce the physical dynamic world to a purely geometrical one."[25] Friedrich Schumann and others had long since protested against the introduction of fictive, mathematically defined elements into psychology (see Chapter 6). Schumann's "observed" elements, however, were themselves geometrical fragments, such as points and lines. The implication was that these unchanged sensations are combined or summarized in different ways by the mind under different stimulus conditions. Köhler argued that this was only Cartesian geometrism in another guise.

Empiricist David Hume, too, had refused to accept mathematically defined essences not found in sensation, such as the length of a line. His sense impressions, like the atoms of classical mechanics, were indivisible. Descartes, however, rejected physical atomism, for the abstractions of his geometry required infinite divisibility to ensure that qualities could be put into a series and measured. It was this presupposition that Köhler meant when he spoke of geometry, and it was this presupposition that physicist-philosophers like Helmholtz had transferred from mechanistic physics into sensory psychology. Yet Köhler could not accuse Helmholtz of being ignorant of physics. In fact, he was attacking two versions of elementism, one formalistic, derived from the imperatives of mathematical analysis and closely linked to rationalism, the other phenomenalistic and empiricist. From the viewpoint of Wertheimer's Gestalt concept, however, the two had much in common; and in any case, by the time Köhler began to study psychology they had become hopelessly intertwined. Köhler opposed both versions of elementism at once, without distinguishing clearly between them.

As Köhler conceded, the tendency to "geometrism" and "summative thinking" is part of ordinary life, not an artifact of the mechanical worldview alone. The things of daily experience, after all, are arranged geometrically, have fixed, often linear borders, and do not tend to dissolve into continuous flows and processes.[26] Thus, the argument against elementistic "geometrism" could be interpreted moderately or radically. One could either expand empiricism by accepting the existence of "suprasummative" entities and relations alongside our normal, everyday "geometrical" experience, which is what twentieth-century empiricist philosophers have generally done. Or one could assert that process, flow, and distribution are *both ontologically and epistemically primary,* and that arrangements of fixed objects exist only within them. This is the position that Köhler, Koffka, and Wertheimer took. For them the term "Gestalt" applied both to objects and to their relatedness in contexts; moreover, they argued that, because Gestalt processes underlie the latter, these result in the former. This is what Köhler meant when he said that Gestalt laws "determine" what we are to regard as Gestalten. It was a bold leap, both in physics and in psychology, from the assertion that there are numerous exceptions to "geometrism" to the claim that we must proceed from these exceptions in order to develop either a coherent theory of perception or a unified philosophical worldview.

Psychophysical isomorphism

When Köhler discussed specific theories about the brain processes corresponding to perceived Gestalten, he argued that there were two choices. Older theories assumed that mutually isolated sensory processes correspond to isolated conduction pathways in the nervous system (see Chapter 6). Köhler conceded that the idea of numerous physical systems corresponding to locally independent sensory or psychophysical processes "is not an impossible thought." Indeed, Wertheimer's theory of the brain processes underlying apparent motion was quite compatible with this view. However, physiological observations reveal not only isolated pathways from the retina to the visual area of the brain, but also numerous transverse functional connections among the nerve fibers at all levels. In addition, psychological observation discovers numerous "suprageometric" characteristics of the perceived world, such as the anisotropic structure of visual space, with its characteristic asymmetries between up and down, or right and left. Thus, Köhler argued, from the points of view of both psychology and neuroanatomy it is not in the least "inexact" to propose "a second possibility which makes just as much sense and has much greater physical interest." This was the idea that "the somatic field could just as well be one physical system."[27]

If either possibility were really to be carried out, Köhler acknowledged, then it must be shown that there is an *"objective similarity* between the Gestalt characteristics of psychophysical events and those of the phenomenal field – *not only in general,* in the sense that we are dealing with Gestalten in both cases, *but in the specific character of every Gestalt in each individual case."* Wundt had loosened the bonds of strict psychophysical parallelism in order to ensure the independent value of the psychical. Johannes Müller had regarded psychological events as representations (*Abbilder*) of events in the nervous system. G. E. Müller had given that theory systematic form in his "psychophysical axioms" and had applied it to color perception. But "we mean something different and more radical even than this," Köhler asserted: *"actual consciousness is related in each case to corresponding psychophysical events according to (phenomenally and physically) real structural properties, not merely connected to them without an objective relationship."* This was the postulate that Köhler later called "psychophysical isomorphism." Toward the end of his life, he restated it thus: "Psychological facts and the brain events that underlie them are similar in all their structural characteristics."[28]

Most writers who asserted a qualitative difference between the physical and the psychical realms argued that even the most exact knowledge of the brain would tell us nothing about conscious experience. In a footnote attached to his statement of isomorphism, Köhler opposed this claim: "A brain observation is in principle conceivable, which would recognize in Gestalt, and thus in the most essential characteristics, something physical similar to that which the subject phenomenally experienced." He conceded that such an experiment was "nearly unthinkable" at the time from a practical point of view, but this did not detract from its possibility in principle.[29] Decades later, after he had emigrated to the United States, Köhler attempted to carry out just such tests (see Chapter 22).

At first glance, all this reads much like Mach's isomorphism; Mach, too, advocated "seeking" brain events that corresponding to sensory processes (see Chapter 4). The title of the chapter in which Köhler's postulate appeared quoted Goethe's declaration, "For what is inside, that is outside" *(Denn was innen, das ist aussen)*, which Richard Avenarius had already cited in his own version of empirical parallelism. For Avenarius, this poetic *aperçu* underscored his claim that immediate experience is intimately dependent on events in the organism, specifically in a part of the brain. One student of the matter has even implied that Köhler "profited" from Mach and Avenarius without citing his sources, in order to avoid offending Stumpf. But Köhler had already cited Mach in his dissertation, and he cited Mach's history of mechanics several times in *Die physischen Gestalten*. The immediate source for Köhler's isomorphism was G. E. Müller's psychophysical axioms, which had been distilled from ideas of Mach and Hering (see Chapter 4). Köhler derived his postulate by transforming Müller's axioms on the basis of Wertheimer's Gestalt epistemology.[30]

More important than the issue of sources, however, are the fundamental differences between Köhler's isomorphism and that of Mach. In the first place, Köhler's was an isomorphism of "objectively related" structures, not of incidentally correlated point patterns or "components." This difference was rooted in Wertheimer's epistemology, which was completely opposed to Mach's sensationalism. It also led to different conceptions both of psychology and of brain action from those of Mach. On the basis of these conceptions Köhler did something Mach never did – he tried to construct a theory of brain processes that could account plausibly for perceived Gestalten in vision.

According to the "zone theory" of vision then advocated by G. E. Müller and Johannes von Kries, initial reception occurs in the retina. The result is then conveyed along conducting pathways without further alteration to the occipital lobe, located at the rear of the brain, where interaction and association processes take place. In keeping with the idea that "the somatic field could just as well be one physical system," Köhler presented visual Gestalten as the result of "one Gestalt process" in which "the whole optic sector from the retina onward" is involved, including transverse functional connections among conducting nerve fibers. However, only the processes in the occipital lobe are accompanied by consciousness. Köhler gave several reasons for choosing this model, but the most important one was surely his belief that applying it to specific problems would be "simpler."[31]

Such a claim seems strange at first. Would it not be simpler to examine processes in single organs, or at least in anatomically similar regions of an organism, rather than to take on the whole visual system at once? But the problems Köhler had in mind demanded Gestalt processes of considerable extent. In addition to the phenomena of seen movement and stationary Gestalten, these included the subjective geometry of the visual field, mentioned previously, the ability of the apes on Tenerife to "see" objects as means of reaching a goal, and what Köhler called "the dynamic, directed processes" involved in their insightful problem solutions.[32] Apparently he chose a hypothesis sufficiently broad to be able to account eventually for all these structures as well.

His choice had dramatic consequences. In the case of vision, it meant radically reconstructing Hering's model of the "inner eye" (discussed in Chapter 4) and expanding his notion of seen objects to encompass the entire field of things and their ordered arrangement. This eliminated both the retinal image as a fixed, two-dimensional picture transmitted point for point from the retina, then transformed or processed in the occipital lobe, and also any notion of "local signs" or other retinal elements as cues for depth. For Gestalt theory, the three-dimensional world that we see is not constructed by cognitive processes on the basis of insufficient sensory information. Rather, it appears complete as the product of nervous processes occurring in the three-dimensional optic sector, conceived as a vastly expanded version of Hering's "inner eye" (see Chapter 4). Köhler and Koffka continued to use the term "retinal image," but they meant nothing more by it than the arrangement of incoming light rays upon initial reception. In addition, in Köhler's model there was no projection of simple sensations onto the cortex in the way proposed by Helmholtz. Instead, "*For Gestalt theory the lines of flow are free to follow different paths within the homogeneous conducting system, and the place where a given line of flow will end in the central field is determined in every case by the conditions in the system as a whole.*"[33] In contemporary parlance, Köhler had described the optic sector as a self-organizing physical system.

Köhler insisted that such views do not prescribe featureless continuity in the cortex, but are perfectly "compatible with rigorous articulation." Nor was it necessary to claim that brain processes must somehow look like perceived objects. Although he rejected the notion of a fixed, two-dimensional retinal image, Köhler made it clear that the arrangements of incoming light rays on the retina are "in general not physical Gestalten at all, but summative geometrical manifolds, albeit physical ones. . . . Seen Gestalten are therefore not reducible to an image [*Abbild*] of the physical Gestalten of the environment." Instead, he suggested that the psychophysical correlates of circles, for example, would also be symmetrical, but not necessarily circular. He spoke here, as Wertheimer had before him, of "functional," as opposed to "geometrical" similarity.[34]

Köhler tried to show that his postulate was practical by applying it to the figure–ground phenomena first reported by Edgar Rubin in 1914. Rubin had found that the essential differences between figure and ground are that the figure has form qualities, while the ground has none, and that the figure appears to have "thing" character, while the ground has only the quality of undifferentiated "material" (*Stoff*). He demonstrated the difference most impressively with pictures in which, after steady fixation, the part of the picture that was at first seen as the figure suddenly becomes the ground, or vice versa. The best known of these is the goblet figure, which has since come to be associated widely with Gestalt psychol-

Figure 6. Figure and ground (Rubin 1915/1921).

ogy (Figure 6). Rubin himself interpreted these results as the products of a "habitual attitude." Köhler put it rather differently. In his view, the most important condition for seeing a thing, or Gestalten of any kind, is "being set off against" relatively homogeneous surroundings. Thus, for him, figure and ground are not products of habit but "two very concrete and phenomenologically real modes of existence [*Daseinsweisen*] of the optical."[35]

G. E. Müller had already suggested in 1896 that the physiological basis of color vision might best be understood in terms of the reversible chemical reactions described by Walther Nernst. Köhler, who had studied with Nernst in Berlin, generalized this to assert that all excitations in the "somatic field" follow the laws established for reversible chemical reactions, although they may not actually reverse their course. Köhler knew that such reactions produce electric currents.[36] He hypothesized that when a small, white figure such as a circle is exposed on a homogeneous gray background, the result on the retina will be two sets of chemical reactions with a corresponding "leap" of electrostatic potential along the boundary between the two stimulus regions. If equal amounts of electricity are involved on both sides of the boundary, this quality will be displaced over a larger area in the region corresponding to the background than in the region corresponding to the disk. It is this difference in current density, transported along nervous pathways by osmotic diffusion and functionally reproduced in the cortex, that helps visible things attain "their lively phenomenal existence."[37] Thus far the schema appears to be little different from that of G. E. Müller, for the primary interaction occurs in the retina.

Next, to account for the relational character of perception that he had demonstrated in his 1918 paper on "structural functions," Köhler applied his schema to the Weber–Fechner psychophysical law (discussed in Chapter 4). According to Köhler, current densities depend not only on the size of the stimulated areas, but also on objective differences between the two color reactions in the retina. If the color of the ground, for example, appears below the stimulus value at which subjects notice a difference in the figure's color, then the Gestalt produced by the original figure–ground relation is not seen. In electrolytic systems, electromotive forces must attain a minimum value before there is current flow at all. Thus, Köhler reasoned, it ought to be possible to translate one threshold concept into the other by assuming that the electromotive forces in the brain would have specific, measurable values corresponding to the psychological threshold values.[38]

Nernst's theory of galvanic chains had been derived from the theory of potential differences between solutions of unequal concentration. Köhler used Nernst's calculations from the former theory, because, he said, they are identical reaction types, differing only in speed. Working out the calculations, he found that the potential difference does not depend directly upon the difference of ionic concentrations, but upon their ratio, according to the equation $\phi_1 - \phi_2 = $ const. log c, where ϕ is the potential and c the concentration. This was the same type of equation as the one Fechner had derived for his version of Weber's law (see Chapter 4). The relation held within a range of concentrations of 1 : 50. The Weber–Fechner law, too, holds for achromatic colors (black and white) within the same brightness range, which is adequate for ordinary seeing. The brightness ratio between black paper and newly fallen snow, for example, is approximately 1 : 40. Thus, Köhler concluded, Weber's law is applicable in the same way both to the threshold differences for visual phenomena and to the physical-chemical processes he supposed to underlie them.[39]

Köhler had turned the tables on those who argued that we need more complete knowledge of the brain's structure and function before we can set up reasonable hypotheses about the physiological correlates of perception. Instead of bemoaning the fact that the brain was terra incognita as far as higher functions were concerned, Köhler used the situation to his advantage, as a license to set up a hypothetical-deductive model of the kind normally applied in theoretical physics. Köhler was not only claiming that there are formal analogies between physical and psychological structures, but employing structurally "similar" physical processes to *explain* psychological ones. Commentators have appropriately called this style of explanation "physicalism."[40]

Köhler's conception of physical law: Order and beauty in nature

By employing hypothetical-deductive reasoning in this way, Köhler showed that he had a conception of the nature of physical law, and of permissable modes of theory construction in natural science, closer to that of his teacher Max Planck than to the inductivism of Mach. For Planck, however, "the Real" was a name for a mathematical synthesis of the "invariant" laws of nature, abstracted from immediate experience, to be achieved in the future. Köhler sought his invariant realities in the here and now. Like Planck, and like his teacher Carl Stumpf, he clearly accepted the epistemology called critical realism – the view that there is an external world which, however, is not immediately accessible to us in all its particulars. For Stumpf, however, physics and psychology use different methods because psychological objects are directly experienced, while those of the external world are not (see Chapter 2). For Köhler, the distinction was evidently not so clear. He even used Stumpf's phenomenological style of psychological observation to convince himself of the reality of lines of force. For contemporary physics, he said, there is "nothing fictive" about these except their discrete and individual appearance in standard field diagrams. Yet Maxwell's illustrations in the *Treatise on Electricity and Magnetism* make "a pleasing impression on the eye." If one takes good pictures with enough curves and closes one's eyes a little, a "nearly continuous" shaded image (*Schattierungsbild*) appears. Figure 7 reproduces one of the examples Köhler cited in support of his claim. "This way," he said, "one makes it more visibly clear [*anschaulicher*] that something *physically real,* and not the presence of analytic functions alone, is concentrated around the conductor in such a characteristic way."[41] This was phenomenological physics of a literal kind, indeed.

The origin of such intuitive certainties, Köhler thought, "must lie deeper." All of the examples Köhler had offered of physical Gestalten were equilibrium processes, such as the equalization of osmotic pressures in two solutions by the migration of ions across the boundary between them, or the spontaneous distribution of charged particles on conductors. Toward the end of his book, Köhler claimed that all directed processes governed by the second law of thermodynamics fulfill the Ehrenfels criteria. Indeed, Maxwell's field diagrams showed, he maintained, that we can predict "from a purely structural point of view" the movements of conductors and magnets, and the groupings of their corresponding fields, in the direction of "increased evenness of distribution, simplicity, and symmetry."[42]

This was a qualitative version of the tendency, described by Planck, of all processes in physical systems, when left to themselves, to achieve the maximum level of stability, which is synonymous with the minimum expenditure of energy, that prevailing conditions will allow. This can be understood only on the basis of

Figure 7. Uniform magnetic field distributed by an electric current in a straight conductor. James Clerk Maxwell, *Treatise of Electricity and Magnetism* (1873), vol. 2, Figure XVII.

the second law of thermodynamics, that is, the entropy principle. Translated into Köhler's language, the amount of energy in a system will be "as small as the Gestalt conditions permit." Heat currents, for example, are not structures in the same sense that distributions of charged particles are, but "unordered energy"; yet each molecule in the flow is still dependent upon all the others. In extreme cases, a point is reached at which no displacement can occur without affecting all the other parts of the flow. At such points the second law refers to the entire system, and "the law for the system prescribes what occurs in the parts, not the other way around." Since the processes in the brain and in the nervous system are also physical, Köhler reasoned, it followed that they, too, must be directional in the same sense, though admittedly the Gestalt conditions would be far more complicated than those generally holding in physics.[43]

Köhler recognized that the existence of such maximum–minimum processes, even in the brain, says little about their relationship to phenomenal Gestalten. The relation between the "energy concentration" of such processes and their "form of distribution" must still be explained. How, he asked, do such processes look?[44] Köhler attempted to answer the question with an example from hydrostatics, taken from Mach's *Science of Mechanics*. Mach recounted experiments in which the physicist van der Mensbrugghe dipped a square wire frame into a solution of soap and water, then placed a loop of moistened thread on the soap film. When the film inside the loop was punctured, the film outside contracted until the thread formed a circle in the center of the liquid surface (see Figure 8). Noting that such "minimal surfaces" could take particularly pleasing forms, Mach remarked that equilibrium states evidently have something to do with symmetry and regularity. His explanation was that in symmetrical systems "every deformation that tends to destroy the symmetry is complemented by an equal and opposite deformation that tends to restore it." Thus, in such cases "a maximum or minimum of work corresponds to the form of equilibrium symmetry," and regularity becomes "successive symmetry."[45]

Köhler argued instead that regularity is prior to symmetry, and boldly suggested that physical systems in general tend toward end states characterized by "the simplest and most regular grouping." In such situations a quantitative change, a decrease in net energy, has a qualitative result, a change in the distribution of the components in a specific direction. This he called "tendency to simplest shape," or toward "the *Prägnanz* of the Gestalt," alluding to the principle already enunciated by Wertheimer, rather vaguely, in 1914. As an example he cited an observation from a provocative source. In his *Theory of Colors,* Goethe reported that the afterimages of rectangles tend to recede from the periphery, and "he noticed that in square images the corners become blunted until at least an ever-diminishing, round image floats before the eye."[46]

Here, as so often in the history of science, the logic of a theory's discovery was different from that of its justification. The *aperçu* that had led Köhler to write his book came not at the beginning but at the end of that book. It was as though Köhler had gone through such an extraordinary intellectual effort simply to make this single idea seem more plausible. As he recounted it in 1938: "When Wertheimer formulated his principle [of *Prägnanz*] in psychology I happened to be studying the general characteristics of macroscopic physical states, and thus I

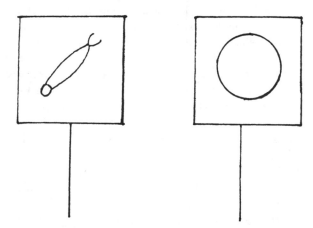

Figure 8. Maximum–minimum distribution: string on soap film. Ernst Mach, *The Science of Mechanics,* 6th ed. LaSalle, Ill. 1960, p. 481. Reprinted by permission of Open Court Publishing Co.

could not fail to see that it is the psychological equivalent of Mach's [maximum–minimum] principle in physics."[47]

In his introduction for philosophers and biologists, Köhler said that his purpose was to help psychologists "to learn to see Gestalten in physics." For others this statement may have been a metaphor, but not for him. For his teacher Max Planck, the unified worldview to which physics aspires refers to a reality outside us, depicted only formally by systems of equations which are testable in experience but are not direct representations of that experience. For Köhler, the structural essence of physical Gestalten, expressed in their mathematical laws, is the same as that of psychological Gestalten, even though no such laws have yet been derived for them. Indeed, for him, understanding nature was something intimately bound up with our intuitive appreciation of its beauty, as much an aesthetic as a pragmatic enterprise. As he later noted, Wertheimer's principle of *Prägnanz* "coincides with a direction often recognized in simplest aesthetics."[48]

Besides the citation just quoted, Köhler also mentioned Goethe in his discussion of the mathematics of physical Gestalten. To underline what he took to be the ontological significance of the fact that similar series of differential equations can cover a wide variety of physical facts, he called them "affinity series" (*Verwandtschaftsreihen*). This was an allusion to Goethe's novel, *Wahlverwandtschaften,* translated into English as *Elective Affinities,* which tells a story of human passion in the language of magnetism and chemical attraction. Köhler thought that Goethe would have been pleased to learn of the existence of these series.[49] Allusions like these were not merely ornamental; they indicate as much about Köhler's scientific beliefs as his learned citations of mathematical physics, few of which reappear in his later work. Actually, both tend in the same direction – the beauty of natural order conceived as a unity is immediately evident to those who would "learn to see" it. For this view the proper term is not Berkeley's *esse est percipi* – to be is to be perceived – but *percipiamus essentia* – we see the essence *in* what exists. Goethe believed, like Spinoza, that reason does not give order to chaotic appearances, but reveals itself in their midst. Köhler's version of Gestalt theory was a true descendant of this belief.

Kurt Koffka was overjoyed at the prospect of not having to be defensive about "real Gestalten" any longer. When he read the proofs of the book at Köhler's request, he later recalled, it seemed to him like "a revelation." In the review he then wrote, he said that Köhler had accomplished "a wonderful unification of two areas of knowledge," and added: "The future will show what this means for science and – let this be expressly emphasized – also for philosophy."[50]

12

Wertheimer in times of war and revolution: Science for the military and toward a new logic

During these years, Max Wertheimer pursued his ideals in other ways. When war was declared he was in Prague. In the early months of the war, Max Brod later recalled, many intellectuals of the Prague circle met regularly in the Cafe Arco to discuss, among other things, their opposition to the conflict and the steps they could take to end it. During one of these discussions, Brod and Franz Werfel thought of publishing an appeal for peace in the leading newspaper of a neutral country. Naively taking Italy's declared neutrality at face value, they decided for Milan's *Corriere della Sera*. To seek support for their idea they went to the philosopher and political activist Thomas Masaryk, whom they thought to be of "realistic, humanistic" and "liberal" rather than narrowly nationalistic views. They asked the slightly older Wertheimer to go with them, because he was a "rising young colleague" of Masaryk's in philosophy who might be heard with respect.[1]

As a newly mobilized regiment paraded noisily outside, the trio ascended the stairs to a small office behind a house on Prague's Wenceslas Square, where Masaryk's party newspaper had its offices. Despite the hectic atmosphere, Wertheimer was able to present their idea to Masaryk. He seemed to listen patiently for a while, but then interrupted curtly. Referring to a German-speaking Jew who had just had a Czech woman arrested for making a critical remark about the parade outside, he said, "You should rather see to it that your countrymen cease their provocations." It was a rude awakening for the young intellectuals. "He was right, of course," Brod later remarked; "but he had spoken as a Czech nationalist," who saw the defeat of the Central Powers as a way of achieving Czech independence.

Psychological science for the military

Wertheimer's humanistic and internationalistic views did not prevent his responding to the call to do research on acoustical problems for the German army. The

187

First World War saw the emergence of the technological battlefield, and it also marked a turning point in the interaction of technology and basic science. In numerous ways scientists demonstrated the usefulness of basic research by employing laboratory instruments and techniques to solve military problems, from the development of sound-ranging devices in physics to that of poison gas in chemistry.[2] Psychologists in Germany participated in this process by adapting psychophysical measurement to skills testing for the selection of communications specialists, pilots, and drivers. The focus of these efforts was on the "human factor" – on the human organism as a functioning part of a fighting machine. Practitioners in the new field called "psychotechnics" searched for machine operators whose skills were best suited to the tasks in question – whose reactive idiosyncrasies, that is, interfered least with efficient functioning.[3]

The problem that led to Wertheimer's research was the location of enemy artillery over long distances in the midst of hundreds of roaring guns.[4] For reasons still unexplained, the Germans did not develop mechanized sound-ranging apparatus for this purpose, combining microphones and oscillographs, until the end of the war. At first, the army made do with listeners posted at different locations in the field to record the times they heard cannon fire; its location was then determined mathematically by using the differences in reported times as a basis for triangulation. There were two obvious problems with this method: variability in listeners' reaction times, and lack of clarity about the relationship between time differences and localization. Before sound-measuring troops were formally integrated into the German army in October 1915, it was necessary to unify and make reliable the techniques they would be trained to use. For this purpose the Scientific Measurement Department of the Prussian Artillery Testing Commission, headed by Breslau physicist Rudolf Ladenburg, engaged the cooperation of the Berlin psychological institute in the spring of 1915. The interdisciplinary group that carried out the research included physicists Max Born and Felix Stumpf (a son of Carl Stumpf), as well as the experimenting psychologists Hans Rupp and Kurt Koffka, in addition to Wertheimer and his friend Erich von Hornbostel.[5]

Wertheimer and Hornbostel used apparatus that was similar to that which Köhler had used for experiments on sound localization. This consisted of two tubes, one for each ear, connected with two sound sources – usually telephones – located several rooms apart. Observers were then asked to report whether a given sound came from the left or the right, or from in between – that is, from either side or the front. But Wertheimer and Hornbostel changed the design in one significant way, by making the tubes variable in length. They did so in order to test an original hypothesis – that the decisive parameter for the localization of noise (nonperiodic vibrations of short duration) was not intensity and phase differences, as for ordinary sounds, but time differences. Their initial tests confirmed this hypothesis in full.[6] However, localization with this apparatus was not exact enough for military purposes. Hornbostel and Wertheimer therefore broadened

Zu der Patentschrift 301669

Fig. 1.

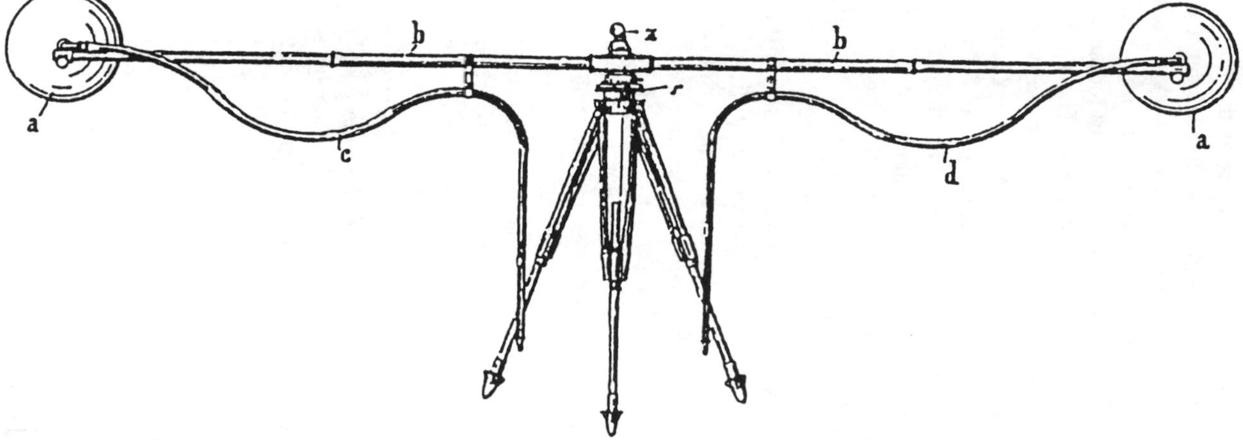

Richtungshörer Hornbostel/ Wertheimer 1915

Figure 9. Diagram of the Wertheimer-Hornbostel “Directional Listener” (Patented 1915). Erich Moritz von Hornbostel and Max Wertheimer, “Vorrichtung zur Bestimmung der Schallrichtung,” Reichspatentamt, Patentschrift Nr. 301669, Klasse 74d, Gruppe 5, patented as of 7 July 1915, issued 28 September 1920.

sound intake with saucerlike devices on the receiving ends of the tubes. The completed device consisted of two funnels or microphones a fixed distance apart mounted on a tripod and connected to the earpieces with tubes or hollow cables (see Figure 9). The operator either turned the apparatus on the tripod until he localized the sound medially, or kept the apparatus stationary and measured the subjective angle of deviation from the instrument's midpoint. The apparatus, formally called a "directional listener" (*Richtungshörer*) but informally dubbed the "Wertbostel," was completed by the end of May 1915 and patented in July. After extensive field tests, it was introduced in the summer of 1916.[7]

After the war, versions of the directional listener were used in aircraft location and in radio research for another fifteen years. The instrument proved inadequate for its original purpose, however, because there was far too much noise in artillery battles to operate it efficiently. A modified version for underwater use was of interest to the German navy, motivated as it was by fears of American submarine and torpedo detectors. To assist in developing the modifications, both Wertheimer and von Hornbostel were engaged as temporary officers of the Artillery Testing Commission from October 1916 for the duration of war. Wertheimer made eight trips to Kiel and Travemünde for this purpose between February and August 1917. From the end of June to the middle of July 1917, he was also sent to the Belgian coast to help develop acoustical warning devices against incoming invasion forces.[8] The modified device was used defensively by intelligence ships; offensive uses were still in the testing stage when the war ended.

As Carl Stumpf wrote in 1918, "I myself would have shaken my head in disbelief if someone had told me what sorts of things would be done in my institute in these war years."[9] A military problem had directly stimulated research aimed at integrating and controlling the "human factor" in person–machine interaction; in the process, basic psychological science had become an instrument of military technology. Despite the antiwar idealism he had shown in Prague when hostilities began, Wertheimer participated in this process. Whether he ever reflected directly on the style, purpose, or moral implications of the science involved is not known. Stumpf invoked the war research done in his institute when he campaigned successfully for its expansion in 1920 (see Chapter 13). The most important result for the Gestalt theorists was the indication that "invariant" relationships were discoverable in hearing as well as in vision (see Chapter 15). As will be clear shortly, however, Wertheimer's involvement did not lead him to redefine his science in a technological mode.

"Conclusion Processes in Productive Thinking": Toward a new logic

In 1916, while still engaged in military work, Wertheimer transferred his teaching rights from Frankfurt to Berlin. There he lectured on general philosophy, logic, epistemology, and pedagogy, and also took over the beginners' exercises in experimental psychology on a temporary basis when Rupp was called to do military research in Austria.[10] At this time, he met Albert Einstein. Their similar political and moral outlooks cemented a friendship that lasted until Wertheimer's death. In these years they had the first of many conversations about the development of relativity theory, which Wertheimer later cited in a chapter of his last work, *Productive Thinking*.[11] He developed the central theme of that book in this period, in an essay entitled "Conclusion Processes in Productive Thinking," written for Stumpf's seventieth birthday celebration in 1918 and published in 1920.

The thrust of the argument was similar to that of Wertheimer's 1912 paper on number concepts. Here, however, the object of attack was no longer western arithmetic, but Aristotelian logic. The question he posed at the outset was, "What happens in genuinely productive thinking," when a recognizable advance in knowledge has been achieved? Traditional logic, epitomized in the classical syllogism "All men are mortal: Socrates is a man : Socrates is mortal," purports to guarantee new knowledge by invoking the rule that the conclusion may not be included in either the major or the minor premise. Thus, the fact that Socrates is a man, or even that he is mortal, is not included, at least formally, in the statement that "All men are mortal." But the knowledge we obtain by obeying such rules, Wertheimer alleged, often seems "empty" and "dead," "like the work of a clerk in a registry."[12]

In Wertheimer's opinion, J. S. Mill was correct to contend that statements like "All men are mortal" are only "inductions in disguise." We cannot really say such things without knowing something about individuals like Socrates. But even if we grant, on whatever basis, that there can be statements that contain all the characteristics of a subject, whether we know them or not,

> then *naturally no "new" knowledge in the genuine [prägnant] sense of the word is possible. . . .* From the point of view of our problem, this is a logic for the good Lord [*eine Logik für den lieben Gott*], or, more exactly, for a scholar who basically already knows and has seen through everything and only orders and adjusts the form of the presentation. But for truly forward-moving knowledge this kind of view is simply crooked [*schräge*].[13]

Wertheimer argued that there are other logical operations besides "the one with the knife," that is, analytics, and "the one with the sack," the logic of classes. To discover them, he called for a naturalistic logic focusing on "real processes, as they actually – fortunately – occur in life" and avoiding "the traditional fixation on mere validity relations." In particular, such an approach would take as its object not an abstract entity "all men," which possesses certain characteristics by definition, and the equally abstract figure "Socrates," which allegedly possesses some or one of these characteristics, but "the Socrates whose known and directly determinable characteristics are actually given." A change in these "given characteristics," he suggested, could produce a kind of feedback effect upon the premises with which one began, a restructuring of the situation that leads in turn to "the grasp of the inner state of affairs."[14]

Wertheimer gave a vivid indication of what this process is like in the following anecdote. Cajus and his friend Xaver are members of the executive committee of their club, but they find the meetings dull and have stopped attending, except for the annual meeting at which the statement of accounts is discussed. One day Cajus returns from a trip and finds a note reporting a unanimous decision reached at a recent executive committee meeting. He becomes angry and resolves to telephone his friend, but reads further and finds: "The decision was reached at the annual accounts meeting, which was held earlier than usual this year. The statement of accounts was approved after a long report from Xaver."[15]

What happens in such situations? Consistent with the vocabulary he had begun to develop in 1912, Wertheimer called the process "recentering." This was something quite different from Mill's inductions, in which repeated experience of a given state of affairs leads us to make a general statement about it. The example just given might seem at first to fit Mill's framework; a negative instance leads to a revision of a previously derived conclusion. For Wertheimer, however, this was a rather poor picture of what was actually going on. In fact, the major premise – in the story this would be "Xaver and I agree" – is "turned upside down" (*umgekrempelt*). "For a moment two premises are there side by side, then suddenly a 'click', a 'snapping together' [*Einschnappen*] [occurs]. What a process! (What?! Xaver was there?! Xaver too – is that *possible? –* aha *– so* . . .)" Thus "something new" is seen in the subject, "and the old concept revolutionized."[16]

Such examples, and Wertheimer's emphasis on visual metaphors, are reminiscent of Köhler's "intelligence tests" with apes. In all of these cases thinking is closely interconnected with perception and action; and in all of Wertheimer's examples the process of change has the same characteristic, discontinuous form that it appeared to have for the apes. First there is an attempt to continue using a familiar concept or behavior pattern, which is blocked, then a pause for active, perceptually aided reflection, then the sudden discovery of or shift into the new conception or behavior.

Such recenterings can also occur on less emotionally charged ground, for example in the solutions to difficult mathematical problems. Consider the task of finding the sum of the external angles of the polygon shown in Figure 10. One method would be to divide the figure into triangles constructed from a midpoint, then to compute the external angles from them. If we eliminate this possibility and look at the situation from a more general point of view, Wertheimer claimed, we can see the external angles as a series of right angles constructed from the sides of the figure, as shown (Figure 10a), and a series of "turning angles" (*Drehwinkel*), indicated by the symbol δ (Figure 10b). To complete or "close" the figure, we need to compute these. But if we realize that a complete "turn" equals 360°,

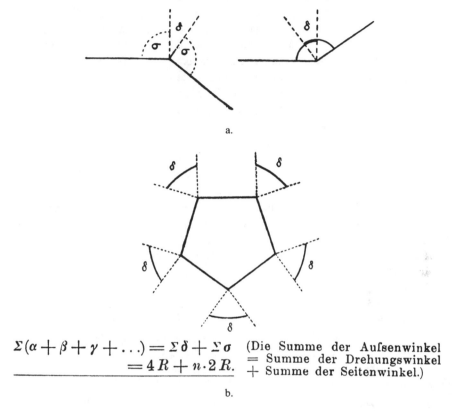

$$\Sigma(\alpha + \beta + \gamma + \ldots) = \Sigma \delta + \Sigma \sigma$$
$$= 4R + n \cdot 2R.$$

(Die Summe der Aufsenwinkel = Summe der Drehungswinkel + Summe der Seitenwinkel.)

b.

Figure 10. Solving the external angles of a polygon. Max Wertheimer, "Schlussprozesse im produktiven Denken" (1920), in *Drei Abhandlungen zur Gestalttheorie* (Erlangen: Verlag der Philosophischen Akademie, 1925), p. 179. Reprinted in Max Wertheimer, *Productive Thinking*, rev. ed. (New York: Harper & Row, 1957), p. 151. Reprinted by permission of Harper Collins Publishers.

"suddenly the situation is clear." Here, too, a shift in our view of the object yields an "insight" that leads to more than a technically correct solution of the problem. *"In such processes the concept that I have of a thing is often not only enriched, but changed, improved, deepened."*[17]

Such shifts occur frequently in the study of history, Wertheimer claimed, as the way we view a historical figure or event radically changes when we learn new facts about it. The history of science, he added, also provides many examples of intuitive recentering resulting in significant theoretical progress, for example Galileo's revision of the concept of inertia, or the young Gauss's "discovery" of the binomial theorem. With the help of Thomas Kuhn, the term "Gestalt switch" has become a common description of such radical changes in scientific think-ing, though it appears that Kuhn did not get the terminology directly from Wertheimer.[18] In any case, Wertheimer wanted more than an evocative descrip-tion. Such accomplishments as Galileo's are often thought to be products of fortunate intuition or of irreducible genius, he noted, but accident or "mere psy-chology" are not the whole explanation. Rather, he suggested, we are dealing here with shifts in "the qualitative hierarchy [*Eigenschaftshierarchie*] of a concept"; and he insisted that "the formal moments involved" can be grasped "in specific laws."[19] This, too, was part of his hope of developing a naturalistic "Gestalt logic," comparable to the "experimental logic" that John Dewey proposed at almost the same time.

Unfortunately, Wertheimer gave no examples of the "specific laws" he had in mind. Another problem was how to account for the urge, need, or will to solve a given problem. For this Wertheimer shifted from structural to functional language, and the question became, "on what aspects of S [the subject] must I concentrate: Or: *how must I apprehend S sub specie the task* [Aufgabe] *here before me?*" This was the terminology of the Würzburg school. Külpe and his co-worker Otto Selz had shown that such problem-solving operations are subject to empirical laws, and thought that this would open the way to solving thorny logical and epistemo-logical problems, such as that of induction. Wertheimer's attitude toward his examples and their philosophical significance was similar. One could claim, he said, that his examples are only cases in which the same object appears in different ways. In the case of Cajus's former friend Xaver, however, the change was more radical than this; "an abyss" separates Xaver before and Xaver after the recenter-ing process.[20]

For a mathematician or a logician, there is no reason why the sum of the external angles of a polygon cannot be found in the manner Wertheimer had rejected. From that viewpoint, Wertheimer's implication that his solution offered a better grasp of the essence of the figure might be interesting psychologically, but it was logically irrelevant. Another distinction Wertheimer seemed to have blurred was that between meaning and significance, or sense and reference (*Sinn* and *Bedeutung*), a commonplace in Continental logic and central to Gottlob Frege's logic of propositions. In the example of Cajus and Xaver, we can say that in sentences including the name "Xaver" that word has a different referent or mean-ing before and after the recentering process, and that the sentence therefore has a different sense. But the logical rules for the construction of such sentences do not change on that account.

Why then did Wertheimer insist that a proper grasp of the processes he described would have any logical significance? The answer is Wertheimer's Ge-stalt epistemology, according to which we directly apprehend real, structural-functional relations – in this case between an "object" named Xaver and Cajus's beliefs about that object's connection with him – that make a difference for the practical logic of our thinking and acting. Philosophy, he implied, ought to concern itself with such real relations, structures, and processes, and not with formal relationships alone. Accordingly, his "major premises" are all empirical statements. However, his version of naturalism could bring him into epistemological difficulties analogous to those Köhler encountered in his work with anthropoids. In the case of Cajus and Xaver, the person named Xaver actually remains the same. If Wertheimer had meant to say that it was a different Xaver in fact, and not only for Cajus, he would have fallen hopelessly into solipsistic idealism. He was therefore quite correct to say that it was "the concept of a thing" that changed, not the thing itself. Still, he offered no way of designating "Xaver$_1$" and "Xaver$_2$" such that it might be possible to formulate and codify any new form of inference that would be needed to capture the process by which new beliefs about Xaver are acquired. Logicians might well concede that such shifts are important enough in practical life, and that the processes behind them might be fit subjects for psychological research. But without such a formulation, or hints about how one could be developed, it was bound to be unclear just what significance these processes were supposed to have for logic.

Wertheimer was criticizing the widespread assumption that Aristotelian logic offers not only rules for the construction and proof of propositions, but also an adequate descriptive guide to actual thought processes. Mill had attacked this assumption by contending that universal statements are inductions in disguise. Many psychologists then substituted Mill's inductive logic for that of Aristotle as a paradigm for thinking. In reaction against Mill, in turn, formalist logicians like Frege sought to separate logic from psychology altogether. The strategy Husserl adopted in his *Logical Investigations* for keeping the two together was to seek and describe "experiences of truth" (see Chapter 5). Yet he accepted Frege's priorities, saying that his aim was an account of the validity, not the genesis of propositions. Wertheimer opposed all of these positions. He defended a naturalistic standpoint against formalist rationalism; his epistemology allowed him to develop a naturalism more encompassing than Mill's, and a psychological realism more immediate than Husserl's.

From intellectual to political revolution

This combination of an open-ended program with high hopes for the future was also characteristic of Wertheimer's other activities and associations in this period. Among his associates at the time was a student, Karl August Wittfogel, who was then active in a dissident left-wing faction of the *Wandervögel* youth movement. He later described his views as "a mixture of Karl Marx and [Chinese sage] Lao Tzu"; Wertheimer became "first a teacher, then a friend." It was Wertheimer who introduced Wittfogel to Käthe Kollwitz, who had begun to speak out openly against the war. Einstein's involvement in socialist and pacifist groups is well known. Whether Wertheimer participated directly in his friends' antiwar activities is not known; but such associations contrast strongly with the nationalistic, even annexationist stance taken by the vast majority of German professors.[21]

However, as we have already seen in the case of Wertheimer's involvement in military research, allegiance to liberal humanist ideals did not necessarily imply opposition to the entire established order. Wertheimer and Einstein made this clear in an incident that occurred shortly after the collapse of the monarchy and the outbreak of revolution in November of 1918. Soon after the revolt began, a student council (*Studentenrat*) was formed at the University of Berlin, modeled on the workers' and soldiers' councils that had been organized across the country. One of the Berlin student council's first acts was to depose and arrest the rector and other university officials. Einstein was asked to negotiate with the students, because his political views were thought to give him some influence with the radicals among them. He invited Wertheimer and Max Born to accompany him to the Reichstag building, where the council met. That the three men had already become companions is suggested by a photomontage Born made of their portraits and sent to Einstein as a postcard in July 1918; in his response, Einstein called it "the clover leaf" (Photo 9).

According to Born's later account, before they could present their business at the council session, the chairman asked Einstein for his opinion of new regulations for students that had just been proposed:

> Einstein thought for several minutes, and then said something like this: "I have always thought that the German universities' most valuable institution is academic freedom, whereby the lecturers are in no way told what to teach, and the students are able to choose which lectures to attend, without much supervision and control. Your new statutes seem to abolish all this and to replace it by precise regulations. I would be very sorry if the old freedom were to come to an end." Whereupon the high and mighty young gentlemen sat in perplexed silence.[22]

Such views must indeed have been puzzling, coming as they did from someone who was known for his pacifist convictions, and who had reassumed German citizenship only reluctantly when he was appointed professor in Berlin.

The council decided that it had no authority in the matter and sent the petitioners to the newly appointed president of the Republic, Friedrich Ebert, who then wrote a few words in their behalf to the appropriate minister. In contrast to Wertheimer's earlier attempt to intervene in political affairs, in Prague, this venture was a complete success. "We left the Chancellor's palace in high spirits," Born wrote, "feeling that we had taken part in a historical event and hoping to have seen the last of Prussian arrogance, the Junkers, and the reign of the aristocracy, of cliques of civil servants and of the military, now that German democracy

Photo 9. Max Wertheimer, Albert Einstein, and Max Born. Photomontage 1918. By permission of Albert Einstein Archives, Jewish National and University Library.

had won."[23] Evidently Einstein and Born believed that democracy and the German ideal of *Wissenschaft* could be combined, that the authority, status, and "freedom of science" that had once been a welcome gift from above could be retained under the new regime. Perhaps Wertheimer shared such hopes and also Einstein's opinion of academic freedom.

What was the connection between Wertheimer's political stance, insofar as we may conclude anything about it from the evidence available, and his philosophical position? His concern for the highest achievements of creative thinking, which continued throughout his life, could easily be construed as elitism. However, Wertheimer did not limit his examples to the thought of great scientists. In his work on number concepts he had already made it clear that for him the solutions that ordinary people discover to the mathematical problems posed by everyday life are determined by the same psychological processes, and are in every way as admirable, productive, and beautiful as those allegedly higher achievements.

Wertheimer's philosophical naturalism was informed by an inseparable combination of allegiance to moral and aesthetic values and the commitment to discover them in life, not to impose them on life from above or without. In this view, the so-called "higher" values can no longer be the exclusive property of a privileged academic elite. In Germany, that elite had identified itself for decades with its social role as servants of the state. Wertheimer's approach offered a way of showing that this is not necessary, that allegiance to "higher" values, rightly understood, is fully compatible with a commitment to democracy.

Wertheimer stated his personal philosophy quite clearly in a brief contribution he made in 1918 to a small volume dedicated to the Jews of Prague, which also included contributions from Max Brod and other members of the "Prague circle":

These are also my wishes and hopes for the youth of Prague: as it was expressed long ago in the Jewish will, that God should live in every daily deed, in the powerful present [*im kräftigem Diesseits*]. That seems to be coming again often; and much from Prague: heartfelt work in the midst of reality, toward life [*mitten im Wirklichen, dem Leben zu*].[24]

Clearly, for Wertheimer there was no opposition between philosophical naturalism and religion, but his Jewishness had little in common with orthodox piety. The idea of daily life as a ritual ordained by God's commandments, characteristic of traditional Judaism, had become a commitment to "the powerful present" much like the Jewish existentialism that Martin Buber was beginning to develop at just this time.

Wertheimer expressed similar hopes in the language of German philosophy at the end of his *laudatio* for Stumpf's seventieth birthday, quoting what Stumpf had written in memory of Lotze the year before:

Forge ahead on the reconstruction of a worldview, do not let yourselves be put to shame by the physicists, *arm yourselves with all the weapons of natural science in order to proclaim nonetheless and yet again the supreme position of the spiritual and the good in the world;* that is what humanity expects from you, if it is to awaken to new life from its most difficult crisis as from a terrible dream.[25]

Gestalt theory was a revolt from within, despite its radicalism. In the course of their scientific training, Wertheimer, Köhler, and Koffka absorbed and fully accepted the terms of the challenge to experimental psychology in its German academic environment. Like other experimenting psychologists, they hoped to find empirical solutions to long-standing philosophical problems. Many of the leading philosophers of their day rejected precisely this hope, contending that the essential qualities of consciousness are beyond the reach of experimentation. In response to this challenge, the Gestalt theorists tried to develop an approach to psychology that would overcome philosophical dualism, a scientific worldview that would support progress in both fields without rejecting experimental methods. Theirs was a common effort, and each of them developed his thinking in close relation to that of the others. However, because Koffka, Köhler, and Wertheimer had different personal and intellectual styles, their contributions to that effort differed.

Max Wertheimer had a profound metaphysical vision that formed his research at every juncture. That vision generated the epistemology that was the core of Gestalt theory. He showed concretely what that vision revealed first in his essay on number concepts, then experimentally in his demonstration of the *phi* phenomenon. As he explicated the new view's philosophical implications in lectures, he began to develop "Gestalt laws," including the law of *Prägnanz*. He described his method as a search for the "essence" of the given. In fact, it was Stumpf's experimental phenomenology taken even more seriously than even its creator had done. Going far beyond Stumpf's methods, he added a hypothetical-deductive model that posited structured "whole processes" in the brain to account for the "dynamic realities" he was discovering, and claimed for it the same ordering and predictive significance for psychological research as hypotheses have in natural science.

Wolfgang Köhler was a sophisticated natural scientist with a philosophical bent. Though outwardly cautious and careful to present his ideas and findings in a way that could be understood and accepted by older colleagues, he did not hesitate to draw bold conclusions. When he called his proposal to discard the conventional distinction between peripheral or physiological sensations and central or psychological operations upon sensation a "tentative suggestion," his apparent caution was a way of gaining a hearing for new ideas. However, as he transformed Stumpf's experimental phenomenology in his work with animals, showed that concepts from classical empiricism were an inadequate basis for the construction of evolutionary hierarchies, and used his training in natural science to extend Wertheimer's Gestalt viewpoint from the psyche to the external world, he showed that he was in many respects the most radical thinker of the three.

Kurt Koffka was an experimental psychologist in a social setting not exactly designed for him. In his work with his students in Giessen on apparent motion, he began to show how Wertheimer's model of theory and practice could be developed into a program of cumulative research. At the same time, he developed, defended, and applied Wertheimer's approach with a will to system. In his polemic against Vittorio Benussi, he brought out the implications of the new view for the psychology of perception. In doing so, he extended the Gestalt concept from perception to behavior, attributing both to "structured whole processes" in the brain. With this the foundations of Gestalt theory as a psychological system were essentially complete. Köhler's observations on apes seemed to provide independent confirmation of Koffka's systematic constructions. It was partly in order to deal with these observations that Köhler transformed G. E. Müller's psychophysical axioms into a psychophysical isomorphism of organized structures, and developed a holistic theory of brain action to account for them.

Gestalt theory's answer to the attack on psychology in Germany as a philosophical discipline was to claim that it applied only to a psychology based on sensationalist and empiricist assumptions. They did not accept the distinction between the genesis and the validity of ideas as a dividing line between scientific psychology and philosophy, but proposed to reconstruct psychological theory instead. With their new epistemology prescribing order *in,* rather than behind, the appearances, the Gestalt theorists offered a way to justify psychology as both an empirical and a philosophical discipline, and thus to accomplish the task at which their predecessors had failed, while upholding "the supreme position of the spiritual and the good in the world." Like other philosophers and scientists, experimenting psychologists had assumed that natural science had to be based on mechanistic assumptions. Köhler's extension of Wertheimer's epistemology to physics also meant, as Koffka proclaimed, that "the opposition between mechanism and [psycho-] vitalism has been overcome."[26] The error of the vitalists, Köhler had said, was to define form and order in such a way that only living matter could possibly possess them. Seen in the different light cast by Gestalt theory, the need for multiple ontologies or universes of discourse disappeared, and the unity of science, life, and mind could be established.

Because philosophy in Germany was dominated at this time by the theory of knowledge, and experimenting psychologists had concentrated on problems of sensation, perception, and cognition, the Gestalt theorists grasped the situation to which they responded in the first instance as a problem in the psychology of knowledge. Accordingly, the roots of the new worldview they offered also lay in that field, and the early development of their theory was largely, though not entirely, confined to it. Whether their new epistemology could also solve other psychological problems, or support a worldview that academic philosophers and other intellectuals in Weimar Germany would accept, will be considered in the following chapters.

Part III

The Berlin school in Weimar Germany

13

Establishing the Berlin school

The period from 1920 to 1933 marked the high point, though not the end, of Gestalt psychology's theoretical development, its research productivity, and its impact on German science and culture. How did Gestalt theory, once created, grow and develop? Given that modern scientific research requires laboratory space and equipment, it is important to show how the Gestalt theorists acquired positions from which they could mobilize such resources, before considering the effect this had on the growth and development of their approach, or the ways in which they projected their innovations into the wider disciplinary and cultural fields.

The founders of Gestalt theory were rewarded for their boldness with career advancement. The most prominent promotion was Wolfgang Köhler's appointment to succeed his teacher Carl Stumpf as professor of philosophy and director of the Psychological Institute in Berlin in 1922. This event, along with the founding of the journal *Psychologische Forschung* in that city in 1921, gave the Berlin school of Gestalt psychology its name. Kurt Koffka helped establish a new psychological laboratory in Giessen in 1919, where he remained, with interruption, until he left permanently for the United States in 1927. Finally, in 1929, after productive years with Köhler in Berlin, Max Wertheimer became professor of philosophy and codirector of the laboratory in Frankfurt, where Gestalt theory began. Though their teaching styles were quite different, all three proved to be capable organizers and directors of research. They and their students had the means to develop the core ideas of Gestalt theory in new directions. As they did so, the situation of their science began to change.

Psychology in Germany during the Weimar period

The First World War was a turning point for academic psychology in Germany. As a result of the use of psychophysical methods in the selection of drivers and antiaircraft observers, and in the development of specialized equipment such as the artillery range finders on which Wertheimer and Hornbostel worked, the field emerged from the war institutionally strengthened. New positions dedicated to applied psychology were created in institutes of technology and commercial academies. Professors such as Walter Moede at the Technical Academy in Berlin and Fritz Giese at the Technical Academy in Stuttgart offered their services to government and industry. University psychologists, particularly Karl Marbe in Würzburg and William Stern in Hamburg, were also leaders in applied work; they and others created departments for applied psychology in some university institutes. But the dominant aim there remained what it had been before 1914 – philosophically and pedagogically relevant research.[1]

Retaining such purposes did not exempt academic psychology from the political struggles of Weimar Germany and interwar Austria. The two most prominent new institutes founded in German-speaking universities in that period, in Jena and Vienna, were products of school-reform programs that led to bitter political and academic-political battles.[2] Such political involvements and the emergence of applied psychology complicated the identity and status situation of a science that had begun in Germany with basic research on problems derived from philosophy and sensory physiology. In 1921, Society for Experimental Psychology chairman G. E. Müller sent a letter to university faculties without psychological institutes and to the education ministries of the German states. Pointing to psychologists' war service, he requested that new chairs be created for the field and that existing institutes be better funded – to little avail. In a gesture reminiscent of the philosophers' protest of 1912, the Society's executive committee published a statement in 1929 against the filling of "chairs previously reserved for the special support of psychology . . . by representatives of pedagogy or pure philosophy." The petitioners contended that the rise of "the previously unknown profession of practical psychologist" and the continuing need for psychologically trained teachers actually called for an increase, not a decrease, of professorships in psychology. However, rather than redefining or redirecting their work toward technological goals, they insisted that the linkages of psychology with philosophy had grown stronger, not weaker: "Less than ever can the theory of mind, of values, and of culture do without the exact study of mental life."[3]

By this time, a plethora of alternative psychologies had arisen to deal with the problems of modern youth, sexuality, character, or personality diagnosis and therapy – practical social problems so broad and so difficult to grasp with standard laboratory methods that there was widespread talk of a crisis both in general psychology and in psychotechnics by the end of the 1920s.[4] In a 1930 report to a teachers' association, William Stern warned of the danger in such a situation. If psychology continued to lose ground in the universities, the schools, social welfare agencies, and courts would "cobble together inadequate psychologies for their own purposes"; and the "constantly growing people's movement" propagating "a more or less dilettantish lay psychology" under headings like graphology, "physiognomics," phrenology, "character analysis" and so on would be allowed to continue uncontrolled by independent, critical scientific examination.[5] Apparently, experimenting psychologists in Germany and Austria tried to have it both ways – to retain the tenuous academic and cultural standing assured by their connection with philosophy, while securing and extending state support by claiming control over the training of professional psychologists. To achieve such ends, some appeared willing to loosen their commitment to the primacy of experimentation. The 1929 declaration's authors admitted that earlier experimental methods had been "rigid" and stressed that other approaches were now available. In the year their petition appeared, they also changed their group's name from Society for Experimental Psychology to its more inclusive present title, the German Society for Psychology.

Psychologists had become immersed, however ambivalently, in a struggle for expert authority over matters of the soul in a modernizing society, which had begun in Germany before 1914, but became incomparably more intense in the Weimar era. It was against this complex background of continuity and change in psychologists' view of their field and psychology's social standing in German-speaking Europe that the Gestalt theorists established themselves as a scientific school. In their case, at least, continuing affiliation with philosophy by no means hindered productive research in experimental psychology. Indeed, the superior theoretical sophistication that such a linkage could bring was an advantage.

Berlin

Architecturally, at least, experimental psychology in Berlin benefited from the German revolution. In the spring of 1920 the Berlin institute moved to new quarters in a wing of the former Imperial Palace, near the university's main building (see Photo 10). This was the work of Carl Stumpf, not the Gestalt theorists. During the war, Stumpf had had to reduce his own space in the insitute's quarters temporarily to make way for the military research discussed in Chapter 12. In 1919, he involved himself in discussions about the use of imperial property after the Kaiser's abdication. Placing heavy emphasis on the wartime contributions, he helped draw up ambitious plans for expanding applied work, possibly incorporating a laboratory that had been supported privately by the Society for Experimental Psychology into an expanded university institute. At one point he even proposed Bellevue Palace, near the Tiergarten, now the Berlin residence of Germany's president, as a potential location. Instead, in April 1920 the institute became the only university facility in the Imperial Palace.[6]

The move more than doubled the institute's size; Stumpf wrote that it had "more than twenty-five rooms," compared with ten in the Dorotheenstrasse. At the same time, its budget increased more than six times, from 4,400 to 28,200

Photo 10. Imperial Palace, Berlin, 1935. (The Psychological Institute was on the 2nd and 3rd floors on the right). Landesbildstelle Berlin.

Marks, a figure almost as large as that for the Physical Institute in 1920 (30,274 Marks). One might suspect that the sudden jump in the Berlin institute's budget was an adjustment for the severe inflation of the time. Ministry correspondence shows, however, that the increase was intended to cover the cost of maintaining a much larger physical plant, particularly of heating the high-ceilinged rooms. Along with the increased budget went a rise in personnel. In May 1920, at a meeting in the Ministry to coordinate efforts in applied psychology in the interest of rationalizing industrial production, Stumpf proposed the creation of a new department in this field with an associate professorship, two assistants, and a budget of 10,000 Marks. A more modest department was founded in 1922, and headed by long-time assistant Hans Rupp, who had been made associate professor for labor psychology the year before. With that the staff rose from one paid and one unpaid assistant to three full-time assistants (see Appendix 1, Table 3). The Berlin institute thus became one of the largest and best-supported psychological research establishments in the world. In Germany, its importance was exceeded only by that of the Leipzig institute, which had thirty-five rooms and five permanent assistants by 1930.[7]

In November 1920, the board of the Albert Samson Foundation voted to grant Wolfgang Köhler a salary until he could secure a position. At the meeting at which this was decided, Stumpf said that there was hope that Köhler would find "a state position" in the near future, but did not specify what he meant.[8] In August he had already resigned the institute directorship, complaining of weakening eyesight, and named Köhler as his representative, while retaining his professorship. Though he submitted a physician's letter to attest his disability, he had another motive for proceeding as he did. The new education minister in Prussia, Carl Heinrich Becker, had proposed to introduce mandatory retirement at age sixty-eight for professors – the age now in force in Germany – as part of a comprehensive university-reform plan. At seventy-three, Stumpf was already well past that limit. Since the institute directorship was not a teaching position, he could designate a representative on his own account. He wrote to a Ministry official that he had Köhler in mind as his eventual successor, and wished to make him better acquainted with the Philosophical Faculty and the institute staff.[9]

In addition to Köhler's youth and relative lack of teaching experience, one other obstacle stood in the way of Stumpf's plan – the tradition of appointing to Berlin professorships only people who had already held chairs elsewhere. That barrier fell in August 1921, when Köhler was named to succeed G. E. Müller in Göttingen, who had had to resign because of the newly enacted retirement rule. Köhler went through all of the corresponding formalities, but by the end of the month he had already been appointed to represent Stumpf both as professor of philosophy and as institute director. Though he received his salary from Göttingen, he seems to have been there only once during his one-semester tenure. This convenient arrangement also allowed Müller to continue in his old position until a true successor could be found.[10]

After these maneuvers, Köhler's formal appointment to succeed his teacher in February 1922 was a foregone conclusion. The Philosophical Faculty had already placed him first a year earlier, far ahead of Karl Bühler. In addition to his outstanding research and his proven administrative skill, the Faculty emphasized Köhler's interest in general philosophical issues such as the mind–body problem and his ability to teach the history of philosophy and natural philosophy as well as psychology. His letter of appointment included natural philosophy among the topics he was to teach. When a Ministry official mistakenly called the position "the professorship for psychology," the Faculty responded that it regarded "the change of designation as objectively unjustified" and asked "that the previous title be retained."[11]

The Ministry evidently preferred other candidates. In the statement to the Faculty just cited, the official in charge of university affairs wrote that he wished to appoint Munich professor Erich Becher to the chair. In his view, "the lack of inner organic unity in the field of experimental psychology" meant that "a generally recognized and established scholar" was needed for this important position.

The Faculty acknowledged Becher's reputation, but pointed out that he had published little experimental work. Though the professorship was and should remain one of philosophy, "the leadership of the Psychological Institute requires a personality who is wholeheartedly dedicated to experimental research and leads actively by example. The greater experience in this respect is clearly on Köhler's side." Thus, the Faculty maintained one traditional prerogative – that of deciding its own membership with minimal interference – by invoking the relatively recent tradition of experimental research in Berlin.[12] A traditional criterion of a rather different sort may also have come into play. William Stern later said that he was invited to Berlin to negotiate for the chair, and was offered the position if he would clarify "a small formality," that is, if he would renounce his Judaism. This he refused to do.[13]

With Köhler's selection to head a greatly expanded institute, a new stage in the history of Gestalt psychology began. It might be called the stage of establishment and reproduction, for the control of institutions to train younger scientists gave Gestalt theory the contours of a scientific school. Yet fundamental dimensions of continuity with the past were evident from the start. The head of the laboratory that would give the Berlin school its name had been appointed in the established manner, as a protegé of the incumbent. Moreover, he had presented himself, and had been accepted, both as an experimenting psychologist and as the sort of "philosophically educated man, filled with philosophical interests" that Wilhelm Wundt had demanded in 1913.

The subsequent history of the Berlin institute reveals further dimensions of continuity in the midst of change. For example, the institute's department of basic or general psychology received two assistants, that of applied psychology only one. Student enrollment in the theoretical department far exceeded that in the applied courses. Thus, even as Berlin outranked all other German institutes, except Leipzig, the primacy of basic over applied research that Stumpf had established remained secure. Psychology's double identity as natural science and branch of philosophy also remained unchanged in the institute's teaching. Between 1921 and 1929, institute staff offered 37 percent of the courses listed in the university catalogue under the heading "Philosophical Disciplines" (Philosophy, Psychology, Pedagogy). Among them, in addition to standard psychology courses, were such titles as "Natural Philosophy" and "The Present Philosophical Situation," given by Köhler, "Logic" and "Introduction to Philosophy," given by Wertheimer, and "Theory of Knowledge and the Natural Sciences," given by Kurt Lewin.[14]

Analysis of the institute's enrollment, too, reveals aspects of continuity with the past. In 1928, Köhler reported that "more difficult admissions requirements" had been introduced, reducing the size of some institute courses. Such restrictions had not existed during Stumpf's tenure, because there had been only a few doctoral candidates. During Köhler's tenure, from 1922 to 1935, 33 dissertations were completed – nearly double Stumpf's annual productivity (see Appendix 2). In fact, however, all that had happened was that the self-selection that had previously obtained was replaced by selection through the institute staff – the elite character of the place remained unchanged. Autobiographies that the doctoral students submitted with their dissertations support this claim. Of the 31 students for whom data were available, 25 came from the upper middle class – they said their fathers were merchants, officers, higher officials, physicians, lawyers, engineers, or Gymnasium or university teachers. Interesting also is the high number of women (15, or 48.4 percent) and foreigners (9, or 29 percent) – much higher percentages than for the university as a whole. Add to this the unusually long average duration from completion of secondary school to the doctorate (9 years), and the picture of psychology at the University of Berlin as an elite field appears complete. As one woman who earned her degree in this period says, psychology in Berlin was not a "bread subject" (*ein Brotfach*), but "a luxury subject" (*ein Luxusfach*).[15]

Those admitted to the rank of doctoral candidate could count on excellent facilities. One student reports that every candidate had one to two rooms for his or her research. Remarkable, too, was the atmosphere that arose during long years of study in a small group, with students often serving as subjects for one another, and the consciousness of being involved in important developments in the field. An atmosphere of privilege radiated from the building itself. Former students all remember the impressive architecture. One recalls that entrance was by way of a grand staircase. Another surmises that the institute must have been in the former quarters of the royal ladies in waiting, because there were full-length mirrors on many of the walls.[16] We may or may not accept former assistant Wolfgang Metzger's nostalgic image of the institute as a "lost paradise." It is nonetheless clear that, in a time when German universities accepted more students than before from the so-called new middle classes (middle-level employees) and even from the lower middle classes, and when psychology in Germany began to move toward professional as well as scientific status, the dominant self-concept in the Berlin institute was that of a scientific elite. This was also clear from Köhler's reaction to the German Society for Psychology's protest statement in 1929, discussed above. In a letter to Wertheimer, Köhler called this move "a plebian revolution" (*eine Pöbelrevolution*) driven by the self-interest of applied psychologists, recommended taking action against it and even considered resigning from the Society.[17]

Such attitudes did not mean that feudal or authoritarian structures dominated within the institute. Metzger later recalled that the assistants in Berlin participated in administrative affairs and meetings "as discussants of equal standing." Other former students and assistants offer more differentiated memories. Köhler, one remembers, was "definitely the full professor (*Ordinarius*)," but gave his co-workers Lewin and Wertheimer considerable freedom. Lewin and Wertheimer were "more open" personally than Köhler, another recalls, but Köhler was always friendly and courteous, and was very close to his own students. Still another student remembers that Wertheimer "had a free and prestige-free interaction with his students, considering the existing situation in Germany. He carried on impromptu, lively, and sometimes very long discussions which lasted far into the night, even in his apartment."[18]

Recollections of Köhler's and Wertheimer's different teaching styles reinforce this mixed picture. Metzger's memory of a course called "Psychological Exercises for Beginners" in winter, 1922, provides a revealing thumbnail sketch:

> One [Köhler] was younger, blond, slim, tall, straight, and with a cutting sharpness of thinking; the other [Wertheimer] somewhat smaller and older, with long hair and deep, kind eyes, with an astounding boldness and wealth of theoretical and experimental ideas. He always made new objections to his own suggestions and required us, his listeners, to consider again and again what one could do to examine these objections and support or refute them. Again and again he also proposed new experimental variations and encouraged us to make predictions about the results.[19]

Students apparently got from such teachers a feeling for science as an active process of discovery in which they, too, could participate, and not a collection of cut and dried procedures to be learned by rote.

Yet another dimension of continuity needs to be stressed. Supervision of doctoral research was close, befitting the reproductive dynamics of a scientific school. Work on the final write-up of the dissertation could be so intensive that the product was essentially cooperative. As one former student recalls:

> Work with Wertheimer was hard. I do not think he ever proposed or accepted a subject for a thesis unless he felt that it led into the very front line of scientific attack, and then when [his own] or someone else's student sent him a paper . . . he would return it, the margins . . . covered with his neat, microscopic shorthand notes which were so beautiful to look at, but which meant so many blows against . . . a shock-proof argument or a crucial experiment. . . ."[20]

The Berlin institute attracted numerous students and scholars from abroad throughout the heyday of Gestalt theory. Visitors included Kanae Sakuma from Japan and G. Usnadze as well as Alexander Luria from the Soviet Union, but exchange with the United States was particularly intensive. It began when Koffka came to Cornell University as Jacob Schiff Visiting Professor of Education in 1924–1925, at the invitation of Robert M. Ogden, who had been a subject in the experiments he had done in Würzburg in 1909. His visit overlapped for several months with Köhler's stay as visiting professor at Clark University in the spring and summer of 1925. Shortly thereafter, Köhler's student Karl Duncker went to Clark University with the support of the Abraham Lincoln Foundation, and Wolfgang Metzger went to the University of Iowa to work with Carl Seashore on audition. These visits were soon followed by the arrival in Berlin of American students funded by the National Research Council and other sources, including J. F. Brown, Donald K. Adams, Karl Zener, Donald McKinnon, and Jerome Frank. Berlin was not the only psychological institute in Germany with international visitors, but it was clearly most impressive for their number and the length of their stays.[21]

Giessen

The situations of the other two founders of Gestalt theory in the 1920s were not so prestigious as that of Köhler in Berlin. But each offered opportunities, and tribulations, of its own. Kurt Koffka had already earned the right to teach at the University of Giessen in 1911, with the aid of former Würzburg school member August Messer. Messer had succeeded Karl Groos as professor of philosophy in 1909, with the support of psychiatry professor and experimenting psychologist Robert Sommer. In the fall of 1919, after Koffka returned from military research in Berlin and in the navy, he and Messer requested 5,000 Marks to establish a psychological institute; the Philosophical Faculty approved the expenditure in November. The funds came from the Society of Friends and Supporters of the university and the budget of the Philosophical Seminar. In 1921, Koffka became Department Head for experimental psychology and pedagogy, with the title and salary of a "non-budgetary" associate professor. In that year he published a book applying Gestalt theory to psychological development, a topic with obvious relevance for pedagogy (discussed in Chapter 14). Conditions for research on the spot appeared good, but university records say nothing about an annual budget for the institute, aside from the salaries of Koffka and his assistant, Hans Georg Hartgenbusch.[22]

Even under these conditions Koffka conducted considerable research. The experiments for the first monographs in the series he edited, "Contributions to the Psychology of Gestalt and Motion Experiences," begun in 1913, were done in Sommer's laboratory. Eleven of the fifteen series items published between 1919 and 1926 came from the new institute. At least as significant was the change in Koffka's teaching after the institute was founded and he became an associate professor. Between 1912 and 1921, he and Messer accounted for 31 percent of courses offered in the "Philosophical Disciplines" (Philosophy, Psychology and Pedagogy). Half of these were psychology courses, but Koffka's offerings also included "Introduction to Philosophy," "Introduction to the Theory of Knowledge," "The Philosophy of Pragmatism," "Introduction to Aesthetics," and exercises on Berkeley, Bergson, Mach, and Avenarius. After 1921, only three of Koffka's courses could be called philosophical. In the summer of 1922, the Philosophical Faculty was divided into natural science and humanities sections, and the experimental psychology courses appeared under both headings. From fall

1925 on, they appeared under natural sciences only. Thus, Koffka, in contrast with his colleagues in Berlin but in harmony with his own temperament, utilized his position in Giessen to assure greater autonomy for psychology as a natural science.[23]

Through this teaching Koffka gathered around him a small but loyal and hard-working group of students. A former student later recalled him as an "extraordinarily intense" teacher intent on demonstrating the value of experimental methods and formulating topics for doctoral theses. "In the seminar discussions," however, "he was liable to press one hard; he was no forbearing examiner." Koffka compensated for this stricter side with his interest in art and travel, and especially with his hospitality. He enjoyed literary and musical evenings with students at his home, at which he and his students read Christian Morgenstern's poetry together and discussed Ibsen, Stefan George, and other writers.[24]

Still, Koffka was not entirely satisfied. Already in 1922, at Robert Ogden's invitation, he had published a full account of the Gestalt view on perception in the American journal, *Psychological Bulletin;* Ogden's translation of his book on development appeared in 1925. In that year, Koffka, whose English was excellent, wrote to Ogden about the insufficient support for his laboratory in Giessen and mentioned the possibility of taking a position in America. In response to an alleged offer from Cornell, the Hessian Ministry of Education applied in 1925 to have his position upgraded to a permanent associate professorship; but the responsible official doubted that the Finance Ministry would approve.[25] Koffka returned to Giessen for only one year, then accepted a second visiting professorship in America, at the University of Wisconsin, in 1926–1927. While there he received offers from that university and from Smith College. The latter offer was arranged by Seth Wakeman, a professor of education at Smith and a student of Ogden's. Koffka chose the women's school partly because the cultural atmosphere of a small New England college town attracted him, but mainly because of the unprecedented research opportunity offered by the William Allan Nielson professorship. This included five years free of teaching duties and a new laboratory built to his specifications, with two assistants of his choice and an operating budget of $6,000. The proferred salary of $9,000 made him, according to his own quite accurate estimate, "one of the highest paid professors in America." He thus emigrated to the United States in 1927 mainly for professional reasons, long before such a step became politically necessary.[26]

Koffka's position in Giessen was offered to Max Wertheimer. In its search for a successor, the Giessen faculty solicited recommendations from the most prominent psychologists in Germany and some physiologists. Though opinions varied widely, the support for Wertheimer was impressive – ten of the thirteen professors surveyed ranked him first. David Katz, for example, called Wertheimer "one of the researchers who are determining the development of modern psychology." William Stern of Hamburg wrote that by appointing him "your university would compensate for an injustice done him by the German universities." Even Erich Jaensch of Marburg, a well-known critic of Gestalt theory, opined that, although Wertheimer was not the sort of all-round talent one might wish, "as a mind" he clearly outranked all other candidates.[27]

Revealing in their own ways were the testimonials by Koffka and Köhler. As Koffka acknowledged, "I myself owe him the best in my development." After voicing equally high praise, Köhler's letter addressed the personal and academic-political issues with characteristic frankness:

> [Earlier] . . . Wertheimer placed no value in a university position, since he wanted to be independent and had independent means. His property has disappeared in the inflation, he has since married, has two children, and no proper income besides a stipend of not even 2,000 Marks. When he becomes ill, the income from lecture fees is gone. . . . Of course his race [sic!] is involved. But I should think that every German who is proud of Lessing and "Nathan the Wise" should disregard this viewpoint in cases of outstanding giftedness. We count Einstein and James Franck as Nobel prize-winners, but we do everything to make it difficult for outstanding Jewish colleagues.[28]

Unfortunately, the Faculty's choice had been based on the premise that the position would be a permanent associate professorship, with good prospects of appointment to a "personal" (nonbudgetary) full professorship for a suitable candidate. In February 1928, the Ministry informed the Faculty that the permanent position had been stricken from the budget for the current year. In June the Ministry offered Wertheimer the position at Koffka's rank, with a base salary higher than Koffka's had been (8,600 Marks versus 7,400 Marks), and a guarantee that the state would make up any difference between his total income, including lecture and examination fees from students, and 12,000 Marks – a reasonable amount at the time, even for a full professor. But the offer came with only a promise, and no guarantee, that the position would be upgraded the next year. Wertheimer refused late in July.[29]

Frankfurt

By then, better things were at last on the horizon for Wertheimer. As early as 1918, when he had applied to have Wertheimer's teaching rights transferred from Frankfurt to Berlin, Carl Stumpf had had only the highest praise for his research and pedagogical skill both in philosophy and psychology, and had added, "I would consider him ripe for a professorship of philosophy." He received the title of nontenured associate professsor in 1922, but this was the result of a reform measure introduced in Prussia for instructors six years past their habilitation, and did not include a salary. In that year, Albert Einstein wrote on behalf of his friend at least twice, at Köhler's request, to recommend him for positions in Kiel and Cologne. The Kiel position was available because of Moritz Schlick's departure for Vienna, where his colloquium later became a meeting point of the so-called Vienna school of logical empiricism. As Einstein acknowledged, Wertheimer may be less suitable "from the point of view of the theory of knowledge" than Schlick's candidate, Hans Reichenbach. Still, he "is not an adherent of ossified word philosophy (Kant Society) but a lively human being, who thinks and experiences for himself, and in this sense is also able to have a liberating effect on young people."[30]

Apparently Wertheimer was not considered for these positions. Instead, he received short-term teaching stipends in 1924 and 1926. Yet even after he gained experience in administration by substituting for Köhler as professor and institute director during the latter's trip to America in academic year 1925–1926, he still had no offer of a full, or even a permanent associate professorship.[31] Perhaps this was due in part to anti-Semitism, as Köhler wrote to Giessen. Yet Jewish psychologists such as William Stern and David Katz had already advanced to full professor at universities in Hamburg and Rostock, respectively. Wertheimer's short list of publications and his failure to produce any major work in systematic philosophy would have counted against him in any case.

The situation changed after Kurt Riezler became provost (*Kurator*) of Frankfurt University in 1927. Riezler had been secretary to wartime Chancellor Bethmann-Hollweg, served in the German embassy in Moscow toward the end of the monarchy, and directed the office of Reich President Friedrich Ebert during the German revolution. In the 1920s, he turned away from politics to philosophical and literary reflections on free will and the "crisis of reality" in modern physics (see Chapter 17). Prussian Education Minister Becker named him to the Frankfurt position in order to create a new kind of university. Riezler planned to make Frankfurt a center for interdisciplinary discussion and research, "a kind of Princeton in Germany," as a friend and colleague later wrote.[32] To that end he aggressively pursued specific appointments. When philosopher Hans Cornelius retired in 1928, Riezler tried to have Wertheimer appointed in his place. The wrangle that ensued showed how difficult it could be to sort out relations among the so-called philosophical disciplines – general philosophy, psychology, and pedagogy – at smaller German universities. The Philosophical Faculty cooperated at first, and even listed Wertheimer on an equal footing alongside Martin Heidegger and Karl Jaspers, because "his investigations [on Gestalt theory] are of fundamental importance for the clarification of epistemological issues." The Ministry, however, replied that Wertheimer appeared better qualified for the chair of "philosophy, especially psychology" in the natural sciences faculty, from which Friedrich Schumann was about to retire.[33]

The Natural Science Faculty had been consulted throughout these discussions, but when they recommended candidates for Schumann's chair in July, they proposed not Wertheimer, but their colleague Adhemar Gelb and Erich Jaensch of Marburg together in first place, followed by David Katz. In doing so they claimed to use the very criteria that Wundt had enunciated fifteen years earlier. Each of the three men, but especially Gelb, they wrote, was "psychologically and philosophically a fully educated person, capable of thinking biologically and [with] thorough training in the design and execution of exact experiments." Wertheimer clearly also met these conditions, and the faculty was "well aware of his high significance for the recent development of psychology" but believed that he would be "better employed in a chair of general philosophy." Told by Riezler that the Ministry would not appoint him to the philosophy chair, the faculty expressed "doubt whether, in spite of all his significance for experimental research, he is a suitable head of a large institute; neither his interests nor his capabilities seem to lie in this direction."[34] As Gelb made clear to Wertheimer, however, such personal considerations were secondary. In fact, the issues that had created tension between philosophers and psychologists since before the turn of the century were still very much alive: "The Ministry was afraid that by giving philosophy to you and psychology to me Frankfurt would be without 'proper' philosophy, a fear surely nourished by Jaspers, but also by others."[35]

In the resulting, quite unusual compromise, Wertheimer came to Frankfurt with positions in both faculties. He received the natural sciences chair, becoming codirector with Gelb of the Psychological Institute. He also became codirector of the Philosophical Seminar with the new appointee in the philosophical faculty, philosopher of religion and social theorist Paul Tillich, who was evidently proposed by Minister Becker himself. As part of this agreement, the Psychological Institute's budget rose 40 percent, from 1,800 to 2,500 Marks, and the Ministry provided an additional 1,000 Marks for books.[36]

With that a galaxy of academic stars had assembled in Frankfurt. In addition to bringing Tillich, Wertheimer, and sociologist of knowledge Karl Mannheim to the city, Riezler also supported the association with the university of the privately financed Institute for Social Research, including an endowed professorship for its young director, critical social theorist Max Horkheimer. The Psychological Institute shared library space with the Philosophical Seminar, which made interaction an ordinary occurrence. Students of the time vividly remember interdisciplinary faculty colloquia, affectionately called the "wisdom" or "truth seminar" (*Weisheits-* or *Wahrheitsseminar*) on such topics as the concept of truth in various disciplines, which often continued late into the night at one of the nearby cafes. In addition to these gatherings and to frequent concourse with Riezler, Wertheimer was friendly with Horkheimer and enjoyed good, though more distant, collegial relations with Mannheim.[37]

With its improved budget and personnel situation, the psychological institute soon became a productive research center visited by scientists and students from abroad, such as Edwin Newman and Willis D. Ellis from the United States. In 1931, Gelb departed for a long-awaited full professorship in Halle, with support behind the scenes from Wertheimer and Köhler. Gelb and Wolfgang Metzger, who came from Berlin to take over one of the assistantships, offered regular laboratory courses. As one student recalled, study in natural science was recommended to all students; and Wertheimer's teaching included both philosophical and psychological topics, as it had in Berlin.[38]

Student–professor interaction in Frankfurt also followed the Berlin model. Once admitted to candidacy, students generally had considerable freedom: "Often one believed he [Wertheimer] had left you alone. But in reality he was waiting for you to be productive and show ideas," one student remembers. Once this happened, the real work began:

> Shortly before the Christmas vacation in 1932 . . . Wertheimer called me with my papers [the first chapter of the dissertation] to his apartment up in the attic room which was his study, far away from the noises of the apartment. He went through my draft word for word – at least that is how I remember it – and I wrote, and rewrote and showed [it to him] again and so forth.[39]

Only four dissertations were completed during Wertheimer's brief tenure. He had had serious health problems – digestive difficulties complicated by heart trouble – since the early 1920s. In February 1932, he suffered a heart attack, and the following semester he took leave for treatment. He may also have seen other problems on the horizon. Metzger later recalled being advised on his arrival to complete his habilitation quickly, if he wished to do so under Wertheimer's sponsorship: "Certainly you have not forgotten that I am a Jew. The Prussian state elections are next March, and you do not know whether I, Wertheimer, can still sponsor you after those elections." In a 1930 letter to Köhler, however, Wertheimer named Metzger as first choice for the assistant position and mentioned the possibility of rapid habilitation, but without political overtones.[40]

Psychologische Forschung

The journal in which the members of all three institutes published most of their work was founded on May 15, 1921, in a contract between the editors – Kurt Koffka, Max Wertheimer, Wolfgang Köhler, Kurt Goldstein, and Hans Gruhle – and the Julius Springer-Verlag, a well-known Berlin publisher in the field of natural science.[41] Goldstein was professor of neurology in Frankfurt and coeditor with Adhemar Gelb of a series of studies in psychopathology that presented results closely related to Gestalt theory (see Chapter 16). Gruhle, a psychiatrist and philosopher in Heidelberg, may have been included to represent psychiatry and thus complete the range of the journal's planned coverage. All of the editors except Wertheimer had had contacts with the publisher in the past. Koffka, who became editor in chief, had been a contributor to a Springer journal, *Die Natur-wissenschaften,* since 1914. Köhler was in the process of bringing out the book version of his anthropoid research with the same publisher.[42] After Koffka's departure for the United States in 1927, Wertheimer became editor in chief. In 1929, when he was named codirector with Wertheimer of the Frankfurt Psychological Institute, Adhemar Gelb, who had already been working with Goldstein for more than a decade, joined the board.

In their opening statement the editors described the journal's intended scope briefly and clearly: "This journal will serve psychology in all of its breadth, including the relations it has, or should have, with other sciences." Apparently *Psychologische Forschung* was not originally intended to be the organ of Gestalt theory, for the editors wrote that, "The acceptance of articles will depend on achievement, not on allegiance to a school."[43] They clearly tried at first to fulfill this aim. In addition to a programmatic essay by Wertheimer, the first volume included contributions by Berlin ethnologist Dietrich Westermann, Prague neurologist Arnold Pick, and comparative psychologist and ethologist Theodor Schjelderupp-Ebbe, in addition to work from psychological institutes outside Berlin. Later volumes included dissertations from Hamburg, Heidelberg, and Göttingen. In addition, articles appeared by internationally known psychologists and philosophers such as David Katz, J. B. S. Haldane, Hans Reichenbach, and Moritz Schlick, all of whom were connected in some way with one or more of the editors, but were not members of the Berlin school.

The intention of covering all fields of psychology was also carried out. Contributions appeared on the psychology of thought, ethnological psychology, and psychopathology. After 1927, the dissertations of the series, "Studies in the Psychology of Action and Emotion," edited by Kurt Lewin, took up increasing space. Still, the psychology of perception, especially vision, and related psychophysical issues clearly dominated. Of 270 original research articles in the journal's first 22 volumes, 151 dealt with such topics. Over the years, research reports from Berlin, Giessen, and Frankfurt, along with writings by the Gestalt theorists themselves, took up the vast majority of the journal's pages. This had a decided advantage for students at the three institutes. The author's honorarium was the first payment most received for their work, and the publisher's reprints obviated the need to pay for printed dissertation copies to submit to the university.[44] Thus, the editors' declared intentions gradually lost force, and the self-presentation of Gestalt theory became more important than that of the discipline as a whole. A reply from Leipzig soon followed – the journal *Neue Psychologische Studien* (New Psychological Studies), founded in 1926, which was intended from the beginning to be the organ of the so-called Leipzig school of "holistic psychology" (*Ganzheitspsychologie*) (see Chapter 18).

Located in three of the only fourteen psychological institutes in German-speaking universities, including one of the most prestigious in the world, and with a journal of their own, the Gestalt theorists had the means of extending and enriching the reconstruction of psychological discourse that they had proposed the decade before. They also had decided advantages in the struggle for impact on German psychology and culture.

14

Research styles and results

Sixty years ago, in a pioneering study of the social conditions of scientific knowledge, Ludwik Fleck described scientists as members of "thought collectives" (*Denkkollektive*) unified by commitments to common "styles of thinking" (*Denkstile*).[1] The Gestalt theorists and their students constituted a "thought collective" of a specific kind – a scientific school, united by the hierarchical teacher-student relationships typical in German university laboratories, and also by shared beliefs about the aims and procedures of psychological science. Some of these beliefs – in the efficacy of certain instruments as controlled stimulus-producers, and in the words of trained or naive observers as adequate data sources for the study of conscious phenomena – were shared by most experimenting psychologists in Germany at that time. The Gestalt theorists gave these instruments and observations new meanings by mobilizing them in support of new claims. Other beliefs, to be outlined below, were specific to the laboratories in Berlin, Giessen, and Frankfurt.

When the Gestalt theorists and their students acted on this complex framework of shared beliefs and practices, the result was the social construction of some impressive scientific facts. The phenomena generated – or, more precisely, allowed to happen – in their laboratories made the culturally resonant words "Gestalt" and "whole," as well as equally resonant terms borrowed from physical science, such as "field," "system" or "frame of reference," come alive in the laboratory. In the process, the Gestalt category acquired a range of meanings that it had not had before. The terms just listed thus functioned as what Kurt Danziger has called "generative metaphors," organizing research within particular laboratories while simultaneously connecting them with wider scientific and cultural discursive fields.[2]

219

Methodological commitments

What held the Berlin school together in the first instance was a deep commitment to Max Wertheimer's Gestalt epistemology, and to the implications for research practice said to flow from it. Wertheimer first stated these principles publically in a manifesto-like article that appeared in volume 1 of *Psychologische Forschung* in 1921. There he called for descriptions of conscious experience in terms of the units people naturally perceive, rather than the artificial ones assumed to be in agreement with proper scientific method. Implicit in conventional psychological descriptions, he alleged, is what he called a "mosaic" or "bundle-hypothesis," the assumption that conscious experience is composed of units analogous to physical point-masses or chemical elements. By making this assumption, psychologists constrain themselves to link contents of consciousness piecemeal, building up so-called "higher" entities "from below" with the help of chance associative connections, habits, hypothesized "functions" and "acts," or a presupposed unity of consciousness. In fact, however, such "and-sums," as Wertheimer delightfully called them, appear *"only seldom, only under certain characteristic conditions, only in very narrow limits and perhaps even only in approximation."* Rather, *"the given is in itself 'formed' ('gestaltet'): given are more or less completely structured, more or less determinative wholes and whole-processes,"* each with its own *"inner laws."* The constitution of parts in such wholes "is a very real process that changes the given in many ways." In research, therefore, "proceeding 'from below to above' would not be adequate, but rather the way 'from above to below' is often required."[3]

Few psychologists held such extreme atomistic or associationistic views any longer, but Wertheimer emphasized that the issue was not "textbook opinions," but rather "what is actually done" in the laboratory.[4] Yet he gave only indirect indications of the procedure he opposed, or of what he proposed to put in its place. In a 1925 paper, Koffka tried to make this methodological stance more explicit. Since there is no guarantee that sensations remain unaltered when not attended to, he argued, sound method requires giving up not introspection, but the analytical attitude that had dominated research using introspection. That attitude treats conscious contents as though they were material things that remain unchanged regardless of the situation in which they appear. "But, whatever the mind is, it is not a mosaic of solid unalterable things." It was therefore more appropriate to try to record and explain phenomena as they appear to both trained and naive observers using ordinary language. This meant, for example, that a line seen by itself and as the side of a figure are, psychologically speaking, "two different things." Koffka acknowledged that this could make psychological observations seem even less reliable than they already did, but asserted that this was the only way to account for the apparent objectivity of natural perception, and thus to end scientific psychology's "remoteness from life."[5]

It is important to emphasize that this "naive phenomenology," as Köhler later called it, was strictly a methodological convention. The Gestalt theorists were not trying to revive the epistemological doctrine called naive realism, which holds that we see the world as it "really" is. Köhler and Koffka made it abundantly clear that the world that appears to us is not a literal copy of the external world, and that there are events "beyond the phenomena," that is, in the brain, which are responsible for the structured appearance that the world has in consciousness, as well as the specific forms that learning and behavior take. For all of the Gestalt theorists, however, the primary task was to explore psychological reality as it appears, and to determine the conditions of its appearance as well as the principles of its organization.

However, despite Koffka's fond hope of ending scientific psychology's "remoteness from life," the Gestalt theorists did not substitute field observations or clinical studies for laboratory experimentation. Rather, they advocated specific modifications in then-standard laboratory procedures. Instead of being asked to fixate on a single point or part of a stimulus array for minutes or even hours at a time, observers were allowed literally to open their eyes wider, to take in more of the field. The purpose was to enable observers to perceive the relatedness among parts or dimensions of stimulus fields, rather than creating situations in which only atomistic sensations were possible. Just as Köhler had criticized Edward Thorndike for designing experiments in such a way as to guarantee animal stupidity, and had designed situations in which apes could exhibit "insight" in problem-solving, he and his colleagues tried to create situations in which Gestalt phenomena, or the relational character of perception itself, could present themselves in experience, rather than being made impossible by the rules of "proper" method.

An event that had occurred years earlier in the Frankfurt institute indicates more precisely what was intended. Among the investigations Friedrich Schumann reported in a 1912 paper on visual perception was one by Wilhelm Fuchs on the long-standing issue of whether two colors could be seen simultaneously, one behind the other, an important problem in depth perception. At first Fuchs obtained no positive result. As Schumann later reported, however, the situation changed in 1913, when Wertheimer and another "strongly visual" observer — probably either Köhler or Adhemar Gelb — came in. Both got positive results, but complained that observation was strenuous because only segments of the colored figures had been presented. When the figures were presented as wholes, the phenomenon appeared immediately, "constantly closed and without any discontinuity." Schumann himself failed to make such an observation. Astoundingly, he admitted that "many other people have a much more pronounced sensory life than I do," due perhaps to his "strongly critical attitude and the effort to obtain greatest exactitude." In this case, he had to acknowledge that "a critical attitude toward various subjective phenomena is not conducive to the perception of a larger complex as a unified whole."[6]

Described in this way, Berlin school research style seems like a radical departure from the ideals of repeatability and impersonal objectivity central to modern science. The discrepancy disappears when one realizes that the Gestalt theorists located objectivity not in any impersonal procedure or in the use of any particular apparatus, but *in the phenomena themselves*. Sought were phenomena, or invariant relations among phenomena, that were so impressive that they automatically acquired an aura of objectivity, despite being artifacts created under laboratory conditions and then referred back to nature. The Gestalt theorists searched for invariant principles of order and meaning believed to inhere in phenomena as experienced under particular stimulus conditions, not correlational or other contingent functional relationships between independent and dependent variables. With this emphasis on inherent order and meaning (*Sinn*), Gestalt experimentation expressed fundamental opposition to technological conceptions of mind.

Sometimes Wertheimer emphasized this opposition when he spoke of "good" or "pure" phenomena, those which most clearly and impressively exemplify the structure of experience in a given situation. As one student of Wertheimer's in Frankfurt put it:

> A Gestalt theoretical experiment was geared up so that it would work in 100 percent of the cases, and if it did not work, well throw it out the window. If only eight out of ten subjects would do it – I don't think Wertheimer used more than ten subjects, and he certainly told us to use ten subjects – and it doesn't work in at least nine, forget about it. That was the methodology.[7]

The proportions are a bit exaggerated, and not all Berlin school studies used so few subjects, but this statement nonetheless captures the heart of the matter.

Despite Wertheimer's highly charged injunction to proceed "from above," rather than "from below," or the Gestalt theorists' loving references to Goethe, there was nothing mystical about this. The goal, rather, was to achieve something like the crucial or decisive demonstrations that had also been sought by the founders of modern science, experimenters like Galileo and Newton – empirical exemplifications of invariant, universal laws of nature. For them, as for their illustrious predecessors, "decisive" demonstrations were dramatic devices intended to achieve a persuasive, didactic effect – to convey the inherent orderliness of nature in immediate, phenomenal form.[8]

Though the Gestalt theorists were committed to starting from "naive" phenomenological description, they did not pursue only qualitative and eschew quantitative research. Berlin school dissertations frequently presented quantitative results, usually in tabular form, sometimes including statistical curves. But the tables and curves almost always referred to the variance among perceptions *within,* not among, individual subjects, or to trends for all subjects taken together and averaged. Not the presence or absence, but the kind of quantification was the issue. The Gestalt theorists' goal was not to discover covering laws that would account for the variance in different subjects' responses under given conditions. Kurt Danziger has found that *Psychologische Forschung* had the lowest proportion of studies with data referring to group rather than to individual performance of the major German psychology journals for the years 1920 to 1935.[9] Gestalt experimentation thus opposed the administrative, classifying style of knowing expressed in the search for covering laws and individual differences and embodied in intelligence and personality testing – a style of knowing that was already becoming increasingly characteristic of psychology in America at the time.

However, such aims and procedures were quite compatible with the social organization of psychological research that still prevailed in Germany in the 1920s. Despite the obvious theoretical differences between Gestalt theory and Wundt's psychology, Berlin school research conformed in certain ways to what Danziger has called the "Leipzig model" of experimentation (discussed in Chapter 1). In Berlin, as in Leipzig, small numbers of investigators shared a common research agenda, usually that of the institute director, and often served as subjects for one another. This type of experimenter–subject relationship was well suited to the social status of Berlin school and other experimenters as members of an educated elite, and to the task experimenting psychologists in Germany had set themselves – generating philosophically relevant information about cognition. In this respect, continuity outweighed change.[10]

Even so, the Gestalt theorists' research styles had certain features that made their approach stand out. Three kinds of hypothesis testing were involved. Predominant were carefully constructed attempts to allow "good" phenomena to happen; but even these demonstrations were presented as confirmations of Wertheimer's claim that "the given" is *gestaltet*. They thus became challenges to the theory-laden descriptions offered by other schools, as well as to common explanations of perception, thinking or behavior in terms of past experience, peripheral processes, or distributions of attention. A second mode of hypothesis testing came closer to the normal science activities that go by that name. This employed standard experimental techniques as well as quantitative inferences from parametric or threshold measurements as evidence for the existence of "invariant" relations within and across sense modalities. Finally, in work on successive comparison, learning, and memory, Köhler and his students developed and tried to test hypothetical models of the brain events presumed to underlie or cause specific perceptual and cognitive phenomena and relationships. They thus attempted to realize in their own way the hope held out by Hering, Mach, and G. E. Müller decades earlier – that psychological facts could become guides to the brain events or processes presumed to underlie them.

Thus, the members of the Berlin school did not limit themselves to a single research procedure, and certainly did not confine themselves to phenomenological description. Rather, they worked through the rich possibilities of several exemplars of the sort discussed by Thomas Kuhn – complex models of theory-laden procedure, often developed initially before 1920. In addition, each of these research styles was enacted in several problem areas. Indeed, precisely the ease with which Gestalt terminology and research styles could be extended to so many different kinds of phenomena lent plausibility to the Berlin school's claims to general validity. Moreover, both the methods and the issues at stake developed over time, thus exemplifying what Andy Pickering has called the "temporality" of scientific practice.[11] Space permits only brief discussion of research on selected topics that best illuminate the way these exemplars operated, and the ways in which research results and theory fed back on one another over time.[12]

Perceptual organization

Wertheimer offered evocative examples of what he meant by working "from above" instead of "from below" in 1923, when he presented a full account of the "Gestalt laws" or tendencies that he had announced in 1914. Alluding to the multiple brightnesses Oswald Külpe had professed to count in his textbook of 1893 (see Chapter 4), Wertheimer wrote: "I stand at the window and see a house, trees, sky. Now on theoretical grounds I could try to count and say: 'here there are . . . 327 brightnesses and hues.' Do I *have* 327? No, I see sky, house, trees; and no one can really have these '327' as such." Moreover, "the particular combination in which I see it [them] is not simply up to my choice." In other words, the perceptual field does not appear to us as a collection of sensations with no meaningful connection to one another, but is organized in a particular way, with "a spontaneous, natural, normally expected combination and segregation" of objects. Wertheimer's paper was an attempt to elucidate the fundamental principles of that organization.[13]

Most general was the law of *Prägnanz*. This states, in its broadest form, that the perceptual field and objects within it take on the simplest and most impressive structure permitted by the given conditions. More specific were the laws of proximity, similarity, closure, and good continuation (see Figure 11). The law of proximity states that a row of dots with equal space between them, for example, will be seen as a series of pairs, while the same number of dots with only slight changes in the spacing will be seen as a single group of three (Figure 11a). The law of similarity states that if an array of equally spaced black and white dots is shown in alternating vertical and horizontal rows of the same color, observers will see verticals or horizontals, as decided by the similar grouping; "to see the opposite organization clearly and simultaneously in the entire figure is usually impossible" (Figure 11b). The tendency to closure is clearest with adjoining figures. If a circle and a square touch at only one point, for example, they will be seen as two distinct figures; but if they overlap, or if one encloses the other, the result may be different. Finally, in images such as the last in Figure 11c, Wertheimer

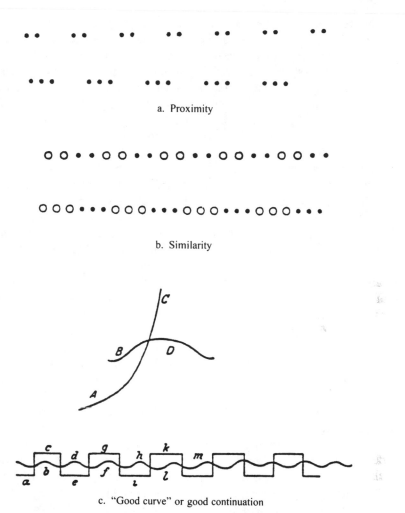

a. Proximity

b. Similarity

c. "Good curve" or good continuation

Figure 11. Gestalt "laws" – organizing tendencies in vision. Max Wertheimer, "Untersuchungen zur Lehre von der Gestalt," *Psychologische Forschung,* 4 (1923), pp. 304–305, 308–309, 322, 323, 325. Reprinted by permission of Springer-Verlag.

claimed that observers typically see the continuous lines AC and BD, and resist the changes of direction necessary to see the organization AB, CD.

G. E. Müller had mentioned the existence of such tendencies in perception as early as 1904, but had said only that they make the perception of stimulus patterns easier (see Chapter 6). Wertheimer maintained that they are determinative for figure, and by extension for form perception in general. He recognized the powerful effect of observers' attitudes and mental set, but by this he understood primarily a tendency to continue seeing the pattern initially seen, even under changed conditions. Nor did he deny the influence of previous experience, such as habit or drill, but he insisted that these factors operate only in interaction with the autonomous figurative forces at work in the immediate situation. How these interactions work in different cases was an empirical issue. Those who would attribute all the tendencies just described to previous experience must "demonstrate concretely for each of these cases and for each of the factors the actual past experiences and times of drill that are involved," a job that "is nowhere as simple and smooth as the answer first seems to suggest."[14]

Wertheimer had not elaborated a finished theory, but had converted the culturally resonant term "Gestalt" and the claim that the given is *gestaltet* into a complex research program for relating the static and dynamic dimensions of perceptual organization to one another. For example, he suggested, one could determine in which cases one Gestalt tendency "defeats" another. Or one could find out whether these effects hold in the same way for both simultaneous and successive, or for stationary and motion perception. Such research could easily extend to other sense modalities besides vision; the law of proximity, for example, clearly plays a role in rhythm. Wertheimer did not exclude quantitative techniques from his program. Indeed, he insisted on them, inviting readers to demonstrate the law of proximity to themselves by varying the distance between dots or dot groups until they found the thresholds for particular groupings. But he made it clear that such measurements should be undertaken only in conjunction with detailed phenomenological description to discover what ought to or meaningfully could be measured.[15]

Much of the research in Berlin, Giessen, and later in Frankfurt amounted to unpacking the empirical and theoretical implications of this paper. The initial steps were usually disarmingly simple demonstrations. In Giessen, under Koffka's direction, Friedrich Wulf had already attempted to demonstrate the applicability of the law of *Prägnanz* to memory before Wertheimer's paper appeared. Specifically, Wulf offered evidence for the proposition that memory traces tend to organize themselves spontaneously over time in the direction of simpler forms. This challenged the view of G. E. Müller and others that either "directional images" or learned, schematic reactions are responsible for the simplifications observed. After showing a series of twenty-six figures drawn on cardboard to six observers for 5 seconds each – usually in their apartments, rather than in the laboratory – Wulf presented them again 30 seconds, 24 hours, and one week later, and then asked the observers to redraw the figures, or to complete partial figures. He found two basic modification tendencies, which he called "leveling" and "sharpening" (Figure 12). In some cases, that is, observers smoothed out distinct features, while in others they exaggerated specific features, such as pointed edges. Sometimes observers altered the figures to approximate a known structure, which Wulf called "normalization." More often, modifications were based on the structure of the

Figure 12. Leveling and sharpening in figures drawn from memory. "V" = presented figure; "w" = reproduction. Friedrich Wulf, "Über die Veränderung von Gestalten (Gedächtnis und Gestalt)," *Psychologische Forschung,* 1 (1921), p. 337. Reprinted by permission of Springer-Verlag.

figure itself. Though he noted the basic difference between "comprehensive" and "isolating" cognitive styles that Vittorio Benussi had already distinguished fifteen years earlier (see Chapter 6), Wulf found that leveling and sharpening occurred in both types of observers. Primary in all cases was the tendency to "better" figures; cognitive styles specified only how that tendency was expressed.[16]

Research on the role of "Gestalt laws" in color and brightness contrast also began with simple demonstrations, but soon proceeded to more sophisticated, apparatus-driven designs. The aim was to show that brightness, color, and form are essentially interrelated, not separable dimensions of experience, because color and brightness contrast play central roles in the constitution of the objects we perceive. In Berlin, Wilhelm Benary employed an experiment devised by Wertheimer to test the law of *Prägnanz* on a phenomenon of brightness contrast. He exhibited a gray triangle in different locations with respect to a black surface, as shown (Figure 13). What Benary called the "belongingness" (*Zugehörigkeit*) of the triangle to the larger figure affected observers' judgments of relative brightness in the two cases. The triangle located inside the larger, black triangle looked brighter than the one outside the cross – exactly the reverse of what would be predicted by a "summative" theory, Benary claimed. He saw this as proof that contrast phenomena are instances of the law of *Prägnanz,* because not local contrast effects, but the function of the gray patch in relation to the organization of the scene as a whole, is responsible for the differing amounts of contrast produced by the different configurations.[17]

Such claims were hardly likely to stand without evidence based on carefully controlled measurement, and Koffka and his students attempted to provide it. In 1923, for example, Koffka himself offered experimental proof that achromatic (black–white) color contrast depends not on the absolute amounts of available

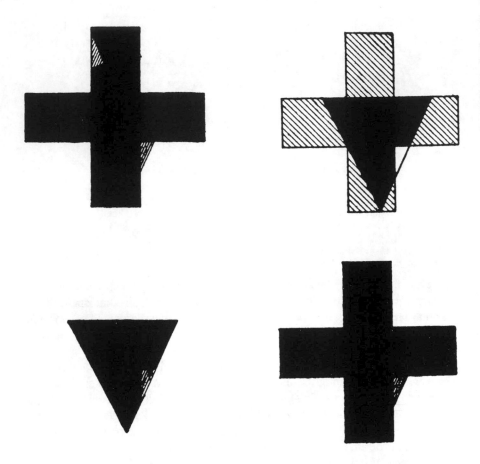

Figure 13. Figural "belongingness" and brightness contrast. Wilhelm Benary, "Beo-bachtungen zu einem Experiment über Helligkeitskontrast," *Psychologische Forschung,* 5 (1924), p. 131. Reprinted by permission of Springer-Verlag.

light but on what he called "stimulus gradients." Decades earlier, Ewald Hering had accounted for Mach bands – subjective gray stripes seen between alternating black and white rings exhibited on a rotating disk (see Chapter 4) – with a contrast theory which stated that, due to retinal interaction, the brightness along the darker side of the boundary between two rings appears greater, while the brightness along the lighter side appears less. Thus, Hering's theory also invoked stimulus gradients, but only for interactions among single pairs of bands. Koffka altered the setup by placing larger, differently shaped patches rather than equally sized strips on the disk, and by carefully calculating the brightness levels so that they no longer varied continuously in a single direction. Experimenting on himself, he was able to eliminate Hering's "boundary-contrast" effects completely. He claimed that the phenomena he obtained cannot be explained by adding up series of point-to-point retinal interactions alone, but had to take account of the "homogeneous" or "inhomogeneous" organization of the entire stimulus array.[18]

In a 1927 Berlin dissertation, Susanne Liebmann pursued this line of investigation further by relating chromatic color to principles of organization, specifically to the figure–ground phenomenon originally studied by Edgar Rubin (see Chapter 11). In 1923, Adhemar Gelb and Ragnar Granit had demonstrated that thresholds for seeing a given color were lower when it was regarded as figure than when it was seen as background. The theoretical significance of this for Gestalt theory was clear – Köhler had already argued in *Die physischen Gestalten* that the physiological counterparts of just such threshold differences are responsible for the way that things appear to us in the world (Chapter 11). Liebmann conducted similar tests controlling for both color and brightness. When she had observers adjust the illumination of colored figures toward the illumination of a neutral ground, she found that the figures began to lose sharpness and definition, until they became only vague and oscillating blotches. At the time, Liebmann argued that such phenomena refuted explanations of optical illusions in terms of visual acuity or other peripheral factors alone. Koffka later claimed that the "Liebmann effect" showed that color alone was not as strong a "field force" as illumination – today called luminance.[19]

Perhaps the most spectacular demonstration of the fundamental role of organization in perception came from research that Wolfgang Metzger published in 1930, in which the Berlin institute's architecture played a leading role. In one of the institute's high-ceilinged rooms, a projector with a specially designed set of lenses cast light at an 80° angle from a high tower toward the middle of a wall surface four meters square (see Photo 11). Screens extending from the wall were lit so that their edges would not be seen, and another screen reached from the top of the wall above and past the subject. As the illumination gradually weakened, observers saw first a somewhat concave surface, then a fog which, after some time, became a foggy mass approaching them through space. This was an "extremely uncomfortable experience. . . . The room literally sucks itself around the observer; a kind of disappearance [*Schwund*] of the environment is experienced, not an indifferent reformation. The eyes automatically seek something solid, a resting point that can stop the disappearance." As the brightness increased again, the field became at first lighter – with a sense of relief – then rapidly brighter, receding from the observer. Metzger called this special environment a "homogeneous *Ganzfeld*." In it, he argued, color and illumination were not so clearly separated as in ordinary, thing-filled space. Against the attribution of such phenomena to reduced stimulus intensity alone, Metzger noted that only the illumination – today called the luminance – of the field, not its intensity level, had changed. Thus the homogenous organization of the surface itself was responsible.

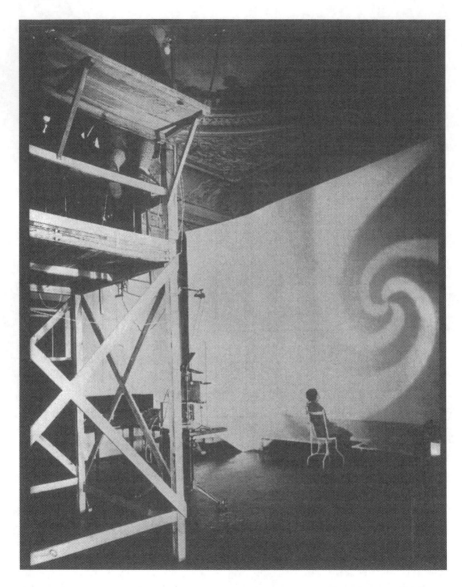

Photo 11. Experiment in progress on Berlin *Ganzfeld* setup. Ullstein Bilderdienst Berlin.

Psychologists today would say that Metzger had shown that a uniformly illuminated field gives no information; the eye needs change in light to function.[20]

Another important claim in Wertheimer's 1923 paper was that habit and drill are secondary to organization. In a 1926 Berlin dissertation, Kurt Gottschaldt tested this claim. He presented five figures individually for one second each to two groups of observers. Both groups were told to memorize the figures so they could draw them later. The three observers in one group saw them only three times, while the eight observers in the other group saw them 520 times. They were then shown 31 new drawings with the initial figures hidden in them in different ways for two seconds each, and asked to say whether anything in them attracted their attention. In both groups, more than 90 percent of responses failed to detect the initial figure. When Gottschaldt instructed observers to search for it, the success rate improved, but only to 31.2 percent of responses in group 1, and 28.3 percent of responses in group 2. Thus, he concluded, frequency of previous exposure alone had no influence on the recognition of embedded figures. Though the instruction, or its resulting mental set, clearly had an impact, this was not enough to overcome "the intrinsic properties of the stimulus object." From the point of view of Gestalt theory, the results were easily explained. The "internal unity" of the figures exposed later was such that the initial figures were "not present at all" in them, psychologically speaking.[21] This is what Koffka meant when he spoke of a line seen by itself and as the side of a figure as "two different things."

Coming closer to real-world perception, Herta Kopfermann explored the role of the Gestalt tendencies in the appearance of plane figures as three-dimensional in research done in Berlin under Wertheimer's direction, but submitted as a dissertation in Frankfurt (see Figure 14). Earlier theories attributed the thinglike appearance of some plane figures either to frequent previous experience of similar shapes as those of three-dimensional objects, or to differing directions of attention. Kopfermann offered evidence to show that in fact certain specific figural factors are primary, and that attention and previous experience are decisive only in ambiguous cases. Thus, only a slight change in the position of one line, as in Figures 14b and 14c, or in a drawing's alignment, as in Figure 14d, suffice to elicit an impression of depth. Kopfermann's work had theoretical significance because the retinal image, too, is two-dimensional; if we can spontaneously see thing-character in two-dimensional objects, then it is not necessary to conceive ordinary object vision as a matter of drawing unconscious conclusions from insufficient sense data.[22]

Motion and organization

Wertheimer had first tried to demonstrate the existence of "dynamic realities" in his 1912 paper on motion (see Chapter 8). Research on motion in Berlin specifically emphasized the linkages with Wertheimer's Gestalt principles of organization. Here again, as in the studies of color and brightness contrast, there was a progression from relatively simple demonstration experiments to more complicated, apparatus-driven designs. In his 1927 dissertation, for example, Jesuit brother Joseph Ternus reversed the claim of Paul Ferdinand Linke (discussed in Chapter 8) that identification of a moving object is necessary to experience motion. He asked instead what kinds of perceived motion are needed to experi-

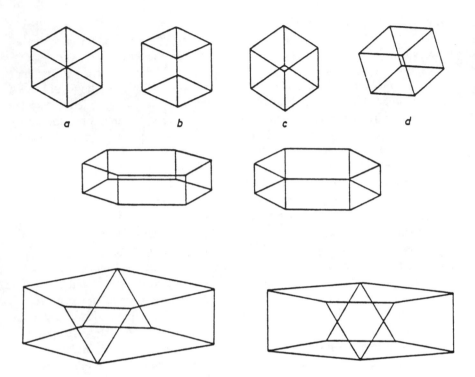

Figure 14. Spatial effects in two-dimensional drawings. Herta Kopfermann, "Psychologische Untersuchungen über die Wirkung zweidimensionaler Darstellungen körperlicher Gebilde," *Psychologische Forschung,* 13 (1930), p. 298, Fig. 1; p. 299, Fig. 7. Reprinted by permission of Springer-Verlag.

ence "phenomenal identity," that is, unified moving objects. When he presented the two dot groups combined in Figure 15a successively, for example, observers saw not individual points, but a moving cross; indeed, they had difficulty finding the common points even when told they were there. Observers reported similar results for Figures 15b–e, but the base line in Figure 15f and the center line in Figure 15g remained phenomenally stable – that is, were seen as the same. What had happened, according to Ternus, was that the common points in the initial figures had changed their function in the figure, and hence were psychologically not the same in the two expositions. Observers saw objects and they saw motion, but their perception of motion did not depend on seeing a single or the same object in motion.[23]

In a spectacular demonstration of both *Prägnanz* and depth effects in motion perception, Wolfgang Metzger used an ingenious setup of his own design, which he called a rotating light-shadow apparatus, composed of four small posts placed

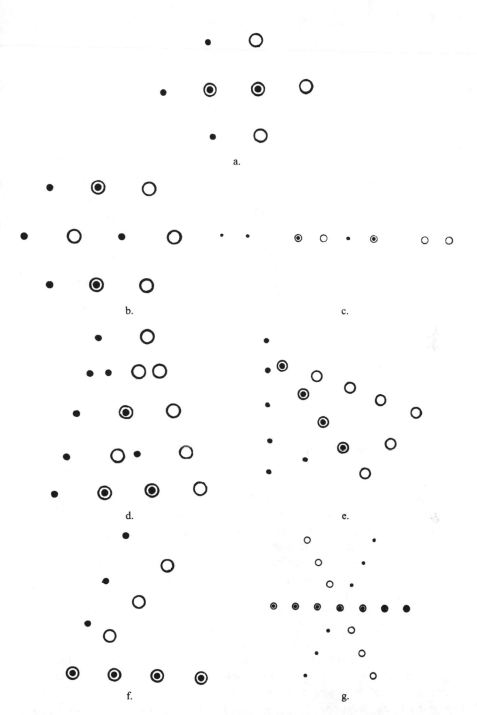

Figure 15. Phenomenal identity (Ternus 1926). Joseph Ternus, "Experimentelle Untersuchungen über phänomenale Identität," *Psychologische Forschung,* 7 (1926), p. 88. Reprinted by permission of Springer-Verlag.

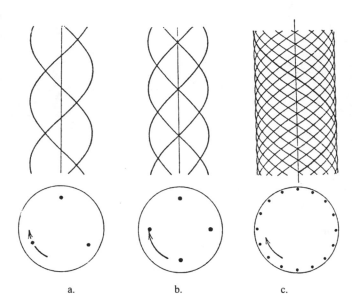

a.   b.   c.

Figure 16. *Prägnanz* in motion perception (Metzger 1934). Wolfgang Metzger, "Beobachtungen über phänomenale Identität," *Psychologische Forschung,* 19 (1934), p. 40, Figs. 46–48. Reprinted by permission of Springer-Verlag.

at varying intervals on a rotating disk, which was hidden from the observer by a translucent window. When he rotated the disk under appropriate lighting, observers reported a powerful depth effect, seeing behind the window not the disc but only four dark round posts in fixed spatial order turning on a vertical axis. Despite the large number of mathematically possible combinations, however, only the patterns in Figure 16 were actually seen. This was a dramatic demonstration of Wertheimer's *Prägnanz* principle in motion perception, but Metzger admitted he could not yet say why the tendency applied in this way.[24]

In work that came between that of Ternus and that of Metzger, Karl Duncker altered both the research modus and the terms of discourse about these issues in his research on what he called "induced motion." This paper, still regarded as a classic study, could well be regarded as an example of what philosopher of science Imre Lakatos has called a progressive problem shift – the enrichment of an established research program, in which propositions generated by specific implications from an overall theory lead to modifications or refinements in the problems considered by that theory – except that the shift was as much practice- as theory-driven.[25] In this case, the shift was to consider the perceiver not as a point at rest located outside the perceptual field, but as a dynamic field part. By "induced motion" Duncker meant what happens when we sit in a train, see another train moving, and then have the impression that the train we are sitting in, too, is moving. He tried to discover what is going on in such situations in the Berlin laboratory. After exploratory observations with light on cardboard, he proceeded to experimental studies in darkened and illuminated rooms. He then eliminated environmental anchorage by working in a darkened room and presenting a stationary point enclosed in a moving rectangle to observers sitting only 100–300 centimeters away. Under these conditions, observers who fixated the stationary point saw it move, and some even thought they were moving themselves.[26]

Mach had studied somewhat similar phenomena decades earlier, and had invoked special "motion sensations" to account for them.[27] Duncker restructured the discussion by combining some remarks from Wertheimer's 1912 paper about the role of the observer's position in motion perception and terminology from relativity theory in physics. The latter he borrowed from G. E. Müller, who had already spoken of "egocentric frames of reference" in perception. Duncker suggested that the functional relation of objects to what he called the "phenomenal frame of reference" was at issue. The question was whether one field part has "only an indirect, or mediated relation" with the surrounding environment, or "an immediate connection" with the surrounding environment, or "an immediate connection" with it.

When we watch automobile or bicycle wheels move, for example, we should see cycloids, that is, the physical resolution of the rotary motion of the wheels and the forward motion of the vehicle. Instead, we can see both kinds of motion at once, though we often also have the impression that the wheel is rolling backward. The reason, he argued, is that the wheel's periphery is part of one reference system with the hub as reference point, while the hub and wheel together form part of another system with the surrounding terrain, or the vehicle and the observer as reference points. The fact that the observer normally plays the part of a third object with the status of reference point explains why resting objects are seen as such even if swept by our moving gaze. The observer's point of regard moves among and is surrounded by the objects of vision; "its localisation is acquired from these objects which thus act as its frame of reference."[28] In the case of moving trains, or in darkened laboratories, the absence of such anchorage explains the resulting disorientation.

In this topic, as in the others discussed thus far, Duncker's phenomenological observations soon led to further work employing quantitative techniques. Examples are studies of visual speed and so-called transposition effects by J. F. Brown, an American National Research Council fellow from Yale University, and by Hans Wallach. Equal motions observed from different distances appear to have nearly equal perceived speeds, although the actual displacement rates of the corresponding retinal images are quite different. When Köhler suggested that Brown find out whether the laws involved are related to the well-known facts of size constancy (discussed in Chapter 4), Brown found that the "lag" in speed constancy was much larger than that between matching sizes under analogous conditions.[29] He then arranged displays to produce equal retinal images, which would appear after compensation as greatly different percepts, by constructing another moving field that was in all respects twice the size of the previous ones. With the two moving fields at the same distance from the observer, Brown then obtained matches showing that the speed in the larger field had to be 1.9 times as large as the speed in the smaller field for the perceived speeds to be the same. This came to be called the transposition principle of speed perception – another example of dynamic self-determination of a perceptual phenomenon (speed) within a reference system. Hans Wallach later worked with perceived direction of motion along similar lines.[30]

Other sense modalities

Central to Gestalt theory's claims to general validity was the ability to discover the kinds of invariant relationships that Köhler, with Max Planck, considered the essence of lawfulness in nature in other sense modalities besides vision. When Wertheimer and Erich von Hornbostel utilized the instrument they invented to determine the direction of artillery and torpedo fire in studies of sound localization (see Chapter 12), they found one such invariant relationship: "the angle at which a sound is heard depends on the time difference with which the same stimulus impacts the two ears." Specifically, the angle of subjective localization increases with deviation from simultaneity, reaching 90° at 630 microseconds; just noticeable differences are greatest near the median, and decrease as the angle approached 90°. In their opinion, this result refuted then widely held views that attributed sound localization to intensity or phase differences, and supported the view that such phenomena are determined centrally rather than peripherally.[31]

Von Hornbostel continued work on directional hearing in the 1920s. In an extensive study comparing hearing with one and two ears, he attempted to show that the latter cannot be constructed from the former. Given similar stimuli and excitations for both ears with respect to frequency, wave form, intensity, and period, observers hear a single, quiet sound localized between the two ears and in front of the head. Increasing differentiation in any of the four variables brings separation of the sound, or diotic (two-eared) hearing. The functional asymmetry appeared analogous to that already discovered by Hering for vision; left-monotic and diotic processes localized to the right appeared stronger. What determines the operation of binaural hearing, von Hornbostel claimed, is not the qualities of individual sound waves or their arithmetic relationship, but the "gradient" or tension between them. These were precise counterparts of the stimulus gradients that Koffka had posited in vision. Perception of such gradients (e.g., 45° to the left) cannot result from an indifferent mixture of localizations to the left and right

Photo 12. Erich von Hornbostel. Courtesy of Professor Michael Wertheimer.

ears, but must be central in origin. He suggested that the physiological processes underlying diotic phenomena are "physical Gestalten" in the sense of Gestalt theory.[32]

Later, in an important handbook article, von Hornbostel presented a comprehensive phenomenology of heard sound, ranging from individual tones to tonal series to groups of tones heard simultaneously. The central claim was precisely analogous to long-standing claims for vision – that this phenomenal order is not strictly parallel to the physical order generated by acoustical research. Single tones, for example, may sound light or heavy, dark or bright, small or large, as well as high or low. Further, tones have colors and brightnesses that vary with the producing instrument. From such relatively diffuse Gestalt qualities von Hornbostel proceeded to Köhler's vowel colors (discussed in Chapter 7). These corresponded in his view to the primary colors in vision, and he ordered them on a triangular graphic, like the color pyramid. From such outstanding points in the tonal series it was an easy step to more musically ordered series and groups of tones. A triad is not a mere sum of two pairs of tones, but a new structure with its own consonance and dissonance. Since such "phenomenal belongingness" persists despite phase differences, and cannot be reduced to intensity alone, it can only be explained by similar "belongingness" in the underlying physiological processes. Psychophysical relationships of this kind need not be considered miraculous, he maintained, if one searched for the yet-to-be-discovered nervous processes in hearing with the help of the psychological facts.[33] The analogy to Gestalt research in vision was clear, but only in his studies of sound localization did Hornbostel try to specify in detail both the phenomenological conditions and the quantitative parameters for such phenomena.

Thought

Particularly indicative of the range and variety of Gestalt-theoretical research – and of the room available for critical modification of Gestalt principles – was Karl Duncker's work on thought. Ostensibly completed under Knight Dunlap at Clark University, where he had gone with Köhler in 1925, Duncker's first paper on the subject was in fact inspired by Wertheimer's 1920 paper on productive thinking and by Köhler's anthropoid work. As Duncker emphasized, "nobody knows better than Wertheimer himself" that his own formulations up to that time "mark off a certain theoretical area, no more."[34] The task was to fill that area with content by specifying more exactly the "act of absolutely singular type" that William James, like Wertheimer, had acknowledged to occur between problem and solution. Here, as in his work on induced motion, Duncker accepted the invitation offered by Wertheimer's open-ended questions. The result was yet another problem shift, in which the original Gestalt perspective was retained but decidedly enriched.

Duncker claimed that Otto Selz's notion of productive thinking as the actualization of learned schemata (discussed in Chapter 9) is too narrowly based on stereotyped situations, such as categorizing. To understand innovative thinking, he argued, it is necessary to study the solution of unfamiliar problems that cannot be solved by applying familiar logical relations. In a preliminary study with five observers, Duncker tried to accomplish this using experimental situations drawn from Otto Lippmann and Helmut Bogen's recently published book, "Naive Physics." Rather unorthodox were his instructions. He asked observers to try to think aloud while working and to draw as much as possible, and assured them that there was no interest in their performance or reaction time, only in their "thinking behavior." They were encouraged to be bold and speak out their mistakes – "You are in charge" – and were also told to stick to the particular conditions, rather than trying to remember solutions. Interaction with the experimenter via questions and hints was expressly permitted – a procedure that was allowable only

Photo 13. Karl Duncker. Courtesy Professor Michael Wertheimer.

because objectivity was being located in phenomena, rather than in an impersonal method.[35]

A sample protocol reveals the processes Duncker found to be at work. The task was to apply X-rays to destroy a tumor inside someone's body, while taking account of the fact that X-rays can destroy healthy tissue at high intensities. The best solution was to use X-rays of medium intensities from different directions so aimed as to concentrate at the tumor's location. Here is one subject's path to the solution, with Duncker's hints in parentheses:[36]

4a. Apply plates to protect the neighboring parts from being destroyed. ("That is not the trouble.")
  b. Send rays through a tube in the esophagus. (Hints: "Draw a cross section of the body; the intensities must be different.")

c. Make the tumor more susceptible by some kind of injection. (Hint: "Treat good and bad tissues differently.")

d. Send weaker rays from two opposite sides so they will meet each other in the center. (diagram) (He saw that this would not help, that the trouble was not yet fully removed. "Why do you let the two radiations go the same path?")

e. Aha! (Time ca. 20 min.)

Duncker emphasized that all the suggested solutions had an "intimate relation" to important subunits of the task, such as doing no harm or destroying only diseased tissue. So long as this was so, even incorrect solutions could be productive. Solution 4d, for example, though bold and intrinsically excellent, is incorrect; but further consideration led from it, with the experimenter's help, to the correct solution.

For Duncker, such protocols showed that productive problem solving is an integrated process that proceeds in definable stages. First and essential is what he called "understanding" or "penetrating the conflict," that is, comprehending what is problematic in the problem situation, which is not the same as merely noticing that there is a conflict. In the second stage, observers represent the conflict to themselves; in the third, they penetrate further into the "conflicting circles," that is, the structure, of the problem situation. In the key fourth stage, observers discover and elaborate what Duncker called the "functional value" of the solution, like Köhler's apes seeing the box that will take them to the food. In the case above, the "functional value" was to weaken destroying power of the rays in healthy tissue while concentrating it where required, on the tumor. Finally, observers realize the solution; that is, they recognize that a particular choice actually fulfills the function required. In Duncker's view, all of these steps are more closely related to perception than to abstraction. More generally, he held that "abstraction is plainly a subsidiary and specific procedure" occurring "only if needed." Though it is obviously necessary to science, it is not needed for thought per se.[37]

Later, in investigations carried out in the early 1930s but not published until 1935, Duncker elaborated on this analysis and applied it to additional, mainly mathematical problems with a larger number of observers. Here Duncker explicitly pulled away from Wertheimer's emphasis on sudden "recentering" of the situation, as he had done implicitly in the earlier study. In his words, "The final form of a solution is typically reached by way of intermediate processes, each of which appears as a solution if directed backward, as a problem if directed forward." Sudden insight certainly occurs, but the problem-solving process is not always as linear or teleological as it would seem from emphasizing such events. Learning from mistakes and making transitions to another stage of considering a problem always means going back to an earlier phase in the process. Thus, to understand "the directional forces" at work, the "theory of 'closure' or '*Prägnanz*' is far too general to be seriously considered here."[38]

Learning, memory and Köhler's psychophysics

Last but by no means least, in work on successive comparison, learning and memory, Köhler – following Wertheimer's precedent of using physiological hypotheses heuristically as guides to psychological research (see Chapter 8) – attempted to create a new model of experimental procedure in order to apply and test his theory of brain action experimentally. Here, too, students did far more than deliver data; they contributed to theory development as well.

The ostensible starting point of this research was a study by J. Borak of Vienna on lifted weights. Borak found that observers noticed small differences between two weights more easily when they lifted the lighter of the two first and the heavier second. As Köhler noted, such discrepancies had already been recorded in the earliest work from Wundt's laboratory, on comparison of sound intensities. His own experiments with the judged loudness of telephone clicks confirmed the predominance of ascending over descending tone intervals; even objectively equal tones are judged more often to be different in ascending than in descending series. When such regularities occur across different sense modalities, Köhler suggested, then it is reasonable to suppose that they are of general importance.[39]

Traditional theories, influenced by associationism, described such judgments as comparisons of present sensations or images with memory images of earlier sensations. But phenomenological investigations by Husserl's student Alfred Brunswig and others had shown that such memory images were often not present. Rejecting Karl Bühler's invocation of psychical "acts" based on sensory "signs" to account for such judgments and Johannes Lindworsky's conception of consciousness as a network of relations into which comparisons might fit, Köhler argued, predictably, that only processes in the nervous system could do justice to the "objective-phenomenal dynamics" at work. His hypothesis was that initial impressions of sounds, light, weight, and so forth, leave not images or "residues," as one recent theory called them, but traces, "quiet pictures" (*stille Bilder*) in the form of energy or electrochemical distributions in specific regions of the cortex, which then undergo structural change as comparison proceeds. Against the criticism that knowledge of the brain was insufficient to support such explanations, he replied that such knowledge is more likely to emerge if we allow our hypotheses about the brain to be guided by the psychological facts.[40] This was a bold effort to rewrite the heuristic psychophysical parallelism already articulated by Ewald Hering and G. E. Müller decades earlier (see Chapter 4).

Later, Köhler applied this suggestion to the perception of relations in general and to memory. Consider, for example, the case of returning from years of travel, seeing someone and remarking spontaneously, "How old he looks!" Köhler argued that such judgments are not products of comparison with a mental image of a younger man, but evidence that immediate experience has the property of referring beyond itself, in other words, that we have awareness of what William James called "states beyond" our actual experience. The same is true in judgments of successive comparison. When we judge successive telephone clicks heard at intervals of five seconds as louder, we only experience one noise. Yet we experience it not "as smooth, isolated and discrete," but rather "it seems to come with a definite direction 'from something in the past.'" Köhler argued that this "something" indeed "determines the direction with which the second noise appears," and that it must therefore be "a fairly direct representation of the first noise." However, because it is not accompanied by a similar experience, the determining something or trace cannot be an event of the same kind as the first experience, but must be physiological. More generally, "it follows that the traces of past experiences are neither an indifferent continuum nor a mosaic of independent parts; rather they must be pictures of past organization." By "picture" he meant not a literal image, but a functional equivalent in the form of a concentration of molecules with properties "corresponding" to those of the original process, that is, with similar organization.[41]

In the early 1930s Köhler's student Otto von Lauenstein took up this idea and developed it in an independent way. Köhler, he noted, had offered two hypotheses to explain successive comparison, but had stated the second only incompletely. The first hypothesis was that successive comparison is the product of a dynamic "transition" from a first to a second event in the brain. Lauenstein accepted physiologist Julius Pikler's criticism that this cannot account for comparison of the first and third items in a series of three impressions, since item 1 has already "transitioned" into item 2. Indeed, if Köhler's physical model of the optic sector is correct, we must assume that the process is always continuous because there are no "empty spaces" in the optic sector. Moreover, Lauenstein's own experiments with neutral tones rather than telephone clicks showed that comparison need not have the "dynamic" leaps or descents ascribed to them by Köhler. Rather, under certain conditions the second tone simply sounds louder – not at the start but in the middle of its appearance – and retains that quality throughout. That is not what would be expected of a "transition."[42]

Lauenstein's alternative was a version of Köhler's second, partially expounded hypothesis, which he called the "sedimentation" theory. He developed this view with the intent of making it general enough to apply to simultaneous as well as successive comparisons. Working from the theory of contour developed in Köhler's *Die physischen Gestalten* (see Chapter 11), he suggested that a "differential apparatus" in the cortex could "measure" the ratios of the phenomenal qualities involved in the form of a potential gradient. We need only ascribe to traces "characteristics that make possible a potential difference similar in kind between it and an excited region" of the field. If the traces remain relatively unchanged, then slight differences in their functional location in the relevant cortical sector would be sufficient to account for – neuropsychologists might now say "to map" – both spatial and temporal differences. As Lauenstein showed, his suggestion was consistent with older experiments by Hollingworth on the so-called "central tendency" – the nervous system's tendency to seek a median level of incoming excitation.[43]

Lauenstein's theory thus opened up the possibility of a physiological explanation for a considerably wider range of phenomena than Köhler had envisaged in 1923. Köhler had already indicated the connection of such ideas to learning in 1929. As he put it, the phenomena of habit formation would seem consistent with the claim that single traces are connected by simple association. But a student of Karl Bühler's had shown as early as 1914 that when observers are asked to read a syllable series they do so "in a definite and rhythmical distribution which usually consists of larger wholes containing subunits." Thus, "evidently a process of organization occurs during learning," especially during the first readings. G. E. Müller had suggested that such results were due to observers' intentional organizing of the material. But Köhler insisted that not all organizing is intentional, and suggested that it might be possible to vary presented syllable series in such a way that "we should be able to deduce from all the rules known in the field of sensory organization corresponding rules for association." Once this ambitious research agenda was underway, "we shall be better able to treat its physiological basis."[44]

This was the purpose of experiments Köhler carried out with Hedwig von Restorff in Berlin and published in 1933. The findings were straightforward enough: Series of nonsense syllables are difficult and unpleasant to learn not because the syllables are meaningless but because the series are unstructured. Observers first learned series of eight pairs of terms, four the same and four different: nonsense syllables, geometric figures, two-digit numbers, letters, and small oblongs of different colors. Four pairs of each sort of object were then presented alternately in combination with one pair each of the others, thus yielding five sets or series. Each series was presented to four or five observers of a total of twenty-two, who were then tested after varying intervals with the standard method of paired associates. The result was the same in all cases – syllables presented singly were remembered far better. Summarized statistically, observers identified 41 percent of the items correctly when syllables appeared four times, and 69 percent when they were presented only once. This seemed to show a definite negative effect of aggregation on retention.[45]

Köhler and Restorff explained this result by hypothesizing that "regions" form in physiological trace fields in a way analogous to the structuring and restructuring of perceptual fields. As a figure stands out from its background, one stimulus object "stands out" in such series. There remained the possibility, however, that each figure could be seen as different from all the others. To test this, fifteen observers learned series of ten items on three different days; they then learned text for ten minutes and were asked to write down as many items as they could remember. On subsequent days, observers were retested with single items from the series isolated against nine items from the other series. The isolated items were retained almost twice as well as the others. Köhler and Restorff argued that in such cases individual traces are absorbed and lose their independence, because each part is connected with every other part. For Koffka, all this meant that "the same laws of organization hold for memory that govern perception," because subsequent events give traces an organization that they did not initially have, as also occurs in perception.[46]

As Erich Goldmeier later put it, association theory and current information-processing models assume that there is no change in the memory trace; on analogy to computer memory storage, change is equivalent to error. But humans are not computers. As Wulf and Restorff showed, traces do change, though accounts may differ as to how or in what direction. In current terminology, processes of some kind intervene between stimulus events and recall. In this "coding" interval, information is lost and structuring or classification gained. The issue is whether this is a self-organizing process or a result of "extrinsic" factors. The associationist view had been derived from work on verbal learning. Against this, the Gestalt theorists emphasized perceptual learning as biologically primary. "The focus on verbal memory ignores the fact that the human brain was not designed for the task of remembering disconnected sentences, words or syllables, or nonsense, least of all verbal nonsense."[47]

Nonetheless, it was clear that traces are not literal "pictures" of stimulus organization, as Köhler had seemed to indicate earlier. Lauenstein admitted that the physicochemical basis for Köhler's original psychophysical hypothesis was "outdated." More important, the recognition that traces change implied the need for an important restriction in strict isomorphism. As Köhler later said, we can assume isomorphism between stimulus correlates and their traces, but "we cannot assume that such isomorphism will be strictly preserved far beyond the time at which the traces have formed." Because traces interact with one another, "the record will tend to revise its own text spontaneously" over time, and "we can only expect the final edition to be a sketchy abstract of a distorted story."[48] In this way Gestalt psychology advanced into new territory – with the active, critical participation of its brightest students, but at a certain cost.

Implications

Looking back in 1935, Kurt Koffka wrote that "Gestalt theory has been rather consistent in its development. It has studied the fundamental laws of psychology first under the most simple conditions, in rather elementary problems of perception; it has then included more and more complex sets of conditions, turning to memory, thinking and acting."[49] Though this statement is correct in a general way, it makes the history of Gestalt-theoretical research appear too linear, as though it had been planned to generate an orderly and comprehensive psychological system. Wertheimer's first research on thinking, for example, appeared in

1912, the same year as his better-known paper on motion perception. Indeed, the main reason for Gestalt theory's extraordinarily rapid growth as a research domain in the Weimar era is that it was driven simultaneously by several different but related theory-laden models for open-ended experimental research programs.

Even so, what the Gestalt theorists and their students demonstrated most often was the applicability of analogies from Wertheimer's principles of perceptual organization to an increasingly wide variety of phenomena – in visual perception and other perceptual modi, then in comparative psychology, learning, and thinking. Duncker's work on frames of reference in motion perception and particularly the studies by Köhler, Lauenstein, and Restorff of successive comparison and memory show that research was not confined to reproductive demonstrations or analogical extensions of Gestalt principles. Rather, both theoretical claims and research results generated new issues, sometimes opening up entirely new topics. The results of that work often led to significant elaborations, even alterations, in theory, and also in semantic repertoire, though not in the basic Gestalt position.

There were, however, problems in the progress. Not all of the research in Berlin, Giessen, and Frankfurt was so rigorously conceived for testing hypotheses about brain function as Köhler's appeared to be. Though the term "theory" usually referred to brain events, there were, as stated previously, three kinds of theory-testing at work. The first two – structured phenomenological observation and the search for quantifiable invariants in the mental realm – were psychological, but not in the sense commonly understood. For the Gestalt theorists, psychological hypotheses referred not to hypothetical mental or emotional processes underlying or producing phenomena, but to principles of self-organization thought to inhere in the phenomena themselves. The third kind of hypothesis was, of course, the step to physiology, compelled in their view by the phenomenal objectivity of organization's results – the impressive "there-ness" of what we experience – as well as the quantitative evidence for "invariant" ordering principles across sense modalities. As critics lost no time pointing out, what had up to then been considered psychological factors – attitudes, attention, and the like – had at best a secondary place in this explanatory logic (see Chapter 18).

The work of Duncker, Lauenstein, and Restorff also raised other, potentially more serious problems. Duncker's research on thinking showed that Wertheimer's metaphors of "centering" and "recentering," though stimulating to research, might not be sufficient to account entirely for productive problem solving as a process. No one seemed prepared to say what the physiological counterpart of that process might be. The same was true of the extraordinarily complex person–environment dynamics discovered in Duncker's work on induced motion. Lauenstein's work on comparison and Köhler and Restorff's on memory showed that analogies from Wertheimer's principles of perceptual organization could lead in promising directions; but whether such analogies would be sufficient guides to the physiological processes at work was an open question. The Gestalt theorists regarded all of these problems at the time as challenges, not as limitations. That testified to the continued vitality of Gestalt theory as a research approach. The evident productivity of the complex, interlocking set of research programs that made up that approach gave the Gestalt theorists the confidence to project their ideas into ever wider theoretical contexts and cultural fields.

15

Theory's growth and limits: Development, open systems, self and society

More occurred in the Weimar period than the application of Gestalt theory to a growing range of research problems. The theory itself evolved as well, as Koffka and Köhler extended the reach of Gestalt language to psychical development, general biology, and the problems of emotion, self, and society. In the process, they enriched their own field of discourse and improved their positioning in the wider culture by linking that field to other theoretical realms with significant cultural resonances. At the same time, they probed the implications of their theoretical commitments and thus tested the limits of their own innovation.

Kurt Koffka on psychical development

Earliest in time of these extensions was Kurt Koffka's effort to present Gestalt principles as adequate foundations for a theory of psychical development.[1] In addition to Gestalt theory, Koffka drew upon a variety of sources. Most prominent among them were Frankfurt neurologist Ludwig Edinger's preformationist view of neurophysiological development and Lucien Lévy-Bruhl's conception of so-called primitive thought. Edinger's conception of "developmental history" (*Entwicklungsgeschichte*) was a pendant to Wilhelm Roux's program of "developmental mechanics" and Hans Driesch's neovitalism (see Chapter 5). Both Roux's program and Driesch's response had originally referred to the development of embryos. Edinger's concept referred to the developmental sequence of species, linking anatomical to behavioral forms in an evolutionary succession. The central distinction, for which Edinger is still remembered, was between the so-called old brain (*Paleoencephalon*) and the new brain (*Neoencephalon*). The former, including everything from the head end of the spinal chord to the olfactory centers, he described as the seat of the reflexes and most instinctive movements. To the subsequently evolved "new brain" he ascribed the functions he called "gnosis" – connecting signals from sensory receptors – "praxis" – noninstinctive

247

movements – as well as simple and complex associations. In these functions, especially in the construction of "memory images from many components," he saw the dividing line between lower and higher animals.[2] Because they lack this ability, the behavior of "old brain" animals is exclusively determined by pre-formed anatomical schemata. The anatomy of "new brain" creatures permits considerably greater flexibility, and the areas lying between and in front of the sensory centers *"clearly increase in size as the animal increases its capacity to guide its observation and activity by intelligence."*[3]

In Koffka's book on psychical development, the distinction between supposed old brain and new brain became "a valuable heuristic principle." Most significant to Koffka was the fact, also emphasized by Edinger, that the human cerebral cortex continues to develop after birth. Evidence about the perception of motion in infants, for example, indicated that it "improves during the course of life" due to growth in the optic sector. But if this is so, Koffka maintained, then "we must think of development in terms of a process of maturation, in the course of which certain regions of the nervous system attain the capacity of forming fixed con-figurations which at first they do not possess." However, he added that maturation is also "dependent upon functional employment" of the organ in question. Thus, for Koffka, development was only partly the realization of preformed anatomical potential; it was maturation aided by practice.[4]

By introducing functionalistic thinking in this way, Koffka could treat learning, too, as "essentially a type of development." Here he differed most clearly with Edinger. For Edinger, new-brain functions were matters of simple or complex associative connections. As Koffka argued, Wolfgang Köhler's demonstration of insight in apes had shown that learning is not only the gradual assemblage of associative connections, but can also occur in a single, "nonheritable achieve-ment." But even such achievements have roots. Just as perception and action become, with practice, an integrated sensorimotor system "uniting phenomena and movements in one total form of behavior," so too in learning "an intelligent (re)construction of the field takes place with respect to the goal."[5] Koffka had adapted Wertheimer's conception of the "knowledge process" to cover the struc-ture of all behavior, both instinctive and intentional. He did the same for the relationship of learning and maturation. Just as the functional employment of an organ aids the development of that organ, so "the maturation of a performance improves with practice," and "practice means *the formation of a [behavioral]* figure, rather than the strengthening of bonds of connection."[6]

Koffka could only extend such metaphors so far because he had accepted Köhler's philosophy of nature. This was most obvious in his discussion of instinct. Citing Köhler's discussion of the second law of thermodynamics, he asserted that there are events in the physical world that "shape themselves" toward a very definite end, the simplest organization consistent with the given conditions. Thus, goal-directedness is not confined to living matter. If we seek an explanation for instinctive behavior, then, we should not attribute it to any vital principle, nor to "an inherited system of connected neurones." Rather, we should "investigate what kind of physicochemical 'closure' produces these astonishing types of behavior, and under what conditions." Koffka admitted that even from this viewpoint "instinct is still a riddle," but "at least it is no longer one that forces upon us the acceptance of psychovitalistic principles."[7]

Since both instinctive and learned responses are present in the same individual, and since both have essentially the same structure, Koffka thought it unnecessary to choose sides in the heredity–environment controversy. A decision is not possible in any case, he argued, without achieving clarity about the nature of experience itself.[8] Having done just that, with Wertheimer's help, he could accept a role for practice in the development of both consciousness and behavior, because he had already subsumed both under a vastly expanded Gestalt concept. Because of his claim that practice could modify or complete the development even of preformed behavioral schemata, Koffka could have been charged with attempting to revive Lamarckism. This was not his aim. Köhler's physical "Gestalt laws" are not acquired characteristics that can be inherited, but invariant structural principles for physical and psychical events, inherited and acquired behaviors. In one sense this style of theorizing was similar to that of Driesch, for both describe development as the realization of an organism's inherent potential. But the location of that potential was fundamentally different. For the vitalists, it was exclusively within the organism. For Koffka, it was in the organism–environment interaction, which works as it does because behavior, reflected experience, and the objects to which both are directed are all subject to the same Gestalt dynamics.

To make this view more plausible, Koffka worked out a number of hypothetical developmental sequences. The most extensive of these was for color vision. This he described as "the gradual construction of new color configurations" by a process akin to that exhibited in Köhler's experiments with "structural functions" in animals. The "most primitive" phenomenon in visual perception, on this account, is the figure–ground distinction already investigated by Edgar Rubin (see Chapter 11). In color vision, the first stage of development would thus be the distinction between chromatic and achromatic colors. This would be followed by the differentiation of reds from blues, or of red from "not red," then of the spectral colors, and finally of intermediate hues. A child who earlier saw only blue and red, for example, would later come to distinguish lilac as "not blue." On this basis the child would gradually distinguish a wide range of color phenomena before learning their common names.[9] Though Koffka offered a variety of observations in support of this hypothetical sequence, he admitted that much of it remained speculative. He did not try to work out a more complete stage theory, of perceptual or cognitive development of the kind that Jean Piaget, Charlotte Bühler,

Oswald Kroh, and others would soon create. In fact, Koffka's schema was little more than a deduction derived from Wertheimer's and Köhler's work and the known facts about the visual spectrum.

In all of this Koffka referred primarily to the development of individuals. Early in the book he raised the issue of ontogeny and phylogeny, and accepted the so-called correspondence theory of Eduard Claparéde and Dewey – the view that "the general characteristics of development are the same for both the individual and the species."[10] But as he proceeded, it became clear that this was insufficient. If his system was to be complete and consistent, and also applicable to pedagogy, Koffka would have to provide explicitly for the social dimension of development. It was here that Koffka drew on the work of Lucien Lévy-Bruhl. Against the widely held view that so-called primitive thought is merely an inadequate copy of Western thinking, Lévy-Bruhl maintained that the consciousness of the simplest societies we know is fundamentally different from our own, because it is determined by the group, not the individual thinker. The core of his argument was the concept of "collective representations" (*représentations collectifs*). The term has often been mistranslated as "collective ideas," but this covers only half of Lévy-Bruhl's meaning. The representations are collective, but they are not only ideas.

According to Lévy-Bruhl, rigid distinctions between intellect and affect, the individual and the group, humans and nature, or even the living and the dead do not exist for so-called primitive consciousness. While the men of the village hunt, for example, the women at home perform rituals required to assure the hunt's success. However, they do this not in order to help the men in any Western sense, but because the rite would be incomplete, and thus unsuccessful, if carried out by the men alone. To account for the coherence of this world, Lévy-Bruhl invoked what he called the "law of participation." The existence of such complex unities, he argued, cannot be explained by claiming, as ethnologists often did, that "primitive" peoples animistically project themselves onto the environment. Rather, reality is more "full" for this form of consciousness, because abstraction hardly exists.[11] Clearly, Lévy-Bruhl's intentions were primarily philosophical, not sociological, ethnological, or even psychological. He drew his picture of "primitive" mentality so starkly in order to impress Western readers with its radical otherness.

Koffka reviewed Lévy-Bruhl's book in the first and only issue of a journal for the psychology of religion that he had tried to launch in 1914 with Wilhelm Stählin, a pastor who had been a subject in experiments Koffka conducted in Würzburg five years before. In their introduction, Koffka and Stählin brought out the evolutionary implications of Lévy-Bruhl's thinking. The psychology of religion, they wrote, must be rooted not in theological propositions or in speculations about the essence of religion, but in biological and historical research. This was a bold position to take at the time, especially for a minister of the established church in Prussia. Specifying the sort of biology they meant, they alluded to the "original lack of differentiation" of human thought. Thus, "everywhere the parts were not present before the whole . . . psychical like all biological development does not proceed along the road of combination, but rather that of differentiation." With this argument, Koffka had accepted Herbert Spencer's view of cultural evolution; for it was Spencer who wrote that all development proceeds "from an undifferentiated, incoherent homogeneity to a more differentiated, coherent heterogeneity." For Koffka and Lévy-Bruhl, however, the original state from which development proceeds may be "undifferentiated," but it is "incoherent" only from a Western point of view.[12]

Koffka integrated Lévy-Bruhl's thinking into his argument in the final chapter of his book on development, entitled "The World of the Child." In general, he simply transferred Lévy-Bruhl's formulations from the so-called primitive mind to that of the child, with one important difference. Both Lévy-Bruhl and Wertheimer expressed their claims as statements about the human mind as such; Koffka presented them in historical form. There is indeed a "child-world" different from that of adults, which persists for a time alongside it. Yet "we are dealing here not with an unchanged child-soul, but with a worldview which continually undergoes a process of transformation." In part this process is one of maturation, during which inanimate objects become distinct from animate ones, objects in general acquire the qualities of things with attributes, and appearance is distinguished from reality. But for Koffka that process is both organic and social: its shape "depends upon the total environment, and above all upon the sociological conditions of this milieu . . . *Man's entire development, including, of course, his perceptions, is dependent upon society.*"[13]

This is obvious for both children and adults in non-Western cultures, where social bonds appear to be much stronger than in so-called civilized society. It is difficult, however, to discover the phenomena and structures of the "child world," precisely because it is "constantly influenced by association with adults and hence [is] not stable enough to show its worth in performance," except perhaps in play. Wertheimer had said that in such a situation it was often understandable that children or members of other cultures resist the demands of teachers from outside.[14] Koffka put it this way:

> The release from [immediately perceived] reality which is possible and easy in our mode of thinking is a specific product of our civilization. The child must go a tremendously long way in a short time in order to learn to think as adults do, in a manner which is not at all natural to him. To lead him along this way, so that his advancement may be vital to him, is the difficult but gratifying task of the teacher.[15]

At the very least, respect was due to children for all that they manage to achieve in their earliest years.

In such passages Koffka seemed to support the idea of the "natural" child, much like Rousseau, the proponents of Deweyite progressive education that was becoming fashionable in the United States, or the advocates of child-centered education who were gaining influence in parts of Central Europe at the time. Teachers appeared here at least to some extent as agents of alienation working on behalf of an adult civilization, with goals fundamentally different from those of children. But the opposition was not in fact so simple. As Koffka recognized, the intervention of parents and teachers in the maturation process is also necessary for the development of culture, and for the development of children within that culture. Thus, children could not simply be left to themselves; even teaching strategies that declare this to be their goal are interventions. Though Koffka sought to explain the development of reason and to make development reasonable, the end result had a tragic cast.

Koffka did not pursue these beginnings further, nor did he produce the developmental social psychology that might have emerged from them. He had asserted that in mental development, "both the phenomena of consciousness and the functions of the organism go hand in hand."[16] But he had not shown in any detail whether or how the functional interaction he posited between the developing individual and society – or its agents, teachers and parents – takes place. Nor was there any consideration of the development of social interactions among children or youths. Without such discussions, the book's usefulness to teachers was limited. More important to psychologists were other difficulties. Though Koffka had packed a wealth of detail into his account, many of his generalizations were deductions from general principles, not inductions from research results. Though they were framed in such a way that they might be tested experimentally, he did not carry out such tests himself. Helene Frank, a teacher who worked at the Berlin institute, made a beginning when she carried out studies of size constancy in children.[17] Koffka had no children of his own to observe, and did not set up a laboratory for child studies; his research interests lay elsewhere. Thus, it was not completely clear how his principles could lead to further work in the field, or how his suggested schemata of development could be made relevant in the classroom. Koffka's hints about the determining role of society were taken up enthusiastically in the Soviet Union by the so-called cultural-historical school led by Lev Vygotsky.[18] In Western Europe and the United States, developmental psychology went other ways.

Wolfgang Köhler on open systems

Fundamental for Gestalt theory as a philosophy of nature was Köhler's attempt to extend his natural philosophy to biology, as he had promised to do in *Die physischen Gestalten*. The context for that attempt was more welcoming in the 1920s than it might have been earlier. In that decade, a number of biologists used analogies from physical field theory to find solutions to the problem of order in organic development, which had been raised by Hans Driesch, without resorting to neovitalism. In 1922, Alexander Gurwitsch employed the term "field" to describe, for example, the regular forms exhibited by the flowers of certain varieties of mushrooms, and distinguish these from the relatively undifferentiated organic structures from which they emerge. In 1926, Paul Weiss presented what he called a "morphodynamics" to account for facts like the regeneration of extremities in amphibians without skeletons by using analogies from engineering, specifically to the functioning of thermocouples. However, rather than imagining the processes involved as self-organizing systems, he attributed them to "generative" processes within the germ cell. In this respect his approach was similar to that of Hans Spemann, the most famous embryologist of the day, who hypothesized that a specific region of the germ cell functioned as an "organizer." All of these proposals resonated with the widespread turn away from mechanistic and toward organicist metaphors in this period.[19]

Köhler had already begun developing his own contribution to this discussion on Tenerife. In a 1919 letter, for example, he suggested that the interaction of eye and eye muscles in response to incoming light could be understood as a complex control apparatus (*Steuerungseinrichtung*) which works to restore balance between organism and environment, much as physical systems tend to achieve equilibrium.[20] In an essay published in 1924 but completed earlier, he began to work out the conceptual basis for such analogies – a distinction between closed, or mechanical, and open, or self-regulating, systems. He began by stating that the problem of order raised by Driesch is the same in perception as in organic development. He maintained, however, as Erich Becher had in 1907, that Driesch's psychovitalism presupposes a physics limited entirely to "machine theory," that is, to the processes described in elementary analytical mechanics. In Köhler's terminology, simple machines, constrained by their rigid "topography," have only one "degree of freedom," while "an inorganic system with several degrees of freedom would behave under different circumstances very differently from a machine, which can do one definite thing or nothing." Hence, the physiological events underlying perception need not be subject in principle to the rigid constraints characteristic of man-made machines.[21]

Applying this thinking to biology in general, Köhler assumed that the tendency of all physical systems to simplicity and regularity proclaimed at the end of *Die physischen Gestalten* applies to processes within the organism, and also to the interaction of organism and environment. As an example he cited the interaction of lens and eye-muscle mentioned previously, which leads to reflexive accommodation toward maximum simplification of the visual field, in this case maximum horopter. He conceded that actual steady states are rare in organic life, but asserted that organisms are always "on the way" to such states. This was another example of the characteristic inconsistency already noted (in Chapter 11), in which a theory emphasizing dynamics posited static distributions as the teleological aim. As Köhler also noted, a theory of such processes would certainly have to address itself to the fact that the embryo apparently gains energy as a result of organism–environment interchange. Nonetheless, he was confident that both the coordination of organs, which is the best sign of order and differentiation in organic development, and purposive behavior in general are instances of the second law of thermodynamics, hence "causally necessary" processes.[22]

Though Köhler had shown biologists and physiologists how they might conceive of certain problematic processes without departing from the ground of natural science, he seemed in doing so to have sacrificed the functional specificity of the biological realm. His frankly physicalistic theory of organic development meant, for example, that interruptions need not necessarily lead to lasting deformation, since the process, once resumed, would still exhibit the same tendency to order as before. Unfortunately, Köhler gave no examples at the time of what he might mean by such a claim, nor did he try to account for the deformations that obviously do occur. Moreover, the idea that all organic processes, including development, growth, and adaptation, are instances of the second law meant that Darwinian evolution is not, or need not be, a product of chance: "Instead of externally directed selection (Darwin), we must consider directly selective factors which . . . will not permit a haphazard origin of forms." Here Köhler agreed with biologists like Oskar Hertwig, who rejected Darwinism because natural selection appears to be based on chance.[23] His theory appeared at least potentially capable of dealing in a sophisticated way with multiple organic solutions to environmental challenges without having to postulate random scatterings. But he did not actually try to account for the enormous variety in natural forms that had been the impetus behind Darwin's theorizing.

Hans Driesch responded to Köhler in a brief article published the next year. His position was straightforward – physical Gestalten do not exist, only "effective units" (*Wirkungseinheiten*) or physical systems. These are sums, though admittedly not and-sums. In fact, and-sums never did exist in physics; mechanics itself relies on determination by systems as units. Driesch admitted that the phenomena Köhler had described, such as the redistribution of charge on a conductor, certainly exist, and that they are affected by the system's boundary conditions. But that was just the point – the boundary conditions are external constraints created by human beings, while organic wholes develop as products of an "inner force." Köhler, he claimed, could not duplicate regeneration experiments in physics. He did not suggest, for example, that an electric current separated into two Leiden jars would produce two smaller versions of its previous distribution. Actually, Köhler had made just such a claim. Even so, Driesch insisted that the physical Gestalten that may indeed exist in the nervous system could "not be wholes of their own force, because such things simply do not exist in the material-physical realm. . . . What exists in physical reality," he concluded, "is surely a totality of place values (*Ortswerte*)," and no more. Thus, the fact that "we phenomenally experience strictly geometric wholes" must be due to a nonphysical factor or force.[24]

In a 1927 essay contributed to a volume honoring Driesch, Köhler did not respond directly to this criticism, but developed his own ideas further. Employing the biologists' vocabulary, he now referred not to the problem of order, but to that of regulation, or what Wilhelm Roux had called self-regulation. Recent experiments by physiologists on the plasticity of the nervous system showed that if a disturbance, such as an injury, eliminated the functioning of individual organs, compensating processes could arise elsewhere in the organism. Such findings, Köhler asserted, were not special cases; "in future we will, no doubt, have to conceive normal functioning, too, in principle, in such a way." Thus, "we must seek new foundations for the theory of normal function in morphogenesis and in physiology." Until now, foundations had been sought almost exclusively in mechanics, because "technology has accustomed us to systems that so limit and prescribe the dynamics of natural forces by means of fixed conditions, connections, pathways, etc." But only "a *theory of systems* seems to be capable of leading us out of the present situation." Köhler then inquired whether there are physical systems that "regulate" in the sense described here, and if so, how we should conceive of them.[25]

As Köhler noted, the fact that closed systems increase in entropy would seem to prescribe a "homogeneity" or "leveling" and preclude the "specific structural differentiation" characteristic of organisms. What he sought, and found, were systems that have observable direction – that is, which "always converge to the same or essentially the same total states" regardless of initial conditions or their paths to those states – but which do not increase in entropy. The characteristic motion of a pendulum, for example, is distributed evenly in space, even though this requires it to move past the equilibrium line and against gravity. The ultimate result of such motion, however, is a static equilibrium. Now, a pendulum is "a system with only one degree of freedom," in the sense described. The orbit of a planet is analogous in a way, since the planet's acceleration varies with its distance from the sun. In both cases there is no obvious capacity for regulation, because the motion configurations depend on the bodies' initial positions and velocities. Thus, "there is no unique final state toward which a system moves, independent of the initial constellation."

But directed forces are at work in such systems, Köhler argued; if inertia were eliminated, they would reach equilibrium. Such is the case when systems show entropy increase, for example under the influence of friction. Morphogenesis, being "rather a slow process," must have a high level of friction. Köhler suggested that the development of organic form is analogous to the behavior of electric currents in liquid conductors, where a buildup of residues on the electrodes ultimately ends current flow. Such systems do "regulate" – that is, there are inorganic systems that "transform themselves in the direction of potential decrease." If conditions permit them to reach equilibrium, the same end state is reached from quite different beginning states. Köhler maintained that this is the only dynamic principle of self-regulation in inorganic nature. Hence, if we seek a natural-scientific basis for organic self-regulation, we must look here – "No other way is open."[26]

Köhler himself noted the limitations of these paradigm cases and indicated additional open questions. "It does not follow" from these considerations, for example, "that organic regulations depend on the same principles" as the physical processes described. Thus, in an implicit concession to Driesch, he admitted that "we could still perhaps be forced to have recourse to other principles than those of physics." Returning to the issue of organic development already posed in his earlier paper, he also acknowledged that "the concept of a system that has been used up to now is perhaps not to be considered completely parallel to the concept of an organism," since energy exchange has not yet been considered. The system concept might better be applied only to "the organism together with its environment." In the case of morphogenesis, the organism-environment unit might then be seen as a closed system, and the developing embryo as "*an open system*" segregated within it.[27]

These ideas, particularly the concept of "open systems," were recognized at the time as significant contributions to theoretical debate in biology. In 1928, for example, hydrobiologist Richard Woltereck cited Köhler when he used the term "ecological Gestalt-system" to describe interactions among plankton algae, insects, and young fishes in lakes and ponds resulting in an equilibrium in numbers as well as a harmony in spatial and temporal (seasonal) distribution among the populations.[28] Ludwig von Bertalanffy, creator of "general systems theory," later acknowledged that Köhler was one of the first scientists to apply the concept of physical systems to biology, and wrote that he first encountered Köhler's ideas in meetings of Hans Reichenbach's Society for Scientific Philosophy in Berlin. In volume 1 of his *Theoretical Biology* (1932), he employed terms from the vocabulary of Gestalt theory, such as "machine theory" and "summativity," as well as "and-sum," in his critique of mechanistic thinking in biology. Indeed, when he stated one of his primary theses – that acknowledging the existence of structure in living material need not lead to giving up natural science – he quoted a review by Koffka of one of Driesch's books. In the second volume, an attempt to develop a mathematics of biological systems published ten years later, he used theoretical distinctions, such as that of reversible and irreversible processes, and also many of the same equations that Köhler had employed in *Die physischen Gestalten*.[29]

Köhler and Wertheimer on the perception of self and others

Equally important for Gestalt theory's claims to general philosophical and cultural significance was its extension to the emotions as well as the perception of the self and of other people. Köhler, at least, was initially reluctant to attempt this, particularly the extension of Gestalt principles to social life. As he wrote from Tenerife in December 1919:

> My fear was always: Just do not create the appearance of philosophy of the bad sort! . . . Perhaps others will now throw in a quick extension to social life [*Volksleben*]. But I cannot do that myself; without exact, sharp thinking through, the basic idea thus easily loses definition, and if I would already direct the Gestalt viewpoint toward the intermediate thing between molecular disorder and the very directed forces that one calls "group, folk, state," I would be somewhat at a loss. Even the mutual influence of people on one another is a problem of which only approximate things could be said, and what a problem![30]

Köhler ventured some initial remarks on emotion in his observations on primate social behavior (discussed in Chapter 10). Most impressive in several respects was what occurred when Köhler appeared at the door of the animals' enclosure wearing a cardboard copy of the mask of a Cingalese plague demon: "Instantly every chimpanzee, except Grande, had disappeared. They rushed, as if possessed, into one of the cages, and as I came nearer, the courageous Grande also disappeared." Here, apparently, was an immediate emotional response to a physical object. He said it was "too facile" to explain this as a reaction to the unknown; for other unknown objects do not have the same effect. Nor was it necessary that frightening objects resemble actual enemies:

> Can it not be that certain shapes and outlines of things have in themselves the quality of weirdness and frightfulness, not in consequence of earlier experience, and not because any special mechanism in us makes them so ad hoc, but because, granted our general psychical makeup, some shapes necessarily have the character of the terrible, exactly as others have the character of grace, or clumsiness, or energy, or decidedness?

Köhler and Koffka later called such immediately perceived emotional qualities "physiognomic" characteristics. The term originally came from the eighteenth-century theologian Johann Kaspar Lavater, who applied the term to the character qualities people saw in the black-on-white profile portraits fashionable in Goethe's time. Koffka and Köhler took it from Hamburg psychologist Heinz Werner's research on what he called "physiognomic" perception.[31]

Max Wertheimer pursued the issue of expression and physiognomic characters in one of his Berlin seminars. As Fritz Heider recalls, he maintained "that each person has a certain quality that Wertheimer called his *radix* – the Latin word for root." This quality expresses itself in a person's appearance, handwriting, mode of dress, movements, talking, and acting, and "also in the way he thinks, what kind of outlook he has, and if he is a scholar, in the kind of theory he builds or adopts." Others recall that he concluded this very popular course with a "physiognomic game," in which he would play melodies on the piano and ask the audience to identify the people or characteristics being portrayed, which they did with a frequency far beyond chance. Such phenomena were bound to cast doubt on conventional attributions of such perceptions to previous experience or conclusions by analogy. As Wertheimer put it in a 1924 lecture, "It is really an imposition on an unsophisticated person to ask him to believe, when he sees another person shocked, terrified or angry, that what he has seen are just certain physical data . . . simply linked externally to what happens psychically in the other person."[32]

In a 1929 paper and in his book, *Gestalt Psychology,* published in English the same year, Köhler addressed the epistemological foundations of the issue directly. Philosophers and physiologists alike have long wondered why perceptions, which so obviously occur "inside" us, are attributed to objects and events outside us. In fact, Köhler argued, Mach and Hering had already sketched an answer to this question decades before. The issue loses its urgency when we realize that our very notion of "outside" is as much a percept as the directions up, down, left, or right, and that "my body," too, is part of this same organized perceptual field, although its inner states generally are not. The so-called "projection theory" thus loses any relevance, since there is nothing to project. Instead, Köhler maintained, the questions must be, what accounts for this given structure of the perceptual field, and under what conditions do the relations among its dimensions change?[33]

As Köhler put it in *Gestalt Psychology,* "In visual experience the pencil there is external to that book and at a certain distance from it." Now, in a very specific, descriptive sense, "My hand is in the same visual field as an experienced object." According to Gestalt theory, we know, at least in a general way, what the physiological basis of the separation between the pencil and the book is. If we apply that model consistently, we can conclude that "There is not the slightest reason why the hand, as a visual object, should be treated in a manner different from the pencil or the book. . . ." On this showing, "No new hypothesis is needed to explain why 'I' am external to those objects and they external to me. . . . I have only to become aware of the fact that, as a visual experience, this 'I' depends upon processes in a definite circumscribed part of my brain as a physical system, no less than do any other objects in the field."[34] Köhler had attributed the phenomenal ego — James's "empirical self" — to a physical rather than a psychological source.

This standpoint made it possible for Köhler to take a position similar to one that the phenomenological philosopher Max Scheler held on a completely different basis — the position that the givenness of both ourselves and other people in experience requires no conclusions by analogy or mystical empathy feelings, but is as immediate as any perception we have. "Since the behavior of others is given to us only in perception," Köhler wrote, "our understanding of others must first of all refer to this source." In a later edition of *Gestalt Psychology,* he acknowledged that even common observations will not guarantee us knowledge of another person's inner states.[35] Nonetheless, he evidently thought that such a view would render the long-standing philosophical conundrum of intersubjectivity — the problem of confirming that others' perceptions are the same as our own – as much a "pseudoproblem" as the projection of sensation. It also made the development of a social psychology on Gestalt lines at least conceivable.

Wertheimer articulated his position on that possibility most fully in an unlikely place – a 1924 article on the psychology of paranoid delusions, ostensibly authored by psychiatrist Heinrich Schulte but actually dictated to him by Wertheimer. In explicit contrast to Freud's focus on frustrated libidinal drives or to explanations focusing on inherent character, emotional, or intellectual weaknesses – all factors located within the individual – Wertheimer asserted the primacy of the social in mental illness. As he put it in his own inimitable way, "A specific situation calls for a specific We." Certain persons, that is, enter a state of "We-crippledness" when they come to feel or notice a "chasm" between themselves and others. In response, depending on personality type or external conditions, individuals may produce a "surrogate equilibrium" by means of delusional ideas, thus constructing powerful images of "the others" as hostile persecutors or pursuers. This reestablishes that individual's sense of relatedness with the group, but in a dysfunctional manner. As an example he cited a Tarter prisoner of war, who developed paranoid delusions when he was put in a hospital where no one spoke his language. In such cases, there was no somatic or personality defect, and the symptoms subsided when the patient was removed to a less threatening milieu. Wertheimer did not deny that somatic factors may also be involved in paranoid delusions. But he clearly emphasized the psychological dimension, and within this the claim that not "society" in the abstract, but specific social situations and individuals' active responses to them together determine individual attitudes and behavior. Of course this was quite consistent with his general Gestalt perspective; in an era dominated by deterministic and hereditarian accounts of mental illness, his insistence on active agency even for the insane was particularly notable. However, though Wertheimer could easily have extended such an account to more common social situations, he did not do so.[36]

Köhler took some tentative, programmatic steps in this direction in 1929. The phenomenal self, he wrote, is usually "the most interesting member" of the perceptual field, but this is not always so. Yet, even when other people replace the self as the "center of our total field," that field retains two fundamental features – its bipolar structure centered around the self–environment nexus, and its "dynamical directedness" toward one or the other pole. Such structures reminded him of "cases in physics in which either lines of force, or a process with a definite direction, develop between two parts of a field." He thought that the comparison "may contain more than a superficial analogy," because "what we experience as our 'self' depends first of all upon this inner situation of our organism as a physiological system." Perhaps such fields of force originate between the people and their environment, including other people. He admitted that not all directed attitudes are as obviously linked with inner organic states as, for example, hunger or sex. But the important point was that social "grouping [is] dependent upon these directed attitudes." Indeed, "grouping based upon social forces is much more interesting than purely visual organization." When four people sitting around a table decide to play bridge, for example, they suddenly become two groups of two. "A great many of the problems of social psychology, some of them much more important and serious than these, will acquire a new aspect as soon as we consider them in such a concrete manner."[37]

Implications for Köhler's psychophysics

Evidently Köhler was trying to combine Wertheimer's ideas, his own physiological model, and some concepts of Kurt Lewin, who was pursuing a similar line of thought in Berlin at the time (see Chapter 16). But it was Karl Duncker's work on frames of reference in motion perception, also published in 1929, that he cited when he later acknowledged the fundamental challenge these issues posed for his psychophysics. As he noted in a 1933 paper, the processes in the brain corresponding to the phenomenal "self" will necessarily be "different from those corresponding to 'outside' experiences," because of the bipolar structure of the person–environment field itself and the concomitant role of directed attitudes in perception. But the "central retina" (*fovea centralis*) and the psychophysical correlates of the phenomenal ego are united by "a very strong link," as shown by the fact that when we say "we look," we mean that we have a visual field organized around this central region. Drawing on Duncker's research, Köhler suggested that it is possible "that in eye movements not so much the entity AB changes position as the fixation point and with it the direction of vision, also simultaneously attention, the directedness of the ego to a field part." Needed, then, was a psychophysical account of such shifts as Duncker had shown can occur in the person–environment relation in induced motion. However, he acknowledged, "a more general and complete treatment of these issues can occur only when at least the first foundations have been achieved for a psychophysics of the third spatial dimension, the connection of the different sensory regions and the region of the ego."[38]

Thus, Köhler himself recognized the considerable problems that remained to be solved before he could apply his heuristic postulate of psychophysical isomorphism to the whole of psychology. We have already encountered a similar problem in the discussion of Gestalt research on learning and memory in the previous chapter. Köhler did not shrink from such challenges. To understand why, it is necessary to clarify a fundamental theoretical commitment referred to only implicitly until now.

Early in the century, Max Planck had pointed out some troubling implications for physicists of Ludwig Boltzmann's work in the kinetic theory of gases. Boltzmann was able to predict the behavior of gases in a closed container, but only by making a variety of radical assumptions, for example that the boundaries of the atoms involved were flexible to a certain extent, not rigid as prescribed by Newtonian mechanics. Even then he could calculate only the relative probability of macroscopic states, not the locations or motions of individual atoms or mass points. His assumption of randomness – that the elements involved behave independently of one another – meant that, at the microlevel at least, entropy was "a measure of disorder," as one commentator has put it.[39] This seemed to contradict accepted interpretations of the second law which posited equilibrium, an ordered state, as the end point of motion. Planck tried to resolve the issue by saying that there were two levels of physical reality, each with its own set of laws.[40] This raised serious problems for any notion of unity in nature – and also, implicitly, for Köhler's psychophysics, which posited similar structural principles in both the physical and the psychical realms.

In his 1927 paper on regulation, Köhler entered this debate by contending that "pure statistics alone would never explain the distribution, improbable in itself, of all free electrons on the surface of a conductor." Entropy increase is a necessary condition of such equilibrium states, but is not sufficient to explain why just this distribution and no other occurs. In 1931, he took a more extensive stand on the issue in a paper presented to Hans Reichenbach's Society for Scientific Philosophy in Berlin. The results appeared in *Erkenntnis,* the journal of the logical empiricist movement, even though Reichenbach made it clear to Köhler in a letter that he did not agree with the argument.[41]

As Köhler pointed out, Boltzmann had assumed that in their initial state, most gas molecules in closed containers followed fully predictable, not random paths. In recent work, theorists had increasingly taken randomness as the normal state and ignored Boltzmann's assumption in favor of "laws of chance or the roulette wheel." Köhler thought it was possible in principle to compute the gas molecules' various "system paths" (*Systemwege*) and interpret the results dynamically. If this could be done, Boltzmann's theorem would "not at all require the indeterministic interpretation that has become customary." In any case, he insisted, "there is no need to turn away from the productive presupposition of a unified conception of natural processes." Clearly, Köhler did not reject the possibility of a fully deterministic explanation of molecular motions. He regarded probabilistic explanations as pragmatically necessary, but not as the last word on the subject. In this, as in so much else, he was in full agreement with Planck.[42]

What appeared then and surely still appears to be a backward-looking attitude on determinism makes better sense in light of Köhler's own theorizing and research, which dealt with macro-, not microlevel processes. As he later wrote, the indeterminism of Heisenberg's uncertainty principle obtains only for the latter. The fields surrounding atomic particles are surely "no less 'real' than the particles" themselves. Thus, for a physiological model of brain processes underlying the perception of wholes and their relationships there is no need "to choose between the microscopic and the macroscopic aspect of nature." We need only give up the assumption that reducing macro- to microlevel events is the only possible route to causal explanation. In any case, "only macroscopic structures . . . can be common characteristics of the perceptual and the physical world."[43]

Thus, though Köhler was willing to give statistical mechanics and quantum physics their due, he clearly shared the unease of his Berlin colleagues Planck and Einstein at the thought of a bifurcated physical world, as well as their loyalty to field dynamics and determinism. It was his dual commitment to continuous–field physics and the unity of nature that made it seem natural to him to conceive the person–environment relationship and even the social realm as fields of forces paralleled by hypothetical field activity in the brain – this in spite of the fact that he and his co-workers were discovering constantly how different the psychical order is from the physical. Other scientists who were closely allied with the Gestalt theorists in other respects did not share Köhler's commitment to this view of physics, or of the relations of physics, biology, and psychology to one another.

16

Variations in theory and practice: Kurt Lewin, Adhemar Gelb, and Kurt Goldstein

Working with the Gestalt theorists in Berlin or Frankfurt did not necessarily mean accepting all of their thinking, though it did mean sharing some of their terminology and methods. Like Koffka and Köhler, Kurt Lewin was a student of Carl Stumpf, and throughout the 1920s he worked closely with Köhler and Wertheimer in Berlin. Lewin employed Gestalt concepts in studies of action and emotion, but he did so on the basis of a concept of science that differed significantly from Köhler's. Moreover, in contrast to the Gestalt theorists' relative distance from practical concerns, he tried to recreate laboratory psychology in a way that would bring it to bear on contemporary problems in the workplace and the school. Adhemar Gelb and Kurt Goldstein pursued a medical version of the same project when they carried Gestalt-related thinking and research into studies of brain-injured patients in Frankfurt. But the theoretical conclusions that Goldstein drew from that research separated him from the Berlin school.

All three scientists faced the scientific issues raised by psychology's transition in German-speaking Europe during the 1920s from a largely academic discipline to a science-based profession. Rather than simply transferring existing theories and laboratory methods to other settings, or bringing new kinds of observers into the laboratories, they tried instead to reformulate both the theoretical and investigative enterprises in psychology to suit new purposes. By drawing upon Gestalt discourse and research practices in their efforts, they implicitly probed the Berlin school's potential and its limits.

Kurt Lewin

Like the Gestalt theorists, Kurt Lewin (see Photo 14) accepted the double identity of natural scientist and philosopher.[1] In doing so, he developed an original account of psychological concepts that he hoped would clarify – and establish – their scientific status, while assuring their autonomy from those of other natural sci-

Photo 14. Kurt Lewin. Archives of the History of American Psychology.

ences. He paired that project, however, with another one – to make psychological theory practical. Gestalt concepts were prominent in both efforts; but the political and philosophical commitments he brought to them predated his encounter with the Gestalt theorists.

Lewin was born in 1890 in the town of Mogilno, in the Prussian province of Posen.[2] His father, Leopold, owned a general store, above which the family lived, and a small farm. The family appears to have been relatively comfortable, but theirs was not the upper-bourgeois milieu of the Gestalt theorists. The Lewin children grew up with an awareness of exclusion from both Prussian and Polish social circles, though they were assimilated enough to celebrate Christmas in addition to attending Jewish religion classes at school. Lewin's biographer suggests that his open, extroverted temperament and the relatively nonhierarchical, interactive research style that grew from it can be traced back to his energetic, articulate, and ambitious mother, Recha.

To improve his educational opportunities, Lewin was sent to board with a family in Posen; when his family moved to Berlin in 1905, he entered the elite Kaiserin-Augusta Gymnasium. After one semester each studying medicine and biology at the universities of Freiburg and Munich, he returned to Berlin in 1910. There he was attracted by the efforts of the Neo-Kantian philosophers of science Alois Riehl and Ernst Cassirer as well as those of Carl Stumpf to combine philosophy and natural science. While a student, he also encountered a lively group of socially and politically progressive students with whom he organized evening courses for working-class adults, took long hikes in the country on weekends and, as one participant recalls, had "long discussions about democratizing Germany and liberating women from the conventional restrictions on their freedom." He joined the socialist student movement, becoming an active member of the Berlin Free Student Association in 1910 and attending the association's 1911 national convention. There he met Hans Reichenbach, with whom he would later share intellectual as well as political commitments.[3]

Lewin indicated his desire to connect scientific psychology with practical issues most clearly and forcefully in a monograph entitled "The Socialization of the Taylor System." This he published in 1920, just after the abortive German revolution, in a series entitled "Practical Socialism," edited by his friend from the youth movement and later neighbor, independent Marxist thinker Karl Korsch.[4] The essay reveals both his sympathy with independent socialist thinking and an underlying practical aim of even his academic research – to find a way for scientists and workers to work together to humanize the factory system.

Lewin did not object in principle to Taylorism's attempt to discover quantitative laws of performance that could rationalize production and thus increase output. Instead he criticized capitalism's use of that effort to maximize profit rather than workers' well-being. The fundamental distinction was between labor's "outer" value as a source of profit for the capitalist or a means of livelihood for workers, and its "inner" value as a source of personal satisfaction. Thus, under socialism labor should be distributed according to its "inner value." If workers were assigned to jobs according to their abilities in a cooperative effort involving management, workers, and psychologists, both productivity and job satisfaction would be enhanced. But in 1920 Lewin also wrote that achieving such a distribution of labor presupposes both freedom from the necessity of working for wages and freedom from the ideological values attached to occupations under capitalism. As long as being a bad officer is still "better" socially than being a good laborer, he argued, fundamental change would not happen.[5]

As the prospects for socialism in Germany faded, it became necessary to think about humanizing the workplace and social practice in general even under capitalism. This remained Lewin's aim throughout the 1920s and even beyond. Films he made in working-class areas of Berlin, work with retarded children, service on examination commissions for blind children, and his continued interest in applied research all exemplified this attitude.[6] Though he continued to teach philosophy as well as psychology courses, his position in the Berlin institute was in the department of applied psychology. In a 1928 paper on the textile industry published together with his nominal superior, psychotechnologist Hans Rupp, Lewin stated quite clearly that he wished to consider work as a process, and "man and machine as a dynamic unity," in order "to reshape that process."[7]

Lewin's pluralistic concept of science

But what concept of science would support the psychological expertise required for such a project? Lewin attempted to answer that question at the theoretical level by combining a viewpoint derived from his earlier studies in philosophy of science with terminology from Gestalt theory. In the 1920s, he elaborated what he called a comparative theory of science, which had three outstanding characteristics: it was empirical, pragmatic, and pluralistic – in decided contrast to Köhler's physicalism. Instead of deriving the natural and the human sciences from different kinds of experience, as Wundt and Stumpf had done, or following the Southwest German Neo-Kantians in establishing transempirical norms for science as such, Lewin adopted the approach Cassirer had taken in his book, *Substance and Function* (1910), based on comparative studies of the concepts scientists actually construct.[8]

Lewin's first and most extensive attempt to realize this program was a study of the concept of time series in physics and biology published in 1922. The central point of this book was that the treatment of objects in time series in physics and in biology are not the same. When physicists refer to a particle of matter persisting in a series of instants, this constituted in Lewin's view as much a genetic (i.e., a temporal) series as a reference to the path of energy from a lump of coal to a steam plant and thence to a light bulb. In both cases what he called "genidentity" (*Genidenität*) – continuous existence over time – is attributed to the object in question. When an egg develops into a chicken, biologists also refer to the life history of a single organism, even though the end product might have no molecules in common with the initial stage, except perhaps in the germ cell. In evolutionary history, biologists construct series of species linked by descent, even though there may be no proven material linkages between them. Thus, in both sciences the relations in question are constructed, not observed; but the entities defined as persisting over time differ not according to some essential difference between "living" and "dead" matter, but according to the point of view required by the scientific task at hand. Though it is possible to compare equivalent concepts among these sciences, Lewin concluded, the concepts of biology are not reducible to those of physics.[9]

In the 1920s, Lewin extended this pluralistic, pragmatic analysis to other disciplines. As he argued in a 1925 lecture, because the sciences examine different problems, equivalent objects have different "modes of existence" in them. Iron, for example, is not the same in physics and in economics. The price of iron bars is an important feature of iron to an economist, but is not part of the object "iron" at all to a physicist.[10] In current philosophy-of-science terminology, Lewin argued that the same word has different meanings according to different contexts of use. Also important was Lewin's claim that these conceptual object-worlds develop over time. Sciences are comparable with one another, he insisted, only if we take them at equivalent stages of development.

Lewin was in contact with the founders of the logical empiricist movement, particularly Hans Reichenbach, throughout the Weimar period. In the early 1920s, Lewin and Reichenbach, then instructor of physics at the technical academy in Stuttgart, made intensive, though unsuccessful efforts to found a journal for empirical philosophy of science; for a time Köhler was also involved in the project.[11] Lewin presented an outline of his comparative theory of science at a 1923 conference in Erlangen, attended by Reichenbach, Rudolf Carnap, and others who, in Carnap's words, "had the common aim of developing a sound and exact method in philosophy." Reichenbach and Lewin shared an interest in illuminating the conceptual foundations of science by examining actual scientific concepts rather than a priori categories. Reichenbach praised "The Concept of Genesis" for this reason in the *Psychologische Forschung*. Later he took up the idea of "genidentity" in his discussion of relativity theory. Lewin, for his part, cited Reichenbach in his discussion of "strict law" in science (see below). But he rejected Rudolf Carnap and Otto Neurath's later call for a "unified science" based on physical language. Lewin's conception of science employed physical concepts, but without physicalism.[12]

In Lewin's view, it was possible to derive what Reichenbach had called "strict," exceptionless laws for psychology without reducing psychical phenomena either to physical events or to interactions of psychical "elements." As the appropriate explanatory objects for psychology he proposed "event types" (*Geschehenstypen*), thus taking the type concept from morphology and then-popular typological personality theories and applying it to behavioral units analogous to ideal types in sociology. He expressed the hope that laws referring to "pure" cases of these event types would eliminate what he called "historical-geographical" factors, such as previous experience or heredity, from psychological explanations, just as physicists seek to eliminate "residual factors" (*Restfaktoren*) as much as possible from their laws. Thus, ironically, just at the time when statistical explanation was coming to the forefront in both quantum physics and in American psychology, Lewin opposed the idea that psychology was or ought to be limited to "merely statistical laws," or "mere rules," as he also called them, such as those advanced by Hermann Ebbinghaus for memory.[13]

In his famous paper on the transition from "Aristotelian" to "Galilean" thinking in psychology (1931), which appeared in the journal *Erkenntnis* – the organ of the logical empiricist movement – Lewin disparaged the dependence on statistical laws in psychology by calling it an example of "Aristotelian" thinking. In the so-called Aristotelian physics of Galileo's opponents, the classification of phenomena defined their essence; hence there were different laws for different kinds of motion. Examples of this kind of normative-typological categorizing in psychology include the terms normal and pathological themselves, or "the obstreperous three-year-old," or the rankings pervasive in intelligence testing. In such thinking, Lewin argued, lawfulness is equated with frequency, and unusual events are regarded as unnatural, even incomprehensible. The result is that the validity of psychological laws appears limited to specific populations, for example, to one-year-old children in Vienna and New York in 1928. In contrast, he noted, psycho-analysis had already proposed to derive general laws in psychology by positing dynamic principles of the unconscious that held for both normal and pathological cases. But he criticized Freud and Adler for overgeneralizing from particular types of behaviors to all of psychology, and pointed out that modern physics conceives the physical world as an all-embracing unity, eliminates normative categories and derives universal laws from concrete, albeit ideal cases.[14]

In a "Galilean" dynamic psychology, Lewin wrote, there is no opposition between the general validity of a scientific law and the specifics of individual cases. Instead of computing "the abstract average of as many historically given cases as possible," researchers would focus first on "those situations in which the determinative factors of the total dynamic [event] structure are most clearly, distinctly and purely to be discerned."[15] Lewin proposed to expand the reach of phenomenological observation still further than the Gestalt theorists had already done, by recreating situations in which the structures of ideal-typical person–environment interactions could be made to appear, as it were, in the laboratory. This he envisioned as a necessary basis for deriving formal, ultimately mathematical, descriptions of their dynamics, in a manner analogous to the way in which Galileo had derived the law of free fall from the relationships inhering in an ideal case – that of downward vertical motion in frictionless space.

Such language pointed both to a similarity and to a significant difference between Lewin's concept of science and that of the Gestalt theorists. From Cassirer's discussion of Galileo's procedure in *Substance and Function* he had learned that "the first goal of experimental inquiry" in classical physics "is to gain a *pure* phenomenon," that is, as close an approximation as possible to an ideal case.[16] Lewin's claim that psychological laws could be derived in an analogous way from idealizations of concrete psychological situations met well with Wertheimer's demand for experiments demonstrating the "essential laws" of "good" phenomena. However, the laws he sought were neither physical nor physiological, but explicitly psychological, even though his language was rich in physical metaphors (see below). Though he and the Gestalt theorists agreed that the given was only a starting point, and that the structures that generate or account for the given were not themselves conscious, the postulated functions or forces doing the explaining were different.

Lewin's Gestalt theory of action and emotion

Lewin developed his conception of psychological science in tandem with an extensive research program on the psychology of action and emotion. Like his philosophy of science, that program exhibited both affiliation with and independence from Gestalt theory. Lewin had established his reputation as a researcher with experimental studies of will, an issue discussed in Stumpf's seminar when he began his training. These studies directly challenged associationist theories of volition. Psychologist Narziss Ach, an adherent of the Würzburg school, had suggested in 1910 that the "determining tendencies" stimulated by an experimenter's instruction inhibit subjects' tendencies to recall associative connections they had already learned; the resulting delay in carrying out the instruction would thus be a measure of will. Lewin tested this by asking observers to learn lengthy series of meaningless syllables, then instructing them either to reverse or rhyme the syllables. The prediction was that they would either take longer to complete the "heterogeneous" second task or give wrong answers. To his surprise, he generally got no "inhibitive delay," and only a few errors.[17] From this and further studies, he concluded that associationism is "untenable," because the predicted effects failed to occur even under optimal conditions. The Gestalt theorists later cited these studies in their own attacks on associationism. As Köhler noted, however, Lewin argued that not self-organizing processes in the brain, but an unconscious psychological process – the dynamic influence of the motivation introduced by the experimenter's instruction – was decisive.[18]

After Wertheimer published his progamatic paper of 1921 (discussed in Chapter 14), Lewin worked his way into Gestalt-style research with a 1923 paper on the perception of upside-down figures and a 1925 study with Japanese visitor Kanae Sakuma on depth effects in motion perception.[19] Only after this did he begin to apply Wertheimer's and Köhler's concepts to action in two papers, "Preliminary Remarks on the Structure of the Mind" and "Intention, Will, and Need" (both 1926). Lewin fully accepted the claim Koffka had advanced in 1915, that actions, like percepts and the "knowledge processes" that produce them, are structured wholes (see Chapter 9); his term was "action wholes" (*Handlungsganzheiten*). As he argued in 1928, only by proceeding from such action-wholes can one determine which parts are psychologically relevant for the work process. But he enriched the concept of organism–environment interaction that Koffka had developed in two ways. First, he distinguished the physical setting and the way it appears to the actor, concentrating on the latter. Second, he suggested that the psychical person is itself a complex, "layered" (*geschichtet*) whole. The needs that influence a person's interaction with the (perceived) environment can thus come from different layers. Usually, he focused on the "surface" layer, which he called in places the Ego, following Freud, and in others the *Motorik* because of its intimate connection with action.

Lewin's reference to the psychical person was in keeping with the trend toward personality theories in Germany during the 1920s, and it marks his most important difference from Gestalt theory as originally articulated. In visual perception, physical stimuli can be considered releasers for processes in the optic sector. For the psychology of action and the emotions, Lewin argued, stimuli act as "activators" for processes within the psychical organism, which only then result in action or emotion. Thus, a stimulus is effective "not according to its physical intensity but according to its psychological reality" for the actor. Examples of this are when the arrival of a letter leads to a long journey, or an innocent remark precipitates an angry reaction. Reorganization of the perceptual field may lead to a "restructuring" of behavior, as Wertheimer and Köhler suggested, but this was not the only or even necessarily the primary source of such changes. Rather, "the totality of the forces present in the psychical field controls the direction of the process"; this is not limited to perceived objects and their relations to an actor, but can include objects and needs of which the actor is not conscious.[20]

Thus, in contrast to Koffka and Köhler, Lewin emphasized that referring to the phenomenal form of an action is insufficient by itself. A complete, causal account of actions must also discover the dynamic motivational forces that "carry" them. These he imagined to reside in the Ego or psychical self, which is in turn a functional part region within the psychical totality that includes both the organism and its (psychical) environment. Both the self and the field in which it lives clearly develop over time, but Lewin opposed psychoanalysis' emphasis on the determinative role of early childhood experiences. Though this may be true in a general sense, and specific childhood experiences may be very significant for present psychical processes in some cases, Lewin maintained that such influences are in most cases as extensive as the influences of fixed stars upon physical processes on earth. These undoubtedly exist but are "extremely small, approximately zero." For an understanding of behavior, it is only necessary to concentrate on the structure of the field at a given time – that is in the physical present.[21]

Despite his emphasis on the independent reality of psychological dynamics, Lewin was certainly not shy about using terms from physics as metaphors for psychological processes. He did not call his approach "field theory" until after he emigrated to the United States in 1933, but the expression "field forces" is ubiquitous in these theoretical papers. He even employed Köhler's concepts directly, saying that the person–environment relationship consists of "multiple strong Gestalten" in communication with others, forming a single, "weak" Gestalt. Most famous of these "strong Gestalten" are those related to what he called the "demand character" (*Aufforderungscharakter*) of objects. The roots of this concept are already visible in Lewin's provocative essay, "War Landscape" (*Kriegslandschaft*) (1917), written while he was convalescing from wounds he had suffered at the battlefront. There he observed that objects change their appearance from being "peace things" to "war things" when they become part of a battle zone. To see a house as a source of firewood, for example, would be barbarous in peacetime, but quite normal and maybe even necessary in war.[22] The example he used in 1926 was a "peace thing": a mailbox has a different relation to me when I have a letter in my hand than when I do not. In the former case, the mailbox seems almost to jump out of the environment and announce its presence.

Lewin attributed such phenomena to the presence of what he called "quasi-needs," contending that objects related to them exert greater psychological "forces" or become more prominent in the field than at other times. To account for these he suggested that dynamic "tension systems" emerge in specific psychical "regions"; these function in the same way as the tensions caused by real needs, transforming the psychical environment in accordance with the person's current intentions.[23] Though he spoke only of psychical "energy" and strictly separated this from physical force, he followed Köhler in describing the satisfaction of needs as a process that reestablished equilibrium "at a lower level of tension." In a move reminiscent of Freud's concept of displacement, he noted, however, that quasi-needs, like true needs, can be satisfied in various ways. In the case of the mailbox, for example, substitute satisfaction can be had by having a friend mail the letter. This language sounds physicalistic, and Lewin's pluralistic philosophy of science sanctions the use of such terminology, but with the clear proviso that analogies only are involved. The "tension" in such systems, for example, is not directly measurable, as is the tension in a coiled spring. This allowed Lewin to have the legitimating benefit of reference to physical terminology, without adopting Köhler's more literally physicalistic mode of explanation.

Research as social practice

The group of students who gathered around Lewin in Berlin were already putting these concepts to work as he developed them. In doing so, they worked through a comprehensive research program on the psychology of action and emotion, topics that had long been thought inaccessible to rigorous experimental research. By recreating ordinary-life situations in the laboratory, they came remarkably close to realizing the Berlin school program of reconciling science and life – as well as Lewin's primary commitment to humanizing social practice. Only two of these studies can be examined here.

Nearly all of the studies from Lewin's group illustrated the importance of interpersonal interaction, and so did the group's own dynamics. Instead of the distance and hierarchy between teacher and student that were traditional in German universities, Lewin and his students met regularly for vigorous discussions of all sorts of issues, usually in the Schwedische Cafe across the Schlossplatz from the Institute. At first they gathered on Saturdays, then more often; eventually the group got a name, the *Quasselstrippe,* or "chatter line." One participant characterizes the creative process in the group as "one long discussion."[24] One evening someone expressed amazement at the cafe waiter's apparent ability to remember what everyone had ordered without writing anything down. Some time after they had paid, Lewin called the waiter and asked what they had ordered. He replied indignantly that he no longer knew. This was the stimulus for Bluma Zeigarnik's famous investigation of memory for completed and uncompleted tasks.[25]

Narziss Ach had argued that volition creates a dynamic connection between stimulus and action. If this is so, then the tendency to carry out intentions should not persist longer than the impact of any physical stimulus impinging on an organism. The waiter's behavior showed, however, that at least certain intentions could be retained far longer, without conscious effort. Zeigarnik's work and Maria Rickers-Ovsiankina's study on the resumption of interrupted tasks showed in different experimental situations that intentional acts can occur, in fact are more likely to occur, after a significant delay.[26] Zeigarnik presented a series of mainly manual tasks to a total of 164 subjects, instructing them to complete these as rapidly and correctly as possible. She interrupted them half the time and allowed them to finish the other half. Afterward she asked the observers to recall the jobs they had done, and say which ones had been most or least interesting or pleasant. She found that uncompleted tasks were indeed remembered much better than completed ones – roughly 90 percent better in adults and 110 percent better in children.[27] Lewin's and Zeigarnik's explanation for the results was that the experimenter's instruction and the uncompleted task created competing "tension systems." Because the system established by the uncompleted task was generally stronger, it remained in memory until released by completing the task, especially if interruption occurred past the midway point or near completion. However, Zeigarnik also argued that fatigue led to a "loosening" of the tension system and could even reverse the results.[28]

This study showed clearly how the Lewin group's methods differed from those of the Gestalt theorists. In an unpublished 1918 paper on the training of subjects for self-observation, Lewin had developed a conception of the experimenter–subject relationship as an interactive process. The aim was not to exclude "bad" subjects, but to overcome their performance anxiety and clarify their responses, in order to go beyond sensory functions to the psyche's affective and voluntary dimensions. To achieve this, Lewin proposed that the experimenter vary the instruction and ask and answer questions from subjects, so long as these remained directed toward a "correct [observational] attitude" and avoided any hinting at expected responses. He conceived the experimenter–subject relationship in moral terms; the aim should be sincere collaboration in a common effort. His conception of training correspondingly emphasized positive incentives, the integrity of the experimenter-trainer, and the interest of the student-trainees in the task.[29]

Zeigarnik reported that she behaved differently with different types of subjects. To those who worked to please the experimenter she showed her pleasure; with those who ambitiously strove to succeed, she examined their work closely; only with those who appeared involved in the task for its own sake – often children – did she remain passive. As she later said, varying her behavior this way was not easy; "I was like an actress.[30] Clearly this was not the cool objectivity at work in the Gestalt theorists' experiments. Nor did such behavior conform to the ideal of experimenter–subject interaction Lewin had described ten years earlier, which had emphasized mutual honesty in a common effort. Zeigarnik reported that subjects often objected, sometimes strenuously, to being interrupted, and some suspected a trick. In fact, Lewin's experimenters sometimes resorted to outright deception to elicit certain behaviors.

This was certainly true of Tamara Dembo's dissertation on the dynamics of anger, published in 1931.[31] This study was a direct response to Lewin's call for experimental studies to clarify the specific meaning of, and the forces at work in, the processes theorized about in psychoanalysis. To be tested in this case was the claim that frustration could release emotions and aggressive wishes stored up from childhood memories. Dembo's experimental design was again a "Galilean" procedure, creating a situation in which the dynamics of anger might express themselves directly. The setup created a radical change in the experimenter–subject relationship, from one of social equals to one based on power. It involved an actual struggle between subject and experimenter, who deliberately frustrated subjects' efforts to complete the assigned task, then prevented them from leaving the room. The tasks – throwing rings onto bottles, or finding a way to grasp a flowerpot outside a prescribed area in the room – appeared sufficiently challenging to arouse interest, and subjects joined in the fun at first. But as it became clear that the mutual interchange essential to good games was not in evidence – for example, when they got no further in the flowerpot game but the experimenter continued to insist that there was a solution – a variety of reactions ensued, some quite irrational. Some subjects, for example, began to daydream; others imagined they were out of the room or thought about other plans. Ultimately rage broke out that even extended to attacks on the experimenter.[32]

Dembo listed more than forty categories of responses, a variety far too great to be reducible to associations of "feeling elements" and volition. Instead, she tried to develop what she called a topological explanation, emphasizing the emergence of inner and outer "barriers" between subject and solution, or between subject and the outside world, in interaction with the "firmness" and "loosening" of tension systems within the person. Thus, some particularly controlled or inhibited subjects would express their anger very quietly at first, then break down completely as their intrapersonal scaffolding collapsed.[33] In 1922, Lewin had already presented topology as a geometry for defining space–time relations independent of any measuring system. By the late 1920s, he had transformed this abstruse branch of mathematics at least programmatically into a device for the representation of psychological field forces, with the implicit hope of moving eventually to a process-oriented rather than a performance-oriented concept of measurement. To realize that ideal, he required a way of generating formal representations of concrete psychological situations, as well as "the concrete structure of the psychological person and its internal dynamic factors." Dembo's study was the first to integrate this programmatic claim with experimental results.[34]

Social fields of force in the life world

Lewin considered Dembo's work as fundamental as that of Zeigarnik, because for him this was a study of change in the structure of psychological fields in conflict situations. The practical relevance of work on conflict was especially obvious in the field of education, and Lewin immediately emphasized the point in a 1931 talk to a group of Montessori educators entitled "Education for Reality." There he brought together for the first time two terms that would become central to his later thought – the psychological "life space" (*Lebensraum*) and "social fields of force."[35] The first was a colorful name for the individual's psychological environment, where Lewin had long located the dynamics of action and emotion; the second stood for the obvious but until then little explored claim that other people, or our images of and feelings about them, are important parts of that environment.

In his talk Lewin summarized the aim of all pedagogy succinctly as "the extension of the psychically present life-space of the child." This meant enabling children to grow and mature, to stretch both their imaginations and their actions toward goals lying ever further away from the needs of the immediate present in which infants live. Deciding how to accomplish that task meant striking a balance between freedom and intervention. For Lewin this was not a question of worldview but a practical matter of organizing social field forces – or allowing them to organize themselves – in the (psychologically) real world. He agreed with the basic Montessori idea of allowing children free choice of goals, but pointed out the difficulties in realizing that program. Inevitably, even in a progressive school, the needs and wishes of adults are among the forces at work in the child's psychical environment. In authoritarian settings everything is oriented to these wishes. This has the advantage of clearly structuring the child's situation in the present, but when parents' or teachers' authority becomes weaker, usually in puberty, the child can lose his or her identity altogether. Lewin thus opposed advocates of rigid hierarchy and discipline in education or child-rearing on practical grounds: "*objectivity*" (that is, a healthy separation between reality and unreality) "*cannot arise in a constraint situation; it arises only in a situation of freedom.*"[36]

From his early essay on Taylorism to this talk to teachers during the Depression, Lewin's aim remained consistent – to humanize social practice in the factory and the school as well as in the laboratory. Koffka's remarks on society and education (see Chapter 15) and the pedagogical examples in Wertheimer's work on thinking suggest that the Gestalt theorists were in sympathy with this goal; but they did not devote their energies to achieving it in the research arena to the same extent as Lewin did. He clearly made the ideological and social power dimensions of this issue more explicit than anyone else in the Berlin school. In another 1931 paper on the psychology of reward and punishment, for example, he stated that the effectiveness of parents' efforts to discipline their children depends on "ideology," by which he meant shared moral values and beliefs. Arbitrary punishment when trust had reigned before can cause a "revolution in the ideology of the child." In such remarks Lewin acknowledged that parent–child or teacher–child relations are ultimately about power and its limits, and that many constituents of the psychological field "are determined by social rather than psychical facts."[37] Unfortunately, he did not elaborate this argument in any detail, nor did he experiment with groups as units or consider the impact of national cultural differences on education until after he left Nazi Germany for the United States in 1933.[38]

In 1935, introducing American readers to his Berlin students' research, Lewin penned an eloquent testimonial: "The fundamental ideas of Gestalt theory are the foundation of all our investigations in the field of the will, of affection [sic], and of the personality."[39] Nonetheless it should be clear from what has been said why, as one American who studied in Berlin put it, "Whether the research of the Lewin group was consistent with Gestalt theory was an important topic of discussion" at that time.[40] The unity of perception and action was one of the cornerstones of Gestalt theory. Thus, it was only natural to transfer Gestalt principles to the study of action and emotion. Yet precisely the pluralistic theory of science, and the emphasis on the independent reality of psychical forces, that made it easier for Lewin to bridge the gap from perception to affect, and from psyche to society, separated Lewin's approach from Köhler's more stringently physicalistic concept of science. Though much of his terminology obviously came from Köhler's field theory, Lewin consistently made it clear that he saw these terms as analogies only. On Köhler's isomorphism he remained agnostic. Köhler confirmed that sense of distance even as he warmly supported Lewin's application to extend his assistantship in 1929. He wrote to the Ministry that Lewin had become not a new proponent of Gestalt theory, but "the driving force of a new and valuable research approach [*Forschungsrichtung*], which he has introduced and whose further development depends completely on him."[41]

Gestalt theory, neuropathology, and the organism as a whole: Adhemar Gelb and Kurt Goldstein

Adhemar Gelb (see Photo 15) began his career with a dissertation for Riehl and Stumpf on Gestalt qualities (discussed in Chapter 6). In it he accepted Stumpf's position that these are the results of "summarizing" psychical functions. After serving as an unpaid assistant in Berlin for one year, he took a paying position in Frankfurt in 1912, replacing Koffka. Though he was not a subject in Wertheimer's motion experiments, he clearly discussed the new Gestalt concept with its creator. In 1914, he reported experiments on space and time perception to the Society for Experimental Psychology that he presented as examples of Wertheimer's "Gestalt laws." Wertheimer referred to Gelb's results in the discussion remark at the same meeting in which he announced those laws (see Chapter 8).

Photo 15. Adhemar Gelb. Psychological Institute, University of Frankfurt am Main.

Later that year Gelb took on a new assignment, for which other work of Wertheimer's helped to prepare him. In a villa near Frankfurt, neurologist Kurt Goldstein, then first assistant at Ludwig Edinger's university institute for neurology, had organized a hospital for brain-injured soldiers with private financing. To assist in both diagnostic assessment and research, Goldstein established a psychology department and recruited Gelb to staff it. The clinic's primary purpose was practical — to treat the men in order to return them to the front, or to determine their appropriate civilian employment and disability payments. This approach was consistent with the ethically grounded, social-welfare view of eugenics that Goldstein had expressed the year before in a lecture on "race hygiene." Since rehabilitation was the primary aim, "not only medical, but also psychological, pedagogical, and occupational measures" were necessary.[42]

Goldstein had another reason to cooperate with a psychologist. Since the 1870s neurological research in Germany had been dominated by a paradigm associated with Carl Wernicke, a leading researcher on the speech disorder called aphasia. In this view brain disorders were disturbances of simple perceptual or motor functions conceived as psychological reflex arcs, the traces of which were stored in "cerebral centers" located near the brain zones identified at the time with higher functions. Researchers following Wernicke's lead proceeded to identify many such disturbances, with names like agnosia, agraphia, alexia, and acalculia, culminating in Hugo Liepmann's delineation of a syndrome he called apraxia, a disturbance of sensorimotor coordination. Goldstein had participated in this tradition by writing a case report on a patient in Liepmann's clinic in 1909; but his ambiguous results indicated the situation was not as cut and dried as Wernicke's paradigm had made it seem.

By this time, criticism from Sigmund Freud, Pierre Marie, Konstantin von Monakow, and others had begun to take its toll. In particular, the fact that many patients could compensate to some extent for lost functions indicated that one brain region could take on the function of another, thus undermining strict localization. The issues paralleled those being raised in psychology at that time – the relation of peripheral and central, "lower" and "higher" brain centers – more broadly, the efficacy of explaining pathology "from below," as disturbances of fixed reflex paths to specifically localized centers. The parallel was not incidental. Wernicke's paradigm had been based on what he took to be "commonly accepted laws of experimental psychology – just the "laws" of association that then came into question. No wonder Goldstein wanted to work together with a psychologist.[43]

Max Wertheimer probably learned of this debate while working with neurologists Arnold Pick in Prague as well as Otto Pötzl and Sigmund Exner in Vienna. He entered the discussion directly when he tested motion perception on one of Pötzl's patients in 1911. Pötzl had reported that the patient, who had damage in both occipital lobes, saw a moving lamp as a number of lights. Wertheimer confirmed the finding with his hand-held slider apparatus (described in Chapter 8).[44] This seemed to support the claim that such functions were not localized part-operations, but said nothing about what they might be instead. Either some coordination of "centers," the corpus callosum – the tissue connecting the two halves of the brain – or, alternatively, Wertheimer's "whole processes" in the cerebral cortex itself might be involved. Goldstein's patients presented an opportunity to research such issues.

Particularly significant in Gelb's research was a modified version of the tachistoscope Wertheimer had used for his motion studies. Instead of having subjects look through a viewer, as Wertheimer and Schumann did, he projected figures, letters and words onto a screen placed about one meter away from the patient, between the patient and the apparatus (see Photo 16). He used the apparatus as a diagnostic tool to supplement standard tests of attention, concentration, and memory; Goldstein claimed it was also helpful in reading therapy.[45] Its most important use, however, was for finer analysis of specific functional disturbances. For this it was necessary to expand the boundaries of laboratory subculture by

Photo 16. Experiment with brain-injured patient, ca. 1917. Kurt Goldstein, *Die Behandlung, Fürsorge und Begutachtung der Hirnverletzten* (Leipzig, 1919). Reprinted by permission of Verlag Ambrosius Barth.

making the often uneducated patients into collaborators, training them to report what they saw without causal attribution. Apparently this was not easy, but Gelb reported that they soon got the knack of self-observation, producing results of impressive quality. He and Goldstein reported only a few cases in the first volume of a projected series, but claimed, using Wertheimer's language, that these were "pure" cases of the phenomena in question, and therefore of fundamental theoretical importance.[46]

Paradigmatic was the case of "Sch.," first reported in 1918. The patient, who was twenty-four when the study began in June 1916 had suffered two wounds to the back of the head a year earlier, one of which had penetrated the brain. He could speak and read, but was "psychologically blind" – he could not recognize figures or words directly, but had to trace them first with head and hand movements. Interestingly, he was ignorant of his own method, and thought he read in the normal way. Gelb and Goldstein called his condition "figural blindness," because "the word [or form] as a whole was not present in his perception." In their view motion perception gave a decisive diagnostic test for this condition. The patient saw neither real hand or arm motion, nor motion with alternate presentations of Wertheimer strips on the tachistoscope screen, but only series of "isolated places in space," like the patient Wertheimer had examined in Vienna; even these he saw only when presentation speed was very slow. But he did experience tactile motion, and recognized landscape features by inference. Shown a picture with trees on two sides and a gap in between, for example, he concluded that "it may be a road coming out of the forest because it is brighter" at the opening. Apparently Sch. compensated for the loss of form perception by reorganizing sensory functioning at a lower level, inferring the presence of objects from remaining sensory cues with the aid of touch. Hugo Lissauer had interpreted the syndrome, which he called "mental blindness" (*Seelenblindheit*), in terms of the then-standard distinction between "associative" and "apperceptive" centers, claiming that only the latter had been damaged. Citing Wertheimer's hypothesis that cortical "whole processes" underlay motion perception in normal subjects, Gelb and Goldstein argued that Sch.'s behavior showed loss of the physiological function connected with "Gestalt seeing."[47]

Later, Wilhelm Fuchs, a neurologist who assisted Gelb and Goldstein, presented additional case studies from the Frankfurt hospital that bore still more directly on Gestalt theory. The patients had a condition called partial total hemianopsia or amblyopia – disturbances or destruction in part of one or both visual hemisphere – due to bomb splinters in the back of the head. Fuchs's experiments showed that patients were quite capable of seeing objects as wholes despite their disability. They were seen, however, in locations different from the objective ones. Tests showed that displacement occurred primarily toward the "competent" side. Fuchs reasoned that "unitary whole processes" were involved, for if injured and uninjured parts of the cortex had worked independently, two separate localizations would have taken place. In other cases there were completion effects – that is, patients reported seeing whole figures even when a stimulus figure presented with the tachistoscope crossed both competent and "blind" portions of the retina. He claimed these were examples of "central completion," and argued that greater figural "cohesion" lowered the threshold for it.[48]

It seemed clear that an account based on the construction of such objects from elementary sensations could not explain perception where there was no stimulation. Association from past experience also appeared incapable of accounting for the phenomena; showing the patient familiar images such as those of a dog or a face gave negative results. Fuchs cited studies of motion perception in the eye's so-called "blind spot" by Koffka and his students to show that completion effects were not due to pathology alone. He argued instead that "only certain characteristic and hence coercive figures" produce completion, those for which seeing a part already implies seeing the whole. Walter Poppelreuter, who had obtained similar results with brain-injured soldiers in Bonn, had called the process "totalizing Gestalt apprehension." The name signaled his view that a psychological function, "a special act of apprehension must be added [to sensations] to achieve the apprehension of a form, a Gestalt." In hemianopsia, then, this "totalizing apperceptive function" remains intact, compensating for the damaged tissue.[49]

Fuchs's explanation required no such additional functions. In his view, the patients had developed what he called a "pseudo-fovea" – their visual fields had restructured themselves spontaneously around a new center, in such a way that objects in them appeared to be in normal relation to one another and to their surroundings. In ordinary vision, the physiological fixation point, then called the "central retina," and the psychological center of attention usually coincide. This produces a phenomenon called concentric narrowing of the visual field, a pathological condition already known to both Helmholtz and Hering. But Fuchs's patients never experienced their reduced visual fields as in any way restricted. They had clear perceptions of right, left, up, and down, and thought that they were looking straight ahead, even though they localized objects off center. In such cases, and perhaps in general, Fuchs claimed, "Attention does not organize the subjective visual field; the organization of this field assigns the center of attention."[50]

For the Gestalt theorists these cases showed that perceptions could indeed be built up or concluded from sensations, but only in the limiting cases characteristic of pathological and laboratory conditions. Even in pathological cases, as Fuchs showed, spontaneous reorganization of both the visual and the corresponding cortical fields could go quite far. However, it was hard to claim that the fragmentation of perception in these patients was the same as the unstructured wholeness of children's seeing. Hence it was difficult in this interpretation to link the pathological with the genetically or evolutionarily prior, as was customary at the time and as Koffka had done in his book on mental development (see Chapter 15). In addition, the suggestion that what had been damaged were physiological processes underlying "Gestalt seeing" was too vague to provide much insight into the anatomical structure or the functional characteristics of such processes.

More significant for Goldstein was that the Gestalt theorists' interpretation of these "pure" cases seemed limited to vision. The patients in question also exhibited disturbances of other sensory and cognitive processes. After he succeeded Edinger as professor of neurology in Frankfurt in 1919, he, Gelb, and their co-workers published two dozen more case studies from the hospital for brain injuries. Work on recovery of function in other sense modalities, and on speech and language disorders, led him in 1924 to make a broad distinction between "concrete" functions in aphasics, in which the world consists only of discrete, unconnected perceptions, and "categorical" – later "abstract" – functions involving symbolic ordering and representation of reality, and to characterize only the latter as disturbances of brain functioning as a whole.[51]

By 1927, however, Goldstein had begun to distance himself more explicitly from Gestalt theory. In a major handbook article he emphasized the importance of psychology to neurology in general, but argued that it was unwise for neurologists to let themselves be guided by whatever psychological views happen to be current. The widespread acceptance of associationism and the reflex doctrine had supported localization theory. Now that his and others' research on functional substitution and compensatory achievement were putting all of these doctrines on the defensive, he asked, "should we [therefore] for example become adherents of Gestalt psychology?"[52] Put in that way, the question appeared to answer itself.

Instead, Goldstein proposed what he called an "organismic," or "holistic" method (*Ganzheitsmethode*) to replace both the reflex-localization model and the diagnostics and treatment connected with it. The method consisted of three fundamental principles. The first and most important was the injunction to proceed from a phenomenological overview of the whole organism in a particular setting. Years earlier Goldstein had expressed the belief that not local injury alone, but "its influence on the whole person" was decisive in determining the full extent of patients' impairment.[53] This axiom was no more than good medical practice. But from this practical norm Goldstein drew two far-reaching theoretical conclusions: that the nervous system is "a network," a *unified apparatus* rather than a collection of separate anatomical parts; and that the organism is a "psychophysical unity," a self-actualizing individual which maximizes or optimizes its "preferred performance" in response to "irritation" from without. He later called this doctrine the "biological basic law." Though he spoke of "a mean point" to which excitations or behavior strive to return, not mere survival, and certainly not only the ordered equilibrium of a visual field, but the integrity of a biological individual was at stake. From this he derived his third basic principle: in pathological cases, recovery or reorganization of sensorimotor coordination is an achievement of whole organisms to compensate for "catastrophic" damage to their total functioning; there is only a gradual difference between such responses and minor disturbances in healthy functioning.[54]

Goldstein's view had certain similarities in terminology and in general intention with Gestalt theory. Aside from their common holism, the most obvious of these was his view of tension or "irritation" as a disturbance, and the restoration of equilibrium and balance as the primary goal of action. In later statements he even employed the term "centering."[55] Yet the object to which Goldstein's language referred was not the organism—environment relation as an objective structure, as it was for Kurt Koffka, but the organism and its functioning alone. Ernst Cassirer, Goldstein's cousin, was a frequent visitor to his clinic in the 1920s. For Cassirer, Goldstein's observations of "abstract" and "concrete" functions in speech were "negative proof" of the central importance of symbols in the constitution of a meaningful world; and Goldstein's description of the organism as a self-actualizing "psychophysical unity" echoed the terminology of Cassirer's Hamburg colleague, William Stern. Finally, Goldstein characterized the psychophysical relation much less specifically than Köhler. He treated the nervous system as an undifferentiated totality, whereas Köhler considered at least some systems, like the optic sector, as relatively separate from the whole.[56]

In his major theoretical work, *The Organism* (1934), written in exile in Amsterdam, Goldstein criticized Gestalt theory more specifically than ever before. A "good Gestalt" is not an end in itself, he asserted, but a means to an end, "a definite form of coming to terms of the organism with the world, that form in which the organism actualizes itself, according to its nature, in the best way." Köhler's emphasis on processes occurring within a relatively stable "topography" was an artifact of vision experiments in laboratories, where only one part of the organism is activated. "The whole, the 'Gestalt', has always meant to me the whole organism and not the phenomena in one field." For whole organisms in ordinary surroundings, the struggle for self-actualization is as important as the struggle to exist. From this protoexistentialist vantage point, Gestalt theory was merely "a psychological physiology."[57]

Goldstein's ideas received considerable attention in the 1920s and 1930s. His results and concepts appeared to coincide with neurologist Henry Head's work on aphasia and other speech disorders, Frankfurt physiologist Albrecht Bethe's studies of "neural plasticity" in insects, and Karl Lashley's work on what he called the "equipotentiality" of brain structures in rats.[58] His conception of an organism's mode of existence as a way of "being in the world" and his concept of self-actualization later had considerable impact on existentialist philosophy and psychology. As Ann Harrington has argued, his individualistic use of holistic vocabulary separated him from conservative thinkers who employed organicism as a marker for antidemocratic sentiment in the Weimar period.[59] However, the fact that recovery was rarely complete in his patients implied that there were inherent limits to plasticity. Moreover, it was not clear whether recovery even of complex functions required an undifferentiated kind of full-organism holism, as Goldstein claimed, or only a less rigid conceptualization of individual functions and their anatomical location.

The challenge to Gestalt theory at the time was clear: Could Köhler achieve the broad applicability he wanted for his field theory of brain action without adopting full-brain or organismic holism? Such differences as well as those already mentioned led to a certain tension between Goldstein and the Gestalt theorists in the 1930s. Koffka wrote to a colleague in 1938 that Wertheimer bore "a moral grudge" against Goldstein for his "unfair" depiction of Gestalt theory in *The Organism*. B. F. Skinner recalled that in a 1940 conversation with him Köhler called Goldstein's conception "Gestalt psychology in the colloidal state."[60]

Despite such tensions, all of these approaches fed back into the development of Gestalt theory. Lewin's thinking pushed both Koffka and Köhler to consider more fully the role of the self in experience and to integrate some version of that concept into their system. Gelb's research with brain-injured patients appeared easy to integrate into the Gestalt framework, but Goldstein's conclusions from it showed that this was not necessarily the case. It was possible to respond to these challenges without giving up the basic standpoint of Gestalt theory. But Lewin's work implicitly indicated the limits of that standpoint, and Goldstein's work did so quite explicitly.

17

The encounter with Weimar culture

Surely enough has been said to indicate both the liveliness of Gestalt theory's scientific development in the Weimar period and its multiple links with important theoretical issues of the day, including the relations of physics and biology, psychology and evolutionary history, individual and group. All of these topics had at least indirect connections with broader ideological concerns and debates of that time. It would have been strange if that had not been so, for in that difficult era calls abounded for a new worldview that could somehow make sense of the German educated elite's changed existence.

In the early nineteenth century, the educated middle and upper middle classes in the German states had sworn allegiance to a self-concept that saw in literary felicity the sign of a cultivated personality, and this in turn as the high point of cultural history. Throughout the century, such thinking successfully supported the status of *Bildungsbürger* and gave them their own positive identity. As the century progressed and the newly unified nation's power increased, that once-cosmopolitan identity became increasingly tied to the alleged duality between Germanic *Kultur* and Western, especially French, "civilization" – an opposition expressed with full force during the First World War. But with Germany's rapid industrialization, the status hierarchy that had supported all this had already begun to break down. In the Weimar period, without a monarch at the top and stable prosperity as the glue, many *Bildungsbürger* saw their position threatened. They feared degradation from the status of bearers of culture to that of mere functionaries, which was in fact already happening to some of them. Many saw in the rise of the natural sciences and technology a symptom or even a cause of the fragmentation, objectification, and materialization of culture that they decried. The experience of war and abortive revolution, inflation, and later of the Depression exacerbated what Peter Gay has called a "hunger for wholeness." Fevered holism was most prominent in the chaotic early years of the republic and during its final agony after 1929. The relatively stable years between were the heyday of modernism, so-called "new objectivity," and technology worship in German culture. Science, too, prospered in that period; by 1928, financial support for basic science exceeded prewar levels. But even as many educated Germans participated in these modernizing trends, many others saw their fears of fragmentation confirmed.[1]

Repeated proclamations of crisis in both science and society generally prefaced the calls for "synthesis" and a unified worldview that came from this environment. Indeed, in this period fragmentation and uncertainty appeared to overtake the natural sciences themselves, once proud purveyors of belief in progress. Many of the fundamental conflicts in science that came to public attention in the Weimar period began before 1914. But they acquired particular urgency and wider cultural impact early in the 1920s, and it was in that decade that these disputes came with increasing frequency to be called crises. Crisis talk was not limited to Weimar Germany, nor was it confined to any single realm of thought or political viewpoint. Though intellectual historians have tended to identify the crisis trope with political and cultural conservatism, Marxists, too, employed it, albeit with rather different intentions. It was a shared diagnosis, the predominant discursive norm, of the age; only its interpretation differed with the commentator's point of view.

Crisis themes: From science to social thought

In keeping with the historically all-inclusive concept of *Wissenschaft* in Germany, crisis talk and concomitant demands for a new synthesis or worldview were addressed to all disciplines, not only to the natural sciences. But such demands were especially vehement there, due to the ambivalent standing of natural science as an ornament of prestige and a source of power that also seemed opposed to predominant humanistic values in German elite culture. The conflicts that spurred crisis talk were important scientific issues and not only reproductions of ideological disputes; but it proved difficult to keep the two dimensions distinct, for precisely the appearance of fundamental conflict and disunity in scientific discourse made science as a whole vulnerable to crisis talk. As the terms fragmentation and synthesis already imply, the problems of whole and part, and of the alleged separation of science from "life" or immediate experience, played important roles in nearly all of these controversies. The account here is limited to those that impinged most closely on Gestalt theory.

In mathematics, the clash in the 1920s was between proponents of symbolic and so-called intuitionistic conceptions of number. The conflict's outlines paralleled those of prewar disputes among Planck, Mach, and others over the status of mathematics in theoretical physics. The purveyor of the intuitionist challenge, the Dutch mathematician L. E. J. Brouwer, based his work on an analysis of mathematical continua similar to Mach's view of physical continua. For Brouwer, mathematical series, such as that of real numbers, are not fixed orders, as set theory would have it, but continuous wholes stretching to infinity. Because continua cannot be composed of parts, he argued, no judgment is possible that a number with a certain character (e.g., an even number) exists or not unless we actually count through the entire series to confirm it, which is impossible by definition. Thus, he concluded, if mathematics is to maintain any contact with our intuitive experience of number, it must reject the traditional laws of logic, in this case the law of the excluded middle. To save the logical coherence of mathematics, the Göttingen theorist David Hilbert developed an "axiomatics," which treated both mathematics and logic as complex symbol systems, their connection with reality guaranteed only by what he termed "metamathematics." With Planck, he firmly believed that only such symbol systems could remove the last vestiges of "anthropomorphism" from science, thus guaranteeing the invariant lawfulness of nature, and with it the unity, objectivity, and general validity of science. In a 1921 essay entitled "On the Foundational Crisis of Mathematics," mathematician and philosopher Hermann Weyl, a disciple of Husserl and a strong proponent of intuitionism at the time, spoke in frankly political terms of a "threatening dissolution of the regime (*Staatswesen*) of analysis" and proclaimed "Brouwer – that is the revolution!" Hilbert replied the next year, "No, Brouwer is not, as Weyl thinks, the revolution, but only the repetition of a *Putsch* attempt with old means." The rhetoric was milder later in the decade, but the term "crisis" and the basic issue of the connection, or lack of connection, of mathematics and immediate experience remained.[2]

Better known to the educated public was the upheaval brought to physics by relativity theory and then quantum mechanics. Among the earliest uses of the phrase "crisis of science" was in the title of a scurrilous antirelativity pamphlet circulated by experimental physicist Johannes Stark at a conference on relativity in Bad Nauheim in 1922. There and later the opponents, mainly experimental physicists, denounced relativity and quantum mechanics as mathematical artifacts with no empirical basis, or more viciously as subversive "Jewish" science. Others, notably Planck, Einstein, and Wilhelm Wien, tried to preserve the ideal of a unified causal world-picture including relativity and quantum theory, while a self-styled avant-garde including Weyl, Hans Reichenbach, Max Born, and Werner Heisenberg confidently proclaimed, in Heisenberg's words, that "quantum mechanics establishes definitively that the law of causality is not valid," or, as he wrote elsewhere, "has no object." Observing the debate, theoretical biologist Ludwig von Bertalanffy concluded that the uncertainty principle meant that a fundamental limit had been reached: "We must – due to the Heisenberg relation – grasp the electron and the light quantum that illuminates it as a unified entity, a "Gestalt" that cannot be broken down any further."[3]

All this led Kurt Riezler, among many others, to write of a "crisis of reality." By this he meant both the inability of mechanical, deterministic physics to come to grips with the phenomena of biology, psychology, or human history, and also the overthrow of deterministic physics itself by relativity and quantum mechanics.

Thus, just as reality's multiple dimensions could not be subjected to unified mechanical order, so physics had failed to subject the external world to the unified symbolic order of differential equations. The flexible use of mathematical tools in quantum mechanics heralded, in Riezler's opinion, not the acceptance of multiple realities but a return to subjectivity, not the hopeless fragmentation of the world-picture but the recognition that "physical reality is not the absolute." The world-picture of physics depends on something that its laws cannot comprehend or measure – the fact that "the manifold given to us prior to all thought is in no way chaotic. It is not composed of individual elements, sensations. . . ." But this perceived order is in turn a "mirror image" of our "forming understanding," which necessarily creates a conceptual system on which the objective reality (*Sosein*) of the world depends. Hence, for Riezler, as for Weyl, conflict in science justified a reaffirmation of idealism.[4]

Evidently, what were exciting innovations for some scientists were clearly taken by others, and also by educated laymen, as good reasons to reassert the primacy of humanistic ideals. The perceived fragmentation of nature and physics' loss of deterministic certainty were also taken as metaphors for the social realm, as fears for social and political stability found expression in worries about the coherence of science and its relation to "life" – meaning the experience of *Bildungsbürger*. Such a situation enhanced the appeal of holistic social thought.

Particularly prominent during the chaotic early years of the decade were the "organic" conceptions of the state put forward by Austrian conservative philosopher and social theorist Othmar Spann and German philosopher of biology Theodor von Uexküll. Von Uexküll based his theoretical biology on a provocative, scientifically quite useful, application of Kantianism to organic life, according to which each organism lives in an environment (*Umwelt*) but perceives only that portion of it relevant to its functioning (*Merkwelt*). Analogously, his "biology of the state" assigned to each occupation its own limited functional role in state and society, so that changing one's job or going on strike are betrayals of the organic community. In this wealthy landowner's perspective, the newly installed parliamentary system was comparable to "the tapeworm that has killed a noble war horse, then tries to play the horse itself." Spann offered a theory of traditional estates (*Ständestaat*), in which society was conceived as a "superorganism" (*Überorganismus*), or an "organic whole" (*organische Ganzheit*), properly organized as a hierarchical structure of guildlike organizations, each ruled by the one above it. This romanticized version of medieval society or of Plato's ideal state Spann offered as a "third way" between "liberalistic" capitalism and Marxism.[5]

A somewhat different, extremely influential, view came from self-styled philosopher of history Oswald Spengler. The first volume of *The Decline of the West* had the subtitle, "Gestalt and Reality," which referred to what he took to be two fundamentally different methods. One, the "exact, deadening procedure of modern physics," abstracts from the flow of becoming and, by trying to subject its objects to mechanical, numerical order, kills them. The other, "morphological" or "formative," method "operates in the realm of moving and becoming" by taking the objective creations of a culture, its institutions, architecture, and so on, as signs of its "soul" or "life," specifically of its stage in the cycle from vigorous birth to decay and death. Such terminology reproduced the by-then commonplace distinction between Germanic *Kultur* and Western "civilization" that was part of the standard vocabulary of conservative thought. According to Jeffrey Herf, however, in contrast to the conventional cultural pessimism with which he is usually identified, Spengler's language was a call for action. He argued that the time was ripe for a revival of productive "formative forces" (*Gestaltungskräfte*) to "overcome" the decayed spirit of urban *Geist* and *Geld;* and he included engineers and factory workers alongside peasants, soldiers, and artists among the likely bearers of such renewed productivity.[6]

The notion of a "third way" became common among both rightists and leftists during the Weimar era. Just as some philosophers sought to overcome dualism, political intellectuals claimed to seek a new synthesis that would overcome old antagonisms. Most often such syntheses amounted to little more than traditionalist or more radical folkish conservatism in a new package.[7] But such terminology was by no means limited to conservative thought. The categories of fragmentation and totality were central to the discourse of Western Marxism in the 1920s. For Georg Lukacs, as for Ernst Bloch, the latter concept sustained hope after the 1917 revolution in Russia. Historical development, Lukacs thought, would inevitably lead to a normative totality – the self-realization of class consciousness in a classless postrevolutionary communist society. After the failure of the revolution to spread beyond Russia and the relative stabilization of capitalism in the mid-1920s, however, the location of totality shifted to culture and the academy. The reorganization of the Frankfurt Institute for Social Research under Max Horkheimer in 1931, for example, was an attempt to overcome perceived fragmentation and overspecialization with interdisciplinary research, and thus, in Gilian Rose's words, to establish "Marxism as a form of cognition *sui generis,*" with "no universal class" as privileged carrier.[8] In one respect, the efforts of these Marxist thinkers paralleled those of the natural scientists discussed above. They, too, tried to establish a realm of intellectual transcendence while nonetheless retaining some sort of connection with reality.

The practical challenge of "humanistic" psychology

Calls for a new synthesis or "third way" came most frequently from philosophy, alongside theology the traditional meaning-giving discipline, to which psychology remained institutionally and intellectually connected. In this context, a resurgence of older attacks on natural-scientific psychology was to be expected, and in these debates holistic vocabulary and the culturally resonant term Gestalt were quite prominent. Now, however, more urgently than before, the opponents offered fully developed alternative approaches to psychology based on theories, or typologies, of personality; and they emphasized not only the theoretical, but also the practical superiority of their views over those of the experimentalists. Examples of this trend were the challenges raised by Karl Jaspers and Eduard Spranger.

Jaspers, a former psychiatrist, had taught philosophy and psychology in Heidelberg since 1913. In his influential text on general psychopathology, he criticized the then-dominant observational and diagnostic practices of Emil Kraepelin, which were based in part on Wundt's experimental techniques. The "sick personality," he asserted, must be described for itself, not classified according to a preestablished schema. This could best be achieved by combining the physician's empathic "understanding" of the patient with the patient's own self-observations. Ultimately, explanatory and intuitive methods should supplement one another, since each can grasp only part of mental life. In this methodological pluralism Jaspers was akin to Max Weber, whose influence he later acknowledged.[9]

In the 1920s, Jaspers offered a psychology of "worldviews" that was clearly intended as an alternative both to metaphysics and to empirical psychology. It was right for philosophers to study worldviews, he wrote, because they are "something whole and universal," and "concerned with the whole one calls philosophy." Specifically, Jaspers's method involved a typological use of the Gestalt category. That is, he presented a systematic account of types of attitudes toward and pictures of the world. Each of these he said was organized around a "center," from which "derived forms" (*abgeleitete Gestalten*) branched out. Examples of world attitudes were the active, contemplative, mystical, and enthusiastic. Types of world-pictures included the sensorispatial, *seelisch*-cultural, and metaphysical. The mechanical world-picture, which Jaspers regarded as a variant of the sensorispatial, was a particular target of polemic because it was so often overextended:

> Thus the psychiatrist enjoys 'brain mythology,' the psychologist a mass of theories of the extraconscious (*Ausserbewusstem*). One puts such great value on measurement, counting, experimentation, that this all becomes an end in itself, and one no longer knows why this is all being done. One consoles oneself with the thought that this will all be fruitful sometime. Thus the mechanical world-picture has a deadly effect on all intuitions by forcing them into itself and thus impoverishing them.

In contrast to the experimentalists' deterministic-causal ideal, Jaspers insisted that the truth of his scheme could be assessed only by its "intuitive evidence," by which he meant not rational argument or historical documentation but conceptual clarity and a kind of recognition by the reader of what was intended. Though he refused to say which worldview one should have, he claimed to provide material for such decisions.[10]

Spranger was a philosopher of education and a self-styled disciple of Dilthey. But while Dilthey had hoped to integrate experimental results of "explanatory" psychology into a more broadly based "descriptive" psychology (see Chapter 5), Spranger hardened the battle lines, asserting that natural-scientific and "humanistic" (*geisteswissenschaftliche*) psychology are fundamentally opposed. Like Dilthey, Spranger accused experimental psychologists of reducing mental life to an interplay of essentially meaningless hypothetical "elements," and protested that such explanations would never do justice to the experience of meaning (*Sinn*) characteristic of human psychology. "I demand the return of the word 'psychology' for the science of meaning-filled life [*sinnerfüllten Leben*]," he wrote. Only on such a basis could a pedagogical theory be developed that would restore to German youth the idealistic values of former days.[11]

Spranger called his alternative "structural psychology." This rested on a system of six ideal personality types, each of which he linked with a specific world-view, or "form of life," which came to the fore in different historical epochs. Individuals could thus be described as mixtures of the theoretical, economic, aesthetic, social, religious, or power type, with one type predominating over the others. This scheme clearly derived from Dilthey's notion of historically "typical" cultural individuals, and his method was similar to Max Weber's procedure in sociology. But Spranger's intent was practical as well as theoretical. He had been appointed to a chair in Berlin by Prussian Education Minister Carl Becker in 1919 to consult on educational reform in Prussia precisely because he favored reviving Humboldt's ideal of *Bildung* as personal self-cultivation. In a 1921 pamphlet he opposed the use of standardized tests in Gymnasium admissions, "for the individual is viewed here in the end as something measurable and graspable in numbers, not as a structural principle of the soul." Although he favored encouraging gifted but less well-off young people to attend university-preparatory schools, his position ultimately meant retaining the privileges of *Bildungsbürger* – in this case, of Gymnasium teachers.[12] His typological personality theory also gave them a conceptual tool for exercising that privilege.

From outside the university came yet another challenge, from proponents of so-called "scientific graphology" and "characterology," led by Ludwig Klages. With the help of handwriting analysis, Klages and his followers claimed to discover people's true inner lives behind their "masks of courtesy"; for the expressive movements in handwriting represented nothing less than the soul's self-expressive power of "formation" (*Gestaltungskraft der Seele*). The resemblance to Romantic philosopher Carl Gustav Carus's belief that the outline (*Gestalt*) and expressive movements of a person's body were keys to that person's character (noted in Chapter 6) was not coincidental; Klages brought out new editions of Carus's works to emphasize the affinity. In his view, handwriting analysis was better suited to selection of personnel for employment than either academic psychology's tests, which he dismissed as trivial, or the methods of industrial psychotechnics, oriented as they were to merely external achievement. In a 1921 essay on "the essence of consciousness," Klages claimed that our speaking of loud or soft colors, of bitter feelings or sweet love — that is, the fact that we can create and understand such meaning-filled mixtures of sensory and emotional description — testifies to the synaesthetic, formative power of *Seele*. On the metaphysical plane, he propounded an opposition between this formative, expressive "soul," the realm of experience, images, and cultural objects, and "mind" (*Geist*), the radically distinct world of causally connected material objects. Leftist critic and theorist of mass culture Siegfried Kracauer had considerable fun with graphology in a satirical piece for a Frankfurt newspaper, but such ridicule did nothing to reduce its popularity. Precisely the mobilization of conceptual resources from German Romanticism in the service of a practical professional role was the secret of its success.[13]

Unifying all these approaches was a shared mode of discourse marked by a preference for intuition and experiential immediacy over experiment and alienating fragmentation. But neither the discourse of power that had enhanced the appeal of natural science throughout the world nor claims to expert authority were absent from this discussion. Privilege was accorded, rather, to those traditionally endowed with cultivated or trained intuition – *Bildungsbürger.*

The Gestalt theorists' response

All of the leading schools of experimental psychology in the Weimar period offered their ideas at least in part as answers to such ideologically freighted demands for meaning, or to calls to reconcile science with "life" in theory and in practice; and Gestalt theory was no exception. Inititally, the Gestalt theorists had addressed their claims to a limited audience of experimenting psychologists, philosophers, and natural scientists interested in epistemological issues. In the intense atmosphere of Weimar culture, their views took on new significance, and it was in that period that they presented their ideas to broader publics and with increasing confidence as the germ of a new worldview. A brief look at their political and intellectual affiliations and those of their students will help to understand the direction their answer took. While the vast majority of German professors and students rejected the Weimar regime as the very embodiment of political disintegration and social fragmentation, evidence about the Gestalt theorists' political views yields a different picture.

In contrast to most German-speaking academics, Max Wertheimer was a free-thinking internationalist and a friend of Albert Einstein, who held similar views (see Chapter 12). In September 1922, Einstein asked Wertheimer to represent him at meetings of the League of Nations' Commission for the International Organization of Intellectual Workers to be held while Einstein was traveling in East Asia. In the postwar years, scientists were slow to restore international cooperation, owing in part to the bitterness of war propaganda. German scientists in particular were concerned that international scientific relations not be restored at the price of German national pride. Einstein's pacifism and his involvement in peace groups as early as 1917 was well known, and caused many to question his appropriateness as an international representative of German science. Anti-Semitism was by no means absent from such considerations. Thus, it was particularly significant when Einstein wrote to Wertheimer, "I know only one whose free and objective spirit I trust in every respect, *and that is you.*" Though he assured Wertheimer that his participation was "a purely personal affair," he noted that the Prussian Education Ministry wished to be informed about the proceedings afterward; but what he told them would be up to him. Wertheimer was reluctant to accept, because he was a Jew and now a citizen of Czechoslovakia. He suggested that it would be better to send a Christian with better language skills instead. This response seems to have irritated Einstein, but does not diminish the similarity of the two men's views.[14]

Wolfgang Köhler fit into this progressive picture less obviously. In a 1920 letter from Tenerife, he referred to the libel suit then being brought by Center party leader Matthias Erzberger against opponents who had accused him of treason for signing the Versailles treaty, and expressed astonishment at what appeared to him to be the new Republic's lack of order and discipline, or a common sense of values:

> What Erzberger is accused of is the cancer of this country [that is, Spain], and when a nation (again like this one) accustoms itself to knowing and bearing such things, then what results is what is understood as "parliamentary government" = deals by the party leaders, alongside which any objective effort for the public good is only demanded by completely naive people.
>
> Otherwise I am impressed by the unbelievable slowness and weakness of the executive. What help are so many laws and decrees, if one does not force the people to follow them? Does democracy mean the same as a general slovenliness [*ein allgemeines sich gehen lassen*]? Many seem to mean that, and beneath complaints about 'militarism,' etc., hides resistance against a life that proceeds – militarily or not – anyway *somehow strictly.* But I have been away for six years and have not personally experienced the deprivation of war, so I should not make any judgments.[15]

The references to "objective efforts for the public good" and to "strictness" were hallmarks of Prussian political culture. Though Köhler did not favor Bismarckian Bonapartism, his evident frustration with what appeared to him to be the Republic's lack of forcefulness corresponded to the later views of the German People's Party of Gustav Stresemann, whose followers cooperated with the Republic more out of realism than conviction. In a 1925 letter to Cornell University psychologist E. B. Titchener, Köhler appeared more positive about the regime. He confessed that he had not foreseen the victory of Paul von Hindenburg in the

Republic's second presidential election, and speculated that many people had voted for him as "the most popular figure . . . whereas they would not hesitate for a moment to defend the democratic constitution against tendencies of conservative restoration. Therefore, I hope, nothing of great importance will be changed in Germany." Asked to place the Gestalt theorists politically, one former Berlin student who knew them all linked Köhler with the party of Stresemann, Koffka with the Social Democrats, and Lewin with the independent, left-wing Social Democrats. Of Wertheimer another said, "he was a partisan of democracy." Of Kurt Koffka a former Giessen student wrote, "When after the assassination of Walter Rathenau the Reichsbund of avowedly Republican teachers was founded by Professor Ludwig Hüter, Koffka immediately joined it."[16]

The assistants and graduates in Berlin for whom information is available appear to have been left of center politically. As already noted (in Chapter 16), Kurt Lewin was a friend and sometime neighbor of the independent leftist Karl Korsch; his 1920 essay on the "socialization" of Taylorism appeared in a series edited by Korsch entitled "Practical Socialism." Though he refrained from using such language later, his active commitment to science as an agent of social change continued. A former student later described him as one of those Weimar intellectuals who, though not a Communist, "approved of the Soviet Union."[17] Wolfgang Metzger and Rudolf Arnheim, to give additional examples, wrote reviews of current psychology for the *Sozialistische Monatshefte* (Socialist Monthly) from 1923 to 1929. Their articles publicized the Berlin school's research, including Lewin's work, but also covered events, publications, and topics throughout the field, including psychoanalysis, characterology, Couéism, the psychology of women, and the idea of "joy in work" (*Arbeitsfreude*). Although these essays generally made no direct reference to politics, the place of publication was a message in itself.[18]

Here and there, the writers also gave hints of Gestalt theory's potential political implications as they saw them. In a review of a book by Gina Lombroso on the psychology of women, for example, Arnheim opposed her equation of maleness with egotism and femaleness with altruism. Men could be altruists, too, he argued, if only for the sake of a "thing," such as science. Lombroso fails to acknowledge that altruism can apply to anything except people, and she is thus an example of her own theory. What is needed, Arnheim concluded, is a wider definition of intelligence that would recognize that quality "less in the proper ordering of abstract thoughts than in the reasoned grasp of a concrete situation," as in Köhler's anthropoid work. By such a definition, "no inferiority of women could be contended."[19]

The Gestalt theorists continually emphasized their allegiance to natural-scientific methods and thinking. As Wertheimer asserted in his programmatic paper of 1921, the question "to what extent that which exists has meaning (Sinn)" is "first of all *a matter of fact*."[20] Thus, it was only consistent that the Gestalt theorists' chief public affiliations outside academia in the Weimar period were with publications and societies dedicated to philosophical reflection on science. Kurt Koffka served on the editorial board of the journal *Annalen der Philosophie,* edited by pragmatist Hans Vaihinger, from its founding in 1918. Both Köhler and Lewin published in the journal *Erkenntnis,* the successor to the *Annalen* edited by Hans Reichenbach and dedicated to philosophical reflection on natural science. In the late 1920s and early 1930s, they were prominent participants in the gatherings of the Society for Scientific Philosophy, best known today as the institutional organ of the so-called Berlin school of logical empiricism, in which Reichenbach played a leading role. The list of lecturers at the society's meetings shows that the group was open to a variety of viewpoints, including nonpositivist positions like that of Gestalt theory. Lewin lectured in 1930 on the transition from "Aristotelian" to "Galilean" thinking in psychology; Köhler spoke the same year on "The Current Situation of Empirical Philosophy" and later on "The Current Situation in Psychophysics." When Reichenbach took over the chairmanship in the early 1930s, Lewin joined the board. Though the Society took no political positions, controversy flared when Lewin's friend, independent Marxist Karl Korsch spoke on "Empiricism in Hegelian Philosophy" in October 1931. After this the Society was briefly forbidden to advertise its lectures at the University of Berlin.[21]

Even without such evidence it would be clear that Gestalt theory in the Weimar years was more than a research orientation. The clearest indication of Gestalt theory's relation to the intellectual ferment of Weimar culture was Max Wertheimer's lecture on December 17, 1924, to the Kant Society in Berlin, entitled simply "On Gestalt Theory," which was published the next year. There he emphasized that Gestalt theory had emerged from concrete research, but suggested that it could also be a response to "a problem of our times." Tentatively, but nonetheless hopefully, he offered Gestalt theory as a new world view that would overcome philosophical and psychological dualisms.[22]

Wertheimer described the "problem of our times" not with the categories of cultural pessimism and alienation, but as a fundamental discontinuity between science and concrete human experience. Anyone observes, for example, that "a pupil grasps the point"; but when one goes to texts in psychology or pedagogy for enlightenment about such a phenomenon, "one is shocked at the poverty, aridity, unreality, the utter triviality of what is said." Terms like "concept formation," "abstraction," "generalization" or "judgment" offer the illusion of rational control; others, such as "intuition," "happily give free scope to the imagination." But "these terms merely name the problems." The same difficulty, Wertheimer maintained, "has appeared everywhere" in science. "For a long time it seemed self-evident, and very characteristic of European epistemology and science, that the scientist could only [view phenomena] first as an aggregate, as something to be dissected into elementary pieces," before establishing lawlike relations between them. "Briefly characterized, one might say that the paramount presupposition was to go back to particles," then "to analyse and synthesize by combining the elements and particles into larger complexes."[23]

There were two common solutions to this dilemma, according to Wertheimer. One separated scientific method and vital human experience with an attitude of "truly grandiose resignation." Another, more academic strategy separated the natural sciences from the humanities, thus accepting the widely held "prejudice" that natural science is necessarily mechanistic and "piecemeal." Employing a familiar metaphor of the period, Wertheimer stated that his approach offered a third way:

> Gestalt theory poses the radical question, is it at all the case that when I hear a melody I really hear the sum of individual notes, at least as a primary basis? Is it not perhaps the other way around, that what I have before me, what I also have in place of the single tone, is a part determined by the structure of the whole?

Anticipating the criticism that such views are unscientific, even irrationalistic, Wertheimer queried, "Is it not possible that a certain viewpoint, certain fundamental assumptions concerning scientific method and approach, have become widespread and have achieved prodigious maturity without being at all a necessary feature of genuine scientific method?"[24]

The use of probing, open-ended question marks marked a distinct difference from the dogmatic style of other critics. Wertheimer emphasized that the implications of such claims could not be discussed in "generalities," because the concepts in question, such as whole and part, "suffer particularly from being considered as cataloguing material rather than an aid to the penetration of the given." Instead, he sketched the implications of the Gestalt viewpoint for specific research problems in psychology, such as the threshold concept and color perception. To this he added suggestive remarks about the self and society. Thus, for example, "man is not only a part of a field, but a part and member of his group"; the notion of a separate ego only "develops under very special circumstances." Hence, just as elementary sensations are "a late cultural derivative" (as Koffka had already argued in 1921 – see Chapter 15), so too is possessive individualism. It is more natural, he implied, for people to work together for a common goal than to be in opposition to one another.[25]

Finally, after pointing to the implications for physics and theoretical biology already brought out by Köhler, Wertheimer spoke eloquently of the impact Gestalt theory could have in philosophy. He waxed most inspiring on the mind–body problem and the alleged opposition of materialism and idealism. Like the notion of elementary sensations or the possessive individual, he argued, the separation of the physical and the psychical was a late cultural creation; in reality there is identity of Gestalt structure in the two realms regardless of "the material characteristics of the parts." Here Wertheimer's commitment to aesthetic over technological values in science came through more clearly than it ever had before. Addressing the alleged opposition of living consciousness and dead matter, for example, he expressed wonderment:

> People speak of idealism as opposed to materialism, and they mean something beautiful by idealism and by materialism something gloomy, barren, dry, ugly. Do they really mean by consciousness something opposed to, let's say, a peacefully blossoming tree?. . . Frankly, there are psychological theories and even plenty of psychological textbooks which, although they speak continuously only of conscious elements, are more materialistic, dryer, more meaningless and lifeless than a living tree which has probably no consciousness in it at all.[26]

Wertheimer ended the lecture by inviting his listeners to imagine the world as "a large plateau on which musicians are seated, playing." Rather than a "senseless plurality," in which "everyone acts arbitrarily – everyone for himself," or a probabilistic order, in which "whenever one musician played C, another would play F so many seconds later," the connections between the tones established only "by some blind piecemeal relationship," he proposed a different, more appealing image – that of "a Beethoven symphony, where from a part of the whole we could grasp something of the inner structure of the whole itself. The fundamental laws, then, would not be piecemeal laws but structural characteristics of the whole."[27]

Such a vision had profound implications for Weimar ideological disputes and for the status of psychology as natural science in that context. To attacks by Spranger and other proponents of so-called humanistic psychology against atomism and mechanism in experimental psychology, the Gestalt theorists could reply that theory and research in that field need not possess either characteristic. Going further, they maintained that precisely psychology, suitably reformed, could provide the foundations for a powerfully synthetic worldview, unifying discourses about nature, life, and mind without sacrificing experimental method. Psychology could thus retain the role assigned to it by both Wundt and Dilthey as the fundamental discipline of the human studies, without sacrificing its legitimating links to natural science. As Kurt Koffka expressed it, also in 1925: "The opposition between understanding and explanatory psychology disappears. And it must, because knowing, according to the view expressed here, is not a process foreign to nature, because value and meaning are not domains of reason but are rooted in the great Being of the world itself."[28]

With this program, and in their political and other intellectual affiliations, the Gestalt theorists also aligned themselves implicitly with the minority of academics who supported the first German Republic. Wertheimer's remarks on group life did so explicitly. Many intellectuals in the Weimar period employed the metaphor of the Beethoven symphony in order to emphasize their rejection of the allegedly atomizing influence of industrialization, urbanization, and democratization, and their support for the transcendent unity and elevation that they believed high culture, especially German culture, could and should embody.[29] As already shown, conservative thinkers used the term "organic whole" as a marker for their social theories, and Spengler employed the Gestalt category in support of his cultural pessimist vision. In contrast, Wertheimer used the metaphor of the Beethoven symphony and the term Gestalt rather differently – to enhance the appeal of a new, explicitly natural-scientific and nonnationalistic philosophical anthropology.

The concrete political implications of his message, however, remained ambiguous. If it is more natural to work in concert than in opposition, for example, then what was to be done about the actual political situation in Weimar Germany, dominated as it was by constant conflict? Was it enough to trust in the reforming impact of democratic structures when so many were working openly to undermine them? If possessive individualism was a late arrival on the cultural scene, did Wertheimer mean to mourn nostalgically the loss of "natural man"? Or did he mean to call, however vaguely, for a move forward to a higher stage, perhaps a form of humanistic socialism? Wertheimer left these questions unanswered. Though he realized that some of his remarks raised issues of history and cultural theory, he did not discuss those topics at the time.

Köhler's contributions to this discussion came later in the Weimar period and set somewhat different accents from Wertheimer's. In a lecture entitled "Essence and Facts" (*Wesen und Tatsachen*), presented to the Kant Society in January 1932, for example, Köhler opposed the stark alternative of conceiving the world either as a meaningless "play of atoms and wave vibrations," or evoking transcendental "essences" (*Wesenheiten*) as sources of truth independent of experience, as Husserl did. Like Wertheimer, he offered a third way, which he called "a realism of meaning" (*Sinnrealismus*), based on the claim originally made by Dilthey that meaningful relationships (*verständliche Zusammehnänge*) are immediately present in experience (see Chapter 4). Examples of such experiences are the immediate sense of connection between feelings of coolness and great enjoyment while having a glass of cold beer after a long walk on a hot summer day; or the sudden experience of fright while sitting in a house as an earthquake begins. Here are cases in which intution yields not merely incidental, but rather causal connections. "As a matter of fact," Köhler argued in *Gestalt Psychology,* such experiences tell us "more than any scientific induction could. For induction is silent as to the nature of the functional relation which it predicates, while in the present examples a particular fact of psychological causation was directly experienced as an understandable relationship." This was a provocative reply to those who regarded subjective intuition and objective rationality as opposed. For Köhler, invariance and causal determination need not be ascribed only to the formal laws scientists construct and attribute to nature, but can be experienced directly. There need be no mystery to such experiences, he argued, if they are regarded as expressions of directed field forces in the brain.[30]

In his 1932 address, Köhler envisioned a unified worldview based on such intuitive experiences of causality that would end philosophy's self-limitation to critiques of knowledge and rededicate it to questions of value and practical life. But this was no more than a vague program at the time. He stated his position more fully in his William James lectures at Harvard University in 1935, later published as *The Place of Value in a World of Facts* (1938). There he extended his claim to ethics by claiming that we have immediate experiences of "requiredness" – when we know without reflection what behavior is objectively necessary in certain situations. However, he could not account for ethically grounded differences of opinion about what is required or obviously necessary, or for the all too painful presence of irrationality, amorality, and contempt for law in the real world of the 1930s.[31]

Köhler suggested something more of the complex cultural resonances and the ambivalence of his position in a 1930 essay called "The Essence of Intelligence." In a striking anticipation of current criticisms of intelligence testing, he argued that in typical test situations children were being confronted with "a kind of potpourri of demands" before anyone had first discovered what the nature and determinants of intelligent behavior might be. "Anybody who is not a very naive prisoner of the values of his particular civilization and cultural era, of his country and social milieu, will surely have to realize" that intelligence has been defined to correspond "to those requirements which the present-day school, the present-day city in Europe or America, the present-day middle class consider important." Why were such measuring devices being introduced before science could give them a secure foundation? Köhler's explanation was simple. The appreciation of genuine intelligence is "one of the most difficult and noblest processes of understanding"; fearing that difficulty, administrators substitute a mechanical operation with standardized results. "Anxiety begets bureaucracy. . . . Not all teachers are such unfailing artists of the understanding that we can trust to their judgment. Let us give them the machine! That is the intelligence test."[32]

Köhler's rhetoric here paralleled his earlier polemics against "a conservative system" in his 1913 critique of dualistic assumptions in the psychology of perception (see Chapter 9). At that time he had portrayed himself as a young scientist eager to shake off the bonds of convention. Here, by using "the city," "the machine," and even "the middle class" as signifiers of opposition, he expressed himself in the vocabulary of cultural conservatism. He even appeared to align himself with Eduard Spranger's 1923 rejection of intelligence testing in Gymnasium admissions as an example of "falsely understood democracy" – that is, as a threat to the privileged role of *Bildungsbürger* in elite selection. However, though his essay appeared in a volume on child development with the subtitle "heredity and environment," he rejected the hereditarian claim that some individuals have "special nervous arrangements lacking in others, which automatically produce intelligent performances."[33] Rather, he asserted, such performances depend on objective properties of the physiological medium involved, which can change even in highly endowed people, when they have a heavy cold, for example. Köhler's refusal to adopt technocratic values in science thus enabled him to resist the blandishments of then-fashionable eugenic discourse and left open the door to a claim that Wertheimer could have endorsed – that people of all social groups are capable in principle of intelligent behavior.

Gestalt aesthetics and political commitment: Rudolf Arnheim's film theory

Given the Gestalt theorists' commitment to an aesthetic orientation in science, it was appropriate that Gestalt psychology came into most extensive contact with Weimar culture in the film criticism and theory of Rudolf Arnheim. More clearly than others, his work exposed the problems of trying to realize Gestalt theory's promise as a unified worldview in that tension-filled atmosphere. The conflicts between order and freedom, classical balance and modernistic imbalance, with which he wrestled had not only psychological and aesthetic but also political dimensions and implications, because contemporary debates on film repeated and refracted the crisis talk that was endemic in the culture. Put exceedingly briefly, literati and philosophers saw in film the quintessential medium of urban technological mass culture. To many commentators it purveyed distraction instead of cultivation and exemplified "Americanization," meaning the comodification of culture and its capitulation to hedonism. As in other forms of crisis talk, critics on both the right and the left shared this diagnosis, differing only in the plus or minus sign they attached to it. Spengler and Martin Heidegger feared, while Bertolt Brecht and Walter Benjamin praised the medium's potential. New in the 1920s, however, was the emergence of a sophisticated discussion of the film medium as such and its problematic embodiment of cultural modernity. To that process Arnheim made an original contribution.[34]

Arnheim was born in 1904 in Charlottenburg, then a suburb not yet incorporated into greater Berlin. He attended a so-called reformed Realgymnasium, which offered Latin but replaced Greek with modern languages and natural science. After graduating in 1923, he divided his time at first between his father's piano factory and the university, but gravitated early to the Psychological Institute courses of Johannes von Allesch and Erich von Hornbostel on the psychology of art and music. He earned the doctorate in 1928 with examination fields in psychology, philosophy, art history, and music history. His dissertation on expression seems to have come out of Wertheimer's course entitled "Knowledge of People" (*Menschenkenntnis*), which emphasized the psychological interpretation of ex-

Photo 17. Rudolf Arnheim (photo by Wilfried Basse). Fotoarchiv, Deutsches Filmmuseum, Frankfurt am Main. By permission of Rudolf Arnheim.

pressive gestures, faces, and the like. Approved officially by Köhler but published in the series edited by Wertheimer, it was in fact the result of intensive cooperative labor with the latter. "It was a very personal relationship," he later recounted. "I was often at the Wertheimers' house in Karlshorst [a Berlin suburb] . . . And we would work in the evenings. I would sometimes stay overnight, sometimes babysit."[35]

The dissertation was one of many responses to the challenge to natural-scientific psychology posed by the graphology of Ludwig Klages and others. Arnheim attempted, first, to verify experimentally the existence of the so-called "physiognomic" or graphological "gift," and then to discover the psychological processes involved in it. In careful quantitative studies he was able to show that psychology students of various levels with little or no background in graphology or "physiognomics" could assign handwriting samples or quotations correctly to their authors with an accuracy far beyond chance. In a preliminary experiment first tried by Wertheimer before the First World War, for example, Arnheim found that of over seventy subjects more than half could correctly identify handwriting samples of Michelangelo, Raphael, and Leonardo da Vinci after viewing them simultaneously in enlarged format for twenty seconds. Such results should not be surprising, he claimed, for so-called "physiognomic" phenomena are far more common than we think. For one thing, they are the basis of our ability to adapt to the moods of others in conversation.[36]

Especially revealing was Arnheim's analysis of erroneous attributions. Among the most important sources of error were: employing "external" criteria, for example attributing a handwriting sample in "German" script to Wagner because he was a representative of "true Germanism"; and basing judgments on individual parts of a handwriting sample or a face rather than the whole. In qualitative analyses of portraits and silhouettes Arnheim showed, further, that character attributions are based on specific figural characteristics. For example, when he showed subjects a portrait of a young girl's eyes, then suddenly revealed the whole picture, they often exclaimed "That's a totally different person!" But the response was not so clear when whole portraits were shown, then partially covered. Again, particular facial parts cut out of one portrait and inserted in another produced characteristic changes in the second picture's expression. Thus, he concluded, allegedly scientific characterology or graphology based on schematic "psychograms," or supposedly characteristic features such as a bold nose or sensual mouth, was bound to be inaccurate.[37]

Nowhere in his dissertation did Arnheim work with film images. Nonetheless, his finding that ordinary people do spontaneously what respectable science thought to be magical became a foundation stone of his film criticism. Arnheim began writing art and film reviews for a variety of papers, including the independent left-wing journal, *Die Weltbühne,* while still a student. After receiving his degree he went to work full time for *Die Weltbühne.* As he put it in a recent reminiscence:

> after getting my degree I had my fill of the dry tone [of academe] for a while. What attracted me irresistably to journals like *Die Weltbühne* was the connection of combative social and cultural criticism, elegant language, and pleasure in mischievous little pranks [*Unfug im kleinen*]. Among those of us who made up the core of the group, it was difficult to say whether we were more interested in what we had to say, or in the nearly sensual pleasure of saying it. Both were serious matters to us.[38]

The essential aspects of Arnheim's film theory were already visible in his critiques and reviews, though he did not fully elaborate that theory until 1932, in the book *Film as Art.* The key points, and their relation to Gestalt theory, can be summarized briefly.[39]

Arnheim was a passionate advocate of pure form in film. To him this meant film as a purely visual medium, which was somehow spoiled by the addition of sound or even color. He supported this viewpoint with a careful analysis of film optics, which he insisted is not the same even as perceptual, let alone physical, optics. In film, three-dimensional objects necessarily appear on a two-dimensional surface, with the result that they appear foreshortened. Moreover, the screen limits the visual surface, hence the viewer's frame of reference. The lack of depth also affects the kinesthetic dimension – if rotated in a certain way, the camera can make stationary objects like walls appear to move. Also lacking are the constancies of size and form. And yet the viewer accepts and responds to these potentially disorienting structural features. An onrushing locomotive filmed straight on, for example, has an even more powerful effect than in real life: "The nearer the engine comes the larger it appears, the dark mass on the screen spreads in every direction at a tremendous pace . . . and the actual objective movement of the engine is strengthened by this dilation."[40] Precisely this fact – that "we can perceive objects and events as living and at the same time imaginary, as real objects and as simple patterns of light on the projection screen" – defines the potential for creative use of the film medium, especially montage. Following Soviet filmmaker and theorist Wsewelod Pudowkin's pioneering work, Arnheim maintained that the ideal artistic aim of film is to create simultaneously an illusion of reality and new expressive forms that transcend reality, and thus to "create pictures beautiful in form and of profound significance, as subjective and complex as painting."[41]

To Arnheim, adding color and sound, instead of enriching the medium's expressive possibilities, reduces them in favor of mere verisimilitude. On color, for example, he argued that black and white photography, by reducing everything to the gray scale, made art "sufficiently independent and divergent from nature" to liberate creativity. Because color film is not likely to produce such a "transposition of reality," the filmmaker's options are limited to "the possibility of controlling the color by clever choice of what is to be photographed"; thus the camera increasingly becomes "a mere mechanical recording machine."[42] On sound he acknowledged that "by sheer good luck, sound film is not only destructive but also offers artistic potentialities of its own." In a review of Jacques Feyder's "Anna Christie," for example, he found Greta Garbo's effortful German irritating; "and yet her wonderful, deep voice (which, by the way, is amazingly reminiscent of Elizabeth Bergner's) enriches the image that we have loved for years." But he still concluded that, if used to realize false ideals of naturalism, sound, like color, created only a hybrid, not a pure form.[43]

Direct references to Gestalt theory abound in Arnheim's film theory. He cited Wertheimer's motion paper, for example, to support the claim that stroboscopic motion is in fact "imperceptible montage," and that this is the principle that made it possible, for instance, for Eisenstein to make stone lions roar in *Battleship Potemkin*. Köhler's work on sudden insight resulting from a recentering of the visual field and Wertheimer's paper on productive thinking were of obvious help in his discussion of comedy, since so many film jokes depend on sudden changes in points of view, or in the functions of filmed objects. Finally, in his discussion of expression in acting, he noted, echoing Wertheimer's and Koffka's ideas on civilization, that human expression appears to have "degenerated" in so-called civilized people compared with the much more distinct forms found in animals. "In a good work of art, however, everything must be clear . . . and therefore human expression on the screen must be plain and unmistakable. Hence a film actor must be capable of producing 'pure' expression."[44]

With such loyalties to the autonomy of film form Arnheim swam against the current of his time. After 1925, advocates of fashionable "new objectivity" (*Neue Sachlichkeit*) admired documentary realism in visual art and praised modern technology's potential to reform and improve ordinary life. In a 1925 *Weltbühne* debate on photography, Arnheim warned against overpraise of the technical and an overly simple linkage of artistic realism and social progress. Later he saw an irony in the popularity of Charlie Chaplin's films among the proponents of "new objectivity": "In a Chaplin film no face, no hand movement is true to nature." Still later, he criticized Piel Jutzi's *Berlin Alexanderplatz* for editing out in the interest of heightened realism precisely those aspects of Alfred Döblin's novel that best conveyed the helplessness and despair of the character's situation: "The more the visual instruction of film sharpens our feeling for the true, the natural, the more perceptible it becomes in so-called social films that film people speak of the proletarian as the blind man does of color."[45] Arnheim was clearly not hostile to social criticism in film. Rather, he argued that only films that achieved their full artistic, that is, expressive potential would also have the desired impact on society. His admiring commentaries on the films of Pudowkin and Sergei Eisenstein exemplified this position. Of Pudowkin's *Mother,* for example, he wrote: "If it is true that a work [of art] is better, the more elementary the content it presents, then here is great art: The raging anger against the oppressor, surely a fundamental drive of humanity, cannot be shown more nakedly and concentratedly."[46]

The censorship debate provoked by Bertolt Brecht's film *Kuhle Wampe,* a reportage of working-class life in the Depression, brought out Arnheim's particular mix of aesthetics and politics most clearly. He praised the film as a breath of fresh air, "a piece of naturalness and reality," but noted certain "imperfections." Among these were the fact that "the arguments offered do not follow from the action but are pure conversation," and the filmmakers' refusal to exemplify the workings of the system in a single story. Just these weaknesses, he remarked ironically, "should make the censors more lenient," precisely because they reduce the clarity of the political message.[47] When the Berlin authorities ignored his ironic advice and banned the film, Arnheim joined the protest. In 1929, he had argued that there was no longer any justification for the 1920 censorship law and restriction on youth attendance besides mere inertia and "the petty bourgeois opposition to art of the people who govern us." In the *Kuhle Wampe* case, he altered his diagnosis of the censors' motives: "It becomes steadily clearer that one is determined to forbid all films that are in any way freedom-loving or progressive, regardless of the cost." In the speech he gave at a protest demonstration on

April 13, 1932, he proclaimed that political censorship is not a sign of government power, but of weakness. Vague appeals to freedom of expression are thus of little use: "It is not our business to appeal to the ideas of freedom and the sense of justice of this state. For with that we only help to maintain the necessary fiction that this is anything but a power struggle."[48]

The Berlin officials who banned *Kuhle Wampe* were Social Democrats, not conservatives; Brecht was linked by this time with the Communists. Such language shows that Arnheim, like many other intellectuals of the time, was being drawn into the battle between Social Democrats and Communists that contributed to disunity on the Left, and thus to the fall of the Weimar Republic. But his review of *Kuhle Wampe* made it clear that he did not share the dogmatic perspectives of either the socialist or the communist parties. In both aesthetics and politics he sought a position in the radical middle. He opposed attempts to make film a mere extension of photography or the theatre, but he also opposed progressive intellectuals' claims for the medium as a cultural change agent, aesthetic modernist claims for the potential of the film medium itself to grow and change, or attempts to associate radical stylistic innovation in film automatically with political progressivism.

Despite his awareness of the film artist's potential to change viewers' frames of reference, Arnheim insisted that the ultimate aim of film as art is to restore order and balance to the viewed world, albeit at an enriched, even rarified, plane of experience. As he later commented, "A population constantly exposed to chaotic sights and sounds is gravely handicapped in finding its way. When the eyes and ears are prevented from perceiving meaningful order, they can only react to the brutal signals of immediate satisfaction."[49] This radical middle position between modernist innovation and classicist order was analogous to that of Gestalt theory in psychological science and in Weimar culture. However, Arnheim, rather like his mentor Wertheimer, did not discuss the social and political implications of his thinking in the same detail as he lavished on its psychological and aesthetic dimensions.

After emigrating first to Italy in 1933, to Great Britain in 1939, and then to the United States in 1940, Arnheim developed a full-fledged psychology of art based on Gestalt principles. But his position on film as art changed not at all, for the links with Gestalt theory, especially the loyalty to pure form, clear and *prägnant* expression, and balanced vision were there from the beginning. His work was Gestalt psychology's most substantial contribution to Weimar culture outside the realms of philosophy and psychology. But precisely that contribution pointed to an ambiguity in Gestalt theory's claim to be a new synthetic world view. As Wertheimer had emphasized Gestalt theory's rootedness in science, Arnheim, too, insisted that film appreciation should be grounded in a full understanding of the psychology of expression. Though such a viewpoint enhanced both the legitimacy of their arguments at the time and the solidity of their contributions in the long run, it was unclear how it could respond to the urgent demands of intellectuals and others for reassurance, or radical change, in the here and now.

Rather than howling with the antiscientific wolves, the Gestalt theorists held to the nineteenth-century belief that philosophy ought to be based on empirical science, while radically revising the positivist epistemology on which that belief had originally been based. In a 1923 essay on the intellectual situation of Europe, writer Robert Musil, who had worked in the Berlin psychological institute and knew Koffka and Köhler, referred to Köhler's theory of physical Gestalten as a ray of hope in an era that "has no philosophy, less because it is unable to produce one than because it rejects offers that do not agree with the facts." Whoever "has the knowledge to understand it," he wrote, "will experience how, on the basis of empirical science, the solution to ancient metaphysical difficulties is already implied."[50]

Other intellectuals in Weimar Germany also recognized Gestalt theory's potential. Critical social theorist Max Horkheimer, for example, supported Wertheimer's appointment in Frankfurt with remarkable praise. "The emergence of a new science of life, which strives to unify physiological and psychological investigation," he wrote, "has greater importance for real knowledge than all hasty 'syntheses' of cultural totality [*Gesamtkultur*]. But I have nowhere found a higher concept of the philosophical and general weight of such strivings than in the writings of Gelb and Wertheimer."[51] The independent Marxist thinker Ernst Bloch, however, later wrote of Gestalt theory that its perspective was laudable but too cautious by far for the radical change demanded by the times. Because its results "belong to science," he carefully exempted it from his broad indictment of 1920s phenomenology, especially that of Max Scheler, for combining otherwise sound observations with a Platonistic and "feudal" vocabulary of "order." Yet by discussing the two viewpoints together, he clearly implied the potential for misuse of the Gestalt viewpoint.[52]

Still more severe was the criticism of Marxist students, some of whom were loosely connected with the Berlin institute, in a 1932 pamphlet. In the opinion of one of them, Gestalt theory's epistemology bore a positive resemblance to the materialist viewpoint of Marx and especially that of Engels, since it claimed, in opposition to idealism, that real objects have real counterparts in both brain and mind. However, completely ignoring the work of Kurt Lewin, they asserted that the Gestalt approach failed to treat the social context of individual consciousness, and had no connection whatever to psychotechnics or any other practical purpose: "It keeps itself in scientific isolation and naturally finds no purchasers for its research results and discoveries. It is thus condemned to unproductivity; it has . . . no economic justification for its existence, it is there for its own sake." Thus, it might solve the crisis of psychology, but not that of society.[53]

All of the participants in these Weimar-era debates expected far more of Gestalt theory than it could deliver. It did not, nor could it, resolve the multiple crises of Weimar culture. Like their competitors, the Gestalt theorists, too, accepted without serious examination the assumption that conceptual change alone would suffice to that end. By emphasizing an aesthetic discourse of inherent order and meaning over technological categories of domination and control, they reaffirmed *Bildungsbürger* values. But by remaining loyal to natural science, they also affirmed their belief that those values could be reconciled with modernity.

18

The reception among German-speaking psychologists

Some sense of the connections between the Gestalt theorists' psychological and political views may have played a role in the critical reception accorded Gestalt theory by other German-speaking psychologists. More clearly visible were two closely related but potentially contradictory considerations. An index of the cognitive and social authority of incipient or established disciplines in a given culture is their control over conceptual or methodological tools and institutional resources. Given the relatively scarce resources available to psychology as an embattled specialty of philosophy in the Weimar period, and the hierarchical organization of scientific institutions in German-speaking Europe, the reception of Gestalt theory understandably reflected a competitive struggle among leading institute directors, each of whom offered his own psychology-based worldview.[1] That controversy involved both normative claims about what psychological theories should look like, and broader, ultimately ideological concerns. On the other hand, again due to psychology's relative lack of cognitive and material resources, the discussion also reflected a struggle to achieve some sort of consensus despite the competition. In 1927, Karl Bühler made his contribution to Weimar-era crisis talk when he spoke of a "crisis of psychology," meaning the discipline's failure to achieve agreement on the most basic issues of subject matter, method, and theory.[2] Psychologists' responses to Gestalt theory reflected their desire to overcome that "crisis," and thus to present a united front that would be, or seem, adequate to the demands of the day.

Priority claims and scientific authority

That the Gestalt theorists were indeed in tune with the times became clear when others laid claim to the Gestalt concept. Soon after Wertheimer's declaration of principles in the journal *Psychologische Forschung* and Köhler's appointment in Berlin, both G. E. Müller and Wundt's successor in Leipzig, Felix Krueger,

307

claimed priority for their treatments of the Gestalt problem. Later in the decade, others also pointed to the long history of the issues involved. The priority claims were markers in a wider contest for disciplinary authority and public acceptance, as well as for efforts to establish and enforce discursive norms in matters psychological.

In 1923, while still chairman of the Society for Experimental Psychology, elder statesman G. E. Müller published a full account of his own "complex theory" and compared it with Gestalt theory. The appeal to generational authority was obvious when he claimed that he was only presenting a view he had defended "for decades," indeed "long before any of the so-called Gestalt theorists possessed scientific maturity."[3] In fact, he considerably extended his earlier theorizing on the topic, trying to explain perceived form by combining association and its accompanying physiological processes with psychological processes such as attention, "collective apprehension," and habit. In this attempt, the "coherence factors" to which Müller had alluded briefly in 1903 (see Chapter 6) acquired considerably more weight than they had had before. But their function remained primarily to make the grasping of complexes by "collective apprehension" easier. To these he now added "physiological" coherence factors corresponding to contour, similarity, and proximity, which he located in a hypothetical "formative zone" between the peripheral organ and the higher cortex. Attention creates a "disposition" in this part of the "formative field" in the optic sector, which corresponds to a psychological disposition to attend to similar stimuli. "Collective dispositions" for simple figures are especially strong when they are symmetrical.[4]

Müller thus acknowledged for the first time that the "phenomenological unitariness of a complex" was a significant factor in perception. But he insisted that more complicated "complex" or coherence factors are acquired only gradually in response to "the needs of life," that is, by habit. All these processes were part and parcel of what Müller called a "battle of the sensory stimuli for the narrow consciousness." In this "contest" stimuli "gain favor by pulling in a certain higher activity," such as attention or memory. Some of these formulations are reminiscent of notions of "neural Darwinism" currently popular in neuropsychology.[5] The real problem, however, was to explain the fact that complexes have effects "as a unified whole" different from those of their components. Though he claimed that his complexes were Gestalten in Köhler's sense, Müller emphasized that for him there was no fundamental difference between the excitations for 1 and 0 and those for 10. Such hypothetical entitites as Wertheimer's "whole processes" were "not something real." There were only mutually interacting excitations, not some "whole" process determining them. Moreover, he argued, Gestalt theory cannot explain the reproductive effects of complexes, the effects of successive Gestalten, the influence of experience or the obvious effects of shifts in the direction of attention (e.g., seeing 1 and 0 or 10).[6]

Köhler replied slightly more than a year later with a fifty-six page polemic in

Psychologische Forschung. After summarizing Müller's theory, he lost no time getting to the heart of the matter – the clash of technological-mechanistic and holistic-aesthetic values in science. "Müller's psychology has very little understanding for the inner meaning of psychological events," Köhler wrote, but rather assembles individual "facts or factors" and attributes their connection to "particular chance circumstances." Indeed, Köhler continued, "as human nature is according to Müller, so is his mode of theory construction. . . . That there might be a fundamental difference between the only externally practical and the internally naturally meaningful is just not recognized, I suspect only poorly understood, even condemned as a modern mysticism."[7] Köhler reminded his readers that "not long ago there was a sort of departure from experimental psychology in general and in broadest circles" due to such views as Müller's, and implied that a psychology that upheld the primacy of intrinsic meaning and value in human nature would be more attractive.

More specifically, Köhler accused Müller of creating the appearance of theory by adding new terminology, and of presupposing what needs to be explained. "Collective attention," for example, was obviously not the only, or even the most important explanation for the appearance of forms in consciousness. When I attend to the space between a piece of paper and a stone on a table, no amount of effort will change the objects seen or their relationship to one another. Or, to take the opposite case, we can attend to one aspect of an ambiguous figure and try to produce its complement as hard as we like, but with no result, until suddenly, completely unexpectedly, the other aspect appears as it were by itself. Moreover, as Karl Bühler had already pointed out, attributing form perception to "collective attention" should mean that the rest of the field which is not attended to remains unformed – an absurd conclusion. The same issues arise with Müller's sudden introduction of "involuntary collective attention," which apparently does not require "higher activity." Such neologisms only cover theoretical nakedness. The only remaining explanation for form perception would be Müller's physiological coherence factors. But if collective attention "increases the physiological coherence" of excitations, then it is clear that such psychological processes only strengthen "groupings and unifications that were in principle there already." With that collective attention becomes a secondary factor, and physiological coherence primary.[8]

But just how can collective attention, or any association theory, work as needed? When I read the printed word "Charlottenburg," for example, what happens? Do I see one letter after another, then summate them with the help of a memory image? Or do I see fourteen distinct letters at once and then summate? In either case, Müller cannot possibly require so many acts of attention if his theory is to be at all realistic. The idea that the more letters there are in the word, the less clearly each will be perceived, does not really address the issue. As Köhler made clear elsewhere, this is a fundamental problem for any association theory – it demands too much of ordinary perception. Here Köhler used the case to show that Müller's choice of examples was one-sided. Nearly all of them came from his work on reading and memory, and there he might be right. Two percepts, 1 and 0 or 10, may indeed come from changing the direction of attention. But what about such units as ?, which for Müller would be two elements of the letter R? We do not ordinarily identify such figures, or select them from the wholes in which they appear, Köhler suggested, because they are not "natural parts" of those wholes. This is in fact the point Müller has failed to grasp – the importance of "natural parts" in the perception of "strong" forms. In such Gestalten, the parts can change, and attention can be a factor in such change, but only secondarily and not often.[9]

Müller's rebuttal called the fundamental disagreement in worldview and values between him and Köhler a difference of "conviction" (*Gesinnung*), a word often employed in political debate. Until Köhler and his comrades say more clearly what "inner meaning" is and how one might discover it scientifically, he averred, "evil people will always have the suspicion that such expressions are only phrases with no content, though surely [they are] well suited to win the sympathies of unclear heads of a certain direction." Müller did not say whom he meant, but it was clear that such people belonged beyond the scientific pale. The rest of his rebuttal consisted only of accusations of distortion and of insults. In his rejoinder, Köhler forgave Müller's sarcasm with a touch of irony, quoting a recently published letter of William James as saying that "Müller will live to feel ashamed of his tone."[10] Müller's book was widely cited by other critics of Gestalt theory, but his "complex theory" generated no research.

In a later, particularly bitter exchange, historical reconstruction again became a weapon in a struggle for discursive authority. This time the disputants were nearer to one another in age, and in intellectual pedigree. After Kurt Koffka published a handbook chapter entitled "Psychology" in 1925, Karl Bühler and Otto Selz, former students and, like Koffka, co-workers of Oswald Külpe, accused him of appropriating ideas from their research without attribution and incorporating them into the "new" Gestalt theory. Bühler even claimed that, while "about half" of Koffka's book on psychical development came from his own book on the same subject, his name appeared only "everywhere where Koffka thinks differently." In comparison, "it has something moving about it, how carefully all of Wertheimer's expostulations are registered with the name of their author." The wider territorial issue became clear in Bühler's complaint that "the compositum Gestalt psychology has passed into the family ownership [of the Berlin school]; one will only be able to use it, for better or worse, in order to designate the special features of their theory." A term that had designated a problem area, open to all, was becoming the name of one group's particular approach to its solution.[11]

Otto Selz struck a similar chord, but with different historical overtones. In his own work on productive thinking and his recently expounded "complex theory," he too, like the Gestalt theorists, wished to bridge the gap between nature and mind, but by "including psychology in the biological sciences" rather than invoking physical systems. Instead of actually discussing these issues, Koffka preferred to offer a "historical construction" of his own devising, forgetting, for example, that Graz school members had also spoken of "open complexes" and "Gestalt formation." Scientific development, he suggested, is more continuous than Koffka would have it. "A philosopher might strive for a new philosophy, but it would never occur to a physicist to present a new physics. In psychology it should not be otherwise. Even 'revolutions' do not eliminate the continuity of an advanced science." Selz had implicitly invoked a communal norm – proponents of new theories have an obligation to acknowledge their predecessors and similar-sounding contemporaries, if only to keep the peace.[12]

Koffka did not deign to reply to Bühler, but addressed Selz's objections by arguing that Selz's theory remained a machine theory in spite of its author's intentions. It was no accident, he claimed, that Selz called himself a "psychovitalist." Both he and Hans Driesch accepted the need for some additional entity or factor to account for order in biological processes. Gestalt theory denies this by claiming that self-organizing processes are common to nature, life, and mind. Despite Selz's claim that "reflexlike" cognitive schemata are not the same as associations, he still postulated an additional factor or mechanism ("task awareness") to account for genuinely productive solutions to problems. There is thus "a contradiction between what Selz wanted and what he has achieved." One reason is probably Selz's assumption that insight can only extract from a situation what is already in it to begin with. This eliminates the possibility of productive thinking from the start. New ideas are not "realizations" of potential knowledge, but fundamental restructurings of already existing states of affairs.[13]

Discursive norms and ideology

Most significant about other exchanges were the many indications that the Gestalt theorists had not fully resolved experimental psychology's orientation problem in the cultural context of the 1920s. Wundt's successor in Leipzig, Felix Krueger, for example, disparaged the Berlin school's originality and went on to propose an alternative, quite different "holistic psychology" (*Ganzheitspsychologie*). Taking his cue from Dilthey's notion of lived experience (discussed in Chapter 5), he emphasized the role of feeling and will in the constitution of experience, and added the dimension of time, in order to build a bridge to historically developed "cultural wholes."[14] The Gestalt theorists ignored the dimension of feeling, he claimed; yet emotion is prior to cognition in human development, and it is feeling that gives "depth" to experienced totality. Gestalt theory overemphasizes what he called the "thing" aspect or dimension of experience and "objectifying" the given. Such "premature objectification" is responsible for the Gestalt theorists' flight into "physicalism." They thus fall back onto common ground with the associationist psychology they reject: "Here as there the slogan dominates, 'Psychology without a soul.'"[15]

The antirationalist thrust of this message fit in well with increasing criticism of Western rationalist "civilization" and praise of Germanic *Kultur* among philosophers and conservative intellectuals. Krueger's writings were full of such motifs, in particular criticism of "goal rationality" and the danger it posed for "the internal, enduring values." "The truly creative individuals," he wrote, "are rooted deep in their folk, much deeper than the average big-city person with his incapability of devotion and, with all his cleverness, his helpless submission to slogans." But he criticized Spengler and Klages for being too pessimistic about Germanic culture's potential for renewal, and recommended reading romantic nationalists such as Father Jahn, Ernst Moritz Arndt, and Wilhelm Riehl: "These deep thinkers, because they were also whole men and understood their folk, knew about the growth laws of true community and the core structure of the soul."[16] Though they had been schooled in exact experimental procedure, Krueger and his Leipzig colleague Hans Volkelt combined such evocative images of "depth" with what one commentator has called "methodological irrationalism." Volkelt, for example, wrote that, since "all psychological analysis is nothing other than the determination of wholes by other wholes," empirical psychological laws are ultimately "abstractions" that can only approach concrete reality asymptotically – "the ineffability of individuality, of the final manifold of a soul, remains."[17] Here scientific and cultural-political positions merged.

Intradisciplinary and wider cultural discourses also merged in Krueger's claim that Gestalt theory's "objectification" of the given prevents it from discovering truly psychological explanations. For Krueger, this meant referring phenomena to what he called underlying emotional, motivational, or cognitive "dispositional structures." These were the source of what he repeatedly called the "impulse to wholeness" (*Drang nach Ganzheit*) characteristic of human experience. In places, he presented these structures as regulative concepts, or as a "biopsychological necessity of development." More often, he claimed that they are "the objectively real, structural ground" of phenomena, the essence of psychological being as such. Thus, Krueger's "structure" concept was ultimately a metaphysical category; indeed, for him it was the link between psychology and philosophy. He did not oppose Gestalt theory's ontological claim that wholes are prior to parts in experience, but offered a different location for wholeness – one within the subject – than the Berliners' "objective situation."[18]

Marburg professor Erich Jaensch showed in another way how scientific and wider cultural discourses merged when he rejected Köhler's physicalism as "materialist" in favor of what he called a "biological" – meaning a typological and characterological – approach. References to biology were common in the Weimar period, from the social Darwinism and "race hygiene" of the eugenics movement to Spengler's comparison of human cultural history with the life cycle and resulting prophecy of the "decline of the West." We have already seen that typological concepts were commonplace in the personality theories of so-called humanistic psychology (see Chapter 17). Jaensch's particular notion of biology was connected with his research on so-called eidetic imagery in children and teenagers. These images, he claimed, combine the sensory immediacy of after-images and the structural richness of normal perceptual imagery. He and his colleagues Oswald Kroh and Hans Henning thought that these are the "fundamental images" (*Urbilder*) from which all perception and memory images develop.[19]

Jaensch's criticism of Gestalt theory – expressed in lectures and occasional published remarks since the early 1920s and more systematically in 1929 – began as a vehement defense of the Marburg institute's research program. In a 1921 paper he specifically suggested that work with eidetic imagery "also opens a way to unlock the today much discussed 'Gestalt problem' for psychological analysis. Pointing to analogies from the natural sciences offers no substitute for this." Rather, such a "premature" step to physics "could easily lead to giving up strict research method in psychology, only recently gained at great effort, in favor of playing with analogies in the manner of [idealist philosopher Joseph] Schelling." He was so indignant when Koffka attacked his group's eidetic imagery research in 1923, before the series was completely published, that he refused even to cite Koffka's paper.[20] Jaensch's more systematic critique, like others', began with a historical point. The Gestalt theorists had not acknowledged that their predecessors had already done away with rigid reliance on psychical elements in favor of dependence on relations. Thus Köhler was attacking a straw man. In fact, Jaensch claimed, Gestalt theory was part of a great movement in philosophy away from the sterile opposition between positivism and Neo-Kantian idealism and back to psychological reality. In view of this it was ironic that Köhler's physicalism negated just this movement by ignoring aspects of the psychical realm that make it fundamentally different from the physical, such as will, purpose, and intention, as well as the dependence of perception on the structure of the whole organism, that is, on personality.[21]

Jaensch's position was similar to that of the Leipzig "holistic" psychologists in its emphasis on the "organic" personality and will, and its concept of development as a process of differentiation from a supposedly primary holistic state. Like Krueger and Spengler, he saw in positivism a sign of cultural disintegration to be "overcome" – a favorite word of the period.[22] But by retaining evolutionary concepts he seemed, at least at the time, to remain committed to some form of natural scientific approach; and there was plenty of experimental work done in Marburg. Even when he rejected Köhler's thinking as too physicalistic, Jaensch, like the Gestalt theorists, held on to the hope of a general philosophical anthropology based on natural scientific psychology.

Closer to the heart of the Gestalt theorists' position and concerns were the criticisms of Hamburg professor William Stern and Vienna Psychological Institute head Karl Bühler. Each of these critics sought to unite one of two central aspects of traditional educated middle-class ideology – the cultivated personality as the goal of *Bildung* and language as the stronghold of *Kultur* – with experimental psychology. Stern's critique, expressed most fully in a 1931 dissertation by his student Martin Scheerer, but also in his own writings from 1928 onward, was summarized crudely but effectively in the slogan, "No Gestalt without a Gestalter." The claim was that the perceiving subject was relatively unimportant for Gestalt theory. Stern maintained that continuing references to the dependence of perception on the state of the organism at a given time were not sufficient to account for the specifically human aspects of experience, especially for people's intentions giving meaning to their actions. Nor could analogies from perception account for the experienced and observable coherence of the human personality over time. Stern clearly did not wish to reject the Gestalt idea as such, but to shift its locus from the interaction of organism and environment to the personality. He conceived the person as a "goal-oriented whole," the psychophysically neutral "point of convergence" of the inner and outer worlds, and thus as the starting point of all philosophy.[23] In this his views were similar to those of Krueger and Jaensch, and both cited this point with approval.

However, Stern's personalism was not the same as cultural-conservative subjectivism. Stern's "person" was not a fixing agent for otherwise unstructured or incompletely structured experienced totalities, as in Krueger's thinking. Moreover, Stern's "person" was not the creation of a historically developed culture, but an autonomous individual living in tension with transpersonal unities like the folk, society, or humanity. "Although the person is codetermined by all the transpersonal areas to which it belongs," he wrote, "it draws its individual wholeness not from them, but asserts it over against them as something autonomous and enriches them by embodying those influences in its own being," that is, by "introcepting" societal and cultural values. This appeared to support a liberal-democratic outlook rather than Krueger's German nationalist authoritarianism.[24]

At several levels, then, there were affinities between Stern's thinking and Gestalt theory. As Stern's student Martin Scheerer showed, however, basic sympathy did not mean uncritical support. In addition to the issue of personality, Scheerer's most important criticism was that the Gestalt theorists' natural-scientific pretensions resulted in lack of clarity. As Köhler implicitly admitted, they had no way of distinguishing the concepts of structure, whole, Gestalt, and system from one another. Moreover, Köhler's "physicalization" or "materialization" of consciousness renders the person passive.[25] For Scheerer, as for Krüger, there was a fundamental tension in Gestalt theory between "objectivizing" and "subjectivizing" motifs. It was imperative, he concluded, to reassert "the 'rights' of subjectivity" in psychology by focusing on the intentional character of consciousness and behavior, as well as the problems of representation and signification. "The ground-breaking results and the broad-ranging ideas of Gestalt theory" could and should be made fruitful in that task.[26]

Karl Bühler focused on the nearly complete absence of language from the Gestalt theorists' research agenda, which he attributed to their physicalist bias. He acknowledged that with their proof that "the whole is capable of determining its parts *realiter,*" the Gestalt theorists had made "a historic contribution" to perceptual theory. In addition, Köhler's application of the systems concept to biology had "put the vast array of control (*Steuerungs-*) processes available to living systems in the proper light." But he emphasized, citing Erich Becher's earlier criticisms, that the existence of such systems does not require any parallelism of Gestalten in both the psychical and physical realms. A "one-sided structural monism" would never account for the "many degrees of freedom" or the goal-directedness in both animal and human action. Nor could it deal with the emergence of ever more complex forms and behavior in evolutionary history, particularly human language.[27]

To do this, Bühler proposed a more pluralistic conceptual framework that distinguished the categories of structure, meaning, and value but related each to the others. His earlier work on Gestalt perception led him to modify Brentano's intentional model of consciousness by showing that sensory stimuli are perceived differently in different contexts. In an attempt to give this idea a biological foundation, he developed a theory of perception in which sensations serve as signs or cues, by analogy with the biological concept of signal. Applied to language, Bühler's sign theory – now recognized as a forerunner of modern semiotics – emphasized the representative function of words, particularly the fact that there is no one-to-one, unambiguous relation between things and their representations or signs. Such thinking, Bühler claimed, is better suited to establish the category of meaning (*Sinn*) in psychology than Gestalt theory.[28]

Though language clearly has structure, Bühler argued further, it also has the functions of expression and representation; any complete theory of cognition or language must begin with or account for these. This raised the fundamental issue of how psychological theories ought to look, which Bühler did not hesitate to put in normative terms. There was no need to throw the sensation concept overboard, just because Stumpf and Helmholtz had occasionally gone too far with the idea of "unnoticed" sensations. For the Gestalt theorists, especially Koffka, sense data are clearly "no longer signs," but "the real determinants of things." Yet "form without matter constitutes no reality. Whoever wants in the name of a holistic theory of perception to eliminate the sensation concept must be able to say what other matter he wishes to introduce in its place, to make his wholes capable of existing, or he does not know what he is saying." Such logical problems cannot be solved by invoking self-organizing structures. Moreover, if the Gestalt theorists wish to cast aside the old rationalism, then they must explain emotionally conditioned deviations from the "best" event structure, or the transcendent value of symbols beyond immediate experience. Even the Gestaltists are forced to operate with "attitudes," he noted.[29] Bühler's polemics hid a basic similarity in the two viewpoints. Both were based on the view that a given structural context decides the significance of both perceptual and linguistic components. Nonetheless, Bühler was right to note that it is difficult to see how the Gestalt theorists could account even for the representational function of language, let alone the roles of symbol and myth as defining constituents of culture.

Köhler did have a few words to say about language in *Gestalt Psychology*. After presenting examples of onomatopoeia and citing examples of "expressive" terms, such as "a black mood" or "bright" tones, from the writings of Ludwig Klages, he pointed to evidence from the work of Erich von Hornbostel that in primitive languages, "the names of things and events often originate according to this similarity between their properties in vision or touch, and certain sounds or acoustical wholes." He also cited a study by Usnadze showing that subjects are more likely to apply a certain combination of vowels and consonants ("baluma") to selected rounded figures than to pointed ones ("takete").[30] Provocative as such examples might be, it was not at all clear how one could use them to distinguish "objective" relations from the culturally determined attributional properties of specific languages. Nor did Köhler say how one could account for the variety of such attributions being so much greater than the range of available phonemes with the notion of intrinsic meaning alone.

The reception of Berlin school research

There is currently much interest in the resolution or closure of controversies in science. Bruno Latour has argued that controversies are decided by the more or less effective mobilization of material, personal, rhetorical, and conceptual resources by particular research teams. On this accounting, institutional politics and the formation of tactical alliances with useful nonscientists are more important than reason; laboratories tend to appear as monads, with networks being created mainly among those with common interests. In contrast, Peter Galison speaks of "trading zones," areas of exchange and communication between laboratory (sub-) cultures where the meanings of terms and conventions of measurement can be negotiated while competition continues.[31] This case provides evidence for both viewpoints. The strongly hierarchical structure of German scientific institutions and the relatively scarce resources available to experimental psychology made a Hobbesian struggle among its schools seem inevitable; it was in this context that the term Gestalt became a cultural resource effectively manipulated by a single research school in Germany while still being claimed by the others. However, closer examination of research in other psychological institutes on issues raised or pursued by the Berlin school indicates that it was also possible, though not always easy, for the rival camps to discover common ground.

The more general criticisms of Gestalt theory discussed above did not prevent constructive reception of the Berlin school's research. Yet here, too, ideologically freighted insistence on "the rights of subjectivity" was far from absent. Most widely taken up by far was Gestalt research on motion. Though there were doubts about the existence of the phenomenon Wertheimer had called "pure" *phi* (motion without a moving object), criticism focused mainly on his physiological hypothesis, not on his observations. Vittorio Benussi and Kiel psychologist Johannes Wittmann argued, for example, that in their experiments motion was seen only when both objects were attended to simultaneously and in a certain unified relationship. This emphasis on the role of "apprehension," a psychological process, in motion perception was the basis for a complete rejection of Gestalt theory by Wittmann and his student, Bruno Petermann.[32]

By the late 1920s the controversy over the *phi* phenomenon's existence appeared resolved, at least in Germany, though some researchers continued to deny it. More frequent were attempts to refine its observation and more precisely measure the stages of its appearance. Most significant of these was a thorough study by Wilhelm Neuhaus, working in Narziss Ach's Göttingen institute. Criticizing Wertheimer's alleged "confounding" of the three measurement parameters exposition time, time interval between first and last stimulus, and distance of stimuli from one another, Neuhaus constructed a new apparatus on the basis of Schumann's tachistoscope to investigate the effect of each of these variables independently of the others. The results enriched Wertheimer's description by adding two intervening stages between simultaneity and optimal motion. But precisely this enrichment had critical implications for Wertheimer's physiological hypothesis, in Neuhaus' view. The idea of a "crossing" function is bound to fail with more than two stimuli; moreover, more complicated stimulus constellations presented with the new apparatus produced "unstable" phenomena not explained by any physiological hypotheses.[33]

Hamburg psychologist Heinz Werner carried out a series of studies in "structural psychology" that paralleled those of the Gestalt theorists in certain respects, but ultimately supported William Stern's personalistic viewpoint. In a 1928 study, for example, Werner and a collaborator, using a stroboscope, presented dissimilar figures to subjects accompanied by an audible rhythm. This produced an impression of motion even in subjects that had been unable to have one with optical stimuli alone. The procedure also worked in reverse – that is, unrhythmic beats could produce a collapse of the motion impression, even in subjects who had seen motion optically. This was proof, they said, "that normally not the eye, but the self as a whole sees motion."[34]

In addition to its broader critique of Gestalt theory, the Leipzig school of *Ganzheitspsychologie* made the more concrete point that the Gestalt theorists neglected the emergence, or "microgenesis" (*Aktualgenese*) of Gestalten. In a sweeping review of "experimental results of Gestalt psychology" presented as an invited address to the Society for Experimental Psychology in 1927, Friedrich Sander, a Leipzig associate of Felix Krueger, claimed that study of this process revealed the importance of emotion in perception. In the standard demonstration

of such "pregestalt" experiences, his student Erich Wohlfahrt presented complicated, bright line figures to subjects in a darkened room, then gradually increased
the exposure until subjects said they experienced them as a "final form" (*End-
gestalt*). The process was characterized, he claimed, by whole qualities "difficult
to name and impossible to draw." At first the objects appeared like an embryonal
sac in which a live object twitched and kicked, then as a labile, circlelike figure,
with an interior at first unstructured and characterized by "feelinglike complex
qualities" like "resistance." Finally, more regular forms appeared, but in "jumps"
rather than gradually, as "regions" of the figure underwent similar stages. Only at
the very end did the figures appear as detached from the observers' feelings.
Emotional experiences were thus "not mere accompanying phenomena but functionally essential" to the "pregestalt" experience, proof of a "structural disposition
of the soul to formation (*Gestaltung*)."[35]

Wertheimer's original conception of the "knowledge process" had provided for
the emergence of Gestalten, but "microgenesis" was not a central research issue
for the Berlin school until Wolfgang Metzger took it up in his *Ganzfeld* research
(discussed in Chapter 14). Though he did not say this was his aim, Metzger's work
clearly answered the Leipzig school's allegation that the Gestalt theorists paid
insufficient attention to the emergence of Gestalten and therefore downplayed the
role of "feeling" in perception. In fact, he tried to measure the effect of what David
Katz had called "impressiveness" (*Eindringlichkeit*) versus that of brightness, and
claimed that the former, which he, like Katz, defined as "feeling strongly affected"
is primary, whereas brightness and whiteness are secondary.[36]

Sander's report also revealed another ideologically loaded difference between
the Leipzig and Berlin schools. Instead of Wertheimer's "centering," the
Leipziger emphasized what Sander called the "dominance" (*Herrschaft*) of Gestalten. The clearest example of what he meant was the figure named after him
(Figure 17). The line 1–2 is seen as longer than 2–3, which is in fact the same
length, because the two parallelograms are different sizes. Like the Berliners,
Sander said that "the result of changed Gestalt formation is a changed meaning of
the lines in the new structure." But "the direction of this dependence" is decided
by what he called the "weight distribution" in the figure.[37] Without pushing the
point too hard, it seems plausible to link this difference in vocabulary to the
different political viewpoints of the schools' adherents. The Berliners acknowledged the existence of hierarchies in experience. Wertheimer even referred to one
Gestalt tendency's "victory" over another, but such language was rare with him.
Not so for the Leipziger.

Like the Leipzig and Hamburg schools, Bühler's Vienna students also offered a
more specific research response to Gestalt theory. Examples are the work of
Ludwig Kardos on thing perception and of Egon Brunswik on size and form
constancy. Only Brunswik's work will be discussed here.[38] Brunswik began his
career as a theoretician in 1929, with a careful historical disquisition on the

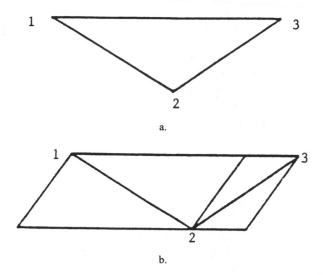

Figure 17. Sander parallelogram (1928). When the isoceles triangle (a) is embedded in (b), the lines 1–2 and 2–3 are no longer seen as the same length.

Gestalt concept, tracing its development from the Greeks and Kant onward, and taking special care to recognize his teacher Bühler as "the actual founder of the psychophysics of Gestalten." The conceptual function of the history was to distinguish the Gestalt concept from those of form, concept, and schema, in order to bring out "the logical particularity (*Eigenart*)" of structural laws as distinguished from normative or causal laws.[39]

In 1933, Brunswik presented a series of studies by his students completed over the past five years supporting his own reconception of the perceptual process. Starting from Bühler's sign and representation concepts, and from Fritz Heider's claim that organism and object are separated by a medium through which perception must be achieved, the issue for Brunswik became the subject's use of sensory cues to infer the qualities, size, or form of distant objects. Brunswik called his view a "lens model," because in it peripheral sensors collect cues and direct them to a central "focal point" where the organism selectively uses them to produce a "central response." The key epistemological distinction was between "intentional" and "intended" objects, by which he meant not only things, but also sensory qualities such as color. According to this conception, subjects try to match "intentional" with "intended" objects, but rarely succeed completely. This was an elaborate way of expressing the well-known fact that perceived size, color, and geometric form only correspond to actual qualities within a narrow range.

The problem had always been to specify quantitatively the relationships between the perceived and the actual qualities. Brunswik claimed that access to systematic measurement could be gained by hypothesizing "intermediate objects" constructed using a variation of standard psychophysical scales. Veridical perception thus became a pole or end point of the scale; in the case of form constancy, this was the frontal-parallel plane. The experimental challenge thus became to measure the range of variation within the continuum of "inferences." Degree of accuracy in these responses was a measure of the organism's achievement, or of what Brunswik later called the "functional validity" of a given act. The idea that subjects can only approximate veridical perception ultimately led, in turn, to a probabilistic theory of perception that is now recognized as an important predecessor to contemporary computational models.[40]

Brunswik's students presented results mainly for variations of attitude and age. Beverly Holaday, for example, tested the role of psychical, somatic, and external conditions in size constancy using cubes seen from an angle; Kurt Eissler did the same for form constancy. Particularly notable was Sylvia Klimpfinger's claim to derive age-graded developmental curves for both size and form constancy. At the end of his programmatic essay, Brunswik raised hopes of extending these findings from individual perception and the development of the constancies to personality type, perhaps even to a theory of the state. As in earlier work on these issues, however, the primary data remained individual adult subjects' judgments of difference between "seen" objects. Notable were Brunswik's claim that only a "minimum of self-observation" had been allowed, and his suggestion that perhaps such methods would satisfy behaviorist criteria for exactitude in psychology.[41]

The similarities between Brunswik's approach and that of Gestalt theory are only to be expected, in view of their common intellectual descent from Brentano and Hering. For both viewpoints, perception is an achievement of the organism. Epistemologically speaking, both take critical realist positions – the perceptual object is not the same as the real object, yet it is also not illusory, but is rather constituted according to empirically verifiable psychological laws. Brunswik's commitment to methodological "behaviorism" would appear to be a major difference from Gestalt theory, but his actual procedure varied little from that of the Berlin laboratory – there was only somewhat less introspection. The most important difference was Brunswik's commitment to probabilism, in contrast to the Gestalt theorists' search for invariant structures in perception. In addition, Brunswik offered no physiological hypotheses to explain his results.

Kurt Koffka responded to the work of Brunswik and his students in his 1935 text, *Principles of Gestalt Psychology*. He focused primarily on the fact that only one aspect – size, form, brightness, or distance – had been measured at a time, with little attempt to deal with their interaction. He also noted that constancy was not perfect, but decreased with angle of orientation. In general, he claimed that such results could be explained by hypothesizing that the perception of objects in other than the frontal-parallel orientation requires "special forces," which are in turn opposed by others pulling it back to the normal. The final organization would be the best balance of these forces. In this view, orientation and shape are mutually dependent.[42]

On the development of the constancies, Koffka cited earlier work by David Katz's students Beryl and Burzlaff, showing that with more than two objects in the stimulus array constancy was present as early as age four. Berlin institute co-worker Helene Frank had found still higher constancy levels than Beryl when objects were placed farther apart. This suggested that younger children are more likely to see excitations of objects located nearer one another as interdependent. Thus, Klimpfinger's curves show only the increasing differentiation of the visual field with maturation.[43]

Institutional rivalry and "the rights of subjectivity"

The reception of Gestalt theory raised substantive issues both in theory and research. But all of the more general criticisms also reflected, both in their vocabulary and in their specific nuances, the most significant concerns and catchwords of German psychologists in the 1920s – holism and meaning, biology and development, language and culture, personality and character. These were also constituting terms of educated middle-class ideology and self-concept in the Weimar period. In their criticisms of Gestalt theory, psychologists pointed out problems with its claims to be both a new, comprehensive approach to mental life and the basis of a unified worldview. In doing so, they responded to the anxiety created by the perceived breakup of the traditional social system that had supported *Bildungsbürger* status by emphasizing the shared values and discursive preferences of the educated elite. Not only in psychology, but elsewhere as well, there were calls to "overcome" supposed oppositions between *Gemeinschaft* and *Gesellschaft, Kultur* and "civilization," or "higher values" and "mere facts." The hope for an overarching "synthesis" was widespread. Psychologists criticized the Gestalt theorists because they claimed to provide such a synthesis, but had not taken one or the other key dimension into account.

The role of institutional competition in all this was undeniable. Even so, certain criticisms were common to nearly all lines of attack. In addition to reminders of the Gestalt problem's long history, the greatest unifier was the rejection of Köhler's physicalism as an insult to the autonomy of the psyche, and thus of psychology. That rejection rested on a widely shared belief that genuinely psychological explanations must invoke processes or entities lying behind and producing phenomena, such as attention, attitudes, or personality types. As Karl Duncker pointed out in his first paper on thinking, such views went back to archempiricist David Hume. He asked: "Are all connections in nature really connections between 'sensible qualities' and 'secret powers,'" as Hume had said? American psychologists seemed to think so; for their entire inventory of explanations consisted of references to such "powers" or third factors, such as frequency, recency, pleasantness, past experience, imitation, and the like.[44] German-speaking psychologists invoked such factors also, but more often they spoke of cognitive or emotional 'powers' such as attention, intention, will, or feeling. Their aim, as Martin Scheerer had eloquently said, was to reassert "the rights of subjectivity." The Gestalt theorists emphasized instead the self-organization of structured situations in organisms and between them and environments. Normative objections to their explanatory approach were inseparable from broader ideological objections to their alleged "objectification" or "rationalization" of the psychical.

Yet the critics were little more successful than the Gestalt theorists in solving a problem that was surely beyond the capacity of any theory. In *The Crisis of Psychology,* Bühler complained of the inability of psychologists to agree on fundamental issues of method and theory, but was confident that a unified science could be created. Which principles would serve as the foundations for such unity remained disputed. Some textbook authors presented the broader, albeit vaguer concept of "structure" as a better common denominator for psychological theory than the more narrowly circumscribed Gestalt concept.[45] But it remained unclear whether invoking "structure" could achieve more than limited theoretical integration within the discipline, much less help aspiring practitioners to find their way in the confusing welter of competing worldviews. Nor was it clear what role such a psychology could play in the training of professional psychologists.

The reception of Gestalt theory among German-speaking psychologists in the Weimar period might usefully be compared with that of relativity and quantum mechanics among physicists (discussed in Chapter 17). The difficulties for relativity and quantum mechanics were at least partly due to the fact that their style of theory construction was not universally accepted even among physicists. This left them fair game for scapegoating in the popular and intellectual press as part of the general "crisis" discourse. Gestalt theory, too, was only partly accepted in its own community, not, however, because its methods were thought too unorthodox, or its ideas too radical, but because its holism did not go far enough. For the proponents of both approaches, secure institutional bases meant that they could do productive research even in troubled times. But the implications of that research did not, and could not, satisfy the ideological demands of the broader public from which the researchers came.

Part IV

Under Nazism and after: Survival and adaptation

19

Persecution, emigration, and Köhler's resistance in Berlin

National Socialism was a new phenomenon in German life, and most Germans reacted to it in much more complicated, vacillating, unclear, and inconsistent ways than we ordinarily assume they did.[1]

It was once accepted that National Socialist "coordination" (*Gleichschaltung*) of all aspects of German life after 1933 proceeded in planned, organized, top-down fashion, with at best token resistance. And it was generally understood that as a result of that process everything worthwhile about German civilization was destroyed or twisted to serve the ends of the new regime, while most important representatives of German science and culture went – or were forced – into exile. Recent research has turned these claims into questions. As a result, the issues of collaboration, persecution, and survival for science under the Swastika have been recast. In particular, this work has revealed a variety of linkages between different sciences and professions and the Nazi state – a variety consistent with historians' views that that state was less of a totalitarian monolith than previously believed. Important factions tried to create Nazified versions of some disciplines, such as "German physics," and allied themselves with ideological watchdogs in the party. But these efforts faced strong opposition from others who emphasized the practical value of pure research, and constructed their own rhetorical connections with Nazism's projects or more concrete alliances with various state or party units. Particularly in the natural and engineering sciences, such efforts gained increasing state support after the promulgation of the Four-year Plan in 1936 made preparation for war a primary aim of economic policy.[2]

Other writers have pointed out that many German scholars' elitist and nationalist attitudes prepared them to welcome or to accept passively the "national revolution" promised by the Nazis. Although some of these "conservative revolutionaries" later turned aside in disappointment, many provided ideological support to the regime throughout its history.[3] Still other scholars have gone further, showing that some disciplines, most notably psychology and eugenical biology, enhanced their professional status considerably under Nazism by offering practical skills to the regime, the former in officer selection for the *Wehrmacht,* the latter in the service of more infamous "selection" programs. In these cases, proclamations of ideological loyalty to Nazism were frequently forthcoming, but were not of central importance; for the services offered contributed directly to central aspects of the Nazi project – world domination and the destruction of the allegedly unfit. Such examples run counter to the "deprofessionalization" that historians have observed in other occupations, such as law and education, under Nazism.[4]

The view remains widespread that the Nazis deliberately destroyed psychology, because they viewed it as subversive.[5] Perhaps the general tendency to equate psychology with psychoanalysis is responsible for this mistaken belief; but in fact, both disciplines, once shorn of their Jewish and leftist members, survived and even prospered under Nazism.[6] Only the case of Gestalt theory will be discussed here; but that history, too, sheds light on the complexities of cultural life under Nazism, on the relationships of science and ideology in the Nazi state, and on the institutional and intellectual situations of science in the postwar German states. This chapter will show that Nazi anti-Semitism and opportunistic pressure for ideological conformity from below, rather than any particular Nazi animosity against Gestalt theory, forced the leaders of the Berlin school into exile, and ended Wolfgang Köhler's extraordinary effort to assure some sort of continuity in Berlin. The following chapters show how the remaining students of Gestalt theory attempted varying combinations of ideological and pragmatic adaptations to their altered circumstances after Köhler departed for America in 1935. These adaptations assured that research and theorizing in the Gestalt tradition continued, albeit in some cases with important modifications; but they could not prevent the Berlin school's increasing marginalization.

The initial challenge and Köhler's response

One of the earliest pieces of Nazi legislation, the so-called Law for the Re-establishment of the Professional Civil Service of April 7, 1933, called for the removal or forced retirement of "non-Aryan" (that is, Jewish) and politically "unreliable" state officials, including professors. As a result, six of the fifteen full professors teaching psychology in German universities lost their posts. Among them were the founder of Gestalt theory, Max Wertheimer, and Adhemar Gelb, a close associate of the Berlin school. The others were David Katz in Rostock, William Stern in Hamburg, and Wilhelm Peters in Jena. Also dismissed was Otto Selz at the commercial academy in Mannheim. Selz and Katz should have come under a provision of the law which exempted Jewish veterans who had served at the front in the First World War. Officials dealt with that in Katz's case by abolishing his professorship; in the case of Selz, apparently, no such legal niceties were observed.[7]

Oral tradition has it that Max Wertheimer had already decided to leave Germany after hearing one of Hitler's campaign speeches on the radio in early 1933. He did in fact go with his family to Marienbad in Bohemia, where he had journeyed the year before for treatment of a heart condition. From there he wrote to the university authorities on the day the Nazi civil service law was promulgated, requesting a leave of absence for reasons of health. Before that letter arrived, however, Frankfurt's Dean of Natural Sciences sent him a curt note informing him that he belonged "to the group of teachers who can expect disturbances in their lectures in the coming semester," and urging him "to make use of your right to take a leave of absence." Within two and one-half weeks Wertheimer had been placed on leave "pending a final decision on the basis of the civil service law." This made him one of the first professors affected by Nazi persecution, and also one of the fifty-two members of the Frankfurt faculty forced to leave their posts in 1933 alone. When the move became public, a leading Berlin newspaper openly regretted the loss of one who "with his economical and content-rich investigations founded one of the most fruitful schools of current psychology."[8]

By this time Wertheimer had received offers from both the New School for Social Research in New York and the Hebrew University in Jerusalem. The latter, for a two-year position, was passed along by Frankfurt colleague Martin Buber. New School president Alvin Johnson wanted Wertheimer, whom he considered "the greatest of the psychologists," to be part of his proposed "University in Exile" for outstanding German scholars who were becoming refugees from Nazism. Wertheimer chose the New School and sailed for New York in September, before the Prussian Ministry had officially decided to dismiss him. When he notified the University of his appointment, he called it a "guest professorship," and University officials replied that they had "no objection" to his accepting such a post. On September 25, however, the Ministry confirmed that as a "non-Aryan," Wertheimer would be forced to "retire." In December, he was denied pension rights because he had not been employed as a full-time civil servant for ten years prior to his dismissal. Wertheimer probably realized that his move to New York would be permanent, but after these decisions he no longer had any choice.[9]

Kurt Lewin, like Wertheimer a Jew of liberal-leftist politics, also emigrated before he could be fired from his position in Berlin. The correspondence on Lewin is an instructive introduction to the institutional side of Gestalt theory's history under the new regime. At the time of the Nazi takeover, Lewin was on leave at Stanford University. In April 1933, when the civil service law was promulgated, he was on his way back to Germany by way of Japan and the Soviet Union. As a decorated war veteran, he was formally exempt from the law. Köhler, who was not a Jew and thus did not fall under the Nazi ban, recommended his continued employment "completely without reservation." However, he noted the possibility that "the race aspect could lead to problems, if Herr Lewin . . . deals with students who know nothing about his participation in the war." He therefore asked the Prussian education minister to extend Lewin's leave, writing that "such a danger would hardly exist next year," and that by then students, once they had learned of his war record, would accept him. A note in the margin of the ministry's copy of Köhler's request reads, "Köhler has absolutely no idea of what the will of the government is."[10]

The Ministry nonetheless extended Lewin's leave until July 31, 1934, but he returned to Berlin only long enough to collect his family before moving on to a temporary research position at Cornell University. In a moving letter dated May 20, 1933, he thanked Köhler warmly for treating him "correctly" throughout, but made it clear that he did not agree with Köhler's assessment of the situation. "Everything within me rebels against the idea of leaving Germany despite all logical arguments," he wrote. And yet:

> The actual loss of civil rights of the Jews has not abated, [but] is increasing daily and will no doubt be carried out completely in the peculiarly systematic German way, whether slowly and methodically, or in periodic waves. . . . I cannot imagine how a Jew is supposed to live a life in Germany at the present time that does justice to even the most primitive demands of truthfulness.[11]

It is unlikely that Köhler had failed to understand "the will of the government," as the Ministry official who commented on his leave request for Lewin thought. He espoused the same line of thinking in his public response to the new government as he did in his private plea for Lewin. That response will be analyzed in some detail here. Köhler was one of the very few German academics who questioned Nazi policies in public. His actions have been assessed mainly in personal and moral terms – the most frequently cited article on the subject is entitled "One Man against the Nazis."[12] Though such an interpretation is eminently justified, it is also necessary to look more closely at the implicit – and explicit – politics of Köhler's position.

The event to which Köhler publicly responded was not his friend Wertheimer's firing, but the resignation of a colleague, Nobel Prize-winning physicist James Franck, in protest against the dismissal of Jewish physicists and mathematicians in Göttingen. Köhler took his stand in a newspaper article, "Conversations in Germany," published in the *Deutsche Allgemeine Zeitung,* a right-of-center journal, on April 28, 1933, only three weeks after the promulgation of the new civil service law (Photo 18). The London *Times* reviewed the article extensively the next day. A slightly abridged translation appeared in the *New York Times* on June 11, in connection with the news of a pending visit of Köhler's to the United States. In the same connection Köhler summarized his views in an interview with the Berlin correspondent of the *New York Times,* which appeared on June 7. He sent a

Gespräche in Deutschland

Von

Prof. Dr. Wolfgang Köhler, Univ. Berlin

Die mächtigen Männer, die eben Deutschland regieren, haben mehr als einmal nach den anderen Deutschen gefragt, nach denen, die bisher abseits stehen und die zu gewinnen sicher lohnte. Wer sie gewinnen will, muß wissen, weshalb sie abseits stehen. Und wenn sie wirklich etwas wert sind, wird offene Behandlung dieser Frage vaterländische Pflicht.

Man kann leicht erfahren, was sie fernhält; denn eben können sie von nichts anderem sprechen. Dabei ist ihnen aller Klatsch zuwider, und ihre Sorge geht mit einem Ernst auf die Nation, der in keiner Partei übertroffen werden kann. Niemals sah ich schöneren Patriotismus als den ihren: Deutschsein ist ihnen eine Aufgabe. Sie wird gelöst, wenn aus dem merkwürdigen, mit sich selbst ringenden Reichtum deutscher Art das Adligste an Gesinnung und Verhalten entsteht. Es ist nur eine Stimme unter ihnen, daß ein Reinigungswerk vor sich gegangen ist, wie es notwendiger kaum hätte sein können; sie bewundern die Wucht des Geschehens, durch welches Deutschland in wenigen Tagen zum erstenmal ein festes Reich wurde, und nicht zuletzt sind sie dankbar für den schroffen Ruck, mit welchem so ungemeine Willenskraft uns alle aus der Bequemlichkeit der letzten Jahre und zu solcher Wachheit aufgerissen hat.

Wie können sie trotzdem traurig sein? Ein Kind vermag zu erkennen, daß sie unter schwerem Druck leben. Sie sind traurig aus Angst um ihre Nation.

Mit politischem Geschehen im engeren Sinn des Wortes hat diese Besorgnis wenig zu tun. Durch unerhörte Konzentration von Macht ist der bisherige Erfolg errungen, und diese Macht geschlossen zu halten, muß für den Führer eine Selbstverständlichkeit sein. Aber wo die eigentliche Politik und die Machtfragen mit der Fülle der Sachfragen verschmelzen, da ist — so sagen diese Menschen — immer noch Deutschland, da ist das unübersehbare bunte Leben von Deutschen in hundert Berufen. Für die Gliederung, für die Standesvertretung dieser Berufe, für die Verfassung deutscher Arbeit hat die neue Staatsführung bestimmte große Ziele, und insofern reicht ihre Politik in jede Sachfrage und in alles deutsche Leben überhaupt. Aber die reicht eben in Sachfragen, und damit ist der erste Gegenstand vieler Sorge erreicht. Keine dieser Aufgaben bis zur letzten kann der Nation gleichgültig sein, und wenn deshalb eine jede in den Händen eines Patrioten liegen muß, so ist schon die nächste Forderung, daß dieser nach Sachkenntnis auf dem betreffenden Gebiet, nach Persönlichkeit und nach Einsicht der beste Mann sein soll, den man im betreffenden Falle überhaupt finden kann. Da nun auf und ab im Lande, in den Fachvertretungen der Berufe, den Verbandsleitungen der Wirtschaft, und bis in die einzelnen Unternehmen ein Mann nach dem andern davon muß, an dessen deutscher Gesinnung so wenig gezweifelt werden kann wie an seiner gründlichen Sachkenntnis und seiner Eignung als Charakter, so höre ich immer wieder die Frage: Warum? Wer sind die Nachfolger? Daß die mächtige Bewegung, die nun herrscht, starke Persönlichkeiten genug für alle wesentlichen Stellen politischer Macht bereithielt, hat sie gezeigt. Daß sie

obendrein auch noch, ferner dem unmittelbar politischen Bereich, für jedes Sachgebiet deutschen Lebens hinreichend viele bessere und beste Männer hergeben könne, für die Verwaltung der Hochschulen, der Berufsverbände, der Einzelunternehmungen in der Wirtschaft, wo immer oben oder weiter unten etwas nach sachlicher Gefordertheit zu führen ist — das scheint meinen Freunden fast ein Ding der Unmöglichkeit.

Ich finde gar nicht, daß diese Menschen etwa neidisch wären. Sie haben einfach Angst um die nahe Zukunft ihres Vaterlandes, wenn in jedem Lebensbereich nicht der schlechthin geeignetste unter allen Männern deutscher Gesinnung, sondern überall der Nationalsozialist des Bereiches die Führung in seine Hand nimmt. Sie wagen nicht zu hoffen, daß dieser immer mit jenem identisch sein werde. Und inzwischen steht ihnen ihr Vaterland näher selbst als die Partei, die schon so Großes am Vaterlande geschaffen hat. Man mißverstehe diese Leute nicht; sie sind keine Aufrührer. Im Gegenteil, nichts wünschen sie mehr als feste Ordnung im Staat, und wenn sie die mächtigen Männer der Regierung um etwas bitten würden, so wäre es weniger Lockerung der Staatszügel als vielmehr gerechte Strenge, und die für alle Deutschen nach gleichem Maß.

Ich fragte solche Menschen, weshalb sie, bei so viel Uebereinstimmung in wesentlichen Punkten, nicht in die Partei eintreten und sich so, je auf ihrem Sachgebiet, zur Auswahl für die wichtigen Stellen verfügbar machen. Die Antwort war immer dieselbe: „Auch vorzügliche Männer haben gehen müssen, anscheinend weil sie nicht zur Partei gehörten. Sie haben das nicht verstanden; denn es geht ja um Deutschland und nicht um eine Partei, und sei es die wertvollste. Es wird uns schwer, einer Partei beizutreten, deren Vorgeben wir in einem so wesentlichen Punkt noch nicht begreifen."

Wenn ich aber weiter frage, so zeigt sich, daß noch eine andere Sorge vorliegt, die freilich ebenfalls aus dem Prinzip der Sachlichkeit um der Nation willen entspringt. Sie betrifft die Rassenpolitik der nationalsozialistischen Partei und des neuen deutschen Staates. Keiner von den Deutschen, die ich meine, leugnet das Vorhandensein eines Judenproblems in Deutschland; die meisten von ihnen glauben, daß die Deutschen das Recht haben, die Zusammensetzung ihres Volkskörpers zu kontrollieren und den zu groß gewordenen Anteil von Juden an der Führung aller wesentlichen Angelegenheiten des Volkes durch weise Regelung zu beschränken. Nur solche Maßnahmen aber können sie innerlich billigen, die nicht auf Umwegen Deutschland schädigen, die nicht plötzlich die Existenz von ganz Unschuldigen zerstören und die nicht die bedeutenden, vornehmen Menschen unter den deutschen Juden schwer verletzen. Denn meine Freunde wollen der These nicht zustimmen, daß jeder Jude, als Jude, eine niedere, minderwertige Form von Menschentum darstellt. Als ich fragte, griff einer nach den Psalmen und las:

> Der Herr ist mein Hirte; mir wird nichts mangeln.
>
> Er weidet mich auf einer grünen Aue und führet mich zum frischen Wasser. —
>
> Und ob ich schon wanderte im finstern Tal, fürchte ich kein Unglück; denn du bist bei mir.

Auch den 90. Psalm las er noch und sagte dann: Kaum hat je ein Deutscher Menschenherzen tiefer erschüttert und getröstet, die in Not sind. Und dies ist von den Juden zu uns gekommen.

Photo 18. "Conversations in Germany" by Wolfgang Köhler. *Deutsche Allgemeine Zeitung*, 28 April 1933.

copy of the interview to the German Foreign Office, with the remark that he had conducted it in Berlin in order to prevent "the otherwise normal distortions and false reports of the foreign press." This remark and its addressee suggest that Köhler was already seeking protectors in the bureaucracy.[13]

The article "Conversations in Germany" appeared on page 2 of the *Deutsche Allgemeine Zeitung*'s national edition – a prominent space often dedicated to opinions considered important but not generated by the staff. Köhler began, appropriately enough considering the newspaper's politics, by saying that it was his "patriotic duty" to ask why "other Germans . . . are standing aside" in these times. These Germans "admire the power of the events by which Germany in a few days has become for the first time a firm state [*ein festes Reich*]." He insisted that "They are not disturbers. Rather the opposite – they wish nothing more than a firm order in the state, and if they were to ask the powerful men of the government for anything, it would be less a loosening of the reigns than for just rigor, for all Germans according to the same standard." But precisely this is what has been missing: "Since up and down the land, in the professional organizations, in the leadership of the economy, and even in individual enterprises one man after the other must leave, whose German patriotism is so little in doubt as their thorough expertise or the appropriateness of their character, I hear again and again the question: Why? Who are the successors?"[14]

Köhler criticized the new policy here mainly on practical grounds. The right-wing German National People's Party was then still in coalition with the National Socialists, and his words expressed, intentionally or not, the concerns of conservative Germans who generally approved of the new "National Revolution," but wondered whether it would harm their own interests. Some may also have worried about some Nazis' apparent lack of concern for proper procedure. Köhler addressed this issue at the end of the article, where he specifically praised the new civil service law, which he said had been "formed with decisive but at the same time careful hands." By this he probably meant the provisions exempting Jewish war veterans, mentioned above, which had been included to mollify the German Nationalists.

On Nazi racist policy, Köhler tried to take a middle position: "None of the Germans whom I mean denies the existence of a Jewish problem in Germany. Most of them believe that the Germans have the right to control the composition of their population and to reduce the proportion of Jews in the leadership of all the essential affairs of the people, which has become too large." However, only those measures should be taken "which do not damage Germany indirectly, which do not suddenly destroy the existence of innocent people, and which do not sorely injure *the significant, superior human beings* among the German Jews. For my friends do not want to agree to the thesis that every Jew, as a Jew, represents a lower, inferior form of humanity." In the *New York Times* interview, Köhler suggested that the government should present its policy as a continuation of the long-standing practice of apportioning high positions in the universities and the bureaucracy according to religion. This "would have a most tranquilizing effect on the outside world, and would win over to our new government the adherence of a multitude of German citizens, including some of the most valuable."[15]

Such statements expressed an ambivalence widely shared by the academically educated elite in Germany toward the Jews. For who could or should distinguish "the significant, superior human beings" among them from the others, except the members of that very elite? Köhler had expressed just this mix of liberalism and elitism years earlier, in his letter supporting Wertheimer's appointment in Giessen (see Chapter 13). According to his own recollection, Max Planck voiced the same view – "that there are different kinds of Jews, worthy and worthless to humankind, among the former old families with the best German culture, and that one must make distinctions" – in his famous interview with Hitler on May 16. Hitler's reply is well known: "That is not correct. A Jew is a Jew – all Jews hang together like chains. . . . I must therefore act against all Jews in the same way."[16]

What mixture of tactics and conviction lay behind Köhler's words? There is no reason to question Köhler's courage, or his sincerity. As he wrote to Wertheimer at the time, "It was simply so that one could no longer do anything with oneself or do any work until that was written." But at the same time, he calmly assessed the article's political impact as "more positive for the author than not," and noted that "nothing has happened to me, although care was taken that the issue came into every hand that it was intended for. In any case, everyone today reads this journal. As for the issue itself, naturally nothing has been achieved by it." Köhler had expected to be arrested for what he had written; he and members of the institute stayed up the night of April 28 playing chamber music, waiting for the police. The fact that they did not come is evidence that, as Fritz Stern has written, "in the first weeks of the new regime the possibility of cautious criticism still existed without the price of martyrdom. It was a period in which the National Socialists . . . were cautious and experimental in their dealing with 'respectable' people."[17]

Foreign readers, however, including some who had worked closely with Köhler in the past, were confused about his intentions at first. Kurt Koffka, for example, wrote to his friend and student Molly Harrower on May 10, 1933:

> The article is extremely well written; if anything could have any effect at all it would be this cautious, and yet brave appeal. What startled me is the introduction, in which he praises the achievement of the New Regime in rather glowing terms. I do not know whether this is just politics in order to give more weight to his defense of liberals and Jews, or whether it represents his own opinion.[18]

Only after talking with Köhler himself one month later did Koffka become convinced of his courage: "I learned a lot about Germany without changing my ultimate judgment about it. Köhler is a very courageous man, much more than one could know."[19]

Similar ambiguities emerged from German readers' reactions to the article. Richard Hertz, an official in the Foreign Office and a nephew of physicist Heinrich Hertz, wrote warmly that Köhler had responded to "a hunger for humanitarian and noble gestures" corresponding to "the surfeit of stupid moves to which we are witnesses"; his words evoked "that knightly and large-hearted Germany, which it is pure joy and most pleasant duty to serve." Such gratitude was characteristic of many readers' responses, including that of James Franck. Köhler's colleague on the Philosophical Faculty, Max Dessoir, however, wrote in a somewhat different vein: "I hope that this warning will have a far-reaching and enduring effect. Naturally there is much blame also on the Jewish side. I like to say that the Jews are the salt of the peoples. Our people was too salty. But it will not do it good if one takes away all the salt."[20]

The battle for the Berlin Institute

Köhler made two other aspects of his position clear after the fall of 1933. By this time the Nazis ruled alone, and on November 3 an order had been issued requiring all university classes to begin with the Hitler salute. On the first day of the fall semester, Köhler entered the lecture hall and gave a half-hearted version of the salute. He then read a statement in which he made it clear that he did not agree with the worldview of the regime, but that he would begin class as he did because this was required by the university administration. Thus Köhler distinguished between his duty as a civil servant and his personal politics. Students greeted his statement with applause. In December, however, members of the National Socialist Students' League raided the Psychological Institute, claiming to be looking for evidence of subversion by Köhler's assistants, Karl Duncker and Otto von Lauenstein. After this incident, Köhler apparently arrived at an agreement with the new rector, eugenical anthropologist Eugen Fischer, that in future he would be informed of such moves in advance.[21]

When the institute was searched a second time, on April 12, 1934, again without advance warning, and authorized in the rector's absence by the dean of the Philosophical Faculty, Ludwig Bieberbach, Köhler protested in the sharpest terms. Once already, he wrote to Fischer the next day, "My authority as institute director was severely shaken," when Nazi students "interfered with my rights as head of house." The new incident is "a heavy attack on my privileges as institute director," constitutes "open interference in institute discipline" and represents "a severe disavowal of me personally in front of the students."[22] This was the style of a member of the intellectual nobility protesting a plebian uprising.

Köhler's stance here was of a piece with his public position. In his protest to the rector, he invoked the power and status that had been a matter of course for

German professors for decades. But that status and power had always been secured by professors' loyalty as civil servants and by their apparently apolitical stance in public affairs. In return for these, the state claimed to guarantee freedom of teaching and research. In the Wilhelmine period the result of this tradeoff had been unprecedented expansion of state-supported science, of which the Berlin institute itself had been a part. In "Conversations in Germany," Köhler stated his conditions for maintaining loyalty to the new state; in his letter to the rector he tried to defend the prerogatives supposedly guaranteed by such loyalty.

The vast majority of Köhler's colleagues did not share his understanding of the relationship between professors and the state. They followed the example of their predecessors in the First World War either by openly proclaiming their loyalty to the new regime without conditions, or by keeping quiet, thus leaving the field to the regime's supporters. Only a few professors besides Köhler dared to act as though the relationship between scientists and the state in Germany were a contract among equals. Three examples are the resignation in protest of Nobel Prize winner James Frank from the chair of physics in Göttingen – the event to which Köhler responded in his article – the equation of the persecution of Einstein with that of Galileo by Max von Laue in a speech to the German Physical Society in September 1933, and the memorial assembly in honor of the émigré Jewish physical chemist, Fritz Haber, in January 1935, held despite a ban by the Minister of Science and Education, Bernhard Rust.[23]

In his letter to the rector, Köhler resigned his position as institute director in favor of his chief assistant, so he would not be obliged ex officio to deal with Fischer until the latter remembered his agreement, apologized for the lapse, reinstated the assistant Otto von Lauenstein, who had been fired as a result of the raid, and disciplined the offending students.[24] Yet, astonishingly enough for those who still adhere to the top-down theory of Nazi "coordination," Köhler's resignation was not accepted; for there were other forces at work.

The students' role

First, the institute's students supported Köhler against the Nazi students, who were members of the National Socialist Students' League. This group had won control over student affairs in Berlin as early as 1931. In May 1933, they had been the primary force behind the infamous book burning, which took place in the square in front of the university.[25] These victories, along with the support of professors like Bieberbach, gave them the confidence they needed to raid the Psychological Institute in December 1933.

An anonymous article in the November newsletter of the Berlin students' organization for the Philosophical Faculty entitled "Has the Psychological Institute Coordinated Itself?" prepared the ground. The author of this piece was Hans Preuss, a psychology student who was then head of the so-called Science

Office of the Berlin university branch of the German Students' League. He later stated that he wrote the article with the approval of the national organization's chief of political education, "in order to make the authorities also aware of the situation in the Psychological Institute." Preuss described the institute "as a stronghold of Communists and Jews," and added that "it was always the case that the German student felt isolated in the institute. It was also obvious that Jewish students were preferred over German students." He then denounced Köhler's assistant Karl Duncker and institute mechanic Hans Haar for "Communist activities." Haar was known to be a Social Democrat. Duncker, who was otherwise the only nonpolitical member of a well-known Marxist family, may have worn an antifascist armband during struggles between Nazi and anti-Nazi students at the University before 1933.[26]

Revealingly, in making his charges Preuss used the generic singular form, "the German student," rather than the plural "German students." The institute students' reply to his article in the same paper showed that they understood that Preuss' actual motives were more personal than political. In answer to the accusation that "the German student" had been rejected in favor of Jews, for example, they wrote: "Students are only accepted in any case according to their scientific achievements. It cannot be denied that a choice is made among the students." Köhler made the same point in a letter to rector Fischer, and also named another criterion for institute membership: "Members of the institute behave decently, openly, loyally, and courteously, or, when they prove unable to manage this, they tend after awhile to depart."[27]

However, with this exchange the institute's students put themselves in a difficult situation, for they were also members of the German Students' League. The ambivalence that could result from such divided loyalties came out after the second raid on the institute in April 1934. In May, twenty-four current and former members of the institute asked the head of the Berlin students' association to help solve "the untenable conflict" in which they found themselves by "restoring trust between these two communities." Notable here were the use of the term "community," a key word from both conservative and Nazi propaganda, to refer to a small group, rather than the German people as a whole, and also the students' demand that the accusations of disloyalty be investigated by the Gestapo. One participant asserted in an interview that these were deliberate tactics to assure that they would be heard.[28]

In early June, the institute students went still further and announced a torchlight parade in Köhler's honor. When the Berlin students' association forbade this exercise, the students appealed to the national organization and also to the Prussian Education Ministry. As one of them later recalled: "We were naive enough to believe that we could build a wall around ourselves and do science inside, while the Nazis busied themselves with their 'folk' [sic!] . . . if only we would appeal to the Ministry, the higher-ups would put the students in their place." This appears to have been Köhler's hope as well. As he put it in a letter to his American colleague and friend, Harvard philosopher Ralph Barton Perry, "I am trying to build up a special position for myself in which I might stay with honor. As yet it seems to work, but the end may come [any] day."[29]

That at least some students took Köhler's public position on Nazi racism seriously is evident from the record of the interrogation to which they were subjected on June 18, 1934. Asked by Berlin Student Leader Richter how German students "could stand behind a professor who does not support the new state and also officially supports the Jew Wertheimer," Ottilie Redslob – daughter of Erwin Redslob, who had been Reich Supervisor of Art (Reichskuntswart) in the Weimer regime – answered, "I know that Professor Köhler in no way denies the Jewish problem. There can be no talk of false racial consciousness." However, as the interrogators went on the offensive, their long-range goals became more obvious. In Richter's words:

> Köhler and his assistants are only the beginning. The whole university is still liberalistic. The university is a school for leaders; for that we can only use National Socialists, who teach National Socialist doctrine. . . . I think that it would be better not to teach number theory for ten semesters than to have it taught by a Jew.[30]

Here was a challenge that went far beyond the status of a single institute. By 1934, however, Nazi students' widening interference in university affairs was beginning to irritate the Ministry, and even to excite opposition among those in the party who valued organization and discipline. It was in this situation that institute students could openly defy their nominal leaders and support Köhler. As they wrote to a Ministry official in June, "as Germans we must demand that the leadership principle be employed in a dignified and clean manner. . . . [Moreover,] we do not believe that it is the purpose of the Führer to deny the possibility of influence to people whose German patriotism is not in doubt."[31]

In July 1934, in an appointment arranged by Köhler, a psychology student, Ilse Horstmann, and assistant Hedwig von Restorff met with Theodor Vahlen, a mathematician and a long-time Nazi who had recently become administrative director of the Ministry. In the statement they brought with them, Restorff damned the Nazi students' attacks as "cowardly" and "unmanly," and evoked the spirit of "comradeship" that had reigned in the institute heretofore:

> The Psychological Institute has the character of a research institute: the "teaching program" is small, and psychology is not a gainful profession. Whoever among the students comes to us must therefore have special interest, be especially capable, and must know that the style of work is determined by the strict demands of science. . . . Now come attacks from without, from people who have no insight into the importance of our tasks

and into this close community life. Is it not a matter of course that all the energies that would otherwise be expressed in work are now engaged for defense?

She closed by asking that institute students and staff be allowed to continue their work in the interest of German science, without ignorant interference from without, and that the Nazi student group be ordered to return "to the area for which it is intended."[32]

After the meeting, Köhler thanked Vahlen for hearing the students and repeated his demands. He would retract his resignation if his assistant von Lauenstein were reinstated, the institute employee (Schmidt, a custodian) who had denounced him were forced to take a leave of absence, the current head of the psychological section of the students' organization were removed and the Ministry made an official declaration of trust in the institute's leadership. The answer came in September. The Ministry officially assured Köhler of its trust, reinstated von Lauenstein and directed that the Nazi students be sharply reprimanded. Seen in context, this was not only a face-saving gesture for Köhler, but a blow by the ministerial bureaucracy against "disorderly" attempts at "coordination" from below. A year later, the Nazi German Students' League, to which the Berlin group belonged, was rendered politically impotent. However, the Ministry's response was clearly not motivated by a desire to aid Köhler's resistance; rather, it was an insistence upon orderly and at least superficially legal procedure, and thus upon the prerogatives of the bureaucracy.[33]

Pressure from abroad

Still more significant than the students' support was the pressure put on the Prussian Ministry of Science and Education by the German Foreign Office. At the end of May 1934, that agency queried the Ministry about letters it had received from abroad, for example from Edgar Rubin in Copenhagen, asking whether the rumor was true that Köhler had resigned. This was only five weeks after Köhler's letter to Fischer, in which he did in fact try to resign.[34]

But such inquiries were not needed to alert the Foreign Office to the situation of the institute. Already on April 17, 1934, four days after the second student raid, a still-unknown author submitted a "secret memorandum" on the institute to the Foreign Office. The writer certainly had a high opinion of Köhler's reputation as well as that of Gestalt psychology, and a firm grasp of the danger at hand in this case:[35]

> Independently of legal considerations the case touches very directly on German cultural policy abroad. The Berlin psychological school (*Gestaltpsychologie*) and its director, Wolfgang Köhler, enjoy a very high reputation abroad and have attracted up to now many foreigners to Germany,

especially from the United States and Japan. . . . Compared with this school the rest of [German] psychology hardly has any significance. Its representatives are rather indifferent for the prestige of German science.

After [Max] Wertheimer and [Adhemar] Gelb were fired or retired as a result of the civil service law and found new positions abroad, Köhler was the only representative of Gestalt psychology in Germany. Up to now, Köhler has refused all offers from America. After this event [the students' raid] it is no longer beyond possibility that he will accept a call from abroad. . . . From the standpoint of German cultural policy abroad, Köhler's departure would be a severe blow. . . . The material and psychological effects would be a burden on Germany's reputation, and not only its scientific reputation, while a foreign country would gain a productive scientist.

A marginal note to this document reads, "we do not want another Schrödinger case." This referred to theoretical physicist Erwin Schrödinger, who had resigned his chair and gone to Oxford in protest against the dismissal of so many of his colleagues.

This was the pressure point that Köhler had already touched, intentionally or not, in his interview with the *New York Times* a year earlier. Apparently Ministry Director Vahlen – a Nazi since 1923 – responded. He informed the Foreign Office of his statement of confidence in Köhler on September 2, 1934, the day he gave it to the university administration, and added the assurance that Köhler would "continue to hold his office."[36] Köhler himself seems to have been involved in some of these moves; surely there is no doubt of his cleverness at bureaucratic infighting. In mid-June, for example, he spoke with Kaiser Wilhelm Society director Friedrich Schmidt-Ott at the suggestion of Max Planck.[37] But it seems clear that pressure from abroad, managed or not, contributed to his remaining in his position for so long.

Denouement

But Köhler's battle was not over. In October 1934 he accepted an invitation from Harvard University to give the William James lectures in philosophy. During his absence, Hans Keller, a former assistant to David Katz in Rostock, was named to the assistant's position vacated by Kurt Lewin, without consulting Köhler. When the Ministry then wrote in February 1935, to ask whether Köhler's resignation request were still current, he replied angrily in the affirmative. But as late as April 1935, he wrote to Karl Lashley, with whom he wished to work in Chicago, that he still held out hope for an arrangement with the authorities.[38]

In May, however, the denouement began, when the Gestapo's Berlin bureau curtly inquired of the Ministry why the assistants Duncker and von Lauenstein, "who had worn the Antifa [antifascist] symbol and acted in a communistic manner, were still in office" at the institute. On June 14, Vahlen, citing "new information," ordered the administrative director of the university to fire Duncker immediately and added that von Lauenstein should already have been released by that time.[39]

With that, Köhler's position became untenable. A year earlier, he had written to Wertheimer that Duncker's habilitation thesis on productive thinking (discussed in Chapter 14) was complete, and that "there should be no more difficulties, so long as they do not come from without." In May 1935, while still in America, he had postponed deciding on an offer from Swarthmore College president Frank Aydelotte. In July, he wrote that the situation had changed: "This measure of the government is morally equivalent to my dismissal." His final resignation, submitted in August, was accepted on September 6 and announced on the 20th, though the formalities were not completed until the 28th. By then Köhler had already begun teaching at Swarthmore, with a visiting professorship supported by the Rockefeller Foundation and the Emergency Committee in Aid of Displaced Foreign Scholars. Interestingly, the resignation was treated as an early retirement, and the university paid his pension until the outbreak of the Second World War. He arranged to have the funds transferred to his first wife, Thekla, and her daughter, signing a statement assuring that the child was "raised in a German manner."[40]

Yet even this did not end the saga of the Berlin institute. Köhler himself cannily summarized his situation in a letter to his colleague, Donald K. Adams: "This is a case of their modern brutality (another man uses this method in order to push me out)."[41] In fact, the men most responsible for Köhler's departure were not Nazi government officials, nor even Nazi students, but two professors: Johann Baptist Rieffert, a philosopher and psychologist who had been the first head of the psychological section of the Reichswehr from 1921 to 1931; and Ludwig Bieberbach, a mathematician later notorious for his attempt to create what he called a "German mathematics." Bieberbach, who was dean of the Philosophical Faculty in the spring of 1934, had authorized the Nazi students' second raid on the institute in the rector's absence. According to a later Ministry investigation, Rieffert and certain student accomplices were responsible for denouncing the institute's assistants to the Gestapo.[42] Their aim, in brief, was to nazify psychology at the University of Berlin.

Rieffert had earned his doctorate under Benno Erdmann in Bonn, then completed his second thesis in Berlin under Carl Stumpf in 1919. After leaving the army psychological service in 1931 due to a conflict with his superior officers, Rieffert had returned to teaching in Berlin, offering primarily courses in "character study" (*Charakterkunde*). He joined the Nazi party in March 1933, and the SA in July; in the interval he and Berlin university rector Eugen Fischer wrote and circulated a statement in support of the governing coalition of Nazis and German nationalists that was signed by 240 colleagues. When the German Society for Psychology accepted the resignations of its Jewish governing board members in

September, he joined the board. One year later he was nominated for a tenured associate professorship in Berlin. Asked to comment, philosopher and Nazi race ideologist Alfred Baeumler praised Rieffert as a skillful characterologist. Carl Stumpf attested Rieffert's "calm, deliberate" manner of thinking and his "reliable character," signing "with the German greeting." In December 1934, roughly the same time that Hans Keller was appointed to the institute staff without Köhler's permission, Rieffert received a "personal" – that is, nonbudgetary – full professorship for character studies.[43]

The final step in this late-blooming career came from Bieberbach, who was still dean of the Philosophical Faculty. On July 5, 1935, Bieberbach requested that Rieffert be named director of the Psychological Institute, "in case" Köhler should resign. After Köhler's departure Bieberbach repeated his request, and also asked that Rieffert be assigned "to convert the work of the institute especially to the tasks of 'race psychology.'" In response, Rieffert was named temporary director of the institute. A new department for "characterology" was to be created for him, with the aim of "researching the psychology of the races and the German tribes."[44]

As described by Regional Student Leader Kiel, who had denounced Duncker and von Lauenstein to the Gestapo at Rieffert's behest, one goal of the characterology section was to develop a "psychology of the Jews" by investigating their gestures, mode of speech, handwriting, intellect, temperament, and so on. For such a task, Kiel wrote, the present director could obviously not be considered, "due to his friendly attitude toward Jews." The project was approved by Walter Gross, head of the Race Policy Office of the Nazi party, and by Eugen Fischer, in his capacity as director of the Kaiser Wilhelm Institute for Anthropology, Human Genetics and Eugenics. Fischer even said that he wished to participate. The Ministry, however, apparently took a skeptical view of the proposal.[45]

In any case, these plans never bore fruit, primarily because Rieffert failed to mention his earlier membership in the Social Democratic Party on the political questionnaire required of all civil servants. He was found out shortly after Köhler's departure, and was suspended from his professorship in April 1936. One year later, after full-scale disciplinary proceedings, he was dismissed from the university. When his appeal failed he was dropped from the party as well.[46]

By that time, however, the dirty work had been done; Köhler and his assistants were gone. Karl Duncker had put himself through the rigors of a Nazi ideological camp (*Dozentenlager*) for university teachers in the summer of 1934 – which considerably strained his already-delicate mental health – only to be informed by the Ministry that he had no prospects for a career in Germany. Von Lauenstein, the son of a Prussian officer, had protested his patriotism in his response to the Nazi students' accusations against him in 1934, to equally little effect.[47] Both men tried for a time to establish themselves in England, Duncker in Cambridge and von Lauenstein at a genetics research center in London. Duncker moved to the United

States in 1938, when Köhler obtained a position for him at Swarthmore. He died there in 1941, apparently by his own hand. Von Lauenstein returned to Germany, where his wife completed her dissertation with Wolfgang Metzger in Frankfurt in 1938. At the outbreak of war in 1939, he enlisted in the army; he was reported missing on the Eastern Front in 1943.[48]

The Berlin institute continued operating with Köhler's chair unfilled until 1942. The assistants' positions were taken and a full complement of courses was offered; indeed, the range of topics increased. During Rieffert's brief directorship the positions that had belonged to Duncker and von Lauenstein were held by Hans Preuss, author of the denunciation in the student newsletter and an SS member, and Robert Beck, who had also taken part in denunciations. Only Beck remained after Rieffert's enforced departure. Hans Keller assumed the temporary directorship, and was also promoted to associate professor, but seems not to have been a party member. According to one student at the time, he taught a methodologically strict experimental psychology on the model of G. E. Müller, with emphasis on memory and developmental psychology. During this period Kurt Gottschaldt, a former student and assistant of Köhler's who had become head of the psychological department of Eugen Fischer's Kaiser Wilhelm Institute for Anthropology, taught courses in developmental psychology and twin research; he became Associate Professor in 1938 (see Chapter 20). In 1936, the university created a new Mathematics and Natural Sciences Faculty, to which Gottschaldt's position was assigned.[49]

The psychology instruction offered in Berlin in the late 1930s was thus a peculiar mix of Keller's pre-Nazi, even pre-Weimar experimental psychology, Gottschaldt's developmental psychology, and "race psychology" offered under the auspices of the Philosophical Faculty by Race Policy Office head Walter Gross and Nazi race theorist Hans F. K. Günther. A subject called "racial soul studies" (*Rassenseelenkunde*) was taught from 1936 by Ludwig Ferdinand Clauss, who, unlike Günther, was an institute associate (see Chapter 21); he had to take his leave in 1942 after being excluded from the Nazi party. Though it is reported that students went or were required to go to Clauss's and Günther's lectures, their actual impact remains unclear. A former student professed to have heard "not a single Nazi sound" in the institute's regular psychology lectures, and the dissertations of the period appear to support his claim. Some of these were experimental projects on traditional topics in perception, for example on depth perception in film; others treated issues in industrial psychology, child development, and twin research. Still others, however, do not conform to this picture, for example a thesis submitted in 1939 on "The Psychology of Anti-Germanism in France." The contrast with former times could not have been greater. American psychologist Barbara Burks reported on a visit in 1936 that only eight or ten students attended Keller's lectures in general psychology, and remarked on "the utter barrenness of the intellectual climate of this former stronghold of the Gestalt school."[50]

At the Berlin institute "coordination" was attempted not from above but from below, not by top officials but by Nazi students in conjunction with careerists trying to take advantage of the chaos created by Nazi policies. Further, though the careerists succeeded in driving out alleged enemies of the regime, they did not succeed in replacing them as planned. Not until the appointment of Oswald Kroh to Köhler's former chair in 1942 did the Berlin institute come fully into line with the course psychology as a whole had already taken elsewhere in Germany.

20

Two students adapt: Wolfgang Metzger and Kurt Gottschaldt

Transformations of psychology under Nazism

German psychologists offered ideological obeisance and practical support to the Nazi regime very early. The German Society for Psychology accepted the resignations of Jewish governing board members David Katz and William Stern in April 1933 before it was clear whether the new civil service law required such a step. Board member Gustav Kafka then resigned in protest, as did Kurt Lewin. During the Society's first congress after the Nazi takeover, in September, 1933, leading psychologists competed to show how their ideas had prepared the way for or were consistent with the thinking of the new regime. Göttingen professor Narziss Ach placed his research on volition into service, quoting Hitler's proclamation that "The will is everything." Bonn professor Walter Poppelreuter, one of the few in the field who joined the Nazi party before 1933, even called Hitler a "great psychologist."[1]

In the forefront of such efforts were the "holistic" psychologists of the Leipzig school and Marburg professor Erich Jaensch, who had been among the most prominent critics of Gestalt theory in the Weimar period. As early as 1930, Felix Krueger had addressed the Leipzig chapter of the National Socialist students' and teachers' association; thereafter he expressed increasingly open sympathy for what he took to be the aims of Nazism. He greeted Hitler in his opening address as chair of the 1933 congress of the German Society for Psychology as the "far-seeing, bold and emotionally deep (*gemütstiefe*) Chancellor." Unlike Wolfgang Köhler, he did not protest the enforced retirement and emigration of Jewish professors, but instead worked behind the scenes to be sure that the psychologists among them were replaced by other psychologists, rather than "pure" philosophers. In cooperation with Nazi officials he made the topics of the Society's next congresses "Race," "Psychology of Community" (*Gemeinschaft*), and "Education for Community." By 1936, however, he had joined the ranks of folkish conservatives disappointed by the difference between their dream of "community" and Nazi reality. Because of a rumor that he had called Baruch Spinoza and Heinrich Hertz "noble Jews" in a lecture, he was forced to step down as rector in Leipzig, take early retirement, and undergo a humiliating check of his ancestry.[2]

Krueger's younger adherents behaved otherwise. In 1937, for example, Friedrich Sander, who had succeeded Kurt Koffka in Giessen in 1928 and then became full professor in Jena in 1933, replacing émigré Wilhelm Peters, presented *Ganzheit* and Gestalt as the "leading ideas of the German movement" and core concepts of a "German theory of the soul." In the same place he went so far as to advocate the "eradication of the Jewish parasitic growth" (*Ausschaltung des parasitisch wuchernden Judentums*) and the compulsory sterilization of Germans with "inferior hereditary stock" in the name of the "will to pure form [*Gestalt*] of the German essence." Before 1933 Sander had already titled the monograph series on perception that he edited "Ganzheit and Gestalt," thus laying claim to the Berlin school's intellectual territory.[3]

Jaensch promoted a different sort of holism from that of the Leipzig school, one that focused on personality and was oriented to the typological style of thinking characteristic of Nazi "race psychology." Earlier, he had distinguished integrated from more labile or "synaesthetic" personality types, but had not presented them as unequal or "racially" determined. After 1933, synaesthetics became inferior and the Jews emerged as important, though not the only representatives of this so-called "Antitype" (*Gegentypus*). This thinking became the psychological basis for Ludwig Bieberbach's "German mathematics," which went still further than Jaensch and posited fixed relations between psychological and racial "types." Jaensch took over the *Zeitschrift für Psychologie* in 1933 and made it into a platform for such work. In 1936, he succeeded Krueger as chair of the German Society for Psychology.[4]

Jaensch's criticism of Gestalt theory grew more virulent and more openly anti-Semitic after 1933. In that year, he brought Köhler's physicalist position into connection with avowed materialistic "tendencies from the East," against which German psychologists needed to construct "restraining walls." For Gestalt theory's "leveling" tendency to subject all wholes to the same laws, Jaensch substituted a series ranging from physical systems to "supraindividual" or societal wholes. If psychology were to become "the protector of a certain ethos," he intimated, it would be well to emphasize that individual personality types are "the most important link" in this chain, and to return to the values of "full humanity, feeling and temperament, especially manly and heroic activism." In 1938, in his notorious book, *Der Gegentypus* (The Anti-Type), Jaensch falsely claimed that the Gestalt theorists had "dominated all psychological professorships" in the past. He further accused them of generalizing to humanity in general the situation of the so-called "lytic S-2" or nonintegrated "synaesthetic," in his new view a mainly Jewish personality type, in which thinking is "separated especially from the life of the will, passive and inactive." He concluded that "Gestalt theory is thus the proper expression of two fundamental characteristics of the past era: the dominance of the S-2 structure [read: of "Jewish" intellectualism] on the one hand and the dead object culture of materialism and physicalism on the other."[5]

Other psychologists who made their careers in the Nazi period pursued different versions of the same double strategy, adapting their discourse and research to what they took to be Nazi expectations while explicitly attacking Gestalt theory. Bruno Petermann, for example, a student of Kiel psychologist Johannes Wittmann, had criticized the Berlin school in the 1920s primarily on scientific grounds, for insufficient emphasis on factors such as directed attention. In the Nazi period he called the Gestalt viewpoint "an automat theory" which ascribed perceived forms to stimulus organization rather than the "forceful directedness" of inner processes. An approach "from within" was better suited, in his opinion, to the "political science" that would fulfill the demands for an "organic worldview" in the Nazi era – a "folkish anthropology" based on a theory of something called the "racial soul" (*Rassenseelenlehre*), binding and forming the individual in a "folkish living whole" (*Lebensganzheit*). In 1938, Petermann succeeded Narziss Ach as professor and institute head in Göttingen, at Jaensch's suggestion.[6]

Rather different uses of the Gestalt concept came from Ludwig Ferdinand Clauss and Ferdinand Weinhandl. In the 1920s, Clauss had already created a field he called "racial soul studies" (*Rassenseelenkunde*), based on the claim that "racial" characteristics were identifiable from specific arrangements of facial features and body postures. In 1937, the year after he became a professor at the University of Berlin, he published a pamphlet based on his lectures entitled "Race is Gestalt" in the series *Schriften der Bewegung,* edited by Philipp Bouhler, head of the Nazi party chancellory. Where the Berlin school had only recently flourished, he asserted that a Gestalt like the outline of a "Nordic" peasant's head, for example, "is graspable neither by reason nor by mere factual knowledge." Such science yields, rather, intuitive "truths of style."[7]

Weinhandl, a student of Alexius Meinong, had developed an alternative, explicitly hierarchical version of Gestalt thinking in the 1920s, in which he argued, in the vein of Christian von Ehrenfels's *Kosmogonie,* that Gestalten could be ranked objectively according to their level or complexity, or "height," and their "purity." At that time, he treated the Gestalt theorists with respect, while emphasizing the historical precedents for their thinking. During the Nazi era he presented the most direct ideological denunciation of the Berlin school. In a talk entitled "Philosophy as Tool and as Weapon" given at an SS training course in 1941, he smeared Husserl's phenomenology and Berlin school Gestalt theory alike as "expressions of the true Jewish spirit."[8]

Clauss's position clearly remained consistent before and after 1933. All of the other psychologists engaged in what can be called ideological adaptations – attempts to reformulate earlier theories to make them appear more acceptable in the new climate, exploiting buzzwords like "whole" and "type." Other leading psychologists favored more pragmatic adaptations to the new regime, but they, too, paid homage to the new discourse. Former Jaensch co-worker Oswald Kroh, for example, redirected his pedagogical research to the goal of a "folkish study of humanity" (*völkische Menschenkunde*) based on the dual criteria of "folkish achievement and political fitness" (*Eignung*). Acknowledging that "wholeness" (*Ganzheit*) has become the leading category of psychological thinking," he claimed that "basically any psychology that proceeds holistically is prepared today to present achievement diagnoses" on the basis of Jaensch's typology or similar frameworks.[9] After Jaensch's death in 1940, Kroh became chairman of the German Society for Psychology; two years later he became Köhler's successor in Berlin.

Such careers were possible because of the increasing employment of psychologists in personnel assessment by the German Labor Front, the Nazi People's Welfare organization and, most important of all, in the military. Psychologists had been hired to help select officers for the army as early as 1925. German rearmament and the reintroduction of compulsory military service in 1935 increased demand for officers and selecting psychologists alike. By 1941, there were between 450 and 500 psychologists in all branches of the military, more than the membership of the entire German Society for Psychology in 1932. In 1941, at the height of Germany's conquests in Europe, a committee of military men, psychologists, physicians, and academics created a diploma examination, the first professional certificate in psychology. The need to train students for this certificate led to the creation of new professorships and the filling of vacant chairs, including that of Köhler in Berlin by Kroh. Psychology had at last emancipated itself from *mater philosophia*, not primarily as a result of initiative from within, but by actively responding to demand from without.[10]

So much for the belief that the Nazis destroyed psychology in Germany. Important for what follows is that this dramatic shift in the status and structure of the discipline went hand in hand with a reorientation of theory and research emphasis that had already begun in the 1920s — a shift of emphasis from the experimental study of cognition toward "characterology," or personality diagnostics. The discourse for which this word stood combined then-respectable approaches to personality like the "layer" theories of Philipp Lersch and Erich Rothacker, which posited different levels for intellect, will and emotion, and basic temperament (*Gemüt*), Ernst Kretschmer's "constitutional" theory, which posited inherited personality types linked with body build, and the far broader, irrationalist speculations of Ludwig Klages, who spoke of a conflict between something called "soul," taken to be essentially Germanic, with something called "mind," essentially Western (see Chapter 17).

Most prominent by far was the system developed by Lersch, and he, too, made his bows toward both holistic discourse and Nazi phraseology. In his most important work, "The Construction of Character" (*Der Aufbau des Charakters*) (1938), he claimed that the unification of the person in all his rational and irrational dispositions is the meaning of life: "Primarily in this connection character study will have to serve the educational tasks of our German present, [for] true and fruitful humanity is only achieved when the processes of thinking and goal-directed willing are bound together with the prerational forces of the feelings and values in that living unity which has its basis in the racial substance." In addition, he wrote that "race, blood, and folk" are "ontic wholes . . . in which the individual is enmeshed and in the unfolding and preservation of which he must cooperate." With the help of such ideas – and because his characterological system was the predominant one in military psychology – Lersch advanced to full professorships in Breslau, Leipzig, and Munich. Apparently he did so without joining the Nazi party, though he did sign a declaration of loyalty by German professors and teachers to Adolf Hitler and the National Socialist state in November 1933.[11]

There was thus no contradiction in principle between such versions of holistic discourse and the practical reason of "characterological" personality diagnostics. In Wehrmacht officer selection, tests of sensorimotor coordination and reaction speed were used alongside expression diagnostics and the interpretation of candidates' behavior in simulated command situations. Sought in this mix of tests and trained intuition were intelligence, control of emotional expression, strength of will, and ability to command – all traditional qualities of the Prussian officer. In this context it was possible, even necessary, to combine what might otherwise be considered premodern discourse and methods with up-to-date test instruments and sophisticated role-playing to help construct an ultramodern fighting machine.[12]

The students of the Gestalt psychologists who remained in Nazi Germany thus faced complex challenges. On the one hand, they confronted ideologically tinged attacks on their teachers' work, combined in some cases with attempts to incorporate parts of it, and the term "Gestalt," into approaches quite different from their own. On the other, they faced the task of demonstrating that Gestalt theory, or their current research, could be relevant to the regime's practical aims. The two most prominent among them, Wolfgang Metzger and Kurt Gottschaldt, responded to these challenges in rather different ways. Their cases, like many others, suggest that the bare alternative between collaboration and resistance is inadequate to the range of behaviors exhibited by scientists under Nazism.

Wolfgang Metzger

Wolfgang Metzger was born in Heidelberg in 1899. Like his mentor Wolfgang Köhler, he was the son of a teacher, in this case at a girls' school, who later became director of a Realgymnasium. When Metzger graduated in 1917 he had wanted to study organic chemistry. Then he was drafted, and lost an eye in combat on the Western Front. When he returned from a French prisoner of war camp in 1920, he studied German literature in Heidelberg, Munich, and Berlin, but then changed his field to psychology under the influence of Köhler and Wertheimer.[13] His dissertation was on monocular motion perception, a topic for which his war wound made him well suited, but he is best known today for his research with the so-called "homogeneous *Ganzfeld*" (see Chapter 14). This work with an artifical situation lacking all texture or contour showed that such structures were necessary for perception to occur at all. The experiment could also be seen as a response to criticism by the Leipzig school that the "Berliners" had neglected to study the emergence of Gestalten. As will be shown, the theme of concessions to Leipzig "holistic" psychology was a constant in Metzger's career.

When an assistantship with Wertheimer in Frankfurt came open due to Adhemar Gelb's departure for Halle in 1930, Metzger was appointed on Köhler's recommendation. After completing his second thesis in 1931, on the problem of phenomenal identity, Metzger remained there after Wertheimer's dismissal. According to a former student, he stayed for "familial" reasons – he had a wife and child, and was expecting another child.[14] Since he was Catholic, not Jewish, and had only his association with Wertheimer and perhaps his earlier reviews in a Socialist monthly (see Chapter 17), but no demonstrable political statements or activity against him, nothing was forcing him to leave, either.

After Wertheimer's departure, Metzger retained his assistantship and became de facto head of the Frankfurt institute. During these years, he later wrote, he did not attend meetings of the German Society for Psychology in order to avoid "cleverly aimed questions" from Jaensch. In Frankfurt, too, his "survival tactic" consisted mainly of "avoiding any personal contact that was not absolutely necessary." He continued his research with the protection of the head of the university's association of Nazi professors, a gynecologist, who in Metzger's words "had a human heart and a sense of the scientific task under the brown uniform." However, the fact that he joined the SA – the *Sturmabteilung*, the Nazi paramilitary organization – in 1933 alters this image of inner emigration.[15]

In 1936, Metzger published a synoptic account of research on the Gestalt theory of perception entitled *Gesetze des Sehens* (Laws of Seeing), since reissued three times and vastly expanded. This work will be discussed further in the next chapter. Important here is that the first edition of the book had no overt political content. Scientific or scholarly literature was generally not subject to pre-publication censorship in any case. But both Metzger's situation and his writing began to change the next year. He joined the Nazi party in May 1937.[16] In the same year he also published a brief article entitled "Laws of Seeing – Applied" in a popular science journal. One example he chose for applying the "Gestalt laws" was a genealogical diagram from a book on eugenics by Saxon education minister Wilhelm Hartnacke (Figure 18a). According to Metzger, the figure's aim was to illustrate the different hereditary outcomes (*Ahnen-Erbe*) possible from one set of four grandparents; but the diagram as printed failed to clarify the relationships at work, because the eye grasps vertical rows first if they are sufficiently separated. Rearranged as shown and thus allowing verticality to work against the "law of

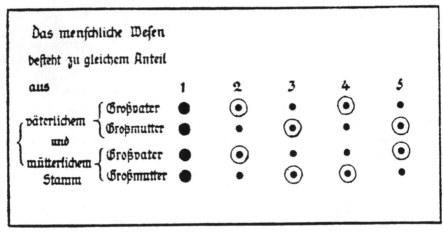

Figure 18. Diagramming heredity (Metzger 1937) Wolfgang Metzger, "Gesetze des Sehens – angewandt," *Natur und Volk,* 67 (1937).

nearness" (Figure 18b), the intended message was immediately obvious.[17] Metzger "warmly recommended" Hartnacke's book, but took no explicit stand on eugenics in this paper. By implication, however, he had offered his services to assure the effective presentation of Nazi population policy.

In the fall of 1937, Metzger was appointed visiting professor in Halle. The incumbent, Adhemar Gelb, had been forcibly retired by the Nazi civil service law.

Metzger's name had been second on the list to replace Gelb as early as 1934, but the negotiations became snarled in a battle between the Philosophical Faculty and Ministry officials trying to force the appointment of Heinrich Schole, an avowed Nazi. The visiting position for Metzger was a compromise.[18] Visiting professors often received the permanent position after the year's trial appointment. However, in conformity with the civil service law, all professorial appointments required a statement from the candidate's chapter of the National Socialist Professors League and the party ideological office on the candidate's political reliability. Leipzig holistic psychologist Hans Volkelt advised that office on appointments in psychology.

Perhaps in order to influence such assessments, Metzger published two short articles in the journal of the local Nazi educators' organization, *Erzieher im Braunhemd* (Educator in a Brown Shirt). The first was entitled "Whole and Gestalt" (*Ganzheit und Gestalt*) – a direct citation of Friedrich Sander's monograph series and thus a nod in the direction of the Leipzig "holists." In it, he criticized the current trend toward "characterology" among his colleagues as too narrow. In his view, statements about the role of tones in a melody, the relationship of character traits in a personality, and the place of particular types of people in society are all based on the same kind of argument. "General" or theoretical psychology thus provided a broader, more appropriate foundation for psychology than "characterology" alone. Metzger had already claimed primacy for "general psychology" in "Laws of Seeing," but without drawing such specific analogies to personality and society. His strategy here was apparently to claim new territory for Gestalt theory by applying it to talk about "the whole person," and thus to establish a claim to both ideological acceptability and practical potential.[19]

It was at this time, too, that Metzger came into conflict with Köhler over the editorial policy of the journal, *Psychologische Forschung.* Beginning with Volume 21 (1935), Köhler had edited the journal alone from Swarthmore. He designated his former assistant Hedwig von Restorff as his representative in Berlin, but apparently did most of the business with the Springer firm himself during summer visits there in 1936 and 1937. As he wrote to Julius Springer, his main interest was to assure that the dissertations completed in Berlin and Frankfurt under his and his colleagues' direction would be published in "a worthy place." He therefore made "certain concessions," among them probably the omission of the other editors' names from the masthead; all except Hans Gruhle counted as "non-Aryans" according to the recently promulgated Nuremberg Laws.[20] For the publishers, international prestige and perhaps foreign income may have been reasons to cooperate in this endeavor.

Indirect evidence indicates that in 1937 Metzger took issue with the continuing appearance of articles by émigré students of Gestalt theory, such as Hans Wallach, in a journal that was still printed in Germany. When Köhler tried to include a paper by Karl Duncker written in English, Metzger inquired of Springer in Berlin whether or not "a very serious error of form" had been made. Köhler then accused him of "interference in the editor's business." As he informed Metzger, he himself had accepted Duncker's essay, "so when you speak of a very serious error of form . . . you are criticizing me." Metzger's attempt to explain his behavior apparently failed. In June 1937, Köhler wrote him an outraged letter accusing him of disloyalty, and sent copies to Koffka and Wertheimer. Mentioning an earlier suggestion of Metzger's that one of the previous editors "be replaced by an unobjectionable person," he continued:

> Not you, but others were of the opinion that there were "objectionable people" among the editors. In situations like that one and the present one, one really does well to take and keep a clear line. . . . I have waited so long in vain for at least a tiny incautiousness to happen to Herr Metzger at least once, out of loyalty to truth and to former teachers.[21]

The quarrel led to a complete break between the two men until after the war. *Psychologische Forschung* continued to appear until 1938, when Köhler requested, over Springer's objections, that publication cease, in part because he did not wish to accept the "compromises" in quality that would be necessary if work from Germany were included. *Psychologische Forschung* did not appear again until 1949, and then with a different editorial board.[22]

If Metzger's moves had careerist intent, they were not successful. He was recommended first for the Halle chair along with Johannes von Allesch, then associate professor in Greifswald. The appointment went to von Allesch, a senior man and a former associate of Köhler's in Berlin with no party affiliation, even though a Nazi Professors' League official acknowledged that his committed Catholicism accounted for "certain inhibitions versus the Third Reich."[23] In 1938, Metzger's name came up in connection with a professorship in Breslau, which went to Philipp Lersch. The next year, he received the title of Associate Professor in Frankfurt. But the turning point in his career came in 1942, when he became professor and director of the Psychological Institute at the University of Münster. In 1972 he suggested that the presence of Cardinal von Galen of Münster, who openly criticized the Nazis' so-called "euthanasia" campaign, might have led to favorable consideration of someone "whose papers were not entirely clean." Still later, he also claimed that Gestalt theory was by its nature "incompatible with any totalitarian ideology," because it dealt with psychological order "that is not imposed from without, but results from the inner dynamics of the situation in question."[24]

Metzger sang a different tune at the time. In 1942, he published two essays in *Volk im Werden,* a journal edited by Nazi pedagogue Ernst Krieck, repeating and expanded his earlier argument for the compatibility of Gestalt theory and National Socialism. These articles are absent from postwar lists of his writings. Of greatest interest among the arguments he advanced is his attempt to reconcile the belief in a "natural order," in which a society's members freely make "the law of the whole" their own without force, with the obvious use of compulsion in the Nazi state. Metzger's response was a psychological justification of the *Führer* principle that was not very different from arguments once offered for absolute monarchs as all-knowing "heads" of their states: Force was needed only for practical reasons, to protect against usurpers or "to repress, cut off [*abkapseln*] and eliminate [*beseitigen*] unrepentent recalcitrants and disturbers of the peace," that is, to compensate for the "blindness" of many to the true needs of the whole. But external force was definitely needed as the political counterpart to the cultural battle against the "spirit of the West."[25]

When this was written, Nazi Germany was at war with the Western powers, and it would be tempting to place this essay in the category of patriotic rather than Nazi propaganda. However, the use of verbs like "confine" and especially "eliminate" are chilling reminders of what was occurring behind the front lines. Wolfgang Prinz suggests that these statements can only be made consistent with Metzger's later claims that Gestalt theory is incompatible with totalitarianism in any form if we remember that he spoke later of Nazism with full knowledge of its evil results, whereas he wrote in 1942 with only its announced goals in mind.[26] Given what had already happened to Metzger's teachers and former friends, this explanation seems weak. It appears, rather, that Metzger employed a mixture of formulae from "holistic" psychology, standard conservative discourse, and terminology of his own in an awkward attempt to come to grips with his situation. As will be shown below (Chapter 22), he later treated the issue of order and freedom in a rather different way.

The work of Metzger's that did most for his career was his theoretical *summa summarum,* published in 1941 and entitled in translation *Psychology: The Development of Its Fundamental Assumptions since the Introduction of the Experiment.*[27] The book had three new editions after the war, and some writers have claimed that the main text was unaltered in all editions.[28] Unfortunately for the self-image of those who studied with Metzger later, this is not true. Four examples will show how he introduced political metaphors to make Gestalt theory appear congenial to Nazism.

On page 1, in a formulation similar to the one already cited, Metzger wrote that the shift to holistic thinking in psychology is "the conclusion of an important stage in the conflict with the spirit of the West, which reached full flower in the seventeenth and eighteenth centuries, and which found its clearest expression in the mechanistic worldview of the French and in English empiricism." On page 2, he added that this "spiritual battle" was now being fought with "the weapon of the experiment." In the second edition (1953), he retained this language, changing only the words "the spirit of the West" to "the so-called modern era." Such language had been standard conservative fare in the Weimar period. At that time, however, Metzger's teachers never presented their thinking in this way. They opposed Descartes and British empiricism because they thought these thinkers were wrong about consciousness, but they did not attack "the spirit of the West" in general. Indeed, they offered evidence from the field theory of Maxwell, surely a "Western" thinker, to support the claim that their thinking was consistent with natural science. It was the Leipzig holistic psychologists who employed this metaphor in the 1920s. Metzger's use of it here was one of many concessions to that school.

On page 51, writing about the "fruitfulness of the atomistic principle in psychology," Metzger maintained, as Wertheimer did, that genuine atomistic states occur, but only as limiting cases, which are relatively unstable. As an example he offered implicit references to England and Weimar Germany:

> a group of individualists can last only under especially favorable conditions, after centuries of undisturbed life together behind the coasts of a completely surrounded island, if prosperity and peace and unity in all essential issues prevail. If one wanted to force the same constitution upon a group that had come together only a short time before, and which is required to live in narrow confines, without natural boundaries, constantly threatened by powerful enemies, that group would collapse in a short time.

In a footnote, Metzger added that this argument came from a book by an English scholar published in 1921, but written mainly before 1914, which concluded that democracy was not an article for export.

On page 189, Metzger wrote about "dominance relations in communal life," citing the famous example of the pecking order in the barnyard, which he called an example of a "centered," or stably structured, set of dominance relations. He then referred to "numerous attempts in past decades to artifically establish uncentered social structures – from the fully 'comradely' marriage to the parliamentary state to the fiction of the equality of dwarf states with leading nations in the League of Nations." All of this shows "that such structures are never stable, but 'snap back' sooner or later according to actual power relations." The references to conservative and Nazi criticisms of Weimar social experimentation and to the European power balance of the 1920s were clearly intentional.

Last but not least, on page 124, Metzger spoke of a fundamental agreement between perceptual theory as he expounded it and "the newest and experimentally best founded form of political theory." Why? Because political theory, like perception, is concerned with the advancement of groupings with the greatest possible firmness and permanence, even under "unfavorable" conditions. The current (that is, Nazi) political theory, like perception, struggles to overcome "arbitrary" groupings, emphasizes the decisive importance of objective relations among cooperating individuals, and realizes that cooperation is best assured when it occurs according to "certain outstanding characteristics of the whole to be constructed." Here Metzger tried to draw an analogy between Nazism's proclaimed goals and Wertheimer's Gestalt laws of perception.

Becoming more concrete, he alluded to the "factor of nearness" in perception, according to which seen objects nearer to one another are perceived as belonging together, whereas similar-looking objects seen further apart are seen as separate. Metzger wrote that this corresponds to the "theory of the enclosed area of settlement" – a reference to the Nazi doctrine of "living space." Further, he alluded to the factor of "similarity," according to which objects that appear similar in shape to others are identified accordingly, even if identity is not exact, and claimed that this corresponds to the "general part of the [Nazi] race theory." He added, however, that the race theory maintains that "the greatest 'firmness' of the group can only be achieved by similarity in genotype, that is, in unchangable aspects – whereas perceptual theory suggests that such structures can be "achieved primarily by characteristics·determined in and by the group itself, or by specific characteristics of its members." Here, where Metzger came closest to explicitly endorsing core teachings of Nazism, he pulled away from a full acceptance of Nazi racial doctrine. But the qualifier was hardly more than a fig leaf. Needless to say, none of these texts reappeared in later editions of the book.

As in Köhler's case, though for different reasons, we must ask: Did Metzger write these words for tactical reasons, or did he believe what he was saying? In response to an interviewer's question, Metzger said toward the end of his life that he had indeed held such beliefs at the time, and denied any careerist motives. He promised to reread his essays from those years and write a statement about the issue, but died in 1979 before completing this task.[29] Some contemporary evidence seems to undercut his later statement. Close reading of the essays and the book *Psychology* indicates that the passages in question could be excised – as in fact they were in the latter case – and the texts would remain coherent. If we are to accept Metzger's claim that he believed what he wrote, we would have to say that both those beliefs and their connections with the rest of his thinking were incompletely "formed" (*gestaltet*) at the time.

Ironically, it would appear that these conceptual gyrations were not really necessary. Michael Stadler writes that in order to preserve the Gestalt tradition intellectually and institutionally, and to defend natural-scientific psychology under difficult conditions, Metzger "had to" weave statements of political loyalty into his writings "occasionally" and conceal his heritage by not citing his former mentors. But once the professional certificate for psychologists was in place and it was necessary in 1941 and 1942 to fill many professorships in a short time, solid scientific qualifications were as important as ideological loyalty. The Nazi party ideological office wrote that it supported Metzger's appointment in Münster because he had done "clean" and "thorough" experimental research, and played down the fact that he was a Wertheimer student. The party professors' organization praised his book *Laws of Seeing* as "an outstanding scientific achievement," and called *Psychology* "comprehensive and also philosophically interesting." The evaluation noted that Metzger had "belonged earlier to a liberalistic social-democratic circle in Frankfurt," but concluded that he had "made an honest change since 1933, so that he counts as politically unblemished [*einwandfrei*] today."[30]

Metzger's response to academic life under Nazism was only one of many instances of anxious adaptation under uncertain conditions. As Reese Conn Kelly has aptly put it, university appointments policy in this period lacked organizational or ideological clarity, and was governed ultimately by "the practical necessity of getting on with the work of the university." However, he adds, "this very unpredictability of the party or the Ministry was often sufficient to cow and corrupt the professoriate, where 'national' and even racial sentiments had not already done so."[31]

Kurt Gottschaldt

The context of Kurt Gottschaldt's adaptation to post-1933 German realities was rather different from that of Metzger's, but his story is equally revealing in its way of the situation of psychological theory and research under Nazism. Born in Dresden in 1902, Gottschaldt, the son of a businessman, studied physics and chemistry in Berlin in addition to philosophy and psychology. At first he was inspired by Köhler's commitment to philosophy based on natural science, and he was a promising student. Koffka and Köhler frequently cited his dissertation on the effect of previous experience on form perception (see Chapter 14), and he received a paid assistantship in 1926. However, Köhler's opinion of his prospects seems to have changed. In a report to the university administration, Köhler acknowledged his competence and cited only unspecified "secondary reasons" for his release in 1929. Gottschaldt recalls that he never felt comfortable in the atmosphere of the Berlin institute, and remembers the sting of having Metzger singled out for preference.[32] He also appears to have been a difficult person.

Gottschaldt then took a position as head of the newly created Psychological Department in the Rhenish Provincial Hospital for mentally abnormal children near Bonn, thus becoming one of the first full-time clinical psychologists in Germany. At the same time he became an assistant at the Pathopsychological Institute at Bonn University, which was headed by hospital director Adolph Löwenstein. The monograph he produced in that setting, entitled *The Structure of Children's Action* (1932), was the most thorough application up to that time of the style of behavior research that Köhler had employed with primates and Kurt Lewin had extended to children. In contrast to methods using "statistical-mechanistic achievement tests," Gottschaldt's approach rested on child-centered or what he called "ego-proximate" (*Ichnahe*) tasks, such as building a tower with wooden blocks. In his view, this ensured that not only the ability to concentrate or to work quickly was tested, but also that the "unified meaningful construction of action-wholes [*Handlungsganzheiten*]," that is, the children's behavioral styles could be observed. Gottschaldt had been influenced by Lewin's ideas in Berlin, but he never belonged to the inner circle of Lewin's students. The monograph appeared as a supplement to the *Zeitschrift für angewandte Psychologie,* not in *Psychologische Forschung.*[33]

For this research Gottschaldt earned the right to teach at the University of Bonn in 1932, with the support of philosopher and theoretical psychologist Erich Rothacker.[34] But when the Nazis came to power the next year, Gottschaldt was denounced for having traveled to the Soviet Union. Like many other Berlin institute members, he appears to have moved in left-wing circles. His first wife is said to have worked for Malik Verlag, a Communist publishing house. In an interview, he recalled that as a student he had been inspired by the speeches of KPD Reichstag member Ruth Fischer, and that he had in fact gone to Moscow and Leningrad; but he added that he had lectured on psychological, not political topics there. He stated that he saved himself from arrest by explaining that his Moscow trip had been strictly scientific, but lost his hospital position; when he tried to continue teaching, student SA members checked his students' identity cards. The university catalogues for the period continue to list him as an assistant; he was kept on at half pay, and Rothacker managed to get him a stipend amounting to the rest. With his future in science looking dim indeed, he returned to Berlin in 1934 and found a job in business, but kept in close contact with his mentor. In 1935, Rothacker recommended him to Eugen Fischer, the founder and director of the Kaiser Wilhelm Institute for Anthropology, Human Heredity, and Eugenics in Berlin, who wanted someone to organize a psychological department.[35]

With that, Gottschaldt came into a field closely connected with one of Nazism's major aims. Fischer's institute was not a product of the Nazi period, but was founded in 1927, with major funding from the state of Prussia, then governed by Social Democrats. The founding was a clear sign that eugenical ideas were becoming important components of Weimar social and health policy.[36] Psychology had been a topic at the institute before Gottschaldt arrived. In 1930, the head of the institute's Department for Human Heredity, Otmar von Verschuer, introduced the term "hereditary psychology" (*Erbpsychologie*) into the research program. Examples of this work are papers by von Verschuer himself on intelligence and Rohrschach test scores in twins, and by his co-workers Ida Frischeisen-Köhler and Marie-Therese Lassen on the heritability of so-called personal tempo, as well as twins' grades and behavior in school. The statistical method used to interpret the results was that of "concordance" and "discordance." Its logic was remarkably simple. If the average difference in the required achievement in dizygotic (fraternal) twins was greater than in monozygotic (so-called identical) twins, researchers attributed at least the "concordant" portion of the results to heredity. This method was similar to those used internationally at the time.[37]

At first glance it would appear that an instrument was thus prepared that the Nazi leadership needed only to take up and use to provide seemingly scientific suppport from both biology and psychology for their own population policy. In fact the transition did not go so smoothly. In 1932, Fischer had criticized the Nazis as late arrivals in the field and rejected their racism. In November 1933, however, Fischer signed the declaration of loyalty by German professors and teachers to Adolf Hitler and the National Socialist state. He eagerly accepted the assignment of assessing the "fitness" of mental patients and other candidates for compulsory sterilization for the hereditary health courts brought into existence by the compulsory sterilization law of April 1933. In a report to the Kaiser Wilhelm Society in June, he asserted that "The institute stands in complete readiness for the tasks of the present state." He clearly saw the Nazis' policies as a golden opportunity to put a eugenical program of population management into practice, and as a pragmatic justification for the basic research already underway in his institute.[38]

When Verschuer accepted a professorship in Frankfurt in 1935, Fischer created a new Department of "Hereditary Psychology" with Gottschaldt as its head. Fischer reported that Interior Ministry official Arthur Gütt, a coauthor of the compulsory sterilization law, supported the founding, and the new department seems to have been well supported. The institute budget for facilities and personnel rose in the year of its creation by 31,300 Marks and 17,000 Marks, respectively, and space for it was included in the Institute's expansive new building, completed in the summer of 1936. A report by Fischer from that year mentions a staff of two assistants, two secretaries, and a nurse.[39]

The department's role in the Institute's research program seems clear enough. The compulsory sterilization law was a program for so-called negative eugenics – preventing the reproduction of those deemed "unfit." A pendant to this was "positive" eugenics, the identification and preservation of biologically "valuable" qualities. As early as 1933, Fischer had written that "new knowledge about the heritability of psychological characteristics is important for positive race hygiene and to influence the biological foundations of culture." From the outset, however, he also emphasized the difficulties of achieving such knowledge. In a 1936 report, he claimed that in the psychical realm there are clearly recognizable, "obviously *inherited* differences" among races, but added that "research into these differences may be one of the most difficult problems in the field of human genetics. The recent plan of the Institute, which stretches over several years, and its new department for hereditary psychology (head: Gottschaldt) are meant for this research. *Step by step and cautiously* a firm foundation for [research on] these extremely important issues is to be laid."[40]

Gottschaldt and his co-workers set themselves the task of applying the methods he had developed in Bonn to twin research. The means to that end were so-called twin camps (*Zwillingslager*) – summer camps for twins, designed to make it possible to observe them in something resembling a natural setting. The first of these was set up on an experimental basis near Berlin with the support of the Berlin region of the National Socialist Welfare Organization. Full-scale camps ran for eight-week periods in the summers of 1936 and 1937 at a children's home on the island of Norderney, in the North Sea. State support made possible a research program of unprecedented ambition and thoroughness. According to Gottschaldt, "all events, experiences, forms of dealing with the world and other people that the children encountered in their daily lives" were carefully recorded in protocols. In addition, psychological experiments assessed "individual functions such as intelligent actions, dealing with specific conflict situations, experiences of success and failure, dealing with rivalries." In another camp observation extended to the nighttime hours, and the twins' "sleeping positions, depth of sleep, sleep disturbances, and the like were observed."[41] Gottschaldt's pride in the creation of an apparatus for total observation was noticeable.

Gottschaldt and his assistant Kurt Wilde, a student of Johannes von Allesch who joined the SA in 1933, and the Nazi party in 1937, reported initial results in both scientific and popular publications. But empirical reports were less prominent during the department's existence than theoretical and methodological positions. In both areas Gottschaldt's thinking showed remarkable continuity with the past. Most significant was the fact that, like Kurt Lewin, he regarded person and environment as a single whole that could not easily be broken up into separable parts. To support this view he cited an example that the émigré Lewin might not have chosen – a dissertation on the psychological situation of prisoners by his Berlin student A. Ohm. As this work showed, even prisoners in interrogation cells – the most controlled environments imaginable – do not all react in the same way to their situation. Thus, "even the environment 'investigative custody' [*Untersuchungshaft*] is experienced in relation to a subject, and *there is no such thing as an objective environment*." Gottschaldt applied this claim to twin research by citing work by Helmut von Bracken, which showed that monozygotic and dizygotic twin pairs exhibit different styles of relating to the environment, and also to one another. Thus, he implied, with twins, too, it was a mistake to assume that psychological "environments" were always the same.[42]

In his criticism of previous research methods, Gottschaldt also showed a continuing commitment to a Gestalt viewpoint and to Lewin's ideas. In an extensive contribution to the authoritative *Handbook of Human Hereditary Biology*, published in 1939, for example, he castigated test psychology, including intelligence testing, along lines already drawn by Max Wertheimer and Wilhelm Benary in the early 1920s and by Wolfgang Köhler in the early 1930s (see Chapter 15). Tests, he argued, provide evidence only about certain specific intellectual functions useful in school. The task becomes getting through the test, and little is learned about the dynamics of learning or behavior. In the same vein, he also criticized analytical approaches to personality assessment emphasizing individual traits as "piecemeal" (*stückhaft*), thus employing Wertheimer's vocabulary. Throughout the article he explicitly cited the writings of Köhler, Duncker, Wertheimer, and Lewin, and of dismissed Jewish professor Otto Selz, thus showing that it was possible to do so without harm in the Nazi period, given the proper context.[43]

A second point of Gottschaldt's methodological critique indicates another source of influence, the so-called layer theories of personality advanced by Lersch and also by Erich Rothacker, his mentor in Bonn. According to these theories, he noted, tests touched only the superficial layers of a personality. Continuous phenomenological observation (*Dauerbeobachtung*) was therefore necessary to reach deeper levels of character and temperament, which Gottschaldt, like other personality theorists, took to be inherited.[44] Of interest here is the way Gottschaldt attempted to combine these two lines of thinking. In his handbook article, on the "hereditary psychology" of giftedness, he described intelligent problem-solving exactly as Köhler, Wertheimer, and Duncker did, as a "restructuring" or even a complete reformation (*Umgestaltung*) of a given situation, which in turn is not a collection of events or individual facts but a continuous process. But he also argued that the person in whom this restructuring takes place is not only part of a concrete situation, as Lewin would have put it, but also a "formed whole" (*gestaltetes Ganzes*), and that intelligent action unfolds through all of the multiple "layers" of the personality – at the emotional as well as the cognitive levels. This view was closer to that of Lersch and Rothacker.[45]

Gottschaldt explicitly opposed "the general and mostly quite unclear holistic conception of typological characterology," and rejected attempts to link styles of form perception to "racial" personality types as methodologically faulty. He also distinguished his position from Leipzig holism, saying that empirical investigation revealed the existence of both "Gestalt contexts" and diffuse "totalities."[46] However, he did not say whether the structure of persons as wholes is different from or the same as that of perceived forms, "Gestalt contexts," or diffuse "totalities." Nor did he make any attempt to relate Gestalt theory systematically to the various "layer" theories of personality then in use. One reason for this may have been that the Berlin school's emphasis on short-term functional-structural analyses was not easy to combine in a rigorous way with biologistic "characterology" or eugenical "race hygiene."

The empirical results Gottschaldt reported most fully during the Nazi period concerned the intelligent achievements of twins studied in the "twin camps." The subjects were 44 monozygotic and 25 dizygotic twin pairs ranging in age from 7 to 18; the average age was 11 years and 3 months. With these children Gottschaldt and his staff carried out a total of 4,076 experimental trials, 2,606 with the dizygotic, and 1,470 with the monozygotic twins, an average of approximately 39 trials per subject. The tasks included standard tests of verbal and logical skills (vocabulary retrieval, conceptual ordering or classification, word and sentence completion tests), and also practical tests of the ordering of objects (packing a suitcase), searching for things in a large field, building, and drawing. The mean difference of the average scores for the dizygotic pairs was more than twice as large as that for the monozygotic pairs. From this result Gottschaldt concluded "that hereditary influences far outweigh environmental aspects in the entire area of intelligent action."[47]

Sophisticated as their statistical approach was for the time, Gottschaldt and his associates still came out more decisively in favor of a hereditarian position than was demanded by their own results. Even before the work began, Gottschaldt painted an optimistic picture of the potential for psychological diagnosis of heredity, especially when paired with neurological studies, and confidently stated that numerous psychological characteristics were inherited. In subsequent reports he asserted not only that intellectual functions are "to a great extent" hereditary, but also quoted Fischer extensively to the effect that inherited racial differences in intelligence were a proven fact. He moderated that position somewhat in the case of women, but downplayed the fundamental methodological problem already being raised by American researchers – the need to control for twins raised apart and twins raised together.[48] Thus, despite his caveats, his work remained a psychological supplement to the Fischer institute's version of the eugenical project.

Gottschaldt, like Metzger, appears to have tried, albeit eclectically, to extend the mode of thinking characteristic of Gestalt theory to new topics, and thus to show that it was suitable for the tasks of the new state. But what was missing from these publications was at least as important as what was said. In the handbook article cited above, for example, he referred to current ideology only once, citing in agreement attacks by his fellow department head Fritz Lenz and Saxon Education Minister Wilhelm Hartnacke on earlier theories of intelligence as examples of "liberalistic intellectualism."[49] He justified his reticence by remarking, correctly, that such topics were outside the handbook's purview, but this did not prevent other contributors from stating their views in favor of "race hygiene," or even advocating enforced sterilization of certain types of criminals.[50] Gottschaldt's emphasis on method gave the impression of scientific care and exactitude. At the same time it justified his restraint in making ideological declarations. Fischer himself took over this task, both in his reports to the Kaiser Wilhelm Society and in public.

Gottschaldt's caution also made it possible to avoid expounding his Department's results too fully. He and his co-workers always emphasized that further research and methodological exactitude were needed to determine the precise interactions of heredity and environment in psychical development. In any case, he added, it was clear that intelligence, for example, was so complicated that even its hereditary aspect could not be carried by any single gene.[51] Such a position could hardly lend much support to eugenicists who wanted to breed a new master race overnight. Gottschaldt never joined the Nazi party. In an interview, he insisted that his emphasis on method was not a tactic, but was a fundamental criticism of Nazi race policy, albeit on practical rather than moral grounds.[52] However, hinting at the difficulties in the way of realizing the eugenical project did not mean rejecting that program in principle.

What impact did all this have on Gottschaldt's career? Fischer seems to have protected his department head. When erstwhile Berlin institute director Johann Baptist Rieffert attempted to denounce Gottschaldt to the Gestapo in early 1936, Fischer demanded proof, and threatened to take him to court for libeling one of his employees if he did not produce it – or so he told Gottschaldt afterward. This corresponds to the general impression at the institute. One openly Nazi staff member later said that "Gottschaldt was a Communist, but Fischer covered for him."[53]

After becoming associate professor in Berlin in 1938, Gottschaldt was proposed at least three times for full professorships. The first two nominations came in 1939, for a chair in Breslau and for Köhler's chair in Berlin; in the second case he was the only candidate. Both of these appointments came to nothing, due to a conflict with Heinrich Harmjanz, the Education Ministry official responsible for university affairs and an SS man. According to a note Harmjanz placed in Gottschaldt's file, when Gottschaldt was called in to begin negotiations for the Breslau position, he proceeded to explain the advantages of his situation in Berlin and the disadvantages of Breslau without waiting for Harmjanz's offer. Harmjanz replied that it was "highly unusual" for a candidate to refuse his first appointment to a professorship; if that was the way he felt, there would be no offer. He termed Gottschaldt's manner "overbearing and arrogant" and opposed his appointment to any professorship at all.[54] In 1942, Gottschaldt's name appeared in second place for the chair in Leipzig as successor to Philipp Lersch. Strongly positive evaluations came from both the Nazi professors' organization and Otmar von Verschuer, who had succeeded Fischer as Kaiser Wilhelm Institute head. Nonetheless, the Harmjanz note remained in his file; and when the first choice, Gerhard Pfahler, refused the position, Gottschaldt did not receive the post.[55] Such an incident could just as easily have happened in the Wilhelmian as the Nazi era.

Much has been made recently of the alleged "compatibility," or "affinities," of certain psychological theories, including Gestalt theory, with Nazism. This issue should be considered under two rubrics, ideological and practical. From the ideological point of view, the implications of these two cases are ambiguous. Many psychologists, including Metzger, tried to rework their earlier theories and offer the revised versions to the new leadership, with at best mixed success. Such attempts say little about an essential "affinity" of those theories with National Socialist ideology – whatever that was. Important for assessing Metzger's case is the absence of key terms such as "will," "folk" and especially "race" from his discourse in this period aside from the single exception discussed previously. Apparently he was willing to maneuver his Gestalt vocabulary toward a version of cultural conservatism that was compatible with Nazism in certain respects, but not in others.

At the practical level, the needs of both the Wehrmacht and of the Nazi labor and social welfare administrations supported the primacy of personality theory and diagnostics during the Nazi period. Metzger's support of "general" psychology against this trend was a defensive position, which he continued to hold after the war. Metzger later integrated "characterology" into his teaching and wrote extensively on pedagogy, thus compromising his strictly basic-research posture (see Chapter 22). To his credit, he did not try to portray his stance on these issues in the 1930s as a form of resistance to Nazism.

Some scholars have argued or implied that in the relations of science and National Socialism, ideological considerations predominated up to 1936 or 1937, but that practical concerns took over thereafter. The fact that Metzger wrote what he did during the war shows that, to him at least, ideology was still important; and ideological debates in psychology continued throughout the Nazi period. Whether the authorities took any notice of these debates in psychology, as they did in physics, is another matter. In Gottschaldt's case, ideology and pragmatics merged so completely in the eugenics program that it makes no sense to posit chronological dividing lines.

Last but not least, there is the moral dimension. The Gestalt theorists' and their students' resistance or accommodation to Nazism can be discussed in moral terms, but only if we try to understand the political conditions of morality in given times. Köhler's strong, Metzger's weak, and Gottschaldt's ambivalent stands should all be seen in the context of the situation of German professors since the Wilhelmine period. Freedom of teaching and research was one of the most important positive features of German academic life. To assure that freedom, professors had the status which Köhler then tried so stoutly to defend. But professors were expected at the same time to be loyal servants of the state. With this ambiguity Köhler, Metzger, and Gottschaldt were all forced to deal. That social situation set the parameters that defined freedom and morality – a lesson that can be extended to other times and places.

21

Research, theory, and system: Continuity and change

The question of continuity and change in psychological research during the Nazi period has often been dismissed with a reference to absurdities like the studies carried out in Erich Jaensch's laboratory of the allegedly "racial" determination of pecking behavior in chickens. "Northern" birds, it was claimed, pecked more slowly, carefully, and successfully at their seed-corn than did "southern" birds.[1] Kurt Gottschaldt's work for the Kaiser Wilhelm Institute for Anthropology shows, however, that "normal" science could also be made to appear serviceable to Nazism's purposes; and there was plenty of such science. Friedrich Sander's students in Jena, for example, continuing research on the "microgenesis" of form perception already begun in the 1920s, examined such topics as the role of "rhythmic Gestalt formations" (*rhythmische Gestaltbildungen*) in raising enjoyment and achievement levels in groups of workers.[2] Such research could easily have been done in the 1920s; in this period it implicitly offered support for the Nazi reorganization of the labor movement.

There was no contradiction between "holistic" psychology of this kind and the turn toward personality diagnostics. Students in Sander's laboratory and in Oswald Kroh's Tübingen and Munich institutes attempted, for example, to prove that "Gestalt" or holistic (*ganzheitlich*) and "analytical" responses to optical illusions correlated with specific personality types. The implication was that standard laboratory exercises, far from being irrelevant to real life, could have diagnostic value, especially when correlated with Ernst Kretschmer's diagnostic tests for "extraverts" and "introverts."[3] The same was true of research in Gerhard Pfahler's Giessen laboratory, comparing Leipzig "holistic" psychology, Jaensch's "Integration Typology," and Pfahler's "Hereditary Character Study" (*Erbcharakterkunde*). Pfahler, a pre-1933 Nazi who advanced to professorships in Giessen and Tübingen after 1933, purported to show psychological differences between so-called Nordic, Falic, and Ostian people and thus legitimate Hans F. K. Günther's widely used system of "race" classification. These and similar studies appeared in mainstream psychology journals, alongside work that could have been published in the Weimar period.[4]

In contrast, both the research program and the conceptual development of Gestalt theory showed remarkable continuity. This was so in part because *Psychologische Forschung* was available as an alternative, protected publication outlet; but that was no longer the case after 1938 (see Chapter 20). The first portion of this chapter explores the ways in which the topics and methods of Gestalt theory persisted or changed after 1933. The second portion analyzes theoretical contributions by Edwin Rausch and Wolfgang Metzger. Both sections consider two questions: Whether and how Nazi-era research and theory reflected the contextual changes just described; and whether this work represented substantial advances in Gestalt theory, or only systematic refinements of established positions.

In a recent essay claiming an "affinity" between "holistic" and Gestalt psychology and Nazi ideology, Wolfgang Prinz has suggested that Gestalt theory in particular was vulnerable to ideological compromise not only because of Wolfgang Metzger's personal predilections, but because it had already reached the limit of its development before 1933.[5] Such claims project subsequent judgments of a theory's ultimate contribution backward onto the history of that theory. They are believable only if there is evidence that the historical actors shared such judgments at the time. Metzger and his students in this period did not act as though they believed that Gestalt theory's fruitfulness was at an end. Their defensiveness was not that of an already rigid and decaying school, but the reaction of an intellectually still-vital but institutionally weakened school to social and political change.

Research remnants – Metzger in Frankfurt

Compared with the prosperity of the Weimar years, the institutional conditions for Gestalt-theoretical research in the Nazi period were considerably reduced, especially after Wolfgang Köhler's resignation in 1935. Many Berlin doctoral candidates managed to complete their work there before Köhler departed. Köhler sent others to finish their degrees with Johannes von Allesch in Greifswald, and still others to Wolfgang Metzger in Frankfurt (see Appendix 2, Tables 2 and 4). The following discussion focuses primarily on the work done under Metzger.

In Frankfurt, Wertheimer's chair was transferred from the Natural Sciences to the Philosophical Faculty, where it was apparently taken by philosopher Hans Lipps. With that the unique coordination of philosophy and psychology characteristic of Wertheimer's tenure quickly ended. According to Edwin Rausch, the philosophy and psychology libraries, which had been in the same room, were divided as early as 1934, because the old arrangement "did not work any more with the new people." At first, Friedrich Schumann came out of retirement to serve as institute director and conduct doctoral examinations. Though he was officially the associate director, Metzger's salary came from his assistantship, which he retained until he left Frankfurt in 1942. The second assistantship, held until April 1933 by Erwin Levy, was taken by Erich Goldmeier after Levy's emigration, and from 1938, after Goldmeier, too, emigrated, by Rausch, who had been a student aide until then. As Rausch recalls, without a full-fledged director the institute "was only a torso, only the assistants hanging in the air." Moreover, many of the remaining students were Jews, like Levy and Goldmeier, or foreigners. The institute budget nonetheless remained largely intact at first. In fact, 2,760 Marks were transferred from the Berlin institute in 1934 – more than the entire annual budget had been during Wertheimer's tenure. This was probably to help support the work of Berlin students sent to Frankfurt. By academic year 1936–1937, the budget was down to 1,530 Marks, a cut of 40 percent compared with 1929, but only a slight reduction from the 1,800 Marks of 1933. Rausch speculates that there was "no attack" on the budget by university authorities because so little was left of the institute after Wertheimer's departure that "it need not be taken seriously any more."[6]

Despite, or perhaps because of, these altered circumstances, it was possible in the 1930s to continue at least some of the lines of work already begun. Indeed, there was hardly any change in research topics, though the range of issues was undeniably narrower than in the Weimar period. Several distinct but related lines of work can be distinguished. In motion perception, Erika Oppenheimer and Walter Krolik followed up on Karl Duncker's classic paper on induced motion (see Chapter 14). In Berlin, Hans Wallach and Ottilie Redslob continued work on perceived motion direction; Redslob completed her work in Frankfurt. A number of studies in both Berlin and Frankfurt attempted to apply the Gestalt approach to the study of touch.[7] Other papers, including two by Köhler and three dissertations begun under him and completed under Johannes von Allesch in Greifswald, followed up on the Köhler-Restorff studies on retention.[8]

In Frankfurt, continuity was assured mainly by concentrating on issues in vision. One organizing question was the role of light and shadow in the perception of objects in depth. Kurt Koffka had discussed the role of brightness differences in the constitution of perceptual objects in Gestalt terms as early as 1923 (see Chapter 14). The topic had become current again with the appearance of a pioneering monograph, "Thing and Shadow," by Vienna psychologist Ludwig Kardos in 1934.[9] In the 1930s, Gestalt researchers pursued the issue in both Berlin and Frankfurt, but on the basis of different exemplars.

In Berlin, Wilhelm Wolff applied Karl Duncker's work on induced motion to brightness perception. He began with older observations of what he called "dynamic contrast" by Exner and Hering, in which, for example, a white field was divided in half and subjects fixated one half while the experimenter altered the brightness of the other. In such cases the seen brightness of both sides changed,

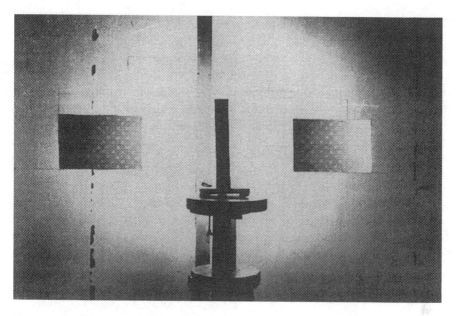

Photo 19. Setup for Turhan's experiment on brightness and depth perception (1935). Wolfgang Metzger, *Gesetze des Sehens* (1936), 3rd enl. ed. (1975), page 537, picture 579. Reprinted by permission of Verlag Dr. Waldemar Kramer.

and when the intensity of the fixated area was increased, the unfixated portion of the field appeared to lose brightness and turn an increasingly darker gray. Noting the apparent analogy to phenomena of induced motion investigated by Duncker, Wolff systematically studied this phenomenon first in a darkroom and then on the Berlin *Ganzfeld* setup. He found that fixation does help bring about change in fixated field parts, as it does in induced motion, and that "enclosed" fields, too, were subject to similar brightness effects, depending on the amount of brightness difference, the speed of induced change and the size of the surrounding field. As he noted, however, a strict analogy to induced motion was not possible, since subjects typically do not perceive their own brightness. Dynamic interactions within perceptual fields could therefore be explained only by referring to inhomogeneities of "trace fields" in the brain. Then "the trace field [would have] the same significance for induced brightness change as . . . the 'subject' (i.e., phenomenal Ego) for induced motion."[10]

In Frankfurt, Mümtaz Turhan, a Turkish student, worked under Metzger's direction on spatial effects of brightness contrast. All earlier work on the subject had related distributions of light and shade to direction of illumination, and Turhan followed suit. But he used two stimulus fields, a surround (*Umfeld*) and an interior (*Infeld*) placed behind it, each illuminated independently (see Photo 19). Using simple concave and convex surfaces, he discovered that a change in the direction of illumination alone could produce drastic alterations in the apparent distance of objects from the observer. Sometimes the interior field appeared to penetrate the surround with concave curvature; with objects held in between the two surfaces and appropriate rhythmic changes in illumination, the experimenter could make one field appear to jump in front of or behind the other.[11] Turhan explored the parameters of this phenomenon by varying the pattern of the interior field (landscapes, shadow strips, and waves), by altering the exposure time, and by reversing the position of interior and surround. In general, he found that with opposing brightness direction or with increased pattern complexity, "forces" work to "push" one surface in front of the other. With illumination in the same direction, subjects tended to see both fields on the same plane as parts of a single surface. All this appeared independently of observers' knowledge or conscious effort.

To explain these results, he suggested that seeing objectively separated surfaces on a single plane under certain conditions, even with differing structure of interior and surrounding fields, was a case of the general Gestalt tendency to closure, which was in turn a tendency to "simplest" distribution, or *Prägnanz*. But to explain what was seen when the two fields do not "belong together," he needed to add further "tendencies." The phenomena seen under opposing illumination could be explained by postulating forces that unify the "illumination field" and produce homogeneity of surface coloration, a tendency to attribute brightness differences to illumination and not to color, and finally a tendency to see "surface unity" with equal illumination. All of these, he added, were relevant for understanding camouflage, since they suggested the possibility of converting opposing to equal illumination and thus avoid being seen. Turhan's proposed explanations, however, remained at the phenomenological level; like Metzger, and in contrast to Köhler's student Wolff, he offered no physiological theory.[12] This concentration on the phenomenological level and reticence about either underlying brain events or any causal explanation were standard features of research under Metzger's guidance in this period.

Perhaps the most interesting research Metzger directed in Frankfurt was Erich Goldmeier's study of judgments of similarity in perception, published in 1936. His starting point was the problem originally raised by Harald Höffding and Ernst Mach in the 1890s: How do we know an object or feature is the same as one we have seen before; or, how do we recognize forms as the same even when they are presented in different positions? (See Chapter 6). In a series of carefully planned experiments, Goldmeier asked a total of more than 100 observers (between 10 and 25 per trial), all but 6 of whom did not know the purpose of the study, to compare successively presented drawings with a standard figure. The results refuted Mach's claim that similarity is attributable to partial identity – the number of equal

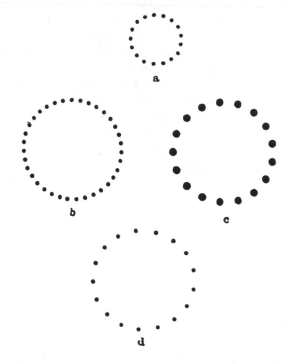

Figure 19. Phenomenal similarity (Goldmeier 1937). (The figures c and d, respectively, are proportional enlargements of the a figure. The b figure is nonetheless judged more similar to a.) Erich Goldmeier, "Über Ähnlichkeit bei gesehenen Figuren," *Psychologische Forschung,* 21 (1937), p. 165, Fig. 15 a–d, 17 a–c. Reprinted by permission of Springer-Verlag.

or identical parts – and also the claim advanced by Mach, Ehrenfels, and others that similarity is based on identity of relations or mathematical proportionality among figures. Consider the following example against the latter hypothesis: in Figure 19, the dot circles c and d are proportional enlargements of a, while in b the number of dots is twice that of a; but subjects still judged b it to be more similar to a. Against the hypothesis of identical relations, Goldmeier argued that the functional roles of relations, not their amounts, are decisive. Clearest in this regard were cases in which similarity persists even in distorted figures. All of these determinations were based on quantitative criteria, not phenomenological observation alone.[13]

In Goldmeier's view, his results showed that what is conserved in perceived similarity are "the phenomenal function of the parts within the [perceived] whole," or "the agreement of those qualities which determine the phenomenal organization" of the field in question. New to the list of such qualia were the dimensions of "form" and "material" (*Stoff*). Metzger had just introduced these in his book "Laws of Seeing," borrowing descriptive terminology from Edgar Rubin's work on figure and ground. Goldmeier found that similarity of form properties was best preserved by proportional enlargement, that of "material" properties by keeping their measure constant.[14] Interestingly, he observed individual differences in observers' judgments of similarity, but attributed these to different levels of comprehension of figural content, and denied that age or personality affected the results. He acknowledged that attention span was a factor in the attribution of "material" qualities, but mentioned this only in passing. Most damaging was his admission that he had no predictive theory "to specify in every possible case which values are singular. Indeed, he admitted that "the prospects for such a theory are dim, and probably depend on progress in neurophysiology."[15]

Research in Frankfurt after Wertheimer's departure thus added no fundamentally new problems to the Gestalt theorists' research program, but extended work on already existing issues, with certain terminological and methodological refinements introduced by Metzger. Most important among the former were terms like "Gestalt center of gravity" (*Gestalt Schwerpunkt*) and references to the "weight" of qualities or properties in perceptual fields – subtle moves in the direction of the Leipzig school. The small number of observers in most studies might be seen as a result of the Frankfurt institute's relative impoverishment, but Goldmeier's case counts against such a view. Also, again with the exception of Goldmeier, publications from the first Frankfurt years were not distinguished by concentrated efforts to respond to substantive criticism of Gestalt theory by other German psychologists. This changed as soon as Metzger went to Halle in 1936 and joined the Nazi party the next year.

An example of the change was a dissertation on memory for form based on work done at Halle, the first study done under Metzger's direction published in a journal other than *Psychologische Forschung*. Koffka's student Friedrich Wulf had claimed in the early 1920s that subjects drawing previously seen figures from memory exhibited "leveling" and "sharpening" tendencies, both governed by a tendency to produce the simplest figures (see Chapter 14). Criticisms of this claim had come from both German and American researchers. The latter, most prominently James J. Gibson and Leonard Carmichael, denied any autochthonous change in the figures and attributed observed alterations in them to their assimilation to one another or to known objects. German authors such as Bruno Petermann and Jaensch student Hans Eilks claimed that Wulf's results were artifacts of a single personality type, or of the fact that his observers had all been members of the same psychological institute. In his dissertation, Metzger's student Siegried

Sorge corrected for these possibilities by using naive observers in larger numbers – rural school children of allegedly "integrated" personality types, according to Erich Jaensch's classification system. He tried to avoid assimilation effects by not using "thinglike" names for the test figures, and by making their structures as different from one another as possible. In addition, he tested the effect of systematically varying the position of the figures.[16]

Predictably, given the larger number of observers employed, Sorge found a greater variety of simplification and restructuring tendencies than Wulf had observed. These included making figures more vertical, introducing right angles, losing fine structuring, closing gaps, and straightening curves or making straight lines curved. He also noted "moderate" assimilation of parts with similar functions in a figure, and a tendency to form the spaces between parts of figures into "subwholes" (*Unterganzer*) – another term borrowed from Leipzig *Ganzheitspsychologie*. But he insisted that none of this refuted Wulf's basic claim, for in every case the same kind of factor was involved – simplification in a specific direction. He also acknowledged the impact of personality type, but only on the amount or speed of simplification, not on its presence or absence. Like other Metzger students, Sorge did not discuss physiological or any other possible causal explanations for his results.[17]

Even the few cases discussed here reveal certain general tendencies in Metzger-guided research: to expand and refine the list of Gestalt factors; to emphasize phenomenology and avoid even the appearance of testing Köhler's psychophysics or any other hypotheses beyond those deducible from Wertheimer's "Gestalt laws"; to respond more sensitively to criticism, albeit without yielding on the priority of Gestalt factors over others; and to point out the potential practical relevance of the work, without undermining the primacy of basic research. The parallels with Metzger's own story as recounted in the previous chapter seem clear. Did this work represent genuine advances in Gestalt theory, or only refinements in an already established position? Phenomenological and conceptual enrichments like Metzger's "form" and "material" properties, or "Gestalt centers of gravity," are different from expansion. Goldmeier's and Turhan's research clearly did not bring about the innovations that Karl Duncker and Otto von Lauenstein had achieved before 1933. Edwin Rausch's work, to be discussed below, was clearly a genuine attempt at theoretical change, but it had no impact on research in this period. In the terms set out in Chapter 14, Gestalt researchers in the 1930s emphasized Wertheimer's strategy of research domain expansion by metaphorical extension over Köhler's hypothesis testing.

What Metzger taught his students was systematic phenomenology of perception. Though this by no means excluded the use of quantitative methods, these were aimed, as before, at establishing the parameters of, or threshold values for, specific experiences within individuals. There was still no serious effort to account for variation among individuals, though some investigators expanded the number of naive observers in response to criticism that earlier results might be artifacts of suggestion. Such modifications did not confront the underlying issue – the possible variability of results on the basis of social, personality, or other factors beyond the stimulus situation alone. There was no systematic attempt to relate perception to questions of social psychology or personality, as was happening elsewhere in response to the practical demands being made on psychology in Germany at that time. This issue was faced directly only in Sorge's study, which was done as Metzger joined the Nazi party and sought, however weakly, to align his work with the shifting German mainstream.

In the meantime, psychological research methodology elsewhere in Germany had changed drastically. The field was ceasing to be a scientific adjunct or prolegomena to philosophy and was becoming an instrument of public policy. Partly as a result of this shift, as we have seen, characterological – that is, typological – methods became increasingly dominant, for these were more in line with the discipline's emerging practical functions. A parallel move to personality research and classification occurred in the United States at the same time, though typological methods were not so prominent there. Gestalt-theoretical research might have been reoriented to adapt to this trend by changing either its topics or its methods. The case of Kurt Gottschaldt, already discussed, indicates that it was at least possible and perhaps necessary to do both, but only at a considerable cost in conceptual rigor. In contrast to other schools, Gestalt-theoretical research in the Nazi period did not employ obviously ideological vocabulary. Some of it substantially enriched knowledge of issues raised in the 1920s and earlier. But its limited impact reflected both decreased institutional resources and depleted intellectual capital – experimental research on cognition was not a dominant line of work in German psychology at the time.

Theory and system

The Nazi period saw two major developments in Gestalt theory that have generally been ignored outside Germany: Edwin Rausch's monograph on "summative" and "nonsummative" concepts, and Metzger's theoretical masterpiece, *Psychology*. Rausch studied mathematics at Bonn University before coming to Frankfurt and discovering Wertheimer's lectures.[18] He became a student assistant at the Frankfurt institute after passing his state teacher's examination in 1931. Though he began his dissertation under Wertheimer, he had not proceeded far by 1933. After Wertheimer's departure he withdrew briefly, then resumed work independently, with Metzger as informal advisor and Schumann as formal examiner. In 1937 his dissertation appeared in *Psychologische Forschung* with the forbidding title, "On Summativity and Non-Summativity."[19] Its aim was to develop a more systematic account of the concepts of part and whole, with the aid of the innovations in symbolic logic pioneered by Bertrand Russell, Rudolf Carnap, Giuseppe Peano, and others.

Rausch's procedure is too technical to be explained fully here. In essence, he proposed to improve on Köhler's "too global" application of the Ehrenfels criteria to both "sums" and "summative groupings" in *Die physischen Gestalten* by presenting a single, simple criterion for summativity: "invariance," the nonalteration of both subtracted parts and the remaining ones on separation of some of them from a grouping. This criterion corresponded with Köhler's example of a pure sum – a collection of stones, one on each continent, the removal of one of which brings no change in the others (see Chapter 11). The broadest example of this was a geometric or symbolic-logistical space consisting of a "mesh" of "positions." The number of possible "removals" or "subtractions" from such a space was easily calculated by an operation equivalent to simple algebraic permutation. This resulted in a series of topological "trains" – that is, chains of part-series depicted in treelike form. The implication was that objects, or sections of a logical space, that can be treated this way are sums; those which cannot are non- or suprasummative. The number of possible subtractions (or: subtractable positions or pieces) that could be performed without change in either the subtracted or the remaining parts thus became a measure of the "strength" of the sum in question. The inverse of this procedure works for suprasummative concepts, if variance rather than invariance is taken as the criterion – that is, when each part and the remainder also change on subtraction.[20]

Rausch's attempt at conceptual clarification was not free of difficulties, many of which he noted himself. A primary problem was that "a proper ordering of Köhler's concepts in our system [is] not possible," because it is unclear whether Köhler's words refer to a division of geometrical spaces, as his did, or what the difference is between a "sum" and a "summative grouping." Further, he could not yet use his system to clarify the concept of "natural part" frequently used by Wertheimer and Köhler. He did think it would be possible to apply his procedure to qualities as well as to pieces and groupings, and thus eventually to include the quality of transposability essential to temporal Gestalten such as melodies. Most problematic, however, was the fact that a "logical sum" is not an "optical sum." Rausch insisted that "always ontically [i.e., physically or phenomenally] real and not logical-ideal alterations" were meant. Though he insisted that such conceptual manipulations "must be useful" for experimental research, he conceded that he had provided "an overview of a conceptual field, not a factual one, in other words not of realities, but of possibilities."[21]

Despite these limitations, Rausch's work had an immediate impact, albeit outside of German psychology. In an analysis of the Gestalt concept published in 1939, émigré logical empiricist philosophers Kurt Grelling and Paul Oppenheim attempted, in explicit agreement with Rausch, to clarify the notions of "sum," "aggregate," and "complex" in a way that would elucidate the actual content of von Ehrenfels's and Köhler's Gestalt concepts and differentiate them from one another. They called the shape of a body, for example, its "geometric form," but distinguished this from complexes such as groups, which they termed "Gestalt individuals." Both of these terms are clearly compatible with Ehrenfels's notion of "Gestalt quality," but Grelling and Oppenheim maintained that they were different from Köhler's physical Gestalten, which they called "systems of effects" (*Wirkungssysteme*). Because complexes of the former kind can also be systems of effects under certain conditions, they concluded that, from a logical point of view, Gestalt complexes and effect systems were "poles of a single order" of concepts.[22] Such analyses could have saved the Gestalt concept from the recurring charge of vagueness, if they had not been ignored at the time. However, because they presupposed an empiricist standpoint, Grelling and Oppenheim, like the other logical empiricists, failed to engage the epistemological core of Gestalt theory – Wertheimer's claim that Gestalten are immanent in experience, not categories imposed upon experience.

Metzger's book, *Psychology: The Development of Its Fundamental Assumptions since the Introduction of the Experiment,* was considerably more ambitious than Rausch's monograph. The original title was "Gestalt Theory," but he changed it to make clear that his aim was to make Gestalt theory the conceptual foundation of general psychology.[23] To achieve this, he employed a strategy rather different from that of Kurt Koffka's major text of the same period, *Principles of Gestalt Psychology* (1935), which Koffka wrote in America. The aim of both writers was to construct a comprehensive system of psychology based on Gestalt theory, but there were differences in both the structure and the content of their books. A brief comparison of their approaches will illuminate the impact of cultural settings on both the construction and the content of theoretical systems in psychology.

The most obvious difference between the books is the identity of the opposition. Koffka wrote that Gestalt theory opposed both materialism and vitalism, because it rejected what he claimed was their common root, "the strong cultural force in our present civilization for which I have chosen the name positivism."[24] Correspondingly, his primary opponents were those American psychologists most obviously identified as positivists – the followers of E. B. Titchener, for whom experimental psychology meant the classification of sensory contents, and the behaviorists. In contrast, Metzger, as shown in the previous chapter, put Gestalt theory in the forefront of the struggle not against a single "cultural force" in a unitary "civilization," but against "the spirit of the West" in general. Metzger had little need to oppose behaviorism or any other form of "positivism" directly, since few writers took such positions in Germany in 1942. Prominent instead among his contemporary antagonists were nonpositivists who opposed natural-scientific psychology, such as Eduard Spranger, or who criticized Gestalt theory for its alleged lack of biological orientation, such as Erich Jaensch. For support he alluded to Germanic thinkers such as Meister Eckart, Hamann, Fichte, and also to Clausewitz, who had described war itself as a dynamically unfolding, indeterministic Gestalt.[25] The relevance of this reference network will be clear from the depiction of the German intellectual scene in earlier chapters. Moreover, psychology in Germany was still a part of philosophy, and the issues its theoreticians addressed continued to be cast as questions of philosophical anthropology.

The different ways in which the two writers constructed their systems also reflected this fact. Koffka's procedure was familiar from the organization of standard textbooks in psychology. He enunciated general Gestalt principles and then applied them to standard topics, beginning with a detailed account of visual perception, proceeding to a critical reworking of Lewin's work on action and emotion, incorporating research by Wertheimer, Duncker, and Köhler on thinking, learning, and memory, and finally applying Gestalt principles to personality and society. Metzger, by contrast, presented not a conventional textbook, but an attempt to revise the theoretical presuppositions of modern psychology. His hope was that this approach would put an end to the "misunderstanding" that Gestalt theory was merely "a psychophysical theory that seeks to explain the entire psychical realm at any price by means of known physical laws."[26] The reference to Jaensch's and Krüger's attacks on Gestalt theory's alleged physicalism was surely obvious to readers at the time.

The first and most important assumptions Metzger questioned were that the "real" causes of events must be sought only behind, not within, phenomena, and its correlative assumption that only the material (*das Stoffliche*) is real, so that the essential qualities of appearances are those of their material substrates. These he combined and called the "eleatic presupposition," alluding to its roots in Platonic philosophy. Husserl's phenomenology had taken a first step toward "liberation" from such prejudices, he wrote, but the final steps had come from Ludwig Klages's "modern theory of expression" and Gestalt theory. Both built their systems upon the assertion of an independent psychological reality, but Klages's "liberation" was only incomplete, because he denied the existence of real objects as tricks of *Geist*. For this Metzger substituted "accepting the given as it is" – the belief, that is, that psychological reality is to be found within, rather than behind, the appearances. Epistemologically, he argued, it may be sensible and necessary to inquire whether something is "real" or not; in psychology only phenomenal reality matters: "There is only one observable subject, and that is 'I', and only one observable world, that is the immediately presented or phenomenal [*anschauliche*] world." Since this presupposition was not empirically verifiable, its value was therefore justified "only by the greater or lesser fruitfulness of the research" that it generated.[27]

Metzger proceeded to explicate no less than five dimensions of reality. The first was the physical world; the others constituted "the psychologically real," beginning with the experienced world which is our only clue to or link with the first. This corresponds to Koffka's distinction between "geographical" and "behavioral" environments. The third sense of "reality" was based on a distinction between things, objects, events, and acts as experienced (*Angetroffen*) and their "re-presentations" (*Vergegenwärtigungen*) to us as imagined, supposed, expected, remembered, planned, conceived – that is, intended in the broadest sense. Only these, he argued, and not all psychological objects, have intentional character. The fourth sense of reality is evident in the experience of "something" versus "nothing" – for example, of qualityless, "empty" space "between" things, the contents of an unopened box or of unseen sides of three-dimensional objects, or the space beyond the horizon. Reality in the fifth sense manifests itself in the distinction between the (phenomenally) real and the "merely apparent" (*Schein*), as in the distinction between pictures and the objects they represent, or between "really" having an emotion and appearing to have it.[28]

Metzger underlined his desire to present his ideas as philosophically relevant by adding remarks on Kant's "thing in itself." As he noted, Kant had ascribed to the realm of experienced phenomena reality in the third of the senses listed above. This was "in itself definitely a step forward"; yet subsequent research had shown that the structure of the realm of appearance (*Anschauungsraum*) is not absolute, as Kant supposed, but differs among perceivers and situations. The appropriate response to this, he maintained, is not the "nominalistic side-step" common to positivism, Neo-Kantianism, and phenomenology; there was too much agreement among observers about the physical world for that. Instead, it was necessary to "complete" Kant's program by determining scientifically the structures of appearance and the empirical principles that govern them; for "we are not of the opinion that philosophy deals with a different reality and possesses another kind of truth than that of the sciences."[29]

Ultimately, Metzger claimed to replace the presumption of "natural disorder" that had allegedly governed all previous scientific thinking about both nature and the mind with the claim that there exist both in nature and in our experience of nature "inner, natural orderings that are not there by force, but 'in freedom'." Indeed, he went so far as to claim that our "knowing mind" (*erkennender Geist*) contains "no finished forms or arrangements that only need to be called forth, but rather (just in the Kantian sense) only the conditions of their emergence – the soil in which they can grow *thus*" and not otherwise.[30] The strategy he employed to sustain this claim was to convert Gestalt principles into metatheoretical concepts and depict them as names for these intrinsic "natural orderings." His chapter headings were therefore not standard textbook topics, but rather terms from Gestalt-style phenomenology of perception, such as "qualities," "contexts," "relational systems," "centering," "order," and "effects." Vision, hearing, and other senses were not treated separately, as might have been the case in conventional texts; instead, examples from all sensory modalities were employed to illustrate the universal applicability of Gestalt principles.

Of particular interest and originality was Metzger's discussion of psychological frames of reference, or "relational systems" (*Bezugssysteme*), the notion already introduced by Karl Duncker in the 1920s (see Chapter 14). The underlying presupposition under attack was that of psychological space as a collection of empty, indifferent "locations." For this Metzger substituted the principle that all location in space and time, as well as all phenomenal judgment, "is based on relations in more extended psychological regions." There were two problems with this claim. As Duncker (and also Walter Ehrenstein) had shown, the fact of relatedness is ordinarily, "hidden" from immediate experience. Rather, as Köhler had noted in 1929, in ordinary life the "absolute" quality of things is their most outstanding characteristic. To come to terms with this apparent contradiction, Metzger recognized that Wertheimer's application of the word "Gestalt" to both seen objects and the structure of the perceptual field as a whole required modification.[31]

Specifically, Metzger acknowledged that the characteristic membership of regions in a relational system "is correlative to but different from the relation of parts to their whole." The experienced "firmness" of objects is based on that of the relational system of which they are parts, not vice versa. Relational systems have "a specific structure that can be more or less firm" according to "preconditions and boundaries present in the organism" for their construction. These are not rigidly determined in advance, but emerge in interaction with given conditions. Thus, a relational system is "a living whole" (*eine lebendige Ganzheit*), which not only reacts to but interacts with local stimulation and its demands. A "true" part is in a two-sided relation with its whole; a part of a relational system is in a one-sided, open-ended relation with the system as whole. A thing in space, for example, leaves no gap on removal, but a piece from a puzzle does.[32]

The benefits of this modification of strict Gestalt principles were clear. It now became possible for Metzger to get a conceptual grip on the myriad additional "tendencies" he and his students had had to suppose to account for results not explained by simple analogies to Wertheimer's "Gestalt laws." To cover these he posited a "principle of branched effects" (*gegabelte Wirkungen*), which stated that "wherever the experienced field has more dimensions than the stimulus field," an infinite variety of experiences can emerge from the same stimulus constellation, depending on the structure of the environmental situation and the state of the perceiving organism. In the pure case, a single stimulus state can produce infinite pairs of matched effects, so that when one increases the other decreases. For example, the size of the retinal image is the basis for both the perceived size and the perceived nearness of objects to one another. Change in size can thus lead to change in proximity, the opposite can occur, or both may result. Analogously, the form of the retinal image "branches" into the seen form of a surface and its perceived extension (*Ausrichtung*) in space.[33]

More broadly, with this principle it also became possible to portray processes ordinarily considered "psychological," such as attention and attitudes, as relational systems, and thus bring them into the purview of Gestalt theory. Most interesting in this regard were concepts experienced as "absolute" which are actually relational, such as small, large; near, far; above, below; early, late; but also clever, stupid. Such examples suggested to Metzger an analogy to coordinates on a measuring scale. Thus, both phenomena such as light and dark adaptation and the perceptual constancies are examples of shifts in the null point of relational systems. As he pointed out, however, the psychological null point is not the same as a "center," and certainly not the same as the center of attention. It can change with time (as in the disappearance of a musty odor from a room), and can also take time to emerge (as in melodic or rhythmic structures). The links of such thinking with the "central tendency" first researched by American psychologist H. L. Hollingworth in 1910 and also discussed by Koffka were intended.[34]

The addition of personality judgments like "clever" or "stupid" to the list of relational scales exemplified a further benefit of Metzger's conception – the possibility of extending Gestalt theory from perception and cognition to personality. For example, he treated "striving for value in itself" (*Eigenwertstreben*) and "status striving" (*Geltungsstreben*) as poles on a continuum of motivations ranging from doing something "for its [or one's] own sake" to acting only to impress others – a notion akin to the later concepts of inner- and outer-directedness. Metzger claimed, however, that it was "at least phenomenologically wrong" to interpret *Eigenwert* as selfishness or self-centeredness; the actual difference was between ranking relative to other peoples' values and "on a suprapersonal, objective measuring rod." The step from here to the social realm was straightforward. One needed only to distinguish relational systems according to whether the null point lies inside or outside the subject.[35]

The costs of such a reconceptualization were as evident as the benefits. One of these was a certain accommodation to the "eleatic principle" Metzger claimed to oppose. This became clear in Metzger's attempt to confront the challenge Köhler had already, though tentatively, faced – that of "imaging" the external world, the self and other people in the brain. Traditionally, he noted, the issue had been posed as a problem of how events inside the organism are "projected" onto or into the external world. Paul Schilder's "body image" concept, first presented in 1924, was a productive step forward, but the location of that image was problematic. Like Köhler, Metzger argued that this was a pseudoproblem; for the inside-outside relation, too, is not a product of the external world but must somehow be mapped inside the organism. The diagram he provided to clarify his position (Figure 20) even included cortical counterparts to emotional and volitional relations to objects.[36]

The diagram also exposed two fatal flaws in this schema: (1) the psychophysical counterpart of the seen object is depicted as a literal copy; and (2) it is shown

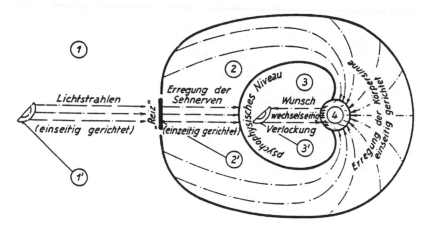

Dependency relation between the physical world, including the organism (= Macrocosm) and the perceptual world, including the experienced ego (= Microcosm). Physical environment (1) with physical object (1'); physical organism (2) with eye (2'); perceived surroundings (3) with perceptually encountered object (3'); body image (4).

Figure 20. Model of "interior psychophysics" (revised from Metzger 1941). Wolfgang Metzger, *Psychologie,* 1st ed. (1941), Fig. 41, p. 271; 3rd ed., Fig. 42, p. 283. Reprinted by permission of Dr. Dietrich Steinkopff Verlag.

as separated from a region labeled "Ego" – the homunculus in person, as it were. In fact, Metzger postulated a fourfold schema, separating both the psychological Ego from the perceptual environment and the "cortical ego" from the physiological counterparts of the objects it perceives. He attempted to save himself from having to postulate applelike images in the brain corresponding to the apples we see by speaking of a "functional" in place of a literal image, as Köhler also did. But he still felt compelled to say that the physical nearness of correlated regions in the central nervous system is what makes the relationship of "body-ego" and "world" so intimate. It also explains why attitude changes initially occur within organisms, not in their relations to external objects.[37]

The problems with Metzger's elaboration of Köhler's concepts were as obvious as they had been in the original. Physical stimuli were still no more than occasions for perception, not their causes, even though Metzger, like Köhler, insisted he was constructing a causal chain. Equally confusing was his insistence on the "Gestalt identity" (*Gestaltidentität*) of the "psychophysical total process and the phenomenal experience of outside and inside." He did not cite Goethe at this point, as Köhler had done, but it was no clearer than before why causal relations should produce such an identity. Metzger later spoke more prudently, but less clearly, of a "causally mediated Gestalt-relatedness" (*Gestaltverwandschaft*), and insisted that the events in question were not Leibnizian monads in preestablished harmony with the external world, but rather two microcosms derived from "the *one* macrocosm of the all-enclosing physical world."[38]

Metzger's account paralleled important contemporary trends in theoretical biology, such as the concept of "homeostasis" elaborated by Walter Cannon and G. E. Coghill, and the work of Erich von Holst on sensorimotor coordination in the central nervous system. But he devoted the most explicit attention to the *Gestaltkreis* theory of Heidelberg physiologist and philosopher Viktor von Weizsäcker, developed in outline by 1933 and published in full in 1940. Melding an extraordinary amalgam of ingredients, including neurophysiology, the philosophies of Joseph Schelling and Max Scheler, and psychoanalysis, von Weizsäcker proposed "to introduce the subject into biology" by presenting the interaction of experience and behavior as a "Gestalt circle," a psychophysical unity in time embracing body and mind.[39]

The *Gestaltkreis* idea was based in part on experiments by von Weizsäcker and co-workers on motion perception with moving observers, which paralleled in significant ways those of Karl Duncker in Berlin (discussed in Chapter 14). In the basic experiment, conducted in a darkened room, the observer sat in a rotating chair, to which a small electric light was attached with a connecting rod. The chair was first set in motion slowly, with the light off, so that observers did not notice they were moving. Only when the light came on did observers realize they were moving; and even though the light was attached to the chair, they did not report it as moving parallel with themselves, but rather in various directions with no obvious order. More important still, the roles of sensory and motor activity changed radically in the two stages, with motor activity dominant in the first, and sensory in the second. This showed to von Weizsäcker's satisfaction that interaction of the visual and vestibular systems was involved, and that sensoricognitive and motor activity cannot be separated. Thus, perception is not understandable without considering the motion of the perceiver, and vice-versa.[40]

At the philosophical level, all this justified for Weizsäcker the claim that the construction of subject–object relationships characteristic of modern biological science ought to be reformulated along lines already followed by relativity theory in physics. In opposition to the rigid coordinate systems of Euclidean geometry and classical physics, Weizsäcker wrote, "biological integration always has only momentary validity. . . . It is thus not properly a system but an ordering of biological acts (*Leistungen*) in the present." Yet, because this ordering occurs over time, in a series of present moments, it "can only occur in contradictions," and hence is "antilogical." The conundrum is resolvable only if we consider the experienced world not as a preexisting, cognized object, but as the emergent product of multiple synergetic interactions of organism and environment.[41] Some aspects of this – in particular the rejection of the constancy hypothesis in favor of a dynamic approach to perception – had obvious parallels with Gestalt theory. Weizsäcker later acknowledged the Gestalt theorists' pioneering work and the influence of Kurt Goldstein on his own thinking, though he claimed he had been unaware of the Berlin school's work when he first developed his theory in the 1920s. But he rejected the Gestalt theorists' claim that variations of the threshold at which forms are seen are themselves the result of Gestalt factors: "Neither the Gestalt makes the threshold nor the threshold the Gestalt. Both cannot be thought of as in causal relation to each other." All that can be determined by experiment are the limiting conditions, not the causes, of formative acts. Gestalt theory had "not given up [deterministic causal] explanation as a research principle in biology decisively enough."[42]

Here was another scientific school claiming to employ the Gestalt concept to overcome both mechanism and vitalism, hence to occupy the same epistemic space as Gestalt theory, but possessing both the institutional authority of physiology and the ideological authority of idealistic philosophy. Metzger pointed out that Köhler had tried to work through a case of dynamic feedback effects, namely the eyes' accommodation to and convergence on moving objects, nearly twenty years earlier (see Chapter 15). He also noted other correspondences between von Weizsäcker's views and those of Gestalt theory, claiming, for example, that the principle of *Prägnanz* was equivalent to Weizsäcker's "tendency to improvise quick solutions according to the principle of geometrical and physical simplification." More fundamental was his argument that it was not possible to consider only person–environment interactions and ignore the psychophysical processes that mediate them. These were "antilogical" only if one accepts mechanical conduction as the sole principle of nervous action, but not if one looks for other kinds of processes in the central nervous system. In any case, von Weizsäcker's "Gestalt circles" are "only a special case of the free play of forces in Gestalt contexts." The term therefore ought to be confined to cases when subjects actively intervene to rearrange their environments.[43]

Metzger's was a full, in places eloquent statement of Gestalt principles and their conceptual foundations. But his book was problematic both as a summary of what Gestalt theory had achieved, and as a response to its critics. His answer to the widespread claim that Gestalt theory neglected the role of properly psychological factors such as attitudes was to assert that "the given" is the psychological. The implication was that such unexperienced entities as "Gestalt centers of gravity" and also attitudes are not causes of what we perceive but part of a larger, self-organizing "Gestalt context" which includes the given. This echoed Koffka's 1915 assertion that the entire organism–environment relationship is a Gestalt. His addition of the "cortical ego" as the presumed location of the experiencing subject only gave attitudes a place to be. Yet he also stated that the organism-environment nexus is a relational system, not a Gestalt. It was not possible otherwise to conceive attitudes and other "subjective factors" as states of an organism, and also account for the fact that such underlying states are not always themselves experienced, even as moods. Neither this nor the emphasis on "letting the things work" with the help of an "objective attitude" accounted for the experience of creating or forming situations as well as having them happen to or in front of a subject.[44] In this way, Metzger reached Gestalt theory's conceptual limits. For this he tried to compensate in part with the terminological concessions to Leipzig "holistic" psychology already mentioned.

Like that of Koffka from the same period, Metzger's book considerably expanded the conceptual range of Gestalt theory. Precisely that elaboration gave Gestalt theory a new, more "finished" look – the look of a system – during the 1930s, which it had not had before. Metzger's bows in the direction of Nazism were not defensive compensations for a worn-out theory, but active, albeit carefully limited, attempts to adapt his thinking to current ideology. The book as a whole was an attempt to make the Berlin Gestalt approach foundational to the emerging realignment of scientific thought and practice in German psychology while simultaneously developing that approach further. He did not do this at the expense of science, but continually pointed to new research issues. Even so, he had only limited success. Though he salvaged part of Gestalt theory's institutional base in Germany, he did not have a major impact on the field as a whole in this period. As Edwin Rausch later recalled, the Frankfurt institute had few students: "The time especially for perceptual psychology was no longer there. The Nazis had other goals."[45]

Contemporary reviews of Metzger's book *Psychology* support Rausch's recollection. The reviewer in the leading scientific organ of the "race hygiene" movement, for example, presented the book as an example of the baneful dominance of "idealistic philosophy" in current academic psychology. Particularly unfortunate was the "complete absence of contact with hereditary psychology" (*Erbpsychologie*). "One has the impression that here the effort to achieve the greatest possible conceptual clarity puts [Metzger] in danger of overlooking essential psychological topics (*Gemüt,* unconscious, and the like) or makes him incapable of penetrating the depths here." In sum, Metzger had missed a chance to fulfill "the demands of the time on science." What "the times" specifically demanded was a psychology that could help "race hygienists" in their selective work by describing "the construction of character" in clear, understandable language. Other psychologists, such as Oswald Kroh, were cooperating with race hygienists such as Ernst Rüdin in such efforts. From this perspective, atomistic and Ganzheit approaches appeared to be brothers under the skin; both were equally removed from "life," that is, the practical demands of the day.[46]

In the Nazi as in the Weimar period, Gestalt theory was criticized for ignoring or dealing insufficiently with key practical issues or ideological catchwords of the time. Despite his frequent references to the Germanic tradition, Metzger's diction remained universalistic and essentially non-Darwinian. Missing in this case were words such as *Gemüt, Tiefe,* and especially "race" or "blood," terms central to the discourse linking psychology, "race hygiene," and the selective functions demanded of social and biomedical science under Nazism. Seen in the context of psychology's development from being a coparticipant in the effort to create an encompassing worldview to an instrument of military and labor policy, or a branch of Nazi "applied biology," Metzger's work was clearly the product of an earlier time.

Metzger offered a rather weak answer to such criticisms in a 1941 essay entitled "Psychology and the Knowledge of People." Writing at a time when psychology was being applied to personnel selection as never before and just when negotiations on the courses to be prescribed for the professional diploma examination were in full swing, he acknowledged that psychophysics and other such book knowledge were "ballast" for pedagogues or other working technicians of mind. Though he criticized the simple-minded use of characterological catalogs in personnel assessment, he implicitly admitted that he had nothing to put in their place by saying that "a beautiful task" lay ahead.[47] This failure was not due to an essential limitation in Gestalt theory. Statements from Metzger's *Psychology,* such as the claim that judgments of character, like perceptual judgments, are relational systems, could have provided a foundation for a theory of personality assessment; but Metzger did not exploit that opportunity at the time.

Gestalt theory began as an original and open-ended theoretical and research program; in Metzger's and Koffka's books it became an increasingly closed system. In both cases, though in different ways, the work of system-building was based on the assertion of broad analogies from perception to other psychological realms. How such assertions could be applied in practice, or even converted into testable research propositions, was at best only unclearly suggested. This brought both men, but especially Metzger, into conflict with the fundamental claim that Gestalt theory's fate would and should be decided by its research success, not its philosophical or ideological acceptability. Of course, this, too, was a philosophical position, in favor of natural science, not philosophical argument, as the arbiter of truth. But research success was and remains dependent in part on the institutional base from which such a position could be sustained – which was and is dependent in turn on researchers' perceived central or peripheral position with respect to societal priorities. As the next chapter shows, that base was limited in postwar Germany as well.

22

The postwar years

Attempts to reconstruct cultural and scientific life in occupied Germany led at first to considerable confusion, exacerbated and often enough caused by political uncertainties. With the founding of the Federal Republic of Germany and the German Democratic Republic in 1949 came gradual reorientations to the cultural imperatives of the newly dominant superpowers. In the process, the compromises into which educated Germans had entered in the Nazi period were too often repressed in both East and West, albeit in differing ways. A complex, conflicted situation resulted, which scholars are only now beginning to explore in depth. Neither the long-held, conventional view of 1945 as a completely new beginning, nor its polemical negation, the attempt to portray the postwar period, particularly in West Germany, solely as a "restoration" of the *status quo ante,* captures the variety of sometimes contradictory programs and results that ensued.[1]

The situation of psychology in the postwar German states exemplifies this varied mixture of continuity and change. In the Western occupation zones and later in the Federal Republic, psychologists gradually reestablished their institutional bases in the universities during the first postwar decade. Only three Nazi professors permanently lost their posts. By 1954, twelve full professors were in place who had held that title before 1945, including Friedrich Sander in Bonn and Oswald Kroh at the newly founded Free University in West Berlin. Exceptions to this trend were Heinrich Düker in Marburg, who had been a political prisoner in the 1930s, and returned émigré Curt Bondy in Hamburg. At the same time, considerable change occurred in the journal landscape. Two reincarnations of the *Zeitschrift für Psychologie* appeared in 1954, one in the West and one in the East, both with the words "applied psychology" added to their titles. Some leading journals ceased publication, like the *Archiv für die gesamte Psychologie.* Others reappeared with new editors. First among the latter was *Psychologische Forschung,* which resumed publication in 1949 with Johannes von Allesch as editor in chief and Wolfgang Metzger on the advisory board, but the name of only one of its previous editors (Hans Gruhle) still on the masthead, and that of Wolfgang Köhler absent.[2]

Nonetheless, the continuity in professorial leadership ensured the persistance of old conflicts in West Germany, such as that between Berlin Gestalt theory and Leipzig holistic psychology. Mainz professor Albert Wellek, for example, a disciple of Felix Krueger and secretary of the German Society for Psychology, created a "genetic holistic psychology" (*genetische Ganzheitspsychologie*) in the 1950s, which linked Krueger's perceptual theory with personality by propounding a fundamental "polarity" of "prerational" and rational forces in the construction of character. This approach proved compatible with the dominant postwar personality concept, Munich professor Philipp Lersch's "layer theory" (discussed in Chapter 20). When the third edition of Lersch's standard text appeared in 1953, the political references it had once contained were carefully excised, and his approach more strongly emphasized the primacy of forces within the individual. In 1947, Lersch wrote in support of personal freedom and independence against the threat of "massification" (*Vermassung*), thus trying to uphold the traditional self-image of the educated elite during the transition to democracy in West Germany, while ignoring the rather different aims that had governed the use of his thinking in the Nazi era.[3]

Such looking away was widespread. Open discussion by psychologists of their behavior in the Nazi period occurred only in the form of complacent reactions to critiques published in Switzerland and the United States. Evidently, as Mark Walker has shown for physicists during the same period, it was important for leading psychologists, at least in the West, to reserve the term "Nazi science" for the apologetic aberrations of a few, in this case Erich Jaensch and his students, in order to dismiss them more easily, while ignoring the more disturbing uses of "normal science" by the many.[4] In keeping with a widespread trend of the period, the general tendency in psychology was decidedly inward looking. In 1949, only 11 percent of citations in experimental psychology were of non-German authors, and the works cited were 25 years old on average; in 1953 the percentage of non-German references was 31, with an average age of 13 years per citation.[5]

In applied psychology, however, trends soon took hold that ran counter to such efforts to maintain continuity. In 1953 and 1955, the first career positions for psychologists in West Germany were created in federal, then in state and local labor offices and unemployment insurance agencies. Educational counseling centers were another growth area for applied work; by 1960, psychologists had replaced physicians and social workers as the leading employees there. The re-establishment of military psychology in 1957, shortly after the Federal Republic's entry into NATO, provided a third source of employment.[6] All of these were sectors that had employed psychologists before 1945. But the emergence of a social market economy accompanied by a welfare state in West Germany led to a subtle shift in preferred forms of psychological diagnostics. The emphasis on trained intuition that had seemed so well suited to elite officer selection in the 1930s continued, but American-style aptitude, intelligence, and personality testing, backed by sophisticated multivariate statistics, appeared better suited to the needs of state employment, welfare, and education administrations.

In response partly to these trends, and partly to the general reorientation of much of West German industry, culture, and society toward the United States, American experimental and statistical techniques made increasing inroads into West German research and professional practice. Articles on personality theory or assessments of particular diagnostic tests made up more than 40 percent of the contributions to the *Zeitschrift für experimentelle und angewandte Psychologie,* the official organ of the (West) German Society for Psychology, in the 1950s. Papers by younger psychologists on recent American approaches to personality and social psychology appeared with growing frequency, but the senior professoriate had a vested interest in the older methods. This was the background to the *Methodenstreit* that raged through the 1950s.[7]

Hamburg professor Peter Hofstätter, who had learned factor analysis and other statistical techniques in the United States and had already tried to import them to Germany with Kurt Gottschaldt's help in the late 1930s, equated these with "empirical psychology" and sharply distinguished them from the ideal-typical procedures of Wellek, Lersch, and their followers, which he labeled "speculative psychology." In reply, Albert Wellek argued that quantitative methods had their place, but should not be applied everywhere for the sake of "pseudo-American" appearances. Though this might have sounded reasonable, he destroyed whatever credibility he may have had with younger psychologists by waxing nostalgic for the "methodological tolerance" and "disciplined freedom" born of unity in a common task which he claimed had characterized German psychology during the Second World War. By the 1960s, the battle had ended in more or less complete victory for the "Americanizers."[8]

In the Soviet zone, psychological institutes were reestablished at Leipzig University, at the Technical Academy in Dresden and, in 1946, at the University of Berlin, renamed the Humboldt University. In the German Democratic Republic, as in the Federal Republic of Germany, the continuation of the Diploma examination of 1941 assured a certain continuity with the recent past and a corresponding emphasis on practical training. In the early 1950s, however, courses in "Marxist-Leninist philosophy" and political economy began appearing in the introductory curriculum, reflecting the East German regime's effort to create a new, Socialist intelligentsia and technocracy. Some psychologists and Socialist Unity Party ideologists tried to go further and impose the Pavlovian orthodoxy that had replaced experimental psychology in the Soviet Union under Stalin, or to give Marxist-Leninist content to pedagogical and social psychology in new ways and thus create a GDR psychology. Others attempted to ally themselves with Soviet approaches more congenial to the German tradition than Pavlovism, such as that of

Sergei Rubinstein.[9] Still others adopted an attitude like that of their Soviet counterparts, who began their publications with simplistic references to Pavlov, but did not mention him thereafter, and expressed only contempt for official doctrine in private.[10] Nonetheless, supposedly "bourgeois" experimental and industrial psychology persisted in "the homeland of Socialism on German soil." A possible reason for this was the continued need for expertise to help sustain and increase productivity in the factories and deal with continuing social problems like juvenile delinquency. Some psychologists may have thought that concentrating on "hard" science would insulate them from the Socialist Unity Party ideologues whose influence was so evident in related fields like pedagogy. But there was nowhere to hide. No planned economy can exist without technical experts – specialists, as the Soviets called them. When experimenting psychologists helped to train such people, or worked with state-owned enterprises to improve productivity by making production processes or their human components more efficient, they accepted at least implicitly the ideological claim that science was a particularly modern and potent form of labor power (*Wissenschaft als Produktivkraft*).

In both German states, psychologists faced the consequences of professionalization, gradually moving away from the grand philosophical ambitions of earlier decades to the more modest aims of specialized research and the practical training of scientists, functionaries, and later therapists. One consequence for psychological theory was increasing pressure to displace the problems of perception and cognition from their formerly central positions. How could Gestalt theory be made to speak to the intellectual and practical needs of these changing, increasingly bifurcated cultures and societies? This question will be answered here by tracing the careers and ideas of Wolfgang Metzger, Edwin Rausch, and Kurt Gottschaldt into the 1960s, focusing on the issue of continuity and change. As will be clear, there was plenty of both, even though Metzger and Rausch worked in West Germany and Gottschaldt worked until 1962 in East Germany. Their common challenge was to establish niches in a new environment.

Wolfgang Metzger in Münster

Considering his earlier career pattern, Wolfgang Metzger's denazification proceeding was surprisingly brief. The work of the Münster institute had come to a stop before the war ended due to Allied bombing; for a time Metzger lived in a nearby village, working as an agricultural laborer and part time as an assistant chaplain. On August 18, 1945, came a visit from American psychologist Heinz Ansbacher, who was conducting a survey of German psychology on assignment with the occupation forces. As Ansbacher later reported, Metzger told him his specialty was "the psychology of perception as applied to camouflage problems." Ten days later, Metzger submitted a report on the subject. This could have endangered him by making him appear to have been an agent of the German military, but the result instead was that his denazification proceeding lasted only a few minutes. On Ansbacher's recommendation, he became one of the first eight professors reinstated in Münster; apparently there was no attempt to investigate either his political behavior or his writings during the Nazi era.[11] This is an example, though a minor one, of the way in which the interests of the occupying forces could decide who was politically acceptable, or what was useful science.

This episode set the stage for Metzger's most important achievement of the postwar era, which was institutional. Largely through his efforts the Münster institute grew from a single room, to a floor of rooms, to an entire building. In 1965, the institute expanded into a second building, becoming the largest in West Germany.[12] But precisely the institutional side of the story reveals a significant aspect of continuity with the prewar history of experimental psychology. Metzger had experienced the advantages of recognition as a natural scientist while working with Max Wertheimer in Frankfurt. In Münster he repeatedly invoked comparisons with biology and chemistry in his early budget requests. However, when a Natural Sciences faculty was created there in 1948, he kept his institute in the Philosophical Faculty, with the humanities. His biographers write that his decision was motivated by a genuine commitment to a definition of psychology's subject matter rooted in the philosophical tradition, which outweighed his allegiance to experimental method. Metzger's reflections on the matter support this interpretation, but they also show that he had reason to regret the move later. One of his proposals for new equipment disappeared into an administrator's file drawer, because the official "regarded its author as megalomaniacal." More wounding still, he was excluded from the inner circle of professors in the Philosophical Faculty, who regarded his field as a "foreign body."[13]

This institutional situation had much to do with Metzger's public defense of experimental psychology in the postwar period. In two essays published in 1952 in *Studium Generale,* a leading organ of academic opinion, he presented Gestalt theory as a humanistic worldview based on experimental science. In the first, he claimed that psychology's true beginnings as a science lay not in the introduction of experimental method, but in the revision of its fundamental assumptions initiated by Christian von Ehrenfels and Hans Cornelius. From the former came Gestalt theory; from the latter, Leipzig holistic psychology. Metzger had to acknowledge that Gestalt theory's research results thus far hardly sufficed to give a scientific account of what common sense already knows to be true about the psyche – that humans experience guilt and remorse, for example. But at least it was no longer necessary to adopt an image of the human that "excludes in advance a genuine and unified understanding of the spiritual [*das Seelische*] and mental [*das Geistige*] in humans and its relation to the structure and functioning of its organism." Thus, Metzger continued to take the philosophy-oriented high road, just at a time when academic psychology in West Germany was becoming more and more a set of problem-solving research and social technologies.[14]

In the second essay, Metzger made it clear that his case for psychology as a natural science with humanistic goals rested on a conception of scientific method that was unusual at the time. There he defined psychology as "the science of the experience [*Erleben*] and behavior [*Verhalten*] of living things and the *nature* [*Eigenart*] *of their world* . . ." The inclusion of both experience and behavior corresponds to the definition in current American psychology texts, but not to the one then accepted in the United States, which spoke of behavior only. Also unusual, then as now, was the claim that experimental psychology strove for "understanding" (*verstehen*) in order to achieve prediction and control. This made sense only because for Metzger "understanding" and causal explanation meant the same thing – comprehending the essential order of phenomena as "inwardly necessary" (*innerlich notwendig*). As he emphasized, such understanding "*can*, but *need not*, assume mathematical form."[15]

This complex middle position had implications for Metzger's exposition of psychology's research methods. He called phenomenological description "an art," just as Friedrich Schumann had done decades earlier; and he carefully distinguished between the "preexperimental," descriptive stage and the "planned intervention" of experimentation proper. But he maintained that there was no opposition between phenomenology and experiment; "each [is] as necessary as the other" for the foundation of a "general" psychology, on which the study of character and personality could build. He did not exclude behavioral observation, diagnostic testing, quantification, or statistics from the methodological canon in principle, but only refused to make such procedures coterminous with experimental method as such. Nonetheless, empirical verification remained for Metzger a fundamental criterion of strict science. That excluded "large parts of contemporary depth psychology and expression theory, as well as so-called humanistic psychology."[16]

In this way Metzger attempted to mediate between holistic or humanistic psychologies based on intuition and American-style behaviorism on the one hand, and between experimental cognitive psychology and personality-centered diagnostics or depth psychology on the other. This mediating strategy was limited somewhat by his own preference for qualitative over quantitative research. For him the ordinal scaling standard in psychophysics and sensory psychology was of little use, since the relational context of conscious experience shifted continually. Though he opposed Albert Wellek's claim that such complexities made strict lawfulness unattainable in psychology, he was still on the losing side in the *Methodenstreit* of the 1950s. Later he opposed the younger generation's devotion to statistics and "Anglo-Saxon dogma" in no uncertain terms.[17] Even so, his commitment to experimental method was not in doubt, and he worked steadily to develop links with American psychologists. In 1954, for example, he attended a Cornell University symposium that reestablished connections between American and German-speaking researchers in perception and helped lay the groundwork for the later resurgence of cognitive science.[18]

That involvement had an ambiguous outcome. Metzger's defense of the Gestalt position against a variety of challenges in the 1950s and 1960s succeeded mainly in illuminating differences in theoretical style, and in separating himself from predominant causal-functional discourse. One example was his attempt to reformulate Wolfgang Köhler's psychophysics in order to make it consistant with contemporary neurophysiology. After his emigration, Köhler had taken up physiological psychology, using EEG's and other methods in an attempt to verify his isomorphism postulate directly. Most interesting were studies done with Hans Wallach and others on so-called figural after-effects. James Gibson had found that a curved line appears less curved after a few minutes' inspection; when it is removed and a straight line shown in the same place, the new line seems curved in the opposite direction. To explain such displacements, Köhler suggested that when direct current flows through brain cells, these become resistant to the flow, which is then deflected elsewhere. He had hypothesized such "satiation" effects in *Die physischen Gestalten* to explain figure–ground phenomena (see Chapter 11). In this case, he suggested that satiation was greater on the concave side of inspected curves than at the ends. Köhler's demonstration of direct cortical currents in cats and humans appeared to support this interpretation. Karl Lashley, Roger Sperry, and others, however, showed, for example, that inserting insulating mica strips into the cortexes of monkeys to block current flow did not prevent them from perceiving forms. For physiological psychologists this ended the matter. Köhler's response and his suggestions for further experiments were generally ignored.[19]

In 1950, Metzger acquainted the West German public with Köhler's results, praising the discovery of direct currents in conjunction with psychical phenomena as a major event in the history of science. In 1961 he was more defensive, trying to make the Gestalt position appear more plausible to psychophysiologists by returning to the "crossing functions" Max Wertheimer had hypothesized as correlates of motion perception and equating them with the "lateral interaction" conceptualized by Hans-Lukas Teuber.[20] In 1964, at a symposium in East Berlin, Metzger adapted still further to current trends by recycling a conception for the psychophysics of intentional action that he had already presented in 1952 as a cybernetic model based on the operations of a servomechanism. As a helmsman acts not directly on a ship's rudder, but on a piece of equipment that operates the rudder, he suggested, so the cortical correlates of voluntary actions are directed not toward the organ in question, such as an arm, but toward its "Body-Ego" (*Körper-Ich*) counterpart in the brain, which takes on the function of the piece of rudder-operating equipment.[21]

This was similar in some respects to Karl Bühler's idea of the perceptual system as a central control apparatus, which he had already advanced in the 1920s (see Chapter 18). Other aspects, however, most notably a concept of memory in which cortical traces of previous experience "resonate" with correlates of current experiences, made room for a less strictly mechanical concept of brain action, and thus anticipated current attempts to develop a psychophysics based on the concept of self-organizing systems, like Karl Pribram's so-called holonomic theory.[22] These ideas connected Metzger with international discussion of the issues, but did not rehabilitate the Gestalt position. Rather, he conceded much to conventional views that presupposed that machine modeling — postulating a mechanism or apparatus — *was* causal explanation.

Revealing in a rather different way of both continuity and change was Metzger's attempt to apply the Gestalt viewpoint to pedagogy. This was due in part to the fact that he was professor of both psychology and pedagogy in Münster. He had already set up an educational counseling center in Münster during the war, and prospective teachers were part of his audience thereafter. It was also consistent with his earlier interests. In the 1920s, while still a student and assistant in Berlin, he had already published articles on the ideas of Alfred Adler.[23] His pedagogical writings had begun in 1941, with an essay on the trainability of "creative forces" published in a journal for industrial engineering. Some of the ideas in that essay reappeared in the first book he published after the war, entitled *Die Grundlagen der Erziehung zu schöpferischer Freiheit* ("The Foundations of Education for Creative Freedom") (1949).[24]

The book's central image was a contrast, attributed to Goethe, between rigid, dead "form" and living, self-organizing Gestalt. Metzger portrayed creative forces as living processes in a person, which needed only to be awakened or allowed to emerge, rather than as "material" to be worked to fit a fixed design, like wood or metal. Seen from this perspective, he argued, using force or adhering to rigid rules with children, students, or trainees will produce only the opposite of creativity. The goal of "fruitful learning" must therefore be "creating in pupils the conditions under which they can achieve insight themselves." The first edition of the book gave only a rough sketch of what this might mean. In the second edition (1962) he offered a fuller outline of the process of productive thinking taken from Wertheimer's posthumous book by that name, which he had translated into German in 1957.[25] Unfortunately, in both editions Metzger presented fewer specific pedagogical suggestions than broad admonishments to avoid mechanical constraints and rote memorizing. The few specific examples of successful pedagogy came largely from the Weimar period: Georg Britsch's theory of art education from the 1920s, Kiel professor Johannes Wittmann's "holistic" method of mathematics teaching, published in 1932, and a similar approach applied to the training of apprentices by Walter Poppelreuter and Wilhelm Arnold in the 1920s and 1930s. Common to all these programs were that instructors helped pupils get the feel of working with particular materials or problems, and then presented clear assignments, but avoided giving rigid rules or stereotyped formulae for carrying them out.[26]

These ideas had obvious relevance to postwar debates among pedagogues and other intellectuals over values. In a 1950 letter to Köhler, Metzger explicitly stated that he wanted to contribute to that discussion, but also admitted that this desire had overcome his judgment that the ideas "perhaps should have been allowed a few years to ripen." The messages he sent were confusing, to put it mildly. The first sentence of the first edition presented his effort as part of the renewal of Germany's "creative forces" in the face of defeat, but one early passage referred to releasing the potential of troops by paying attention to their capabilities instead of following rulebooks. That passage was cut from the second edition. In a sentence he retained, Metzger wrote that the conquest of a weak state or "the destruction [*Ausrottung*] of a defenseless, contemptuously viewed minority . . . cannot properly be called 'struggle'," as the murder of the Jews had been termed by the Nazis.[27]

Such contradictory metaphors of struggle coexisted uneasily with exhortations to quietism conveyed by numerous quotations from ancient Eastern sources, such as Chuang Tzu, the I Ching, or the Zen masters, to which Metzger had been introduced by Shiro Morinaga, a Japanese student who had worked with him in Frankfurt in the late 1930s. Earlier he had enlisted Gestalt theory in a general struggle against "the spirit of the West"; in 1949 he put totalitarianism squarely in the "Western" camp, in order to contrast it unfavorably with the wisdom of the East. Thus, he named "the police state and Taylorism [as] perhaps the most impressive" examples of the West's false belief in the necessity of rigid control, which underlay the machine theory of life and behavior, and added: "The basic belief of the West is false – it sees only one half of Being." He characterized the other half in a quotation from the *Tao te King:* "I act not, and the people becomes good by itself." This concept of leadership was clearly different from the sort of pecking-order dominance that he had idealized in 1941.[28]

Equally problematic was his portrayal of the ego as at best an accompanying stimulus to creative achievements, and at worst a hindrance. Borrowing a phrase from Adlerian theorist Fritz Künkel, he maintained that teachers and pupils should "put off egoistic glasses" and strive for "education to objectivity" (*Sachlichkeit*). This was consistent with the Gestalt tradition. But then he invoked keywords for the aim of education that had a rather different resonance: "self-sacrifice" (*Hingabe*) from Ludwig Klages, "true obedience" from Meister Eckhart, and "self-forgetting" from Chuang Tzu. He even spoke of the need for "an attack on the egotism of the infant," which he claimed was characteristic of Zen.[29] These phrases sounded strangely reminiscent of Nazi-era emphasis on the insignificance of the individual compared with that of "the whole." They were certainly not in tune either with democratic individualism or with the emphasis on "inwardness" (*Innerlichkeit*) then current in West German personality theory. Such confusions were typical of the disorientation in the immediate postwar years, but Metzger retained this language in the second edition. In both editions he also avoided taking any explicit stand on central educational policy debates, such as the retention or reform of the traditional three-tier German school system, or on the role of religion. It was not his purpose to specify the goals to which creative freedom should be directed, he wrote, because he wished to avoid awakening "exaggerated expectations."[30]

Metzger specified the political implications of his pedagogy more clearly in a pamphlet on political education published by the state government of Lower Saxony in 1965. "What political education is or should be," he began, "depends on the political system in which the individual is to take his place." In a democracy, then, the task is to develop attitudes and behaviors that are "completely penetrated by the spirit of the political and social order," affecting not only more narrowly political actions but also daily life "through and through" (*bis ins letzte*).[31] Thus, his aim was to decouple totalistic thinking from its authoritarian connotations and connect it with democratic ideals.

To achieve that result, Metzger advocated "a fundamental change in the public attitude to children" in Germany, to give them less a sense of being ground down under the weight of parental or school authority and more of the feeling of "basic trust" that will make it possible to become independent thinkers and actors. The term "basic trust" came from psychoanalytic ego psychology. But instead of deriving such feelings from the mother–child relationship, he claimed, following Max Wertheimer, that children possess an "original urge to find out things" (*Forschungsdrang*) which produces "as much satisfaction as sucking or being stroked." If this is so, then neither child rearing nor education should restrict children's freedom. Rather, its task is "to stand by the child until it arrives where it basically wants to go."[32]

Metzger tried to reconcile such permissiveness with the institutional requirements of public education by arguing that children's primal urge to discovery was equivalent to a desire "to fulfill the requirements of the objective situation," which was in turn equivalent to Adler's "community feeling" (*Gemeinschaftsgefühl*). Thus, the apparent contradiction between order and freedom disappears, but only in a specific social situation. "The joy of the healthy child in obedience" rests on its "trust that the educator [parent or teacher] knows what must be and what should not be, and that that which he orders is objectively good and necessary." Prohibitions should therefore be restricted only to a skeletal set of house rules and a minimum of safety procedures to prevent life-threatening situations. In support of this he cited a famous study by Kurt Lewin, Ronald Lippit, and Robert White contrasting the dull atmosphere of an "authoritarian" classroom with the "joyful cooperation" induced by "democratic" leadership.[33]

Metzger recognized that instituting such attitudes and policies in Germany would require major cultural changes. "It is much more comfortable to govern a collection of yea-sayers than a multiplicity of people with minds of their own. But to give in to this temptation and defame the uncomfortable thinker . . . is *the sin against the spirit of democracy.*" To reinforce democratic values and make students aware of the difficulties of pluralism, Metzger made recommendations that must have been uncomfortable in the 1960s. German history, for example, should be studied "up to the present" and "not only from native sources." Those who would forbid students and teachers to travel to East bloc countries show only that their own belief in democracy is "*extraordinarily weak.*" Students should learn, further, to consult a variety of press opinions, and to develop a particular distrust of the local press as well as an eye for "managed news." Above all, they should learn that "a constitution, too, is a human creation in which no perfection can be presupposed. A critique of the constitution therefore does not necessarily signify an attack on it." Rather, criticism can result in fuller realization of its intended aims. The state should encourage, not discourage criticism; leadership can awaken "love for this state" and acquire genuine authority only by institutionalizing tolerance in its own behavior, not just by proclaiming it theoretically.[34]

Metzger's sincere endorsement of democratic values was evident in these words. His position contained a far lower dose of stiff traditionalism and bureaucratism than was normal in the Adenauer–Erhard era. But the pedagogy he claimed would support such values rested on a milieu theory, in which external circumstances shaped educational situations at home and in school. Instead of facing the contradiction between the Gestalt theorists' belief in the inherent "goodness" of self-organization and the insufficiency of that belief as a guide to understanding individual or social behavior in the Nazi period, including his own, Metzger accepted the democratic context then in place and tried to strengthen and reform it.

For his multifaceted publications and activities, Metzger earned considerable recognition. In 1960, when Friedrich Sander was forced to step aside as chair of the first International Congress for Psychology held on West German soil, in Bonn, when his Nazi writings were exposed in an East German journal, Metzger took his place. In 1962, when the German Society for Psychology met in Münster, he became its chair, serving until 1964. An honorary degree from the University of Padua in 1965 was the fruit of friendly ties with the intellectual descendents of Vittorio Benussi, such as Fabio Metelli, whose work he had cited since the 1930s. In addition, he maintained contacts with Gestalt-oriented psychologists in Japan and received medals from the universities of Louvain, Trieste, and Prague.[35]

Yet, in spite of his international reputation and the breadth of his interests, Metzger remained in Münster for the rest of his career. In an autobiographical interview, he attributed this to the fact that "after the Second World War psychology, intentionally or accidentally, came fully into the hands of a school not well disposed to me, the achievements of which I am certainly willing to recognize, but whose representatives did not have a high opinion of such a declared supporter of the 'Berlin school'." A further difficulty was the "dogma," which he himself had espoused, that psychology is a humanistic discipline. This led to an official policy recommendation limiting experimental research – and hence advancement chances for experimentalists – to only a few institutes. To remedy the latter problem and to raise the level of basic research, he supported the creation of a Max Planck Institute for experimental psychology in 1960. This was eventually established, but only after he had retired.[36]

Metzger's legacy as institute head was no less ambiguous than the outcome of his career. The Münster institute produced thirty-eight doctoral students between 1948 and 1976, slightly more than the Berlin institute had produced in half the time during its heyday. Some of them rose to prominence later, such as Heinz Heckhausen, who became professor in Bochum and later chair of the German Society for Psychology. Equally impressive, however, was the small number of students who later concerned themselves with advancing Gestalt theory. His biographers write that Metzger "never sought to impose a fixed school orientation" on his students or institute colleagues, due to his "intellectual liberality and his consistent adherence to the Gestalt-theoretical principle of free self-determination of action, limited not by force or barriers but only by the objective demands of the situation." But another former student writes that, although Metzger himself was a popular undergraduate teacher and an understanding and supportive director of research, the general atmosphere in the Münster institute was one of self-satisfied maintenance of traditional views; the progressive, questioning spirit characteristic of the 1920s lived on in nostalgic reminiscence only.[37] Whatever the reasons, the continuity in Gestalt theory's institutional development was broken.

The story of Metzger's postwar interaction with Wolfgang Köhler is a poignant parable of continuity and its limits. Metzger's attempt to reestablish contact with his teacher at the International Congress of Psychology in Edinburgh in 1948 met with an understandably cool response at first, according to a former Berlin contemporary. But their correspondence resumed the next year, and it grew warmer after Metzger wrote to West Berlin in support of Köhler's appointment to succeed Oswald Kroh at the Free University in 1956, then translated Max Wertheimer's *Productive Thinking* into German in 1957. When Metzger asked Köhler for permission to dedicate the second edition of his own book on creative thinking to him in 1961, in honor of his approaching seventy-fifth birthday, he agreed. To accompany the presentation copy Metzger wrote an emotional letter reminding Köhler of a conversation they had had in Berlin, "in which you asked me if I wanted to become a psychologist," and which "determined my entire subsequent life." He expressed the hope "that you are as little disappointed by the results of that conversation as I." Köhler warmly praised the book: "I admire the boldness with which you have freed yourself from the narrowness of customary psychology." Nonetheless, though he traveled to West Germany often, Köhler did not visit his former student in Münster until he came to receive an honorary doctorate in April 1967, only a few months before he died. Before going to Metzger's institute, the old master went to the laboratory awarded as restitution to a returned émigré psychologist, Richard Goldschmidt, whom he hardly knew. With this gesture he clearly indicated both his primary loyalties and the limits of any possible *rapprochement.*[38]

Edwin Rausch in Frankfurt

After earning the doctorate in Frankfurt in 1937 with his dissertation on summative and nonsummative concepts (see Chapter 21), Edwin Rausch became an assistant there.[39] During Metzger's absence in Halle, he temporarily led the institute, maintaining operations with informal working groups. Drafted on the outbreak of war in 1939, he served in the air force psychological service. After the dissolution of army and air force psychology in 1942, he was transferred to the meteorological service because of his training in physics. At first he was able to return periodically to the Frankfurt institute, but later he was posted successively to the Western, Russian, and Polish fronts. He ended up in Austria, where he was captured by American troops. Upon returning from a prisoner-of-war camp in the fall of 1945, he took over his old assistant's position, which he retained until he became associate professor in 1948. In 1954, his position was converted to a full professorship, which he held until his retirement in the 1970s. Although he did not succeed directly to Wertheimer's chair, his career, like Metzger's, assured Gestalt theory's continuous presence in postwar German psychology.

In contrast to Metzger's broad range and willingness to address nonacademic audiences, Rausch devoted nearly all of his publications to extremely exact phenomenological illumination and conceptual clarification of issues from Gestalt theory. The central concern of his theoretical work was to give the treatment of "suprasummative" concepts outlined in his dissertation a psychological basis. An interesting example was his analysis of the categories variability and constancy, published in 1949. Variables and constants are familiar from mathematics and the theory of experimental design; Rausch contended that variability and constancy are not products of abstract thought, but "are already given in simple perceptual processes." Specifically, he combined the vocabularies of the Berlin and Leipzig schools by arguing that such perceptions are "accented" in specific ways, and are thus "a rank-ordering and dominance phenomenon." A cabinet, for example, can be seen as high or low, wide or narrow with respect to the phenomenal dimensions of a room – that is, with the room held constant and the cabinet treated as variable. A sign of such ordering is that observers' statements generally include the prefix "too." Alternatively, the cabinet can be seen as having such qualities "for itself." In such cases, Rausch argued, predicates like high, low, wide, or narrow also imply a two-member relation, this time between two dimensions of one object, with the second member of the pair often left unmentioned. When acquaintences see an adolescent boy with his mother, for example, their statements – about how much he has grown, or how tall he is – often refer to him, though the state of affairs includes both of them. Because the mother is fully grown, she is treated as constant; the son is still growing, hence variable. Here, as elsewhere, psychological relations are not commutative, the way mathematical ones are.[40]

The culmination of this descriptive-analytical effort came in a major essay on the problem of qualities or properties (*Eigenschaften*) in perception, published in 1966. In his 1890 essay on Gestalt qualities (discussed in Chapter 6), Ehrenfels had referred the term "Gestalt" to a particular quality or attribute of certain wholes or complexes, such as melodies. Rausch, following Wertheimer, used the word to describe not a quality but a special kind of whole which has specifiable characteristics precisely *as* a whole. The task he set himself was to develop an exhaustive taxonomy of such whole qualities. He recognized, for example, that ego and environment, like figure and ground, can be conceived either as a relation of one to the other, or as a Gestalt property attributed to one or the other. Applying such considerations to the sphere of "objective" appearances, he distinguished properties of order and construction (symmetrical, closed), textural properties (soft, rough, transparent), or expression properties (proud, peaceful, threatening). From this point of view, whether a given complex is a Gestalt or not is not a yes-or-no decision, but a matter of gradations on a continuum. A sequence of tones, for example, is a Gestalt to the extent to which it can be heard clearly as a melody.[41]

Thus, for Rausch a Gestalt or whole quality does not necessarily disappear when we isolate its parts, as it did for Ehrenfels. In fact, isolated parts have properties of their own that may depend on the features of their new environment and/or on what they retain from their old context. They may be independent and self-contained ("natural" parts), incomplete or "unsaturated," in need of supplementation, "lost" or homeless (*verloren*), or alien (*fremd*). Tones previously in a melody, for example, exchange the part properties they had had for the property, say, of being given as figure, alone, against a background of silence. In other words, true isolation in the phenomenal sphere does not exist. Rausch did not see any of this in a static way, but as stages in ongoing processes. Isolated parts may become unified wholes in their own right, or acquire new part properties as they become integrated into other wholes.[42]

Such considerations led to systematic distinctions among varieties of order given in experience, which Rausch called "stages of *Prägnanz*" (*Prägnanzstufen*). In the 1920s, Wertheimer had already noted that the perception of objects or groupings is subject to threshold phenomena similar to those already known to sensory psychology. As the point in a series at which tones of gradually increasing loudness are heard as louder may vary with specific stimulus conditions, so also may the distance between two dots at which they are no longer seen as a pair. He created the notion of "Prägnanz steps" or stages to apply this observation to perceptual structures in general. Rausch elaborated the idea to a degree that Wertheimer could not have predicted, distinguishing no fewer than seven dimensions of order, each of which he understood as a plus–minus continuum.

The first was that of regularity or "lawfulness," which could range from random collection to meaningful belongingness. The second dimension was autonomy (*Eigenständigkeit*), the degree to which an object exemplifies a prototype from which other objects are derived or to which they may be compared, in the way the rectangle is derived from the square. The third dimension was that of integrity – the degree of completeness or intactness of objects, as opposed to their appearing disturbed or incomplete, with something missing or lost. The fourth and fifth dimensions referred, respectively, to an object's simplicity, meaning its harmoniously articulated and organized appearance, and its diversity – its appearing rich, fruitful, that is, "pregnant" in the English sense of the word. Finally, he distinguished the dimensions of meaningfulness (*Bedeutungsfülle*) and expressivity, based on subjects' knowledge of an object's connections to other objects or reference systems. These were explicitly defined as material rather than purely formal dimensions; they were therefore historical and subject to evolution. This multidimensional scale contrasted considerably in richness and sophistication with Christian von Ehrenfels's earlier attempt to establish the "level" of a Gestalt – its degree of "purity" and richness, or complexity – as a criterion for aesthetic value.[43]

All of these studies, particularly the last, were important contributions to the systematic clarification of Gestalt concepts, and to the reconstitution of Gestalt theory itself as a descriptive science. Rausch's most important empirical work dealt with the phenomenological structure and measurement of so-called geometric-optical illusions, undertaken at Metzger's suggestion in 1938 and 1939. Parts of the study were accepted as a habilitation thesis in 1941; these were published together with additional theoretical material in 1952. The important feature of so-called geometric-optical illusions is that seen figures or features differ from objective geometrical relations. The attempt to measure such differences was already decades old (see Chapter 6). As could be expected, Rausch proceeded from whole figures rather than from their parts, taking as his hypothesis a postulated "tendency to orthagonal orientation," closely related to Wertheimer's *Prägnanz* tendency.[44] He approached the problem of measurement by developing a distinction between "ontogram" and "phenogram" originated by Karl Gerhards. This was based on the idea that the phenogram is a phenomenal "representative" of the seen figure, which can be measured either by comparison with another figure or by reproduction, that is, by drawing series of approximate "figural equivalents." He showed what this meant in experiments with the Sander parallelogram (see Figure 14, Chapter 18), taking phenogram data for all four sides as well as height and width and the length of the diagnonal. He found that observers emphasized or extended the diagonal in some cases, the vertical in others. This confirmed the theory that the illusion resulted from a combination of vertical "raising" (*Aufrichtung*) and horizontal or diagonal "stretching" (*Streckung*) of the figure. Against explanations that associated the variants of the illusion with different personality types, he noted that one subject showed an exaggerated tendency to illusions in both directions.[45]

Rausch's treatment of the "phenogramm" as a phenomenal "representative" made his work on optical illusions broadly similar to present-day cognitive science, for which the nature of such representations remains a fundamental issue. His emphasis on the problem of measurement makes it hard to accuse him of being "merely qualitative," but his approach was obviously not the same as American-style quantification. He did not employ significance tests, consider individual differences in observers' reports, or speculate about the physiological processes underlying perception or cognition. His emphasis on phenomenology and his avoidance of even appearing to test Köhler's psychophysics were characteristic of the trend in Gestalt-theoretical research that had already begun during the 1930s.

On the institutional side, Rausch managed, after a slow start, to reconstruct the Frankfurt institute's operation and to attract a number of students. The initial budget was 1,400 Marks, as it had been in the late 1930s; this was augmented only by a small "American contribution" from an unspecified source in 1949. Many of the studies in the institute's monograph series addressed issues raised by classical Berlin school or Lewinian research. Examples included work on the role of practice in the comparison of groups of points at rest and in motion, on systematic errors in repeated estimations of time intervals, and on the retention of observers' own achievements. The proportion of phenomenology and experiment ranged from an analysis of goal-directed action in sports, which used experiments but rested mainly on extensive interviews with athletes, to work on the relationship of understanding the instructions and test performance, in which phenomenological and experimental analysis were roughly equal, to a dissertation on "activation processes" in learning experiments, which was constructed rather like American-style learning studies of the period. Interestingly, the author of the latter study, Wolfgang Schönpflug, was one of three Rausch students who became professors later; he teaches today at the Free University of Berlin. The other two were Josepha Zoltobrocki in Frankfurt and Manfred Sader in Münster. However, only a few Rausch students pursued the Gestalt tradition in a serious way in their later careers.[46]

Rausch's career was one of calm, patient, and rigorous research, with no branching out into other fields of psychology and no venturing into academic, disciplinary, or general politics. In addition to their personal styles, the careers of Metzger and Rausch also reflected broader changes in the social structure of higher education in West Germany. Metzger hearkened back to an earlier era, when a single institute director was expected, and expected himself, to survey an entire discipline. The days of such generalists were already numbered by 1960. By comparison, Rausch appeared more modern in his complete dedication to a carefully limited program of basic research, though he acknowledged that he could not ignore personality diagnostics in his teaching.[47] But the style of that research looked back in its own way to the past, and he made clear enough that he was concerned more with aesthetic judgments than with the practical assessments of professional practioners.[48] In the context of the 1950s and 1960s, the result was the appearance of a certain narrowness. In Frankfurt, this led not to a pluralism of viewpoints in one department, but to the creation of two additional psychological institutes, one more oriented to developmental and social psychology, the other one of the few university institutes for psychoanalysis in the world.

Kurt Gottschaldt in East Berlin

Kurt Gottschaldt's tale of career reconstruction in the turbulent near-chaos of occupied Berlin is by far the most dramatic of these three stories. He accompanied Otmar von Verschuer when the inventory of the Kaiser Wilhelm Institute for Anthropology was transferred to Verschuer's estate in February 1945, but returned to Berlin in July. In August he obtained permission from Verschuer to remain there permanently, to set up a psychological research station with the support of the health office in the southwest Berlin district of Steglitz, in the American zone. However, the new president of the Kaiser Wilhelm Society, Max Planck, refused to allow this facility to use the Society's name.[49] Several months later, controversy erupted between Gottschaldt and Verschuer over control of the data from Gottschaldt's twin research. In May 1946, as Verschuer was trying to reestablish the institute in Frankfurt and return to his old professorship there, newspaper articles appeared calling him "a National Socialist race fanatic" and bringing to light for the first time his connection with the inhuman experiments of Joseph Mengele at Auschwitz. The information on which these articles were based came from chemist Robert Havemann, who had had himself designated head of the Kaiser Wilhelm Society in Berlin by an official of the new, Communist-dominated city administration. The identity of Havemann's informant remains unclear, but circumstantial evidence suggests that it must have been Gottschaldt.[50] Verschuer at least thought so; he complained bitterly to Kaiser Wilhelm Society Executive Director Otto Hahn that "my efforts to rebuild my [sic!] institute in Frankfurt have been sensibly disturbed by the intrigues and denunciations of Herr Gottschaldt." Ultimately, under pressure from Hahn and Planck, he allowed the twin material to be taken to Berlin.[51]

While the conflict with Verschuer was still in progress, in 1946, Gottschaldt became professor of psychology at the newly refounded University of Berlin, now named after Alexander von Humboldt. The university lay in the Soviet occupation zone, but Gottschaldt later surmised, plausibly, that the fact that he was there and qualified was probably more important than any political connection.[52] Though he remembers that he encountered opposition to basic research from Soviet occupation officials, he was authorized to acquire quarters for an institute in 1946. These were – and still are – at Oranienburger Strasse 18, not far from the ruins of Berlin's now-rebuilt main synagogue. This building, badly damaged and surrounded by rubble, eventually became one of the largest psychological research facilities in Europe. In keeping with the reorganization of East German science on Soviet lines, which emphasized hierarchy and centralized management, Gottschaldt quickly secured a preeminent place in East German psychology. He became a full member of the reconstituted Berlin Academy of Sciences in 1953, director of a research institute for experimental and applied psychology in the Academy and also head of the Advisory Council for Psychology attached to the East German Ministry of Education. To this he added the editorship of the journal, *Zeitschrift für Psychologie,* when it reappeared in 1954.

The old Berlin institute, like the rest of the former imperial palace in which it had been housed, had been largely destroyed by bombs and artillery. By 1954, the new institute had outgrown its predecessor, with forty-one rooms on three floors, and further expansion underway. At that time there were departments for general, developmental, labor, and animal psychology, as well as "constitutional biology." The staff, too, was considerably larger than it had been in the 1920s, including one chief assistant, ten regular assistants, and ten doctoral candidates who did staff work in return for a state stipend – an innovation introduced in 1951 and based on the Soviet model of centralized planning in science as well as in industry. In contrast to Metzger's institute in Münster, Gottschaldt's professorship and the Berlin institute were located in the Mathematics-Natural Sciences Faculty, as they had been since 1938. In 1960, Metzger wrote to a colleague that Gottschaldt's was the only laboratory on German soil that could stand comparison with the best research facilities in the United States.[53]

Most notable, however, in comparison with the past was that basic research was described as the institute's "second main task" after professional training. Students were required to observe and work with at least one of the children in the institute's own childcare center, and also to participate in other internships, called *Praktika.* Among these was assisting in psychological diagnoses at the institute's polyclinic, which carried out hundreds of clinical assessments for courts and welfare offices. The emphasis on practical training reflected both psychology's continuing development as a profession in Germany and Gottschaldt's public priorities, which corresponded with those of the East German state. The unity of theory and practice was central to Marxist doctrine. In his forward to the first issue of the new *Zeitschrift für Psychologie,* Gottschaldt assured his readers that the journal would fulfill its traditional function of reporting basic research results. But he added that it would also take on the role of the former "Journal of Applied Psychology and Character Studies" (*Zeitschrift für angewandte Psychologie und Charakterkunde*); for today, "theory and practice are hardly to be separated in psychology."[54]

Consistent with this new emphasis, Gottschaldt's research in the postwar period concentrated still more than it had before on problems of personality development and social psychology, issues as relevant to the needs of a new Socialist state as they were to administrative agencies in West Germany. Prominent initially was his research on juvenile delinquency, a topic the Gestalt theorists had not treated before, but one of considerable urgency in the chaotic postwar period. Gottschaldt had already confronted the issue in the 1930s in his second job with the "Polyclinic for Nervous and Difficult Children" at the Berlin Children's Hospital. The new study was an intensive workup of 560 cases examined there in 1945 and 1946. Though he insisted on an individual approach to assessment, he developed what he called a "phenomenology" of delinquency's characteristic features: lack of emotion and hence emotional connection to others, related to a negative view of life and oneself; passivity and low energy level; low social order in family, school, or job; disturbed instinctual behavior, from bed-wetting to loose and careless sexual relations; low achievement levels despite normal or higher intelligence. This picture had changed little since before the war; the problem had only grown much greater and spread to all classes. Thus it was "obvious that all of these phenomena of delinquency are connected with the general collapse of social order" in the immediate postwar period.[55]

Consistent with the Berlin school's emphasis on the objective structure of a given situation – and Wertheimer's much earlier analysis of pathological phenomena in individuals (discussed in Chapter 15) – was Gottschaldt's argument that societal circumstances reduced individual youths' criminal responsibility. "Thus – to reconfirm an old result – we are often dealing not with 'delinquent children'," he wrote, "but with 'delinquent circumstances' in the widest sense of the word." Notable in this context was his sharpened position on the heredity–environment issue, one that contrasted with his stance under Nazism: "It cannot be overlooked that in recent decades precisely in the field of applications of human heredity research much misuse has occurred. Even today we encounter chatter [*Geschwätz*] about a general 'hereditary burden' in many welfare agency documents, which easily leads to an inhibiting fatalism."[56]

A related dimension of change was evident in a major paper on twin research published in the inaugural issue of the new *Zeitschrift für Psychologie*. This was a follow-up on seventy of the ninety twin pairs studied on Norderney. Interestingly enough, because some twins lived in the East and some in the West, the research was supported by funding from both the East German State Secretary for Higher Education and the West German Notgemeinschaft in Bad Godesberg, the predecessor of the present Deutsche Forschungsgemeinschaft (DFG). Gottschaldt argued here that the current task of personality theory was to integrate three dimensions: the "concrete situational field" of each person, hereditary "constitution," and a new level – socioeconomic determinants of personality.[57] His report of the follow-up study began by modifying a claim he had already made in 1939.

Whereas he had argued before that identical twins have "concordant" brain stem functions and hence similar basic temperaments, degrees of energy, and "affectivity," while higher cortical functions are more likely to be influenced by environmental circumstances, he now said that higher functions are *"essentially dependent on (the) life-historical, educational, and upbringing situation."* Hence, the capacity for intellectual activity and the ability to get a broad overview of situations varies considerably with living conditions, even in identical twins. New was the finding that differences in "social superstructure" (*soziale Überbau*), including occupational and free time interests, began to appear in puberty. Yet there were some astounding cases in which identical twins living apart chose the same or similar professions. A recent study by a research group in Minnesota presents similar results, apparently without knowledge of Gottschaldt's previous work.[58]

Gottschaldt explored the impact of "social superstructure" in puberty further in the most innovative research he did in the 1950s, on what he called "Persona" phenomena. Though he took the term "Persona" from C. G. Jung, his work explored not issues in depth psychology, but rather the emergence of the separation between what today would be called objective and subjective personality attributions. Specifically, he had his assistants take series of photographs of children from his twin studies (later of others as well), then systematically distort them to present wider and narrower faces. He then had observers match them with their own image in a mirror. Six- to eight-year-olds took it all as an amusing game, enjoyed looking at the distorted photos and had no trouble choosing a "realistic" image of themselves. They became quite anxious, however, when shown distorted photos of near relatives, especially their mothers. Eleven- to twelve-year-olds, by contrast, had already begun to develop a "self-image" different from the "objective" one. They consistently chose photos that made them appear thinner than the mirror said they were – a finding that may be of interest even now for research on psychological dimensions of eating disorders. Gottschaldt claimed that this change in facial "Gestalt" impressions was related to identity formation in puberty, when the subjective importance of social status and bodily beauty increase. He claimed, further, that such images were indicative of societally defined "roles" shaped by propaganda and other media – perhaps a rather strange point to make in a country with a controlled press. This, along with his citation of West German psychologist Peter Hofstätter on these points, suggests a desire to be read in the West as well as the East.[59]

Gottschaldt focused on societal dimensions still more specifically in work on what he called the "we-group," a term taken directly from Wertheimer and Lewin. By this he meant something different from the term "peer-group," already becoming standard terminology in American social psychology at the time. A we-group is not a collection of individuals in similar living or work situations, but a relatively tightly bounded and bonded subunit within such groups, for example a clique, often but not necessarily defined by age. He observed three sets of such groups: 10 groups of about 20 children each in the twin camps of 1936–1937; a single group of 20–22 delinquent children working for one year on reconstructing the Berlin institute building in 1950, and various groups in the institute's child care center. Through systematic extensive observation (*Dauerbeobachtung*) of the kind already employed in his twin research, he differentiated ten different levels of we-group structure and behavioral style. If applied to the workplace, he suggested, knowledge of such groups might help improve productivity.[60]

Gottschaldt had extended the Gestalt concept or analogies from it in clever ways to personality and social psychology, though without citing the Gestalt theorists directly except for occasional references to Kurt Lewin.[61] When speaking of development, however, he translated then-standard personality theories like that of Philipp Lersch into developmental terms. In this manner Gottschaldt built upon previous work and kept current with theoretical discourse in the West, while at the same time adapting to his new location by adding the concept of socially formed "superstructure." As he had done during the Nazi period, he created a sophisticated line of argument that assured him at least some degree of autonomy. It also assured him linkages with West German colleagues and funding through membership in the governing board of the German Society for Psychology and the editorial committee of the authoritative *Handbuch für Psychologie*.

Gottschaldt secured his prominent position and his institute's generous funding without joining the Socialist Unity Party. As in the Nazi period, his publications contained few citations to official doctrine; an isolated exception was his citation of Soviet pedagogical theorist Makarenko's "Pedagogical Poem" on the potential of collectives for resolving the delinquency problem.[62] His students expressed such loyalties more directly. For example, in a lengthy exposition of the ideas of Pavlow and Bykow in the *Zeitschrift für Psychologie* in 1955, Friedhart Klix attempted to reconcile Pavlov's views, Gestalt theory, and Gottschaldt's standpoint by stressing a functional-biological orientation over self-organization.[63] In an attempt to apply Gottschaldt's we-group and Persona concepts to the study of morale and productivity problems in industry, Hans-Dieter Schmidt claimed that, since capitalist exploitation has been "eliminated" by socialism, Western psychological concepts and techniques could be applied in socialized industry with little or no change. They could be "tools or instruments of the [new] ruling class" precisely because they are "socially indifferent."[64]

Such assurances were not universally accepted. In 1957, Schmidt responded to an obviously ideological polemic, in which returned émigré clinical psychologist and Socialist Unity Party member Alfred Katzenstein denounced Weimar-era psychology in general and Gestalt theory in particular as "idealistic," and claimed that the Gestalt theorists had accepted "attractive offers" and gone to America after the advent of Nazism, rather than joining the resistance. One could criticize Gestalt theory, Schmidt replied, but not with such "slanderous" attacks. Admirable as such loyalty was, the controversy was a clear indication that all was not well with Gottschaldt's position in Berlin.[65]

This incident was soon followed by a lecture series organized by pedagogical psychologists at the Humboldt University, the aim of which, as Hans Hiebsch put it, was "to overcome the last inherited bits of bourgeois psychology" and place the discipline in the German Democratic Republic (GDR) fully at the ideological and practical service of the socialist state. Again Gestalt theory was singled out for attack; Ulrich Ihlefeld spoke of "the reactionary core of this seemingly materialistic conception."[66] In February, 1958, the Socialist Unity Party's third conference on higher education called for redirecting research and training to the needs of socialism, expanding knowledge of Soviet psychology, and developing Marxist-Leninist cadres and institutes. To achieve all this, it called for a change in the composition of the Education Ministry's Advisory Council on psychology, which Gottschaldt still headed. According to a later East German account, his joining the governing board of the German Society for Psychology, rather than organizing a separate East German society, was seen as self-promoting collaboration with the "bourgeois intelligentsia" and as willing acquiescence in the Adenauer government's Hallstein Doctrine, which mandated the isolation of the GDR from the non-Communist world.[67]

The same account reports that Gottschaldt's reaction to the party's new program for psychology was to demand that party cadres in his institute transfer elsewhere. Negotiations with the president of the Academy of Sciences proved fruitless. Party members in the Berlin institute transferred at first to the pedagogical faculty; some later went to a new institute in Jena, which had been created for Gottschaldt's one-time protégé, Friedhart Klix. This reduced the staff of the Berlin institute so much that it was no longer capable of fulfilling its teaching duties. Gottschaldt resigned his positions in September 1961, one month after the start of construction on the Berlin wall. In early 1962 he arrived in Göttingen as a full professor. After a brief interval the Academy of Sciences laboratory for experimental and applied psychology which he had headed was attached to the Berlin institute, and Klix took over as institute director.[68] Perhaps all this could be seen as an outcome of the Cold War, were it not for the fact that Klix, a highly competent researcher in the field of cognition who was also a Socialist Unity Party member, continued to maintain links with colleagues in the West, as well as in the Soviet Union, and with the Gestalt tradition. For his part, Gottschaldt remained dedicated to his lifelong theme, the "phenogenetics of human personality." He did a further follow-up study of his twins in the 1960s – this time with DFG funding only, and was still at work on a volume with that title summarizing his twin research when he died in 1991.[69]

Metzger, Rausch, and Gottschaldt attempted either to orient at least some significant concepts and methods from Gestalt theory and research to the intellectual and practical issues posed by new political and social orders in postwar Germany, or to maintain the continuity of the Gestalt tradition while the world changed around them. Their success, as measured by their standing in postwar German psychology, was surely no worse than that of Köhler, Koffka, and Wertheimer in their time. Though their international reputations obviously did not compare with those of the founders, they were surely at least as well known abroad as their German contemporaries. Nonetheless, Gestalt theory was in a defensive position in the postwar period analogous in some ways to the one it had been in during the Nazi period, albeit for different reasons in the two German states.

The remaining representatives of the Gestalt tradition were left behind by the increasingly functional orientation of scientific psychology, which made it biologistic in theory and instrumental in practice. The clearest indicator of this in the West came in a chapter by Norbert Bischof on the epistemological foundations of perceptual theory in the volume of the *Handbuch für Psychologie* that Metzger edited, published in 1966. Strongly influenced by ethologist Konrad Lorenz, Bischof took a frankly evolutionary stand and argued that if the Gestalt position were correct, then organisms oriented themselves to their environments according to "aesthetic," not biologically functional criteria.[70]

In his chairman's address to the German Society for Psychology in that year, Metzger conceded Bischof's point but declared it a virtue:

> Bischof remarked recently that if Gestalt theory is right, then the organism makes basically aesthetic hypotheses about the reality that surrounds it. I would like to say to that, first, that it in fact does so, and second, that it obviously does well by doing so, even when, as can be shown, the aesthetic hypothesis, just because it is aesthetic, sometimes passes reality by.[71]

Metzger, like the founders of Gestalt theory, remained true to an aesthetic rather than a functional, technological, or pragmatic orientation to science. Unfortunately, despite his considerable efforts to show that such an orientation could also have practical value in education, the time was long past when that viewpoint could count on much societal support.

Conclusion

When Wolfgang Köhler died in Enfield, New Hampshire, on June 11, 1967, he had just returned from receiving an honorary degree at the University of Uppsala, Sweden, and similar honors in West Germany. He had been elected president of the American Psychological Association in 1957, and named an Honorary Citizen (*Ehrenbürger*) of the Free University of Berlin in 1962. In the latter year a *Festschrift* in honor of his seventy-fifth birthday appeared in a leading West German psychological journal; another *Festschrift* for his eightieth birthday appeared posthumously.[1] Max Wertheimer was awarded the Wilhelm Wundt Medal of the German Society for Psychology in 1983, forty years after his death. Wolfgang Metzger's honorary degrees have already been mentioned in the previous chapter; both he and Edwin Rausch also were awarded *Festschriften* in leading West German psychological journals. And yet, despite these signs of well-deserved respect in the United States and in Germany, the Gestalt theorists' ideas were ambivalently received.

Remarks by Metzger in the opening article of the volume he edited in the *Handbuch der Psychologie* indicate the defensive position Gestalt theory was in by the 1960s. The growing ascendancy of information-processing conceptions in perception and cognition theory aroused his special wrath. "As though a half century of serious thinking about the problem of elements in psychology had passed without a trace," loose talk about "elements" and "bit numbers" was "simply taken out of thin air," and it was simply assumed without investigation that human cognition and problem solving "must proceed in every case – exactly as in a computer – in the manner of a yes–no game."[2] The advent of the computer had ushered in an era of technology-driven theorizing about the mind, rather like the impact of the telegraph on Helmholtz's thinking about vision. The wave has swollen, not abated, since then. The Gestalt theorists raised central issues and provoked important debates in psychology, theoretical biology, and other fields.

But their mode of thinking and research style accommodated only uncomfortably to the intellectual and social climate of the postwar world.

Two explanations have been given for this outcome. One emphasizes institutional, political, and biographical contingencies. Sheer numbers of adherents do not suffice to assure a school's scientific impact, but they help. For slightly more than a decade the Gestalt theorists produced a significant number of students who understood the intellectual issues at stake and had hands-on experience in the methods deemed necessary to address them. Kurt Koffka, however, was eventually stymied by insufficient funding for his Giessen institute in the 1920s. The Nazis and their would-be collaborators cut off the remaining leaders from their bases in Berlin and Frankfurt while they were still in their prime. The school suffered severe personal blows with the early deaths of Wertheimer in 1943, Koffka in 1941, Gelb in 1935, and Lewin in 1947. In addition, three of Köhler's most outstanding students – Karl Duncker, Otto Lauenstein, and Hedwig von Restorff – died young.

After they left Germany, the founders of Gestalt theory all obtained positions where they could do excellent research, but could not train Ph.D.'s. After his five-year research professorship at Smith College ended in 1932, Kurt Koffka joined the regular faculty there, where he remained until his death. He made six attempts to relocate, but apparently never received an offer. Köhler remained at Swarthmore until his retirement in 1955. Outstanding students, such as Richard Held and Dorwin Cartwright, came to do B.A. and M.A. degrees as well as postdoctoral research with him; his colleagues David Krech, Richard Crutchfield, and Edwin Newman were influenced by his ideas; but he had no Ph.D.'s. Wertheimer prospered by all accounts as a teacher at the New School, but a doctoral program was instituted there only shortly before his death. The lack of American students put the Gestalt viewpoint at a decided competitive disadvantage compared with the neobehaviorist research machine headed by Clark Hull and Kenneth Spence at Yale and Iowa, respectively, in the 1940s and 1950s. Mary Henle remarks with some irony that contemporary cognitive scientists, by taking up issues the Gestalt theorists had once raised, are supplying "the manpower" they had lacked in the past.[3]

The situation in Germany was different. Metzger, Rausch, and Gottschaldt produced more students between them than Köhler, Koffka, and Wertheimer did. There surely would have been still more, had it not been for the Nazis and their helpers. Yet relatively few of their students carried on in the Gestalt tradition. Thus, it seems necessary to consider the impact of broader historical and political contingencies. The most important of these was the superpowers' impact on postwar German culture. We have already seen evidence of this in Kurt Gottschaldt's career. Metzger's polemics against the information-processing approach indicate the impact of Anglo-Saxon theoretical and research styles in West German psychology. Yet German traditions showed considerable staying power.

Moreover, differences persisted between European – not only German – traditions and American approaches. As Finnish psychologist Kai von Feiandt claimed in the 1967 *Festschrift* for Köhler, the latter were still under the sway of behaviorism and focused entirely on performance; the former remained concerned with processes inside the organism and with the role of cognition in perception, accepting the limits on measurement involved.[4]

At least as important as such cultural differences were psychologists' efforts in Germany, as elsewhere, to adapt to new agendas set by a much broader change in twentieth-century life, the professionalization of society. For psychology in Germany, that trend was aided rather than impeded by Nazism. As it proceeded after 1945, the search for empirical principles of subjectivity that had largely dominated academic psychology in its early years became less important than the basic research necessary to support the production and use of instruments for personality diagnostics. Later, it also came to mean creating what Nikolas Rose has called "reflexive technologies" – methods of helping the sufferers of modern life understand and cope with their difficulties.[5]

Gestalt theory, like many other approaches dating from earlier in the century, swam in the wake of this current. In West Germany, the discourse of its leaders was not that of an increasingly technological culture and society, nor generally of those trying to heal its sufferers, though Metzger expressed an affinity for Adlerian psychotherapy, and one Metzger student has created a version of cognitive psychotherapy with his teacher's express endorsement.[6] In East Germany, just such attempts to merge Gestalt thinking with talk of perception as an instrument of adaptation, or of we-groups as more or less functional in the production process, brought Kurt Gottschaldt down, partly because he did not combine it with ritual declarations of allegiance to Marxism-Leninism.

Here is where conceptual issues enter the picture. The strengths and limitations of Gestalt theory helped determine both how well it could live up to its creators' own hopes for a new scientific worldview, and how well their students could adapt to social and cultural change. Perhaps the most difficult, but also the most important issue was that of language. The Gestalt theorists, like most psychologists in Germany, worked mainly with human observers, most of whose reports came in verbal form. Köhler had included the sense of a sentence in his list of Gestalt qualities in 1920, yet studies of language played no role in the development of Gestalt theory. The reason for this is not difficult to discover. In psychologies and epistemologies based on rationalist categories, language constitutes meaning. For Gestalt theory, language expresses meaning that is already there in the appearances, or in the world. This was already clear in Wertheimer's essay on number concepts in "primitive" peoples, in which language was a source of evidence about what was given to subjects in other cultures, not a culturally formed medium that itself structures the given. The work of Roman Jakobsen and the so-called Prague school of linguistics has since made it clear that the Gestalt category, which they adapted from Ehrenfels, is eminently applicable to structural analyses of language.[7] But Wertheimer's epistemology prevented the Berlin school from taking that route.

Other issues that proved fateful for Gestalt theory as a worldview after 1920 were history, personality, and society. In his 1920 essay on productive thinking, Wertheimer had suggested that "recentering" also occurs in our view of historical figures and events (see Chapter 12). The Gestalt theorists' conception of productive thinking was strikingly similar to that of Thomas Kuhn in emphasizing the primacy of dramatic leaps over incremental progress in discovery. But just such leaps are illustrations of what Wertheimer took to be a general human capability, not a historically changing process. As Koffka's book on development showed, the Gestalt theorists did not think that evolution and cultural circumstances were unimportant; but they deemphasized these dimensions in favor of what they took to be more fundamental principles of dynamics. Ironically, this made it difficult for them to grasp the specific character of the changing historical situations in which they found themselves in their own theoretical language.

Equally problematic, though for different reasons, was the Gestalt theorists' treatment of causal explanation and personal agency. The problems they had with providing a causal account of perceptual phenomena by attributing them to field dynamics in the brain have already been discussed (see Chapters 11, 22, and 23). It was also difficult to find a place for an acting, responsible person in a theory that emphasized process, flux, and in particular the "objective" structure of situations. Köhler made a valiant attempt at this after his confrontation with Nazism.[8] Kurt Lewin made an earlier and still more systematic effort to conceive the person as a complex, "layered" whole in a functional interrelationship with both physical and social environments. But for him the person was still part of a dynamic field of forces, which appeared as often to enclose as to be under the control of the Ego. Wolfgang Metzger's conceptual and behavioral gyrations in the Nazi period and after indicate the limits of such situationism.

It would be a mistake to read such results backward into Gestalt theory's earlier history. It was by no means clear in the 1920s that technological-sounding discourses of causation and control would be the ticket to scientific success in German psychology. What was clear was that the scientific community preferred "the psychical" to be located behind, not in the appearances. What many found lacking in Gestalt theory was talk of the cultivated personality, one who is an effective agent in the world by being an inward-looking whole sufficient unto itself, an instantiation of *Seele*. The Gestalt theorists did not join in that chorus, either. Yet its categories remained within the discourse field – an acknowledged part, albeit an ambivalently received one, of the German intellectual tradition.

As early as 1919 Köhler had expressed the fear that lack of rigor would result from any attempt to apply Gestalt theory to personality and social psychology; and his fears were not unjustified. The preferred route to such extensions was analogy or metaphor, and the further the metaphors were stretched, the harder it became to connect them with Köhler's concept of brain action. As the work of Rudolf Arnheim on expression and that of Lewin and his students on action and emotion showed, however, it was quite possible to extend the domain of Gestalt theory into these important research areas, so long as one separated them from Köhler's psychophysics. Nonetheless, systematic application of Gestalt thinking to social psychology was largely an American phenomenon. Best-known of these efforts is the work of Solomon Asch, who studied with Wertheimer at the New School, showing that group pressure could even affect subjects' judgments of the length of lines on a piece of paper.[9]

Equally famous are the many studies by Lewin and his American students of the role of "democratic" and "authoritarian" group atmospheres in the performance and sense of well-being of workers and pupils. Metzger made a point of citing this work in his book on creative freedom as evidence for the superiority of "democratic" groupings. And yet, in Germany at least, there was no sustained attempt to develop a Gestalt-theoretical concept of the social person. The emphasis remained on the interactions of individual organisms with complex environments; other people were simply included on the environment side. It is hard to count this a weakness of Gestalt theory, since the tendency is shared by all other so-called social psychologies. Even so, Carl-Friedrich Graumann contends that concepts from Gestalt theory, such as the role of frame of reference in forming impressions and social judgment of others, form "an unceasing undercurrent" in contemporary German and American social psychology, though their origins have been obscured.[10]

Ultimately decisive, however, was not any limitation of Gestalt theory itself, but a metatheoretical impasse between its theoretical and research styles and those of the rest of psychology. Gestalt theory was and remains interesting because it was a revolt against both mechanism and vitalism that still insisted on claiming the title of natural science. Especially after 1950, its critics increasingly insisted on causal explanations, by which they meant positing mechanisms in the brain or apparatuses in the mind, a procedure that presupposes dualism. Metzger tried to resolve the impasse by claiming that Gestalt principles are not causes of but rather "conditions for the possibility" of experience; but he could not avoid presenting models and positing control functions in the brain.[11]

As sophisticated as the Gestalt theorists were in their appreciation of the way order emerges from the flow of experience, one must ask how such a process philosophy can be reconciled with strict causal determination, as Köhler, at least, plainly wished to do. Kurt Koffka tried to accomplish this feat by insisting that the very principles of simplicity and order that the Gestalt theorists claimed to find in experience should also be criteria for evaluating both descriptions and explanations. For him, the best argument for isomorphism was his desire for "one universe of discourse, the one about which physics has taught us so much."[12] The Gestalt theorists were not the only ones to invoke a criterion of simplicity in this manner. But Koffka and his co-workers never succeeded in convincing their colleagues that it was logically necessary or scientifically fruitful to think that the external world, its phenomenal counterpart and the brain events mediating interactions between them all have the same structure, or function according to the same dynamical principles.

Gestalt theory's most important historical roles have been that of a sometimes unacknowledged stimulant to further thinking, or that of *agent provocateur,* a constant irritant reminding overly enthusiastic or dogmatic scientists of the limitations in their current approaches. In his now-classical text entitled *Vision,* for example, David Marr, one of the creators of computational models of perception and cognition, admits that no mechanisms have yet been discovered in the brain that would be capable of accounting for our experience, say, of "pink Volkswagens going left" as a single percept. Advocates of now-current "parallel distributed processing" and connectionist models claim that they will be able to do this eventually, but are encountering considerable skepticism.

Wolfgang Metzger clearly stated the challenges confronting any such model thirty years ago in what he called the "antinomies" of psychophysics. These were contradictions between (1) the continuous structures of immediate experience and the discontinuous anatomical structures thought to underlie them, (2) the temporal continuity of psychical events and the temporal discontinuity of cortical functions or nervous processes, and (3) the continuously graduated series of psychical intensities and the all-or-nothing law of synaptic transmission.[13] It is fair to say that this threefold challenge remains in force today, as before.

James Gibson has written that the question Koffka asked in his *Principles of Gestalt Psychology,* "Why do things look as they do?" fundamentally reshaped research on perception. Recently, central issues of Berlin school research, such as similarity and perceptual organization, have returned to center stage, although concepts of "top-down" processing offered to deal with the question have at best a questionable relationship to Gestalt theory.[14] Finally, the status of Wertheimer's "Gestalt laws" and particularly of the so-called minimum principle of *Prägnanz* he enunciated remains contested – which is another way of saying the issues involved are still important. Though Julian Hochberg has argued that the principle is too vague to sustain quantitative analysis, William Epstein and Gary Hatfield point out that nearly all perceptual theorists acknowledge that some version of a "minimum principle" must be valid. More recently, Epstein has even suggested that it may be time to "rehabilitate" Gestalt psychology. Although the Gestalt theorists failed to develop a true theory to account for the important phenomena they adduced, he writes, no one else has, either.[15]

A more positive outcome has resulted from the Gestalt theorists' and others' contributions in theoretical biology. Evidence that the challenge Wolfgang Köhler posed when he introduced the distinction between "closed" and "open" systems into that field is still very much alive can be found in the work of Ilya Prigogine, Manfred Eigen, Humberto Maturana, Réné Thom, and others on the theory of self-organizing systems, and the emergence of ordered patterns from chance events. Eigen refers to Köhler directly when he proposes a new treatment of the problem of morphogenesis with the aid of the concept of self-organization. Following Prigogine, Eigen distinguishes two types of structure: "conservative" structures, like crystals, in which a characteristic pattern repeats itself, and "dissipative" structures, like snowflakes, in which patterns apparently never repeat themselves, but are nonetheless products of orderly processes that can be described mathematically. Precisely these characteristics make them suitable candidates for application to morphogenesis. Particularly provocative is Eigen's use of computer simulation to generate potentially predictive mathematical models of "dissipative" processes. But when Eigen asks whether the problem Köhler had initially raised – the relationship of such formative processes to the phenomena of perception – has now been solved, he can only answer, not yet. Neurophysiologists have developed detailed "maps" of the retinal-cortical connections for relatively simple sensory processes, including color contrast; but they have not yet given us an anatomically exact model of any higher perceptual process, let alone of Gestalt perception.[16]

Gestalt theory has been a worthy participant in two central intellectual trends of our time: the revolt against dualism in twentieth-century thought; and the ongoing struggle between science as technological manipulation and control, and science as an attempt to understand and appreciate order in nature. Returning from their time to ours, it seems strange, and sad, that such dichotomies persist – that scholars still feel the need, for example, to consider hermeneutics and natural science as either–or alternatives. Also frustrating is the continuing use of the word "positivism" as a synonym for evil. If positivism means a worldview based on science, then the Gestalt theorists were themselves positivists; yet they insisted that neither science nor the world we experience needs to conform to positivistic prescriptions. The time has long since come to break away from such all-purpose labels, both in intellectual history and in contemporary discourse.

Cognitive psychologist Frederick Bartlett defended a viewpoint rather different from Gestalt theory; but he acknowledged in 1962 that, when Koffka and Köhler first appeared before an international audience, at the International Congress for Psychology in Oxford in 1922, they "produced a profound impression, for the brilliance and scientific integrity of their work, and also for a most attractive capacity for friendship and understanding." To him they "seemed to be opening up a world of promise for psychology, one in which a wholly humane outlook might be combined with exact scientific precision."[17] Koffka summarized that outlook clearly in 1935:

> The acquisition of true knowledge should help us to integrate our world, which has fallen to pieces. It should teach us the cogency of objective

relations. . . . indicate to us our true position in our world and give us respect and reverence for the things animate and inanimate around us.[18]

It remains to be seen whether such an ideal has a chance of being reconsidered in a scientific and cultural world that is still more fragmented today than it was in Koffka's time.

Appendix 1: Tables

Table 1. *Psychological institutes in German universities, 1879–1945*

Founding	University	Founder	Later Director[a]
1879	Leipzig	Wilhelm Wundt	private; officially recognized 1883; F. Krueger 1917; P. Lersch 1936
1886	Berlin	Hermann Ebbinghaus	private; officially founded C. Stumpf 1894; (see Table 3)
1887	Göttingen	Georg Elias Müller	Psych. Dept. of Phil. Sem.; N. Ach 1922; B. Petermann 1938
1889	Munich	Carl Stumpf	private; officially founded T. Lipps 1894; exp. O. Külpe 1913; E. Becher 1916; A. Fischer 1929; O. Kroh 1938; P. Lersch 1942
1889	Freiburg	Hugo Münsterberg	Psych. Dept. of Phil. Seminar; J. Cohn 1901; G. Stieler 1934
1891	Halle	Benno Erdmann	A. Riehl 1898; H. Ebbinghaus 1905; F. Krueger 1910; T. Ziehen 1917; A. Gelb 1931–1933
1894	Breslau	Hermann Ebbinghaus	Psych. Dept. of Phil. Seminar; W. Stern 1905–1916; res. P. Lersch 1937–1939
1896	Würzburg	Oswald Külpe	exp. K. Marbe 1909; C. Jesinghaus 1935
1898	Bonn	Benno Erdmann	Psych. Sem.; O. Külpe 1909 as Psych. Inst; G. Störring 1914; E. Rothacker 1929
1898	Kiel	Götz Martius	J. Wittmann 1922
1904	Frankfurt[b]	Karl Marbe	F. Schumann 1909; M. Wertheimer and A. Gelb 1929; W. Metzger 1933–1941

continued

413

Appendix 1

Table 1. *continued*

Founding	University	Founder	Later Director[a]
1905	Königsberg	Ernst Meumann	Phil. Sem.; N. Ach 1907; officially founded F. E. O. Schultze 1922; G. Ipsen 1936–1938; res. 1941 K. Lorenz as Inst. for Comparative Psych.
1907	Münster	Ernst Meumann	Psych. Dept. of Phil. Sem.; E. Becher 1909; R. Goldschmidt 1919–1933; res. B. Kern 1935; W. Metzger 1942
1911	Hamburg[c]	Ernst Meumann	cont. W. Stern 1916–1933
1912	Marburg	Erich Jaensch	G. H. Fischer 1941
1919	Giessen	Kurt Koffka	to 1927; res. F. Sander 1929; G. Pfahler 1933
1921	Cologne	Johannes Lindworsky	to 1928; res. R. Heiss 1938
1925	Jena	Wilhelm Peters	F. Sander 1933
1930	Greifswald	Johannes v. Allesch	Psych. Dept. of Phil. Sem.; H. Schole 1938
1942	Heidelberg	Willy Hellpach	

exp. = expanded; cont. = continued; res. = resumed
[a]Interim directors not included.
[b]Commercial Aacademy; university founded 1914.
[c]Colonial Institute; university founded 1919.
Source: Ulfried Geuter (ed.), *Daten zur Geschichte der deutschen Psychologie,* vol. 1 (Göttingen, 1986).

Table 2. *Size, budget, and personnel development for six psychological institutes in Germany, 1879–1914*

Institute	Rooms (Year)	Budget[a] (Year)	Personnel (Year)
Leipzig	2 (1879)	600 (1883)	1 Prof. (1874)
	5 (1883)	2,000 (1909)	1 Asst. (1883)
	11 (1892)		1 Mech. (?)
			1 Cust. (?)
			1 Asst. (1894)
			1 Assoc. (1908)
Berlin	2 (1886)	1,000[b] (1894)	1 Prof. (1894)
(see also	3 (1894)	1,400 (1895)	1 Asst. (1894)
Table 3)	10 (1900)	2,400 (1901)	2 Cust. (1895)
		3,400 (1909)	1 Cust.- (1900)
		4,400 (1914)	Mech.
Göttingen	4 (1887)	500[c] (1887)	1 Prof. (1881)
		700 (1901)	1 Asst. (1901)
		1,200 (1908)	
		1,500 (1914)	
Würzburg	4–6 (1896)	280 (1896)	1 Prof. (1894)
		1,200[d] (1909)	1 Assoc. (1902)
			1 Asst. (1904)
			1 Cust. (1906)
			1 Mech. (1910)
			1 Mech. (1912)
			1 Mech.- (1912)
			trainee
Munich	1 (1889)	100 (1889)	1 Prof. (1889)
	? (1913)	4,000 (1913)	2 Asst. (1913)
			1 Mech. (1913)
Frankfurt	? (1904)	1,500 (1905)	1 Prof. (1904)
			1 Asst. (1904)
			1 Mech. (1904)
			1 Asst. (1910)

Prof. = professor; Assoc. = associate professor; Asst. = assistant; Mech. = mechanic; Cust. = custodian.

[a]Budget = annual expenditure for institute operation in Marks, including physical plant, apparatus, and maintenance. Budget figure for Leipzig is for apparatus only.

[b]For additional one-time expenditures, see Table 3.

[c]Primarily plant and maintenance; instruments obtained through private gifts or built by assistants.

[d]From 1906, additional income of 2,000 Marks per year from Leopold Schweisch Foundation.

continued

Table 2. *continued*

Sources: Wilhelm Wundt, "Das Institut für experimentelle Psychologie," in *Festschrift zur Feier des 500 jährigen Bestehens der Universität Leipzig* (Leipzig, 1909), vol. 4, pp. 118–133.

Carl Stumpf, "Das Psychologische Institut," in Max Lenz, *Geschichte der königl. Friedrich-Wilhelm-Universität zu Berlin* (Halle, 1910), vol. 3, pp. 202–207. See also sources for Table 3.

Georg Elias Müller, "Eröffnungsansprache," *Bericht über den 6. Kongress für experimentelle Psychologie* (Leipzig, 1914), pp. 103–109.

Karl Marbe, "Das Psychologische Institut," in *Hundert Jahre bayrisch. Ein Festbuch* (Würzburg, 1914), pp. 106–109.

Oswald Külpe, letter to Carl Stumpf, 22 March 1913, ZStA, Rep. 76 Va Sekt. 2 Tit. X Nr. 150 Bnd. 2 Bl. 216.

Ulfried Geuter (ed.), *Daten zur Geschichte der deutschen Psychologie,* vol. 1 (Göttingen, 1986).

Table 3. *The Berlin Psychological Institute: Size, budget, and staff, 1894–1942*

Year	Size (rooms)	Budget[a,b]	Staff
1894	3	1,000	2 (1 Prof., 1 Asst.)
1895		1,400	3 (+ 1 Cust.)
1900	10	2,400	4 (+ 1 Mech.)
1909		3,400	
1912		4,400	
1920	25+	28,200	7 (+ 2 Asst., 1 Cust.)
1922		33,700	
1925		27,500	
1928		27,500[c]	
1930		23,600[d]	
1935		20,190[e]	

[a]"Budget" refers to annual expenditures for operation of the institute in Marks, including physical plant, apparatus, and maintenence, without personnel expenses or irregular supplements. Figures not corrected for changes in the value of the Mark.

[b]Additional single expenditures for new acquisitions and moving costs: 6,000 M. (1894); 7,300 M. (1900); 5,000 M. (1909); 5,000 (1914). Budgeted expenditure for 1920 move 61,500 m.

[c]Ten percent blocked.

[d]Actual expenditures; budget remained as in 1928.

[e]4,038 Marks blocked.

Sources

Carl Stumpf, "Das Psychologische Institut," in Max Lenz, *Geschichte der königl. Friedrich-Wilhelm-Universität zu Berlin,* (Halle, 1910) vol. 3, pp. 202–207.

Minerva. Jahrbuch der gelehrten Welt, 25 (1921), p. 73.

GStA, Rep. 76 Va Sekt. 2 Tit. X Nr. 150, Bnd. 1–4, *passim.*

BA Potsdam, REM 49.01, Nr. 1662, Bl. 78f.

Table 4. *Research staff of the Berlin Psychological Institute, 1894–1945*

Year	Director	Assistants	Associates
1894	C. Stumpf (oP)	F. Schumann (Pd –1905)	
1900			O. Abraham (Pd)
1905		E. v. Hornbostel (–1906)	
1906		N. Ach (Pd, then aoP – 1907)	E. v. Hornbostel (Pd 1923; aoP 1925–1933)
1907		H. Rupp (Pd 1909; aoP 1921)[a]	
1910		A. Gelb (–1912)[b]	
1912		J. v. Allesch (–1923)[b]	
1916			M. Wertheimer (Pd; aoP 1924–1929)
1919			J. B. Rieffert (Pd –1926; aoP –1934)
1921	W. Köhler (Pd)[c]	K. Lewin (Pd; aoP 1927–1934)	
1922	W. Köhler (oP)		
1923			J. v. Allesch (Pd 1923–1928)
1926		K. Gottschaldt (–1929)	
1927		W. Metzger (–1931)	
1929		K. Duncker (–1935)	
1931		O. v. Lauenstein (–1935)	H. Friedländer (Pd –1933)
1934		H. v. Restorff (–1935) H. Keller (–1936)	
1935	J. B. Rieffert[c] (oP 1934)	H. Preuss (1936) R. Beck (–1943)	L. F. Clauss (Pd)
1936	H. Keller (aoP)[c]		
1937		H.-J. Firgau (–1938)	
1938	W. M. Schering (aoP)		H. Keller (–1944) K. Gottschaldt (aoP – 1945) H. F. K. Günther (aoP – 1943)
1942	O. Kroh (oP)	K. S. Sodhi	
1943		H. Märtin	

Pd = *Privatdozent* (unpaid instructor); aoP = associate professor; oP = full professor.
Sources (in addition to sources for Table 3):
Ulfried Geuter (ed.), *Daten zur Geschichte der deutschen Psychologie*, vol. 1 (Göttingen, 1986), pp. 15–18.
Universitätsarchiv der Humboldt-Universität zu Berlin.

Appendix 2: Dissertations

Table 1. *Berlin dissertations supervised by Carl Stumpf*

Year	Author	Title
1896	Max Meyer	Über Kombinationstöne und einige hierzu in Beziehung stehende akustische Erscheinungen
1898	Kurt Ebhardt	Beiträge zur Psychologie des Rhythmus und des Tempo
1904	Bernhard Groeythuysen	Das Mitgefühl
1905	Hermann Giering	Das Augenmass bei Schulkindern
	Anton Palme	J. G. Sulzers Psychologie und die Anfänge der Dreivermögenslehre
1908	Robert Musil	Beitrag zur Beurteilung der Lehren Machs
	Kurt Koffka	Experimentelle Untersuchungen zur Lehre vom Rhythmus
1909	Herbert Langfeld	Über die heterochrome Helligkeitsverteilung
	Gerhard Gotthardt	Bolzanos Lehre vom "Satz an sich" in ihrer methodologischen Bedeutung
	Wolfgang Köhler	Akustische Untersuchungen I
	G. J. v. Allesch	Über des Verhältnis der Aesthetik zur Psychologie
	Walter Poppelreuter	Über die scheinbare Gestalt und ihre Beeinflussung durch Nebenreize[a]
1910	Adhemar Gelb	Theoretisches über Gestaltqualtitäten[b]
1911	Wilhelm Reimer	Der Intensitätsbegriff in der Psychologie Historisches und Kritisches
1913	Walter Blumenfeld	Untersuchungen über die scheinbare Grösse im Sehraum
	Catherina von Maltzew	Das Erkennen sukzessiv gegebener musikalischer Intervale in den äusseren Tonregionen

continued

Appendix 2

Table 1. *continued*

Year	Author	Title
1914	Alexander Kühn	Über Einprägung durch Lesen und durch Rezitieren
	Richard Liebenberg	Über das Schätzen von Mengen
1915	Percy Ford Swindle	Über einfache Bewegungsinstinkte und deren künstlicher Beeinflussung
	Otto Leeser	Über Linien und Flächenvergleichung
1916	Kurt Lewin	Die psychische Tätigkeit bei der Hemmung von Willensvorgängen und das Grundgesetz der Assoziation
1918	Hans Friedländer	Die Wahrnehmung der Schwere
1921	Ernst Lau	Neue Untersuchungen über das Tiefen- und Ebenensehen

Except where noted, dissertations had Stumpf as first reader only.

[a]Research done in Berlin but submitted and approved for the doctorate in Königsberg. First reader Narziss Ach.

[b]First reader Alois Riehl; secod reader Carl Stumpf.

Sources

Jahresverzeichnis der an den deutschen Universitäten und Hochschulen erschienenen Schriften, vols. 12 ff.

Ulfried Geuter (ed.), *Daten zur Geschichte der deutschen Psychologie,* vol. 2 (Göttingen, 1987).

Table 2. *Berlin dissertations supervised by Wolfgang Köhler, Max Wertheimer,*
and Kurt Lewin

Year	Author	Title
1922	Margarete Eberhardt	Über die phänomenale Höhe und Stärke von Teiltönen (K)
1924	Walter Scholz	Experimentelle Untersuchungen über die phänomenalen Grösse von Raumstrecken (K)
1925	Erna Schur	Mondtäuschung und Sehgrössenkonstanz (K)
	Joseph Ternus, S.J.	Experimentelle Untersuchungen über phänomenale Identität[a]
1926	Richard Meili	Experimentelle Untersuchungen über die Ordnung von Gegenständen[b]
	Wolfgang Metzger	Über Vorstufen der Verschmelzung von Figurenreihen, die vor dem ruhenden Auge vorüberziehen (W)
	Kurt Gottschaldt	Über den Einfluss der Erfahrung auf die Wahrnehmung von Figuren (W)
1927	Bluma Zeigarnik	Über das Behalten von erledigten und unerledigten Handlungen (K)
	Susanne Liebmann	Über das Verhalten farbiger Formen bei Helligkeitsgleichheit von Figur und Grund (K)
	Anitra Karsten	Über psychische Sättigung[c]
	Maria Ovsiankina	Die Wiederaufnahme unterbrochener Handlungen[c]
1928	Rudolf Arnheim	Experimentell-psychologische Untersuchungen zum Ausdrucksproblem (K)
1929	Karl Duncker	Über induzierte Bewegung (K)
1930	Ferdinand Hoppe	Erfolg und Misserfolg (K)
1931	Tamara Dembo	Der Ärger als dynamisches Problem (L)
	Günther Voigt	Über die Richtungspräzision einer Fernhandlung (L)
1932	Otto Lauenstein	Ansatz zu einer physiologischen Theorie des Vergleichs und der Zeitfehler (K)
	Werner Wolff	Selbstbeurteilung und Fremdbeurteilung im wissentlichen und unwissentlichen Versuch (K)
	Curt Berger	Zum Problem der Sehschärfe (K)
1933	Sara Forer	Eine Untersuchung zur Lese-Lern-Methode Decroly (L)
	Sara Fajans (Glück)	Die Bedeutung der Entfernung für die Stärke eines Aufforderungscharakters beim Säugling und Kleinkind (K)
	Wera Mahler	Ersatzhandlungen verschiedenen Realitätsgrades (L)
	Hedwig v. Restorff	Über die Wirkung von Bereichsbildung im Spurenfeld (K)

continued

Appendix 2

Table 2. *continued*

Year	Author	Title
	Meta Jakobs	Über den Einfluss des phänomenalen Abstands auf die Unterschiedsschwelle für Helligkeiten (K)
	Käte Lissner	Die Entspannung von Bedürfnissen durch Ersatzhandlungen (L)
1934	Sara Sliosberg	Zur Dynamik des Ersatzes in Spiel- und Ernstsituationen (L)
	Constantin Calavrezzo	Über den Einfluss von Grössenänderung auf die scheinbare Tiefe (K)
1935	Wilhelm Wolff	Über die Kontrast erregende Wirkung der transformierten Farben (K)
	Hans Wallach	Über visuell wahrgenommene Bewegungsrichtung (K)
1936	Max Simon	Über egozentrische Lokalisation[d]
	Margarete Jucknat	Leistung, Anspruchsniveau und Selbstbewusstsein (L)[e]
1937	Hellmut Bartel	Über die Abhängigkeit spontaner Reproduktion von Feldbedingungen[f]
	Astri Ortner	Nachweis der Retentionsstörung beim Erkennen[f]
	Ilse Müller	Zur Analyse der Retentionsstörung durch Häufung[f]

Except where noted, the first reader is given in parentheses after the title. K = Köhler; L = Lewin; W = Wertheimer.

[a]First reader Eduard Spranger; second reader Max Wertheimer.

[b]First reader Hans Rupp; second reader Wolfgang Köhler.

[c]Research done in Berlin, but submitted and approved at the University of Giessen. First reader Kurt Koffka.

[d]First reader Hans Rupp; second reader Wolfgang Köhler.

[e]Lewin and Köhler listed as "reporters," not official readers; oral examination in July, 1934.

[f]Work begun in Berlin, but submitted and approved at the University of Greifswald. First reader Gustav J. von Allesch.

Table 3. *Giessen dissertations supervised by Kurt Koffka*

Year	Author	Title
1915	Adolf Korte	Kinematoskopische Untersuchungen[a]
1920	Friedrich Kenkel	Untersuchungen über den Zusammenhang zwischen Erscheinungsgrösse und Erscheinungsbewegung bei einigen sogenannten optischen Täuschungen[b]
1922	Friedrich Wulf	Über die Veränderung von Vorstellungen (Gedächtnis und Gestalt)
	Erich Lindemann	Experimentelle Untersuchungen über das Werden und Vergehen von Gestalten
1924	Ludwig Hartmann	Neue Verschmelzungsprobleme
	Adolf Ackermann	Farbschwelle und Feldstruktur
1925	Nikolaus Feinberg	Experimentelle Untersuchungen über die Wahrnehmung im Gebiet des blinden Flecks
1926	Johann (Hans) Georg Hartgenbusch	Über die Messung von Wahrnehmungsbildern
	Adolf Noll	Versuche über Nachbilder
	Paul Kester	Über Lokalisations- und Bewegungserscheinungen bei Geräuschpaaren
1927	Annie Stern	Die Wahrnehmung von Bewegung in der Gegend des blinden Flecks
1928	Alexander Mintz	Über äquidistante Helligkeiten

[a]Published in 1915; degree never awarded due to author's death in World War I.
[b]Published in 1913; submitted and approved as dissertation 1920.
Sources: See Table 1.

Table 4. *Frankfurt dissertations supervised by Adhemar Gelb, Max Wertheimer, Wolfgang Metzger, and Edwin Rausch, 1930–1973*[a]

Year	Author	Title
1930	A. Schwemmler	Über hervortretende Farben (G)
	Herta Kopfermann	Psychologische Untersuchungen über die Wirkung zweidimensionaler Darstellungen körperlicher Gebilde (W)
1931	Ellis Freeman	Untersuchungen über das indirekte Sehen (W)
	A. Fischer	Experimentelle Beiträge zur räumlichen Orientierung (G)
1932	H. Bullmann	Die Phänomenologie des Wachsens und Schrumpfens beim Vergleich von Flächen (G)
	E. Nahm	Über den Vergleich von komplexen geometrischen Gebilden und tonfreien Farben (G)
	Lore Posner	Die Erscheinungen des simultanen Kontrastes und der Eindruck der Feldbeleuchtung (G)
	Wolfgang Hochheimer	Analyse eines 'Seelenblinden' von der Sprache aus. Ein Beitrag zur Frage nach der Bedeutung der Sprache für das Verhalten zur Umwelt (G)
	Wilhelm Sieckmann	Psychologische Analyse des Falles Rat . . . ein Fall von sogenannter motorischer Aphasie (G)
1934	Gertrud Siemsen	Experimentelle Untersuchungen über das taktil-motorische Gerade (M)[b]
	Erika Oppenheimer	Optische Versuche über Ruhe und Bewegung (M)[b]
	Walter Krolik	Über Erfahrungswirkungen bei Bewegungssehen (M)[b]
	Kurt Madlung	Über anschauliche und funktionale Nachbarschaft von Tasteindrücken (M)[b]
1935	Joseph Becker	Über taktilmotorische Figurwahrnehmung (M)[b]
1936	Karl Wiegand	Stroboskopische Versuche mit drei geradlinig angeordneten Lichtreizen (M)
	Mümtaz Turhan	Über die räumliche Wirkung von Helligkeitsgefällen (M)[b]
1937	Erich Goldmeier	Über Ähnlichkeit bei gesehenen Figuren (M)[b]
	Edwin Rausch	über Summativität und Nichtsummativität (M)[b]
	Hans-Georg Spiegel	Über den Einfluss des Zwischenfeldes auf gesehene Abstände (M)[c]
1938	Ottilie Redslob	Über Sättigung gesehener Bewegungsrichtung (M)[c]
	Lotte Lauenstein	Über räumliche Wirkungen von Licht und Schatten (M)
1939	H. J. Schnehage	Versuche über taktile Scheinbewegung bei Variation phänomenaler Bedingungen (M)
1940	Siegfried Sorge	Neue Versuche über die Wiedergabe abstrakter optischer Gebilde (M)[d]

continued

Table 4. *continued*

Year	Author	Title
	H. Keller	(M) (title not yet known)
	H. Schaeffer	(M) (title not yet known)
1941	Eduard Josef Born	(M) (title not yet known)
	Franz Kutsch	(M) (title not yet known)
1954[e]	Egon Becker	Mengenvergleich und Übung (R)
	Kurt Müller	Über die Rolle der Bezugsbereiche bei der Lokalisation. Experimentelle Unteruchungen zum Problem des Ortsgedächtnisses (R)
1955[e]	Kurt Kohl	Zum Problem der Sensumotorik. Psychologische Analysen zielgerichteter Handlungen aus dem Gebiet des Sports (R)
1957	Manfred Sader	Instruktionsverständnis und Testleistung. Untersuchungen über Vorphase und Hauptphase eines psychologischen Prüfversuchs (R)
1959	Alfred Pfeil	Lesen und Benennen im Reaktionsversuch (R)
	Erika Junker[e]	Über unterschiedliches Behalten eigener Leistungen (R)
1962	Kurt Müller[e,f]	Der Aufbau figural-optischer Phänomene bei sukzessiver Reizung (R)
1963	Wolfgang Schönpflug	Über Aktivationsprozesse im Lernversuch. Untersuchungen über den Zusammenhang von Retentionsleistung und elektrischem Hautwiderstand (R)
1965	Josefa Zoltobrocki	Über systematische Fehler bei wiederholter Schätzung von Zeitintervallen (R)
1966	Friedrich Hoeth	Gesetzlichkeit bei stroboskopischen Alternativbewegungen (R)
1970	Udo Schmidt	Figurale Nachwirkungen auf kinästhetischem Gebiet bei Lageveränderung von Objekt und tastender Hand (R)
1973	Paul Tholey	Zur Einzel- und Gruppenleistung unter eingeschränkten Kommunikationsbedingungen (R)

[a]The first reader is given in parentheses after the title. (G) = Gelb; (M) = Metzger; (R) = Rausch; (W) = Wertheimer. The list for Gelb is not complete.
[b]Research begun under Max Wertheimer.
[c]Begun in Berlin under Wolfgang Köhler.
[d]Completed in Halle under Metzger's supervision.
[e]Differs from year of publication.
[f]Habilitation.
Sources: Stadt- und Universitätsbibliothek Frankfurt am Main and sources cited in Appendix 2, Part 1.

List of unpublished sources

I. Archives

American Philosophical Society Archives, Philadelphia
  Wolfgang Köhler papers
Archives of the History of American Psychology, Akron, Ohio
  Kurt Koffka papers
  Kurt Lewin papers
  Donald K. Adams papers
  N. R. F. Maier Oral History
Archiv der Berlin-Brandenburgischen Akademie der Wissenschaften, Berlin
  Albert-Samson-Stiftung, II-XI-121, Bnd. 6, 12 and 13
  Handakte Wilhelm von Waldeyer
Berlin Document Center
  Reichserziehungsministerium (REM) files
  National Socialist Party (NSDAP) Chancellery card files
Bibliothek und Archiv der Max-Planck-Gesellschaft, Berlin
  Kaiser-Wilhelm-Institut für Anthropologie, menschliche Erblehre und Eugenik,
  Bnd. 2399-2406; Abt. II, Rep. 1A, X 97
  Kaiser-Wilhelm-Gesellschaft, Senat
Bundesarchiv Koblenz
  R 73 (Deutsche Forschungsgemeinschaft)
Bundesarchiv Potsdam
Cornell University, Olin Library, Manuscripts and Archives
  Robert M. Ogden papers
  Edward B. Titchener papers
Geheimes Staatsarchiv Preussischer Kulturbesitz, Berlin
  Preussisches Kultusministerium, Rep. 76
    Va Sekt. 2 Tit. X Nr. 150, Bnd. 4 (Universität Berlin, Psychologisches Institut)
  Preussisches Staatsministerium, Rep. 90, Abt. Q Tit. VI Bnd. V Nr. 1
Geheimes Staatsarchiv Preussischer Kulturbesitz, Merseburg (now Berlin)
  Preussisches Kultusministerium, Rep. 76
    REM 49.01, Nr. 1662 (Bauliche Angelegenheiten des Psychologischen Instituts, Universität Berlin)

427

Va Sekt. 2 Tit. IV Nr. 51, Bnd. 17; Nr. 68a, Bnd. 1; Tit. X Nr. 61, Bnd. 6
(Universität Berlin, Philosophische Fakultät)
Va Sekt. 2 Tit. X Nr. 150, Bnd. 1–3 (Universität Berlin, Psychologisches Institut)
Va Sekt. 5 Tit. IV Nr. 4 Bnd. 3 (Universität Frankfurt, Philosophische Fakultät)
Vb Sekt. 4 Tit. X Nr. 53a Bnd. 1 (Förderung der angewandten Psychologie, Berlin)
Korrespondenz Friedrich Althoff, Rep. 92
Harvard University Archives, Cambridge, Mass.
  Edwin G. Boring papers
Jewish National and University Library, Jerusalem
  Albert Einstein Archives
Leo Baeck Institute Library, New York
  Fritz Mauthner collection
Robert Brodie MacLeod papers, private collection, Ithaca, N.Y.
Max Planck Institute for Psychological Research, Munich
  Kurt Gottschaldt papers
National Library of Medicine, Bethesda, Md.
  Lawrence K. Frank papers
New York Public Library, Manuscripts and Archives Division
  Max Wertheimer papers
  Emergency Committee in Aid of Displaced Foreign Scholars
Politisches Archiv des Auswärtigen Amts, Bonn
  Abteilung VI W Hochschulwesen: Deutschland, Bnd. 25–26
Rockefeller Family Archives Center, North Tarrytown, N.Y.
  RG 1.1, Series 216, Box 11 (Wolfgang Köhler)
Gertrud Siemsen memoir, Berlin, 1978
Smith College Archives, Northampton, Mass.
  Kurt Koffka correspondence
  William Allen Nielson papers
Staatsbibliothek Preussischer Kulturbesitz, Berlin
  Korrespondenz Carl Stumpf, Sammlung Darmstädter
Universitätsarchiv der Humboldt-Universität zu Berlin
  Philosophische Fakultät, Dekanat, Nr. 152, 1236, 1439, 1460, 1469–1474
  Universitätskuratorium 8391
Universitätsarchiv Giessen
  Präsidialabteilung, Nr. 24 and Nr. 384
Universitätsarchiv Göttingen
  Universitätskuratorium XVI.IV A.a., 15a (G. E. Müller); 281 (W. Köhler)
Universitätsarchiv Münster
  Psychologisches Institut/Wolfgang Metzger
Universität Frankfurt am Main, Präsidialabteilung
  Personalhauptakte Max Wertheimer
Universität Heidelberg, Psychologisches Institut
  Nachlass Kurt Lewin
Verlagsarchiv des Springer-Verlags, Heidelberg
  Mappen G-118, G-177, K-185, K 188
Max Wertheimer papers, private collection, Boulder, Colo.
Yale University Medical Library, New Haven, Conn.
  Robert M. Yerkes papers, Box 29, Folder 548 (Koehler, Wolfgang) Box 57, Folder 1090
  (Anthropoid Research: Canary Islands)

II. Interviews

Günther Anders	Vienna (Telephone Interview), 5 October 1980
Rudolf Arnheim	Ann Arbor, Mich., 9 August 1976
Rudolf Bergius	Bad Homburg, 11 November 1983
Kurt Gottschaldt	Göttingen, 5 August 1987
Karl Hempel	Pittsburgh, 25 October 1986
Wolfgang Hochheimer	Berlin, 26 May 1980
Anitra Karsten	Frankfurt am Main, 22 February 1978
Richard Meili	Bern, Switzerland, 5 November 1980
Edwin Newman	Cambridge, Mass., 2 October 1976
Erika Oppenheimer Fromm	Chicago, 15 November 1982
Edwin Rausch	Oberursel and Frankfurt am Main, 22 September 1978
Ottilie Selbach (née Redslob)	Berlin, 20 August 1987
Gertrud Siemsen	Berlin, 11 May 1979
Hans Wallach	Swarthmore, Pa., 19 August 1976
Blyuma Zeigarnik	Leipzig, 10 July 1980

Notes

Abbreviations

AAW   Archiv der Berlin-Brandenburg Akademie der Wissenschaften, Berlin
AfgP   *Archiv für die gesamte Psychologie*
AHAP   Archives of the History of American Psychology
APS   American Philosophical Society
BA   Bundesarchiv (Potsdam, Koblenz)
BDC   Berlin Document Center
GStA   Geheimes Staatsarchiv Preussischer Kulturbesitz
HSPS   *Historical Studies in the Physical Sciences*
JHBS   *Journal of the History of the Behavioral Sciences*
MPGA   Max-Planck-Gesellschaft, Archiv
MWPB   Max Wertheimer Papers, Boulder (Colorado)
MWP-NYPL Max Wertheimer Papers – New York Public Library
PAUF   Präsidialabteilung der Johann-Wolfgang-Goethe-Universität Frankfurt am Main
PAAA   Politisches Archiv des Auswärtigen Amts (Bonn)
PsyFo   *Psychologische Forschung*
UAHUB   Universitätsarchiv der Humboldt-Universität zu Berlin
UA   Universitätsarchiv (Giessen, Göttingen, Münster)
ZfP   *Zeitschrift für Psychologie*

Preface

1 I have discussed the development and reception of Gestalt theory in the United States in the following articles: "Gestalt Psychology: Origins in Germany and Reception in the United States," in Claude Buxton (ed.), *Points of View in the Modern History of Psychology* (San Diego, Orlando, New York, 1985), pp. 295–344; "Emigré Psychologists after 1933; Migration, Science and Practice," in Mitchell G. Ash and Alfons Söllner (eds.), *Forced Migration and Scientific Change: Émigré German-Speaking Scientists after 1933* (Cambridge and New York, in press).

Introduction

1 Martin Scheerer, *Die Lehre von der Gestalt. Ihre Methode und ihr psychologischer Gegenstand* (Berlin, 1931), pp. 2 ff.; Michael Wertheimer, "Gestalt Theory, Holistic Psychologies and Max Wertheimer," in Günther Bittner (ed.), *Personale Psychologie: Festschrift für Ludwig J. Pongratz* (Göttingen, 1983), esp. pp. 36 ff. See also Solomon Asch, "Gestalt Theory," *International Encyclopedia of the Social Sciences,* vol. 6 (New York, 1968), pp. 157–175.

2 See, e.g., Aron Gurwitsch, *The Field of Consciousness* (Pittsburgh, 1964), Part Two; Maurice Merleau-Ponty, *The Structure of Behavior* (1942), trans. Alden Fisher (Boston, 1963).

3 Barry Smith, "Gestalt Theory: An Essay in Philosophy," in idem. (ed.), *Foundations of Gestalt Theory* (Munich, 1989), p. 13.

4 Mary Henle, "Wolfgang Köhler (1887–1967)," *Yearbook of the American Philosophical Society* (1968), pp. 141–142; cf. Kurt Koffka, *Principles of Gestalt Psychology* (New York, 1935), esp. pp. 20–21.

5 Fritz K. Ringer, *The Decline of the German Mandarins: The German Academic Community, 1890–1933* (Cambridge, Mass., 1969), esp. pp. 375 ff.

6 Martin Leichtman, "Gestalt Psychology and the Revolt against Positivism," in Allen Buss (ed.), *Psychology in Social Context* (New York, 1979), pp. 47–75.

7 Jeffrey Herf, *Reactionary Modernism* (Princeton, N.J., 1984); Sheila Faith Weiss, "The Race Hygiene Movement in Germany," *Osiris,* 2nd series, 3 (1987): 193–236; Paul Weindling, *Health, Race and German Politics between National Unification and Nazism, 1870–1945* (Cambridge, 1989); Herbert Mehrtens, "Symbolische Imperative: Zum Natur- und Beherrschungsprogramm der wissenschaftlichen Moderne," in Wolfgang Zapf (ed.), *Die Modernisierung moderner Gesellschaften* (Frankfurt a.M. and New York, 1991), pp. 604–616. For a similar argument applied to broader social and cultural trends in the 1920s, see Detlev J. K. Peukert, *Die Weimarer Republik: Krisenjahre der klassischen Moderne* (Frankfurt a.M., 1987).

8 On the vicissitudes of organicism, see Frederick Burwick (ed.), *Approaches to Organic Form: Permutations in Science and Culture* (Dordrecht and Boston, 1987).

9 See: Ulfried Geuter, *The Professionalization of Psychology in Nazi Germany,* trans. Richard Homes (Cambridge and New York, 1992); and my essay, "Psychology in Twentieth-Century Germany: Science and Profession," in Geoffrey Cocks and Konrad H. Jarausch (eds.), *German Professions 1800–1950* (Oxford and New York, 1990), pp. 289–307.

10 Konrad H. Jarausch, *The Unfree Professions: German Lawyeres, Teachers and Engineers 1900–1950* (New York, 1990); Charles E. McClelland, *The German Experience of Professionalization* (Cambridge, 1991); Peter Lundgreen, "Akademiker und 'Professionen' in Deutschland," *Historische Zeitschrift,* 254 (1992): 657–670.

11 For a comprehensive survey, see Barry Smith, "Gestalt Theory and Its Reception: An Annotated Bibliography," in Barry Smith (ed.), *Foundations of Gestalt Theory* (Munich, 1989), pp. 227–478.

12 On textbook histories in American psychology see my essay, "The Self-Presentation of a Discipline: History of Psychology in the United States between Pedagogy and Scholarship," in Loren Graham et al. (eds.), *Functions and Uses of Disciplinary Histories* (Dordrecht and Boston, 1983), pp. 143–188. For a thorough criticism of the errors to which taking present assumptions and disciplinary boundaries for granted have led in both history of psychology and history of philosophy, see Gary Hatfield, *The Natural and the Normative: Theories of Spatial Perception from Kant to Helmholtz* (Cambridge, Mass., 1990), Introduction.

13 For erroneous linkages to Kant, see, e.g., Duane Schulz, *A History of Modern Psychology,* 3rd ed. (New York, 1981), pp. 241 ff.; Daniel Robinson, *An Intellectual History of Psychology* (New York, 1976), p. 312; and George Mandler (ed.), *Thinking: From Association to Gestalt* (New York, 1964). For further examples from standard work in perception rather than textbooks, see Mary Henle, "The Influence of Gestalt Psychology in America," esp. pp. 121 ff. For attempts at correction, see Nicholas Pastore, "Reevaluation of Boring on Kantian Influence, Nineteenth Century Nativism, Gestalt Psychology and Helmholtz," *JHBS,* 10 (1974): 375–390; Hatfield, *The Natural and the Normative.* The primary source of many of these misapprehensions is the discussion of Gestalt theory in Edwin G. Boring, *A History of Experimental Psychology,* 2nd ed. (New York, 1950). Two recent texts take better account of current historical scholarship on Gestalt theory: Thomas H. Leahey, *A History of Psychology,* 3rd ed. (Englewood Cliffs, N.J., 1992); and David J. Murray, *A History of Modern Psychology,* 2nd ed. (Englewood Cliffs, N.J., 1988).

14 Henle, "The Influence of Gestalt Psychology"; Wolfgang Köhler, "Psychology and Evolution," *Acta Psychologica,* 7 (1950): 288–297; idem., "Perceptual Organization and Learning," *American Journal of Psychology,* 71 (1958): 311–315.

15 Kurt Danziger, *Constructing the Subject: Historical Origins of Psychological Research* (Cambridge, 1990), Introduction; Laurel Furumoto, "The New History of Psychology," in Ira S. Cohen (ed.), *The G. Stanley Hall Lecture Series,* vol. 9 (Washington, D.C., 1989), p. 18. On social constructivist and contextualist approaches to the histories of other sciences and in history of philosophy, see Steven Shapin, "History of Science and Its Sociological Reconstructions," *History of Science,* 20 (1982): 157–211; Richard Rorty, J. B. Schneewind, and Quentin Skinner (eds.), *Philosophy in History: Essays in the Historiography of Philosophy* (Cambridge, 1984).

16 For recent attempts to integrate historical and philosophical considerations of objectivity, see Dudley Shapere, *Reason and the Search for Knowledge* (Dordrecht and Boston, 1984); Steve Fuller, *Social Epistemology* (Bloomington, Ind., 1988), Part Two; idem., *Philosophy of Science and Its Discontents* (Boulder, Colo., 1989); Helen Longino, *Science as Social Knowledge: Values and Objectivity in Scientific Inquiry* (Princeton, N.J., 1990); Robert N. Proctor, *Value-Free Science? Purity and Power in Modern Knowledge* (Cambridge, Mass., 1991); Allen Megill (ed.), *Rethinking Objectivity* (Detroit, 1994).

17 These correspond to the three respects in which science is a social enterprise, according to Longino, *Science as Social Knowledge,* pp. 66–67.

18 To this extent, the approach is consistent with, though not derived from, Niklas Luhmann's theory of self-organizing social systems. See, e.g., Niklas Luhmann, *Soziale Systeme. Grundriss einer allgemeinen Theorie* (Frankfurt a.M., 1984); Herbert Mehrtens, "Das soziale System der Mathematik und seine politische Umwelt," *Zeitschrift für Didaktik der Mathematik,* 88: (1987), 28–37. For an example of more flexible constraint talk at the "micro" level, see Peter Galison, *How Experiments End* (Chicago, 1987), ch. 5.

19 Bruno Latour and Steve Woolgar, *Laboratory Life: The Social Construction of Scientific Facts* (Beverly Hills, Calif., 1979); Karin D. Knorr-Cetina, *The Manufacture of Knowledge: An Essay on the Constructivist and Contextual Nature of Science* (Oxford, 1981); Bruno Latour, *Science in Action* (Cambridge, Mass., 1987); Galison, *How Experiments End;* David Gooding, T. J. Pinch, and Simon Schaffer (eds.), *The Uses of Experiment: Studies of Experimentation in the Natural Sciences* (Cambridge, 1989); Andrew Pickering (ed.), *Science as Practice and Culture* (Chicago, 1992); Danziger, *Constructing the Subject.*

20 Timothy Lenoir, "Practice, Reason, Context: The Dialogue between Theory and Experiment," *Science in Context,* 2 (1988): 3–22.

21 Longino, *Science as Social Knowledge,* p. 66.

22 Lorraine Daston, "The Moral Economy of Science," in *Critical Problems and Research Frontiers in History of Science and History of Technology* (papers presented to the History of Science Society, Madison, Wisc. 1991), esp. p. 425.

23 Wolf Lepenies, "Wissenschaftsgeschichte und Disziplingeschichte," *Geschichte und Gesellschaft,* 4 (1978): 437–451; idem., *Between Literature and Science: The Rise of Sociology,* trans. R. J. Hollingdale (Cambridge and New York, 1988).

24 Pierre Bourdieu, "Intellectual Field and Creative Project," *Social Science Information,* 8 (1969): 118; idem., *Homo Academicus,* trans. Peter Collier (Cambridge, 1988); Fritz K. Ringer, "The Intellectual Field, Intellectual History and the Sociology of Knowledge," *Knowledge and Society,* 19 (1990): 269–294. Luhmann's systems model for these interrelationships undervalues the question of power in comparison with that of Bourdieu.

25 Peter Lundgreen, "Differentiation in German Higher Education," in Konrad H. Jarausch (ed.), *The Transformation of Higher Learning 1860–1930* (Chicago, 1982), pp. 149–179; Rudolf Stichweh, "The Sociology of Scientific Disciplines: On the Genesis and Stability of the Disciplinary Structure of Modern Science," *Science in Context,* 5 (1992): 3–15.

26 On the role of conceptual transformations in history of science, see I. Bernard Cohen, *The Newtonian Revolution, with Illustrations of the Transformation of Scientific Ideas* (Cambridge, 1980); Frank Sulloway, *Freud, Biologist of the Mind: Beyond the Psychoanalytic Legend* (New York, 1979).

27 Shapere, *Reason and the Search for Knowledge* (cited in n. 16), ch. 14.

28 Kurt Danziger, "Generative Metaphor and the History of Psychological Discourse," in David E. Leary (ed.), *Metaphors in the History of Psychology* (Cambridge, 1990), pp. 331–356. On exemplars see Thomas S. Kuhn, "Second Thoughts on Paradigms," in *The Essential Tension: Selected Studies on Scientific Tradition and Change* (Chicago, 1977), pp. 293–319; Barry Barnes, *T. S. Kuhn and Social Science* (New York, 1982), ch. 2.

29 On scientific research styles and schools see Mikhail G. Jaroschevsky, "The Logic of Scientific Development and the Scientific School: The Example of Ivan Mikhailovich Sechenov," in William R. Woodward and Mitchell G. Ash, *The Problematic Science: Psychology in Nineteenth-Century Thought* (New York, 1982), pp. 231–254; Jane Maienschein, "Whitman at Chicago: Establishing a Chicago Style of Biology?" in Ronald Rainger, Keith R. Benson, and Jane Maienschein (eds.), *The American Development of Biology* (Philadelphia, 1988), pp. 151–183; Gerald L. Geison and Frederic L. Holmes (eds.), *Research Schools: Historical Reappraisals, Osiris,* vol. 8 (Philadelphia, 1993); Jonathan Harwood, *Styles of Scientific Thought: The German Genetics Community, 1900–1933* (Chicago, 1993).

30 Danziger, *Constructing the Subject,* ch. 4.

Chapter 1

1 See esp. Boring, *A History of Experimental Psychology,* 2nd ed. (cited in the Introduction).

2 See, for example, Joseph Ben David, *The Scientist's Role in Society* (Englewood Cliffs, N.J., 1971), ch. 7, and Frank R. Pfetsch, *Die Entwicklung der Wissenschaftspolitik in Deutschland 1750–1914* (Berlin, 1974).

3 Joseph Ben David and Randall Collins, "Social Factors in the Origins of a New Science: The Case of Psychology," *American Sociological Review,* 31 (1966): 451.

4 Wilhelm Wundt, "Die Psychologie im Kampf ums Dasein" (1913), in *Kleine Schriften,* vol. 3, ed. Max Wundt (Stuttgart, 1921), p. 542.

5 Georg Elias Müller, "Eröffnungsansprache," *Bericht über den 6. Kongress der Gesellschaft für experimentelle Psychologie* (1914), pp. 106–107.

6 For the physiological institute budget, see *Minerva, Jahrbuch der gelehrten Welt,* 23 (1913–1914): 115.

7 Pfetsch, *Wissenschaftspolitik.* For research outside the universities, see Peter Lundgreen et al., *Staatliche Forschung in Deutschland 1870–1980* (Frankfurt a.M., 1986). In order to avoid raising issues of comparative history that would needlessly lengthen the discussion, only universities and psychological institutes within the Reich are treated here.

8 Clemens Menze, *Die Bildungsreform Wilhelm von Humboldts* (Hannover, 1975); Charles McClelland, *State Society and University in Germany, 1700–1914,* ch. 4; Fritz K. Ringer, *The Decline of the German Mandarins* (cited in the Introduction), esp. pp. 110–111; and R. Steven Turner, "The Growth of Professorial Research in Prussia, 1818 to 1848– Causes and Context," *HSPS,* 3 (1971): 137–182.

9 Konrad Jarausch, *Students, Society and Politics in Imperial Germany: The Rise of Academic Illiberalism* (Princeton, N.J., 1982), pp. 147–148.

10 Herbert Schnädelbach, *Philosophy in Germany 1831–1933,* trans. Eric Matthews (Cambridge, 1984), p. 27.

11 Max Weber, "The Alleged 'Academic Freedom' of the German Universities" (1908), trans. Edward Shils, repr. in *Minerva,* 11 (1973): 17.

12 Stumpf to Althoff, 28 June 1895, GStA, Rep. 92, Althoff B Nr. 182 Bd. 4 Bl. 47–48.

13 McClelland, *State, Society and University,* pp. 314–321; Bernhard vom Brocke, "Hochschul- und Wissenschaftspolitik in Preussen und im deutschen Kaiserreich 1882– 1907: Das 'System Althoff'," in Peter Baumgart (ed.), *Bildungspolitik in Preussen zur Zeit des Kaiserreichs* (Stuttgart, 1980), esp. pp. 93–108.

14 Ringer, *Mandarins,* p. 53; McClelland, *State, Society and University,* pp. 239–240, 280, 282, 307; Reinhard Riese, *Die Hochschule auf dem Weg zum wissenschaftlichen Grossbetrieb: Die Universität Heidelberg und das badische Hochschulwesen 1860–1914* (Stuttgart, 1977), pp. 62 ff.

15 Pfetsch, *Wissenschaftspolitik,* p. 52; Peter Lundgreen, "Differentiation in German Higher Education" (cited in the Introduction); Riese, *Die Hochschule,* pp. 158–159.

16 Velma Dobson and Darryl Bruce, "The German University and the Development of Experimental Psychology," *JHBS,* 2 (1966): 74–75. On Lotze's philosophy and research methods, see William R. Woodward, *From Mechanism to Value: Hermann Lotze and Nineteenth-Century German Medical, Philosophical, and Psychological Thought* (Cambridge, forthcoming).

17 von Ferber, *Entwicklung,* pp. 177–178. For an overview of academic careers and the problems of younger scholars, see Jarausch, *Students, Society and Politics,* and Fritz K. Ringer, "A Sociography of German Academics, 1863–1938," *Central European History,* 25 (1992): 251–280.

18 Klaus Dieter Bock, *Sozialgeschichte der Assistentur* (Bonn, 1959); von Ferber, *Entwicklung*, p. 86. On the formation of schools in Neo-Kantian philosophy, see Klaus Christian Köhnke, *The Rise of Neo-Kantianism: German Academic Philosophy between Idealism and Positivism*, trans. R. J. Hollingdale (Cambridge, 1991). For an example from psychology, see David Katz, "David Katz," in E. G. Boring et al. (eds.), *A History of Psychology in Autobiography*, vol. 4 (New York, 1952), p. 196.

19 J. N. Morrell, "The Chemist Breeders: The Research Schools of Justus Liebig and Thomas Thomson," *Ambix*, 19 (1972): 1–46; R. Steven Turner, "Justus Liebig versus Prussian Chemistry: Reflections on Early Institute-Building in Germany," *HSPS* (1982): 129–162; Peter Borscheid, *Naturwissenschaft, Staat und Industrie in Baden 1848–1914* (Stuttgart, 1976); Lewis Pyenson and Douglas Skopp, "On the Doctor of Philosophy Dynamic in Wilhelminian Germany," *Informationen zur erziehungs- und bildungshistorischen Forschung*, 4 (1976): 63–82.

20 Hermann Helmholtz, "Über das Verhältnis der Naturwissenschaften zur Gesamtheit der Wissenschaft" (1862), *Vorträge und Reden*, 5th ed. (Braunschweig, 1903), p. 183. Dietrich von Engelhardt, "Die Konzeption der Forschung in der Medizin des 19. Jahrhunderts," in Alwin Diemer (ed.), *Konzeption und Begriff der Forschung in den Wissenschaften des 19. Jahrhunderts* (Meisenheim am Glan, 1978), esp. pp. 61–64, 95; Arleen Tuchman, "Experimental Physiology, Medical Reform, and the Politics of Education at the University of Heidelberg: A Case Study," *Bulletin of the History of Medicine*, 61 (1987): 203–215; Timothy Lenoir, "Science for the Clinic: Science Policy and the Formation of Carl Ludwig's Institute in Leipzig," in William Coleman and Frederic L. Holmes (eds.), *The Investigative Enterprise: Experimental Physiology in Nineteenth-Century Medicine* (Berkeley, Calif., 1988), pp. 139–178.

21 Avraham Zloczower, *Career Opportunities and the Growth of Scientific Discovery in Nineteenth-Century Germany* (Jerusalem, 1966); R. Steven Turner, Edward Kerwin, and David Woolwine, "Careers and Creativity in Nineteenth-Century Physiology: Zloczower Redux," *Isis*, 75 (1984): 523–529.

22 Rudolf Stichweh, *Zur Entstehung des modernen Systems wissenschaftlicher Disziplinen: Physik in Deutschland 1740–1890* (Frankfurt a.M., 1984), pp. 345–392, 457–471; David Cahan, "The Institutional Revolution in German Physics, 1865–1914," *HSPS*, 15 (1985): 1–66; Christa Jungnickel and Russell McCormmach, *Intellectual Mastery of Nature: Theoretical Physics from Ohm to Einstein* (Chicago, 1986), chs. 4, 10; Kathryn M. Olesko, "Commentary: On Institutes, Investigations, and Scientific Training," in Coleman and Holmes (eds.), *The Investigative Enterprise*, esp. pp. 306 f.; Lewis Pyenson and Douglas Skopp, "Educating Physicists in Wilhelminian Germany," *Social Studies of Science*, 7 (1977): 329–366.

23 Wundt, *Erlebtes und Erkanntes* (Stuttgart, 1920), pp. 293–295.

24 Wolfgang G. Bringmann, William D. G. Balance, and Rand B. Evans, "Wilhelm Wundt 1832–1920: A Brief Biographical Sketch," *JHBS*, 11 (1975): 294. See also Wolfgang G. Bringmann, Norma J. Bringmann, and Gustav A. Ungerer, "The Establishment of Wundt's Laboratory: An Archival and Documentary Study," in Wolfgang G. Bringmann and Ryan D. Tweney (eds.), *Wundt Studies* (Toronto, 1980), pp. 123–157.

25 Boring, *History*, p. 325; Wundt, *Logik*, 3rd ed., 3 vols. (Stuttgart, 1907–1908); *Erlebtes*, p. 313.

26 Wundt, *Beiträge zur Theorie der Sinneswahrnehmung, Einleitung, Über die Methoden der Psychologie* (Leipzig, 1862). For comparison with later statements of Wundt's program, see William R. Woodward, "Wundt's Program for the New Psychology: Vicissitudes of Experiment, Theory and System," in Woodward and Ash (eds.), *The Problematic Science* (cited in the Introduction), pp. 167–197.

27 Wundt, "Das Institut für experimentelle Psychologie," in *Festschrift zur Feier des 500-jährigen Bestehens der Universität Leipzig* (Leipzig, 1909), vol. 4, pp. 131–132.

28 Kurt Danziger, "Wundt's Psychological Experiment in the Light of His Philosophy of Science," *Psychological Research*, 42 (1980): 113 ff.; idem., "The Origins of the Psychological Experiment as a Social Institution," *American Psychologist*, 40 (1985): 133–140.

29 Miles A. Tinker, "Wundt's Doctoral Students and Their Theses, 1875–1920" (1932), reprinted in Bringmann and Tweney (eds.), *Wundt Studies*, pp. 269–279.

30 Bringmann et al., "The Establishment of Wundt's Laboratory," in *Wundt Studies*, pp. 149–150; Métraux, "Wilhelm Wundt und die Institutionalisierung der Psychologie."

31 Émile Durkheim, "La philosophie dans les universités allemandes" (1887), quoted in Métraux, "Wilhelm Wundt und die Institutionalisierung der Psychologie," pp. 92–93.

32 Wundt, "Psychophysik und experimentelle Psychologie," in Wilhelm Lexis (ed.), *Die deutschen Universitäten* (Berlin, 1893), vol. 1, p. 456.

33 Pierre Bourdieu, *Distinctions: A Social Critique of the Judgment of Taste*, trans. Richard Nice (Cambridge, Mass., 1984).

34 Wundt, *Erlebtes*, p. 305.

35 Wundt, "Das Institut für experimentelle Psychologie," pp. 118–119, 121–122; Ben-David and Collins, "Social Factors," p. 456.

36 On specialization in German philosophy, see Julius Baumann, "Philosophie," in Lexis (ed.), *Die Deutschen Universitäten*, vol. 1, pp. 427–449; Herbert Schnädelbach, *Philosophy in Germany 1831–1933*, trans. Eric Matthews (Cambridge, 1984), ch. 3.

37 "Zur Einführung," *Zeitschrift für Psychologie und Physiologie der Sinnesorgane*, 1 (1890), p. 3. Emphasis in the original. See also Manfred John and Georg Eckardt, "Die 'Zeitschrift für Psychologie' in ihren ersten Jahrzehnten," *ZfP*, 198 (1990): 145–163. The journal was divided into two parts in 1908, one for psychology and the other for sensory physiology, but all issues will also be cited in this book for convenience as *Zeitschrift für Psychologie (ZfP)*.

38 "Zur Einführung."

39 Robert Sommer, "Zur Geschichte der Kongresse für experimentelle Psychologie," *Bericht über den 12. Kongress der Deutschen Gesellschaft für Psychologie* (1932), p. 9.

40 *Bericht über den 1. Kongress für experimentelle Psychologie* (1904), p. xxiii.

41 Hermann Ebbinghaus, *Memory: A Contribution to Experimental Psychology* (1885), trans. Henry A. Ruger and Clara A. Bussenius (repr. New York, 1964), pp. 7, 22 ff., 107–108; Kurt Danziger, "The Positivist Repudiation of Wundt," *JHBS*, 15 (1979): 5–30; idem., *Constructing the Subject*, ch. 3. For further discussion, see Chapter 4.

42 Wolfgang G. Bringmann and Gustav A. Ungerer, "Experimental versus Educational Psychology: Wilhelm Wundt's Letters to Ernst Meumann," *Psychological Research*, 42 (1980): esp. pp. 69–71; Siegfried Jaeger, "Zur Herausbildung von Praxisfeldern der Psychologie," in Mitchell G. Ash and Ulfried Geuter (eds.), *Geschichte der deutschen Psychologie im 20. Jahrhundert. Ein Überlick* (Opladen, 1985), pp. 90 ff.

43 J. B. Maller, "Forty Years of Psychology: A Statistical Analysis of American and European Publications 1894–1933," *Psychological Bulletin*, 31 (1934): esp. p. 539. See also Ben-David, *Scientist's Role*, p. 196, Fig. 2. For the number of philosophy professorships, see von Ferber, *Entwicklung*, p. 207. For the number of experimenting psychologists with chairs, see Appendix 1, Table 1, and subtract the name of William Stern, who was not yet a full professor, and also the names of Ernst Meumann and Hermann Ebbinghaus, who both died in 1909. An informal count in Max Frischeisen-Köhler, "Philosophie und Psychologie," *Die Geisteswissenschaften,* 1 (1914): 371, gives fewer philosophy chairs.

Chapter 2

1 For the following, see Carl Stumpf, "Carl Stumpf," trans. Thekla Hodge and Suzanne Langer, in Carl Murchison (ed.), *A History of Psychology in Autobiography,* vol. 1 (Worcester, Mass., 1930), pp. 389–441.

2 Franz Brentano, *Psychologie vom empirischen Standpunkt,* 2nd ed. (Leipzig, 1924), p. 124; *Psychology from an Empirical Standpoint,* trans. Linda McAlister (London, 1973), p. 88. Emphasis mine. Cf. Barry Smith, *Austrian Philosophy: The Legacy of Franz Brentano* (Chicago and LaSalle, Ill., 1994) esp. pp. 35, 42–43.

3 Carl Stumpf, "Reminiscences of Franz Brentano" (1919), trans. Linda L. McAlister and Margarete Schattle, in Linda McAlister (ed.), *The Philosophy of Franz Brentano* (London, 1976), pp. 11, 18; Oskar Kraus, "Biographical Sketch of Franz Brentano" (1919), trans. Linda McAlister, ibid., p. 4.

4 Stumpf, *Über den psychologischen Ursprung der Raumvorstellung* (Leipzig, 1873), pp. 110 ff. Cf. Stumpf, "Zum Begriff der Lokalzeichen," *ZfP,* 4 (1893): 70–74. For Brentano's doctrine of the unity of consciousness, see his *Psychologie vom empirischen Standpunkt,* vol. 1 (1874), ed. Oskar Kraus (Hamburg, 1959), ch. 2. Stumpf acknowledged Brentano's influence on his argument in "Reminiscences," pp. 42–43.

5 Stumpf, "Reminiscences," pp. 33–34; "Autobiography," p. 398.

6 Stumpf, "Reminiscences," p. 43; *Tonpsychologie,* vol. 1 (Leipzig, 1883), p. 43.

7 Stumpf, "Psychologie und Erkenntnistheorie," *Abhandlungen der königlich bayrischen Akademie der Wissenschaften,* I. Kl., 18 (1891), pp. 482, 491, 501–502.

8 Hermann Lotze, *Logik* (Leipzig, 1874), book 3, ch. 4.

9 Stumpf, "Psychologie und Erkenntnistheorie," p. 508.

10 Wilhelm Dilthey, "Ideen über eine beschreibende und zergliedernde Psychologie" (1894), *Gesammelte Schriften,* vol. 5, 4th ed. (Göttingen, 1974), pp. 139, 184, 199, 193.

11 *Briefwechsel zwischen Wilhelm Dilthey und dem Grafen Paul Yorck von Wartenburg 1877–1897,* ed. Sigrid von der Schulenburg (Halle, 1923), Nr. 94 (Yorck to Dilthey, 2 September 1892), Nr. 121 (Dilthey to Yorck, 13 October 1895), Nr. 107 (Dilthey to Yorck, 1 November 1893).

12 For the following see also Ulrich Jahnke, "Promotor des Fortschritts!? Friedrich Althoff und die deutsche Universitätspsychologie," in Bernhard vom Brocke (ed.), *Wissenschaftsgeschichte und Wissenschaftspolitik im Industriezeitalter: Das 'System Althoff' in historischer Perspektive* (Marburg, 1991), pp. 307–336.

13 Ulrich Sieg, "Im Zeichen der Beharrung. Althoffs Wissenschaftspolitik und die Universitätsphilosophie," in vom Brocke (ed.), *Wissenschaftsgeschichte und Wissenschaftspolitik,* esp. p. 288.

14 Philosophical Faculty to Ministry, 13 July 1893, GStA, Rep 76 Va Sekt. 2 Tit. IV Nr. 61 Bnd. 6 Bl. 193–208. Lothar Sprung and Helga Sprung, "Zur Geschichte der Psychologie and der Berliner Universität (1850–1922)," *Psychologie für die Praxis,* 1 (1985), pp. 14, 16–17; Lothar Sprung and Helga Sprung, "Ebbinghaus an der Berliner Universität – ein akademisches Schicksal eines zu früh Geborenen?" in Werner Traxel (ed.), *Ebbinghaus-*

Studien 2 (Passau, 1987). The Sprungs show that Ebbinghaus was not considered seriously for this position, and attribute this to Dilthey's growing antipathy for Ebbinghaus's scientific approach. Correct as this surely is, younger scholars were also commonly expected to get a full professorship elsewhere first, before being considered worthy of an appointment in Berlin; and Ebbinghaus had not yet taken that implicit career hurdle.

15 Lipps to Althoff, GStA Rep. 92 Althoff Nr. 71, Bl. 81, cited in Jahnke, "Promotor des Fortschritts," p. 330; Althoff to Miquel, 10 August 1893, GStA, Rep. 76 Va Sekt. 2 Tit. X Nr. 150 Bnd. 1 Bl. 13–14; Boring, *History*, p. 365.

16 Stumpf to Althoff, 14 September, 28 September, and 9 October 1893; Dilthey to Althoff, 10 October 1893; Althoff to Dilthey, 13 October 1893; Stumpf to Althoff, 14 October 1893; Dilthey to Althoff, 15 October 1893, GStA, Rep. 76 Va Sekt. 2 Tit. IV Nr. 61 Bnd. 6 Bl. 298–301, 303, 305–307, 309–311.

17 Althoff to Stumpf, 17 October 1893; Stumpf to Althoff, 20 October 1893. GStA, ibid., Bl. 314, 317–320.

18 Althoff to Stumpf, 22 October 1893; Stumpf to Althoff, 25 October 1893; agreement Stumpf–Althoff, 12 December 1893, GStA, ibid., Bl. 256, 321–322, 245, 327–329. Althoff to Ministry, 1 July 1895, GStA Rep. 29 Althoff AI Nr. 13 Bl. 77.

19 UAHUB *Chronik der Friedrich-Wilhelm-Universität zu Berlin,* 1894 ff. See also GStA, Rep. 76 Va Sekt. 2 Tit. X Nr. 150 Bnd. 1, Bl. 308.

20 GStA, ibid., Bl. 94–97, 126–132, 252–257, 306–313; Bnd. 2, Bl. 123–126, 159–161, 215–217.

21 Carl Stumpf, "Das Berliner Phonogrammarchiv," *Internationale Wochenschrift für Wissenschaft, Kunst und Technik,* 22 February 1908; idem, *Die Anfänge der Musik* (Leipzig, 1911). For support from private donors, see Stumpf to Ludwig Darmstaedter, 5 and 29 July 1909 and 4 February 1912, Staatsbibliothek Preussischer Kulturbesitz, Berlin, Sammlung Darmstaedter. For other funding see GStA, Rep. 76 Va Sekt. 2 Tit. X Nr. 150 Bnd. 2 Bl. 9–20, 52 ff., 72 ff., 85–87. Each of Stumpf's requests to the Ministry was accompanied by testimonials from Reichstag representatives, mainly from the liberal Freisinniger Verein.

22 Stumpf, "Das Psychologische Institut," in Max Lenz, *Geschichte der königlichen Friedrich-Wilhelm-Universität zu Berlin,* vol. 3 (Halle, 1910), p. 203.

23 GStA, Rep. 76 (Cited note 19), Bnd. 2, Bl. 213 ff.

24 Stumpf, "Antrittsrede," *Sitzungsberichte der Preussischen Akademie der Wissenschaften* (1895), pp. 736–737.

25 Stumpf to Althoff, 23 May 1894, GStA, Rep. 92 Althoff B Nr. 182 Bnd. 4 Bl. 43–44.

26 Stumpf, "Die Wiedergeburt der Philosophie" (1907), in *Philosophische Reden und Vorträge* (Leipzig, 1910), esp. p. 177.

27 Karl-Heinz Ingenkamp, "Das Institut des Leipziger Lehrervereins 1906–1933 und seine Bedeutung für die Empirische Pädagogik," *Empirische Pädagogik,* 1 (1987): 60–70; Danziger, *Constructing the Subject,* p. 131; Monika Schubeius, *'Und das psychologische Laboratorium muss der Ausgangspunkt pädagogischer Arbeiten werden!' Zur Institutionalisierungsgeschichte der Psychologie 1890–1933* (Frankfurt a.M., 1990).

28 Stumpf, "Zur Methodik der Kinderpsychologie," *Zeitschrift für Kinderpsychologie,* 2 (1900), pp. 19–20; Stumpf, "Carl Stumpf," pp. 404–405.

29 Stumpf, "Zur Einteilung der Wissenschaften," *Abhandlungen der königlich Preussischen Akademie der Wissenschaften,* Phil.-hist. Cl., 1907, pp. 3–4, 5–6.

30 Ibid., pp. 26 ff., esp. 28, 29–30.

31 Stumpf, "Erscheinungen und psychische Funktionen," *Abhandlungen der königlich Preussischen Akademie der Wissenschaften,* Phil.-hist. Cl., 1906, pp. 2 ff., 5–7, 30–31.

32 Stumpf, "Erscheinungen," pp. 8–9; idem., "Leib und Seele" (1896), in *Philosophische Reden und Vorträge* (Berlin, 1910).

33 Stumpf, "Erscheinungen," p. 8; cf. James Rowland Angell, "The Province of Functional Psychology," *Psychological Review* 14 (1907): 61–91; John O'Donnell, *The Origins of Behaviorism: American Psychology, 1890–1920* (New York, 1985).

34 Herbert S Langfeld, "Stumpf's 'Introduction to Psychology'," *American Journal of Psychology,* 50 (1937), p. 33. Langfeld states that this account is based on both his own and Koffka's lecture notes. Unfortunately, Koffka's notes are not among his papers.

35 Ludwig Marcuse, *Mein zwanzigstes Jahrhundert* (Frankfurt a.M., 1968), p. 21. Alois Riehl was a Neo-Kantian philosopher who specialized in philosophy of science. His and Stumpf's names lent themselves to a pun on the German expression "mit Stumpf und Stiehl ausrotten" – to destroy root and branch. The German version of the couplet reads: "Die Philosophie gilt hier nicht viel / Man rottet sie aus mit Stumpf und Riehl."

36 Kurt Koffka, *Principles of Gestalt Psychology* (New York, 1935), p. 53.

37 Quoted in Wilfried Berghahn, *Robert Musil* (Reinbek, 1963), pp. 51–52. See also David S. Luft, *Robert Musil and the Crisis of European Culture, 1880–1942* (Berkeley, Calif., 1980), pp. 76 ff., and David Lindenfeld, *The Transformation of Positivism: Alexius Meinong and European Thought, 1880–1920* (Berkeley, Calif., 1980), pp. 56–57.

38 Selected and summarized from UAHUB, *Chronik der Friedrich-Wilhelm-Universität zu Berlin,* 14 (1900) – 28 (1914).

39 For descriptions of the instruments, see William Stern, "Der Tonvariator," *ZfP,* 30 (1903): 422–432, and Friedrich Schumann, "Die Erkennung von Buchstaben und Worten bei momentaner Beleuchtung," *Bericht über den 1. Kongress für experimentelle Psychologie* (1904), pp. 34–40. The donation from Joachim and Stumpf's purchases are recorded in GStA, Rep. 76 Va Sekt. 2 Tit. X Nr. 150 Bnd. 1 Bl. 225–228, 309; Bnd. 2 Bl. 110 ff.

40 Adrian Brock, "Was macht den psychologischen Expertenstatus aus?" *Psychologie und Geschichte,* 2 (1991), 109–114.

41 Stumpf, "Einteilung," p. 25; Schumann, "Erkennung," p. 34.

42 Langfeld, "Stumpf's 'Introduction'," p. 35.

43 Kurt Lewin, "Carl Stumpf," *Psychological Review,* 44 (1937): 189.

44 Quoted from "Feier zu Carl Stumpfs 70. Geburtstag, 21. April 1918," MWPB. Emphasis mine. C.N.V.F. (*conzentrische Gesichtsfeldeinengung*) is concentric narrowing of the visual field, a pathological condition described by Hermann Helmholtz, later thought to be common in hysterics.

Chapter 3

1 See, e.g., David Katz, "David Katz" (cited in Chapter 1), p. 192.

2 Moritz Geiger, "Über das Wesen und Bedeutung der Einfühlung," *Bericht über den 4. Kongress für experimentelle Psychologie* (1911), 29–73. Marbe's remark is on p. 66.

3 Heinrich Rickert, *Kulturwissenschaft und Naturwissenschaft* (1898), 2nd rev. & enl. ed. (Tübingen, 1910), p. 52.

4 Wilhelm Windelband, *Die Philosophie im deutschen Geistesleben des XIX. Jahrhunderts. Funf Vorlesungen* (Tübingen, 1909), p. 92.

5 Ibid., p. 93; see also "Die Erneuerung des Hegelianismus" (1910), in *Präludien. Aufsätze und Reden zur Einführung in die Philosophie,* 7th–8th ed. (Tübingen, 1921), vol. 1, pp. 273–289.

6 See, e.g., McClelland, *State, Society and University* (cited in Chapter 1), pp. 304 ff.

7 On this aspect of Neo-Kantianism's situation, see Thomas Willey, *Back to Kant* (Detroit, 1978) and Köhnke, *The Rise of Neo-Kantianism* (cited in Chapter 1), pp. 269 ff. Both authors are well aware that there were differences of political opinion among the Neo-Kantians, most notably between the left-liberal Marburg school and the more conservative Southwest German school. But the commitment to transcendent values and norms and the opposition to relativism is common to them all.

8 William Stern, "William Stern," in Carl Murchison (ed.), *A History of Psychology in Autobiography,* vol. 1 (Worcester, Mass., 1930), 335–388; idem., *Person und Sache* (Leipzig, 1906).

9 Edmund Husserl, "Philosophie als strenge Wissenschaft," *Logos,* 1 (1910–1911): 319, 320–321; idem., "Philosophy as Rigorous Science," trans. Quentin Lauer, in *Edmund Husserl: Phenomenology and the Crisis of Psychology* (New York, 1965), pp. 118, 119–120.

10 Husserl, "Philosophie," p. 321 and n. 1; Lauer, *Husserl,* p. 120 n. g.

11 Richard Goldschmidt, "Bericht über den 5. Kongress der Gesellschaft für experimentelle Psychologie," *AfgP,* 24 (1912): 96.

12 Clemens Baeumker, "Oswald Külpe," *Jahrbuch der königlich bayrischen Akademie der Wissenschaften* (1916): 73–107; David Lindenfeld, "Oswald Külpe and the Würzburg School," *JHBS,* 14 (1978): 132–141. See also Appendix 1, Table 1.

13 Oswald Külpe, "Psychologie und Medizin," *Zeitschrift für Pathopsychologie,* 1 (1912): 266, 190, 229.

14 Willy Hellpach, review of Külpe, "Psychologie und Medizin," and other articles, *ZfP,* 64 (1912): 440; cf. H. Schwartz, review of Külpe, *Zeitschrift für Philosophie und philosophische Kritik,* 149 (1912): 126–127. On the more recent history, see K. Hauss et al. (eds.), *Medizinische Psychologie im Grundriss* (Göttingen, 1976), Vorwort.

15 Kurt Koffka, review of Külpe, *Deutsche Literaturzeitung,* 33 (1912): 2272, 2274.

16 Wolfgang Köhler, review of Georg Anschütz, "Tendenzen im psychologischen Empirismus der Gegenwart. Eine Erwiderung auf O. Külpes Ausführungen 'Psychologie und Medizin' und 'Über die Bedeutung der modernen Denkpsychologie'," *ZfP,* 64 (1913): 441. Anschütz replied in *ZfP,* 66 (1913): 155–160; Köhler answered on pp. 319–320. Anschutz replied yet again in *ZfP,* 67 (1913): 506; Köhler's "Schlussbemerkung" follows on the same page.

17 Paul Natorp, "Das akademische Erbe Hermann Cohens. Psychologie oder Philosophie?" *Frankfurter Zeitung,* 12 October 1912. Ulrich Sieg, "Psychologie als 'Wirklichkeitswissenschaft: Erich Jaenschs Auseinandersetzung mit der 'Marburger Schule'," in Winfried Speitkamp (ed.), *Staat, Gesellschaft, Wissenschaft: Beiträge zur modernen hessischen Geschichte* (Marburg, 1994), pp. 317 ff.

18 For the numbers, see Karl Marbe, *Die Aktion gegen die Psychologie: Eine Abwehr* (Leipzig, 1913). Some of the signers, such as Ernst Cassirer and Hans Vaihinger, who sympathized intellectually with the experimenters, may have been taken in by the statement's opening disclaimer. For a more detailed account and a translation of the petition, see my article, "Wilhelm Wundt and Oswald Külpe on the Institutional Status of Psychology," in Bringmann and Tweney (eds.), *Wundt Studies,* esp. pp. 407 ff.

19 Wundt, "Die Psychologie im Kampf ums Dasein" (cited in Chapter 1), pp. 528, 533.

20 Ibid., p. 543.

21 See, e.g., Marbe, *Die Aktion gegen die Psychologie;* Moritz Geiger, "Philosophie und Psychologie an den deutschen Universitäten," *Süddeutsche Monatshefte,* 10 (1913): 755.

22 Karl Lamprecht, "Eine Gefahr für die Geisteswissenschaften," *Die Zukunft,* 83 (1913): 18.

23 Georg Simmel, "An Herrn Prof. Karl Lamprecht," *Die Zukunft,* 83 (1913): 233. Fechner's law is discussed in Chapter 4.

24 Ibid.

25 Aloys Fischer, "Philosophie und Psychologie," p. 344; Külpe, "Philosophie," in *Deutschland unter Kaiser Wilhelm II* (Berlin, 1914), vol. 3, p. 9.

26 Edwin Rausch, interview with the author, Frankfurt am Main, 22 September 1978. Rausch states that Schumann told him this in the 1930s.

27 Karl Marbe, "Die Stellung und Behandlung der Psychologie an den deutschen Universitäten," *Bericht über den 7. Kongress der Gesellschaft für experimentelle Psychologie* (1922), pp. 150 ff.

28 Karl Marbe, "Die Bedeutung der Psychologie für die übrigen Wissenschaften und für die Praxis," *Fortschritte der Psychologie und ihre Anwendungen,* 1 (1913): 5–82; August Messer, "Die Bedeutung der Psychologie für Pädagogik, Medizin, Jurisprudenz und Nationalokonomie," *Jahrbücher der Philosophie,* 2 (1914): 183–217; Aloys Fischer, "Der praktische Psychologe – ein neuer Beruf," *Der Kunstwart,* 1913: 305–313. See also O'Donnell, *The Origins of Behaviorism;* Danziger, "Social Origins of Modern Psychology" (all cited in Chapter 1).

29 Franz Hillebrand, "Die Aussperrung der Psychologen," *ZfP,* 67 (1913): 13; cf. Marbe, "Die Bedeutung der Psychologie," pp. 77 ff.

Chapter 4

1 Katherine Arens, *Structures of Knowing: Psychologies of the Nineteenth Century* (Dordrecht, 1989).

2 The term is taken from Schnädelbach, *Philosophy in Germany* (cited in Chapter 1).

3 R. Steven Turner, "Consensus and Controversy: Helmholtz on the Visual Perception of Space," in David Cahan (ed.), *Hermann von Helmholtz and the Foundations of Nineteenth-Century Science* (Berkeley, Calif., 1993); idem., *In the Eye's Mind: Vision and the Helmholtz–Hering Controversy* (Princeton, N.J., 1994), p. 76.

4 Hermann Helmholtz, "The Conservation of Force: A Physical Memoir" (1847), trans. in Russell Kahl (ed.), *The Selected Writings of Hermann von Helmholtz* (Middletown, Conn., 1971), p. 5.

5 Helmholtz, "Recent Progress in the Theory of Vision" (1868), in *Selected Writings,* p. 153; cf. *Treatise on Physiological Optics,* trans. J. P. C. Southall (New York, 1924–1925), vol. 2, pp. 143–146.

6 Helmholtz, *Die Lehre von den Tonempfindungen als physiologischer Grundlage für die Theorie der Musik* (1862), 4th rev. ed. (Braunschweig, 1877), pp. 235 ff., esp. 242–244.

7 Helmholtz, "Recent Progress," in *Selected Writings*, p. 168. The telegraph analogy was first enunciated in *Tonempfindungen*, p. 245.

8 Timothy Lenoir, "Helmholtz, Müller und die Erziehung der Sinne," in Michael Hagner and Bettina Wahrig-Schmidt (eds.), *Johannes Müller und die Physiologie* (Berlin, 1992).

9 Helmholtz, "Recent Progress," in *Selected Writings*, pp. 174–175; cf. *Optics*, vol. 3, p. 4.

10 Helmholtz, *Optics* vol. 3, esp. p. 5; cf. p. 155. The characterization of the logical structure of Helmholtz's "unconscious conclusions" comes from Hatfield, *The Natural and the Normative* (cited in the introduction), p. 167, but he writes that for Helmholtz, "scientific inference is the psychology of perception writ large." For Lotze's theory of local signs and the use Helmholtz and others made of it, see William Woodward, "From Association to Gestalt: The Fate of Hermann Lotze's Theory of Spatial Perception," *Isis*, 29 (1978): 572–582.

11 Helmholtz, *Optics*, vol. 3, pp. 12, 28; on "compounded" tone sensations, see *Tonempfindungen*, pp. 170–176.

12 Helmholtz, "The Facts of Perception" (1878), in *Selected Writings*, pp. 376, 386–387; Hatfield, *The Natural and the Normative*, pp. 199–200. On Helmholtz and Kant see Mandelbaum, *History, Man and Reason* (Baltimore, 1971), pp. 292 ff.; Gerlof Verwey, *Psychiatry in an Anthropological and Biomedical Context* (Dordrecht, 1985), ch. 2; Hatfield, *The Natural and the Normative*, esp. p. 169; Lenoir, "Helmholtz, Müller und die Erziehung der Sinne"; on Helmholtz and Fichte, see R. Steven Turner, "Hermann Helmholtz and the Empiricist Vision," *JHBS*, 13 (1978): esp. pp. 55–58.

13 Emil Du Bois-Reymond, *Die Grenzen des Naturerkennens* (1871), 11th ed. (Leipzig, 1916), p. 41.

14 R. Steven Turner, "Helmholtz, Sensory Physiology and the Disciplinary Development of German Psychology," in Woodward and Ash (eds.), *The Problematic Science* (cited in the Introduction), pp. 147–166; idem., "Paradigms and Productivity: The Case of Physiological Optics 1840–94," *Social Studies of Science*, 17 (1987): 35–68.

15 Stumpf, "Hermann von Helmholtz and the New Psychology," *Psychological Review*, 2 (1895): 2, 6.

16 Ewald Hering, "Der Raumsinn und die Bewegung des Auges," in L. Hermann (ed.), *Handbuch der Physiologie*, vol. 3:1 (Leipzig, 1879), pp. 569–573.

17 Hering, "Zur Lehre vom Lichtsinne" (1874), reprinted in *Wissenschaftliche Abhandlungen* (Leipzig, 1931), p. 6; Turner, "Paradigms and Productivity," p. 51; idem., *In the Eye's Mind* (cited in note 3), pp. 120 ff.

18 Hering, *Outlines of a Theory of the Light Sense* (1905 ff.), trans. Leo M. Hurvich and Dorothea Jameson (Cambridge, Mass., 1964), p. 1.

19 Hering, *Beiträge zur Physiologie* (Leipzig, 1862), p. 166.

20 Hering, "Zur Lehre vom Lichtsinne," sec. 24, quoted in *Light Sense*, pp. 223–224. Emphasis in the original.

21 Ernst Mach, "On the Effect of the Spatial Distribution of the Light Stimulation on the Retina" (1865), in Floyd Ratliff, *Mach Bands: Quantitative Studies in Neural Networks in the Retina* (San Francisco, 1965), p. 269; idem., "On the Physiological Effect of Spatially Distributed Light Stimuli" (1866), ibid., p. 283.

22 Hering, *Light Sense*, pp. 123–124, 166–167, 170. On the relationship of Hering and Mach, see Richard Kremer, "From Psychophysics to Phenomenalism: Mach and Hering on

Color Vision," in Mary Jo Nye (ed.), *The Invention of Physical Science* (Kluwer, 1992), pp. 147–173.

23 The theory was first presented in "Zur Lehre vom Lichtsinne," sec. 27, and the equilibrium idea generalized in "Zur Theorie der Vorgänge in der lebendigen Substanz" (1888). I am relying here on the formulations in *Light Sense,* esp. ch. 4, para. 22–23. For "memory colors" see pp. 6 ff.; on the heritability of traces, see "Über das Gedächtnis als allgemeine Funktion der organischen Materie" (1870) reprinted in Oskar Loerke and Peter Suhrkamp (eds.), *Deutscher Geist: Ein Lesebuch aus zwei Jahrhurderten* (Frankfurt a.M., 1966), vol. 2, p. 371.

24 Anson Rabinbach, *The Human Motor: Energy, Fatigue and the Origins of Modernity* (New York, 1990); David Cahan, "Helmholtz and the Civilizing Power of Science," in idem. (ed.), *Hermann von Helmholtz* (cited in note 3).

25 Helmholtz, "Über das Verhältnis der Naturwissenschaften zur Gesamtheit der Wissenschaft" (1862), *Vorträge und Reden,* 5th ed. (Braunschweig, 1903), p. 183; *Physiological Optics,* vol. 3, p. 31. Hatfield suggests that by the 1870s Helmholz had already begun to modify this militant-sounding mechanistic language and bring scientific creativity and artistic intuition closer together. See "Helmholtz and Classicism: The Science of Aesthetics and the Aesthetics of Science," in Cahan (ed.), *Hermann von Helmholtz* (cited in note 3).

26 Cited in Kremer, "From Psychophysics to Phenomenalism," p. 149; cf. Turner, *In the Eye's Mind,* p. 121.

27 Leo M. Hurvich, "Hering and the Scientific Establishment," *American Psychologist,* 24 (1969): 497–513. For an extensive analysis, see Turner, *In the Eye's Mind,* Part Three.

28 Hering, "Zur Lehre vom Lichtsinne" (1878), repr. in *Wissenschaftliche Abhandlungen* (Leipzig, 1931), pp. 216–217.

29 Mach, "Spatial Distribution," pp. 269–270.

30 Koffka, *Principles,* p. 63.

31 G. E. Müller, "Zur Psychophysik der Gesichtsempfindungen," *ZfP,* 10 (1896), 1 ff.; Stumpf, "Erscheinungen und psychische Funktionen" (cited in Chapter 2), p. 7. Müller cited Mach's essay of 1865, along with similar statements by Fechner, Lotze, and Hering.

32 Franz Hillebrand, "Die Stabilität der Raumwerte auf der Netzhaut," *ZfP,* 5 (1893): 58.

33 Walter Poppelreuter, "Über die Bedeutung der scheinbaren Grösse und Gestalt für die Gesichtswahrnehmung," *ZfP,* 54 (1910): 317–18, 342; idem., "Beiträge zur Raumpsychologie," *ZfP,* 58 (1911): 200–262. See also Walter Blumenfeld, "Untersuchungen über die scheinbare Grösse im Sehraume," *ZfP,* 65 (1913): 242–404. For an account of earlier research on apparent size, see Edwin G. Boring, *Sensation and Perception in the History of Experimental Psychology,* (New York, 1942), pp. 292 ff.; cf. the appendixes by Johannes von Kries in Helmholtz, *Optics,* vol. 3, esp. pp. 390 ff.

34 David Katz, "Die Erscheinungsweisen der Farben und ihre Beeinflussung durch die individuelle Erfahrung," *ZfP,* Ergänzungsband 7 (1911): 7 ff. For Hering's demonstrations and his descriptions of "luster" and volume colors, see *Light Sense,* pp. 8, 12–13.

35 Katz, "Erscheinungsweisen," passim.

36 Ibid., pp. 227, 300, 374 ff.

37 Von Kries, appendixes to Helmholtz, *Optics,* vol. 3, esp. pp. 562, 621, 644 ff.

38 Turner, "Paradigms and Productivity," p. 47; idem., "Vision Studies in Germany: Helmholtz versus Hering," *Osiris,* 8 (1993): 80–103, esp. pp. 96 ff.; *In the Eye's Mind,* ch. 13.

39 See, e.g., Kurt Danziger, "Wilhelm Wundt and the Two Traditions of Psychology," in Robert W. Rieber (ed.), *Wilhelm Wundt and the Making of a Scientific Psychology* (New York, 1980).

40 Gustav Theodor Fechner, *Elemente der Psychophysik* (Leipzig, 1860), vol. 1, pp. 34 ff.; Marilyn Marshall, "Physics, Metaphysics, and Gustav Fechner's Psychophysics," in Woodward and Ash (eds.), *The Problematic Science;* Turner, "Helmholtz, Sensory Physiology" (cited above, n. 14).

41 Wundt, *Grundriss der Psychologie,* (1896), 8th ed. (Leipzig, 1907), pp. 14 ff., 52 ff., 319. For the role of apperception in Wundt's mature system, see Woodward, "Wundt's Program," and esp. Danziger, "Wundt and the Two Traditions of Psychology."

42 Wundt, *Grundriss,* pp. 18–19, 400; idem., "Über die Einteilung der Wissenschaften," in *Kleine Schriften,* vol. 3, p. 1–53, esp. pp. 28, 45.

43 Wundt, "Erfundene Empfindungen," *Philosophische Studien,* 2 (1885): 299, quoted in Danziger, "The History of Introspection Reconsidered," *JHBS* 16 (1980): 249–250; cf. Danziger, "Wundt's Psychological Experiment" (cited in Chapter 1).

44 Wundt, *Grundriss,* pp. 35–36.

45 Wundt, "Über psychische Kausalität," pp. 112 ff.; cf. idem., *Grundzüge,* 6th ed., vol. 3, pp. 755 ff., where he speaks of the "principle of creative resultants."

46 See, e.g., Theo Herrmann, "Ganzheitspsychologie und Gestalttheorie," in Heinrich Balmer, ed., *Die Psychologie des 20. Jahrhunderts,* vol. 2, *Die europäische Tradition* (Zürich, 1979), esp. pp. 576–577. For a different view by a student of the Gestalt theorists, see Wolfgang Metzger, *Psychologie,* 4th ed. (Darmstadt, 1968), pp. 55 ff.

47 Ernst Mach, *Die Geschichte und die Wurzel des Satzes von der Erhaltung der Arbeit* (Prague, 1872), pp. 25–26, 31, 34–35. On the role of Mach's psychology in his rejection of absolute space and time and of atomism, see Erwin Hiebert, "The Genesis of Mach's Early Views on Atomism," in Robert S. Cohen and Raymond J. Seeger (eds.), *Ernst Mach: Physicist and Philosopher* (Dordrecht and Boston, 1970), esp. pp. 96 ff.

48 Mach, *The Analysis of Sensations and the Relation of the Physical to the Psychical* (1886), 5th ed. (1904), trans. C. M. Williams and Sydney Waterlow (repr. New York, 1959). I have cited this rather than earlier editions because this is the one that the Gestalt theorists probably read when they were students.

49 Mach, *Analysis,* pp. 6, 21, 23–24.

50 Ibid., p. 13 and p. 13, n. 1.

51 Mach, *Analysis,* pp. 16–17, 37. For Mach's use of the term "elements" as an escape from solipsism, see Robert S. Cohen, "Ernst Mach: Physics, Perception and the Philosophy of Science," in Cohen and Seeger (eds.), *Ernst Mach: Physicist and Philosopher,* p. 128. The phrase "connecting the appearances" comes from John T. Blackmore, *Ernst Mach: His Work, Life and Influence* (Berkeley, Calif., 1972), pp. 34–35.

52 Mach, *Analysis,* pp. 310–311.

53 Richard Avenarius, *Der menschliche Weltbegriff* (Leipzig, 1891), pp. 4 f. and Sect. 150.

54 "Müsset im Naturbetrachten/ Immer eins wie alles achten;/ Nichts ist drinnen, nichts ist draussen; Denn was innen, das ist aussen; So ergreifet ohne Säumnis/ Heilig öffentlich Geheimnis." Goethe, "Epirrhema" (1820), quoted in Avenarius, "Bemerkungen zum Begriff des Gegenstandes der Psychologie," *Vierteljahresschrift für wissenschaftliche Philosophie,* 18 (1894): 400.

55 Avenarius, "Bemerkungen," vol. 18, pp. 417–418, and vol. 19, p. 13; idem., *Kritik der reinen Erfahrung* (Leipzig, 1888 ff.), sec. 69 ff.

56 Mach, *Analysis,* p. 62.

57 Gereon Wolters, "Verschmähte Liebe – Mach, Fechner und die Psychophysik," in Josef Brozek and Horst Gundlach (eds.), *G. T. Fechner and Psychology* (Passau, 1988), pp. 103–116; cf. Blackmore, *Ernst Mach,* pp. 14–15, 30–31. See also Michael Heidelberger, *Die innere Seife der Natur: Gustav Theodor Fechner's Wissenschaftliche Weltauffassung* (Frankfurt am Main, 1993).

58 Stumpf, review of Mach in *Deutsche Literaturzeitung,* 27 (1886), 947–948; cf. Blackmore, *Ernst Mach,* p. 120 ff.

59 Kurt Danziger, "The Positivist Repudiation of Wundt" (cited in Chapter 1); idem, *Constructing the Subject,* ch. 3.

60 Oswald Külpe, *Grundriss der Psychologie* (Leipzig, 1893), pp. 4, 215 ff. See also Danziger, "The Positivist Repudiation of Wundt" and David Lindenfeld, "Oswald Külpe and the Würzburg School" (both cited in Chapter 1). Külpe did not indicate his own sources clearly in 1893. In 1895, however, after accepting a chair of philosophy in Würzburg, he published a widely read introduction to philosophy in which he acknowledged his debt to Avenarius.

61 Ebbinghaus, *Memory* (cited in Chapter 1), pp. 13–14.

62 Ebbinghaus, *Grundzüge der Psychologie,* vol. 2 (Leipzig, 1897), pp. 7, 549; idem., "Die Psychologie jetzt und vor hundert Jahren," in *3. Congrès Internationale de Psychologie* (Paris, 1901), p. 59. Cf. Danziger, "Positivist Repudiation," p. 229, n. 63.

63 Ebbinghaus, *Abriss,* p. 51. Emphasis mine.

64 William Stern, "Die psychologische Arbeit des neunzehnten Jahrhunderts," *Zeitschrift für pädagogische Psychologie,* 2 (1900): 414.

65 M. Norton Wise, "Mediating Machines," *Science in Context,* 2 (1988): 77–113.

Chapter 5

1 Antonio Aliotta, *The Idealistic Reaction against Science* (1912), trans. Agnes McGaskill (London, 1914); H. Stuart Hughes, *Consciousness and Society* (New York, 1958), ch. 2. Klaus Christian Köhnke, in *The Rise of Neo-Kantianism* (cited in Chapter 1), ch. 7, speaks more cautiously of a differentiation *within* philosophy between Neo-Kantianism and positivism.

2 David Lindenfeld, *The Transformation of Positivism: Alexius Meinong and European Thought, 1880–1920* (Berkeley, Calif., 1980); James T. Kloppenberg, *Uncertain Victory: Social Democracy and Progressivism in European and American Thought* (New York, 1986), esp. pp. 3–4, and ch. 2.

3 Anthony Edward Pilkington, *Bergson and His Influence – a Reassessment* (Cambridge, 1976); on Bergson and the revival of metaphysics in German philosophy, see O. Ewald, "Die deutsche Philosophie im Jahre 1912," *Kant Studien* 18 (1913): 339–382.

4 Henri Bergson, *Essai sur les données immédiates de la conscience* (1889), 96th ed. (Paris, 1961), p. 4.

5 Bergson, "Introduction to Metaphysics" (1903), trans. in *The Creative Mind* (New York, 1946), pp. 213–214, 216. I have altered the translation slightly.

6 Mach, *Analysis,* p. 376; Bergson, "Introduction," p. 207.

7 Bergson, *Creative Evolution* (1907), trans. Arthur Mitchell (1911; repr. New York, 1944), esp. p. 36; Vladimir Jankélévitch, *Henri Bergson* (Paris, 1959), p. 10.

8 William James, *The Principles of Psychology* (1890; repr. New York, 1950), vol. 1, p. 196.

9 Ibid., vol. 1, p. 521 n.; for similar criticisms of Lotze and Stumpf, cf. pp. 522–523 n. For a discussion of the argument, cf. Richard High, "Does James's Criticism of Helmholtz Really Involve a Contradiction?" *JHBS,* 14 (1978), 337–343.

10 James, *Principles,* vol. 1, esp. pp. 258–259, 265, 278–279.

11 Ibid., p. 277. On artistic metaphors in James, see David E. Leary, "William James and the Art of Human Understanding," *American Psychologist,* 47 (1992): 152–160; Jacques Barzun, "William James – The Mind as Artist," in S. Koch and D. E. Leary (eds.), *A Century of Psychology as Science* (New York, 1985), pp. 904–910; Daniel Bjork, *The Compromised Scientist: William James in the Development of American Psychology* (New York, 1983); on James and evolutionary theory see Robert J. Richards, *Darwin and the Emergence of Evolutionary Theories of Mind and Behavior* (Chicago, 1987).

12 Ibid., vol. 1, p. 177. Emphasis mine.

13 Quoted in Nicholas Pastore, *Selective History of Theories of Visual Perception 1650–1950* (Oxford, 1971), p. 245. Pastore recognizes that this is an anticipation of Gestalt theory, but moves (on p. 401, n. 24) to weaken the point by saying that "James actually adopted the mosaic model of the functioning of the nervous system when he explained some fact in detail."

14 James, *A Pluralistic Universe* (New York, 1909), pp. 188–189.

15 For this assessment, see Ralph Barton Perry, *The Thought and Character of William James* (Boston, 1935–1936), vol. 1, pp. 461–462. Not surprisingly, Perry's statement agrees with the evaluation of his friend Wolfgang Köhler, in *Gestalt Psychology,* 2nd ed. (New York, 1949), pp. 198 f.

16 Langfeld, "Stumpf's 'Introduction to Psychology'" (cited in Chapter 2), p. 37.

17 Dilthey, "Ideen über eine beschreibende und zergliedernde Psychologie" (cited in Chapter 2), pp. 144, 193–194.

18 Ibid., pp. 143–145.

19 Ibid., pp. 145, 162, 166–167. Emphasis mine.

20 Ibid., pp. 144, 172. For further discussion see Frithjof Rodhi, "Dilthey's Concept of Structure within the Context of 19th-Century Science and Philosophy," in Rudolf A. Makkreel and J. Scanlon (eds.), *Dilthey and Phenomenology* (Pittsburgh and Washington, D.C., 1987).

21 Ibid., pp. 225 f., 232, 236–237. For Dilthey's earlier views, see his *Einleitung in die Geisteswissenschaften* (1883), in *Gesammelte Schriften,* vol. 1, ed. Bernhard Groethuysen (1922), 4th ed. (Göttingen, 1959), e.g. p. 29. Cf. Rudolf A. Makkreel, *Wilhelm Dilthey: Philosopher of the Human Studies* (Princeton, N.J., 1975), pp. 131–132.

22 Dilthey, "Ideen," p. 157; cf. idem., "Beiträge zum Studium der Individualität" (1896), in *Gesammelte Schriften,* vol. 5, esp. 241 ff. See also Angelika Ebrecht, *Das individuelle Ganze: Zum Psychologismus der Lebensphilosophie* (Stuttgart, 1992).

23 Ebbinghaus, "Über erklärende und beschreibende Psychologie," *ZfP,* 9 (1896): esp. pp. 182 ff.

24 Dilthey's reply is in *Gesammelte Schriften,* vol. 5, pp. 237–240; cf. "Ideen," pp. 171, 173, and Makreel, *Wilhelm Dilthey,* pp. 208–209.

25 Traugott Konstantin Österreich, ed., *Überwegs Geschichte der Philosophie,* vol. 5 (Berlin, 1923), p. 512.

26 Husserl, "Reminiscences of Franz Brentano," in Linda McAlister (ed.), *The Philosophy of Franz Brentano* (cited in Chapter 2), esp. p. 48; Gottlob Frege, "Dr. E. Husserl's Philosophy of Arithmetic," in J. Mohanty (ed.), *Readings in Edmund Husserl's Logical Investigations* (The Hague, 1977), pp. 6–21; Dallas Willard, *Logic and the Objectivity of Knowledge: Studies in Husserl's Early Philosophy* (Athens, Ohio, 1984).

27 Husserl, *Logische Untersuchungen,* vol. 1 (Halle, 1900), 4, para. 22, 9; ch. 7, esp. pp. 124–125; p. 188.

28 Ibid., vol. 2 (Halle, 1901), pp. 7–8. Husserl states that his use of the word "psychologist" is the same as Stumpf's in vol. 1, p. 52, n. 1.

29 Ibid., vol. 2, pp. 75–76.

30 Ibid., vol. 2, pp. 193–194, 196.

31 *Ibid.,* vol. 1, pp. 189–190; vol. 2, p. 200. Husserl acknowledged that these views were similar in some respects to the "theory of objects" that Alexius Meinong was developing at just this time (see Chapter 6).

32 Ibid., vol. 2, esp. pp. 198, 227 ff., 230 ff. For useful discussions of these issues, see R. Sokolowski, "The Logic of Parts and Wholes in Husserl's Investigations," in Mohanty (ed.), *Readings,* esp. pp. 100 ff., and Barry Smith (ed.), *Parts and Moments: Studies in Logic and Formal Ontology* (Munich, 1982), ch. 1. The computer example comes from Barry Smith, "Gestalt Theory" (cited in the Introduction), p. 21; see also *Austrian Philosophy* (cited in Chapter 2), p. 252.

33 Husserl, *Logische Untersuchungen,* vol. 2, pp. 268 ff.; cf. Barry Smith, "Ontologische Aspekte der Husserlschen Phänomenologie," *Husserl Studies,* 3 (1986): esp. p. 118.

34 Husserl, *Logische Untersuchungen,* vol. 2, pp. 16, 18.

35 Katz, Erscheinungsweisen (cited in Chapter 4), p. 30; idem., "David Katz" (cited in Chapter 3), p. 194. In 1911, Katz hastened to add that Husserl had not carried out analyses of color in his courses, and that the type of description involved had already been used by Hering. He may have done this to placate Müller, whose professional relations with Husserl were not of the best. For the relation of Katz's work to Husserl's phenomenology, see Herbert Spiegelberg, *Phenomenology in Psychology and Psychiatry* (Evanston, Ill., 1972), pp. 41 ff.

36 Husserl, *The Phenomenology of Internal Time-Consciousness,* ed. Martin Heidegger (1928), trans. James S. Churchill (Bloomington, Ind., 1964), esp. pp. 44 ff.; Stern, "Psychische Präsenzzeit," *ZfP,* 13 (1897): 325–349; Alfred Brunswig, *Das Vergleichen und die Relationserkenntnis* (Leipzig and Berlin, 1910); Wilhelm Schapp, *Beiträge zur Phänomenologie der Wahrnehmung* (Göttingen, 1910), pp. 17 f., 114 ff.; cf. Spiegelberg, *Phenomenology in Psychology and Psychiatry,* pp. 54 f.

37 Henry J. Watt, "Experimental Contribution to a Theory of Thinking" (1905), repr. in Jean M. Mandler and George Mandler, *Thinking: From Association to Gestalt* (New York, 1964), pp. 189–200; Oswald Külpe, "Versuche über Abstraktionen," *Bericht über den 1. Kongress für experimentelle Psychologie* (1904), p. 61; Ach, *Willenstätigkeit und Denken,* pp. 188f., 192.

38 August Messer, *Empfindung und Denken* (Leipzig, 1908), pp. 5, 80, 91; Karl Bühler, "Tatsachen und Probleme zu einer Psychologie der Denkvorgänge, I. Über Gedanken," *AfgP,* 9 (1907): esp. pp. 346 ff.

39 Külpe, *Die Realisierung*, vol. 1 (Leipzig, 1912), p. 10; cf. Husserl, *Logische Untersuchungen*, vol. 2, p. 714.

40 For a full discussion of the methodological issues, see Kurt Danziger, "The History of Introspection Reconsidered," *JHBS*, 16 (1980): 241–262.

41 Ernst Cassirer, *Substance and Function*, p. 345.

42 Külpe, *Die Philosophie der Gegenwart in Deutschland* (1902), 6th ed. (Leipzig, 1914), pp. 29–30.

43 Külpe, "Über die Bedeutung der modernen Denkpsychologie" (1912), in *Vorlesungen über Psychologie*, ed. Karl Bühler, 2nd ed. (Leipzig, 1922), pp. 311–312.

44 Husserl, "A Reply to a Critic of My Refutation of Logical Psychologism" (1903), in Mohanty (ed.), *Readings*, pp. 3–42; Herbert Spiegelberg, *The Phenomenological Movement* (The Hague, 1960), vol. 1, pp. 119 ff., 150 ff.

45 For a discussion of these criticisms, see Ernst Cassirer, *The Problem of Knowledge: Philosophy, Science and History since Hegel*, trans. Charles Hendel (New Haven, Conn., 1950), esp. pp. 194, 207; on teleomechanism, see Timothy Lenoir, *The Strategy of Life* (Dordrecht and Boston, 1982).

46 Wilhelm Roux, introduction to vol. 1 of *Archiv für Entwicklungsmechanik* (1894), quoted in Garland Allen, *The Life Sciences in the Twentieth Century*, 2nd ed. (Cambridge, 1978), p. 34. For this and the next paragraph, see Allen, esp. pp. 25 ff., 29 ff.; and Jane Maienschein, "Preformation or New Formation – or Neither or Both," in T. J. Horder et al. (eds.), *A History of Embryology* (Oxford, 1986), esp. pp. 79 ff.

47 On Loeb, see Philip Pauly, *Controlling Life: Jacques Loeb and the Engineering Ideal in Biology* (New York, 1986); on the emergence of Driesch's vitalism, see Frederick B. Churchill, "From Machine-Theory to Entelechy: Two Studies in Developmental Teleology," *Journal of the History of Biology*, 2 (1969): 65–86.

48 Eduard von Hartmann, *Die moderne Psychologie* (Leipzig, 1901); Hans Driesch, *Die Seele als elementarer Naturfaktor* (Leipzig, 1903), pp. 52 ff., 62–63.

49 Driesch, *Seele*, pp. 66, 71, 74–76.

50 *Chronik der Friedrich-Wilhelm-Universität zu Berlin*, 22 (1908), pp. 72–73.

51 Erich Becher, "Kritik der Wiederlegung des Parallelismus . . . durch Hans Driesch," *ZfP*, 45 (1907): esp. pp. 416–417, 422–423; idem., "Das Gesetz der Erhaltung der Energie und die Annahme einer Wechselwirkung Zwischen Leib und Seele," *ZfP*, 46 (1908): pp. 97–98; Stumpf, "Leib und Seele" (cited in Chapter 2), pp. 12–13, quoted in Becher, "Erhaltung der Energie," p. 114.

52 Driesch, *The Science and Philosophy of the Organism* (Aberdeen, 1909), esp. vol. 2, pp. 78, 258, 282 ff.

53 On the reception of Driesch's lectures, see Horst Heinz Freyhofer, "The Vitalism of Hans Driesch," Ph.D. Dissertation, University of California, Los Angeles, 1979, pp. 94 ff.; Cassirer, *The Problem of Knowledge*, p. 197.

54 Driesch, *Lebenserinnerungen* (Munich, 1951), pp. 142 ff.

Chapter 6

1 Harald Höffding, "Über Wiedererkennen, Association und psychische Aktivität," *Vierteljahresschrift für wissenschaftliche Philosophie*, 13 (1889): pp. 421–422, 431, 442, 446; idem., *Psychologie im Grundriss auf Grundlage der Erfahrung* (1882), 5th German ed.

(Leipzig, 1914), p. 180. See also Ingemar Nilsson, "Alfred Lehmann and Psychology as a Physical Science," in Bringmann and Tweney (eds.), *Wundt Studies.*

2 Benno Erdmann and Raymond Dodge, *Psychologische Untersuchungen über das Lesen auf experimenteller Grundlage* (Halle, 1898), pp. 148–149, 161.

3 Friedrich Schumann, "Psychologie des Lesens," *Bericht über den 2. Kongress für experimentelle Psychologie* (1907), p. 170.

4 W. Katzaroff, "Contribution a l'étude de la recognition," *Archiv de Psychologie,* 11 (1911): 2–78.

5 Paul A. Bogard, "Heaps or Wholes: Aristotle's Explanation of Compound Bodies," *Isis,* 70 (1979): esp. p. 13.

6 Timothy Lenoir, "The Eternal Laws of Form: Morphotypes and the Conditions of Existence in Goethe's Biological Thought," in Frederick Amrine, Francis J. Zucker, and Harvey Wheeler (eds.), *Goethe and the Sciences: A Reappraisal* (Dordrecht and Boston, 1987), pp. 17–28.

7 Carl Gustav Carus, *Symbolik der menschlichen Gestalt. Ein Handbuch zur Menschenkenntnis,* 2nd enl. ed. (Leipzig, 1858, repr. Hildesheim, 1962).

8 Johann Wolfgang Goethe, *Werke* (Hamburger Ausgabe), vol. 14, p. 1.

9 Goethe, *Werke* (Hamburger Ausgabe), vol. 13, p. 329; Johann Peter Eckermann, *Gespräche mit Goethe in den letzten Jahren seines Lebens* (repr. Munich, 1976), p. 233; Gernot Böhme, "Is Goethe's Theory of Colors Science?" *Contemporary German Philosophy,* 4 (1984): 272, 274.

10 Hermann Helmholtz, "Nachschrift" (1875) to "Über Goethes naturwissenschaftliche Arbeiten" (1853), in *Vorträge und Reden,* 4th ed. (Braunschweig, 1896), p. 46.

11 Immanuel Kant, *Kritik der reinen Vernunft* (1781), ed. Raymund Schmidt (1924; repr. Leipzig, 1979), pp. 174 f.; Johann Friedrich Herbart, *Lehrbuch der Psychologie* (1817), 3rd ed. (Leipzig, 1887), p. 49.

12 Mach, "Vom räumlichen Sehen" (1865), in *Populärwissenschaftliche Vorlesungen* (Leipzig, 1911), pp. 117–123, 119–120; *Analysis* (cited in Chapter 5), p. 109 n. 1.

13 Mach, "Sehen," pp. 121–122.

14 Mach, *Analysis,* pp. 105 ff., 285.

15 Christian von Ehrenfels, "Über Gestaltqualitäten" (1890), repr. in Ferdinand Weinhandel (ed.), *Gestalthaftes Sehen* (Darmstadt, 1960), pp. 12, 17, 18–19; cf. idem., "On Gestalt Qualitites," trans. in Barry Smith (ed.), *Foundations of Gestalt Theory* (cited in the Introduction), pp. 83, 89, 90–91.

16 Ehrenfels, "On Gestalt Qualities," p. 93. For the concept of one-sided versus all-sided dependence relations, see Kevin Mulligan and Barry Smith, "Mach and Ehrenfels: The Foundations of Gestalt Theory," in Smith (ed.), *Foundations of Gestalt Theory,* pp. 124–157; Smith, *Austrian Philosophy* (cited in Chapter 2), p. 247.

17 Ehrenfels, "On Gestalt Qualitites," esp. p. 108.

18 Ibid., pp. 105–106; cf. Smith, "Gestalt Theory," pp. 15–16.

19 Ehrenfels, "On Gestalt Qualities," p. 117, n. 5. On Ehrenfels as poet and Wagnerian, see Reinhard Fabian, "Leben und Wirken von Christian von Ehrenfels: Ein Beitrag zur intellektuellen Biographie," in idem. (ed.), *Christian von Ehrenfels: Leben und Werk* (Amsterdam, 1986), pp. 11 ff.; Gerhard J. Winkler, "Christian von Ehrenfels als Wagnerianer," ibid., pp. 182–213.

20 Quoted in Geoffrey G. Field, *Evangelist of Race: The Germanic Vision of Houston Stewart Chamberlain* (New York, 1981), p. 216; on Chamberlain as an associate of Ehrenfels, see p. 103. Field is careful to note the ideological diversity of the Wagnerian cult, and distinguish Ehrenfels from its racist members.

21 Reinhardt Grossmann, "Structures versus Sets: The Philosophical Background of Gestalt Psychology," *crítica: revista hispanoamericana de filosofia,* 9 (1977): 3–21.

22 For surveys, see Boring, *History,* pp. 441 ff., and Herrmann, "Ganzheitspsychologie und Gestalttheorie" (cited in Chapter 4), pp. 578 ff. The views of James, Dilthey, and Husserl are discussed in Chapter 5.

23 Alexius Meinong, "Zur Psychologie der Komplexionen und Relationen" (1891), in Meinong, *Gesamtausgabe,* eds. Rudolf Haller et al. (Graz, 1968 ff.), vol. 1, esp. pp. 287, 294–295. For Meinong's earlier philosophy, see Lindenfeld, *Transformation,* pp. 92 ff.

24 Stumpf, *Tonpsychologie,* vol. 2, p. 228; Hans Cornelius, "Über Verschmelzung und Analyse," *Vierteljahreschrift für wissenschaftliche Philosophie,* 16 (1892): 417; Meinong, "Beiträge zur Theorie der psychischen·Analyse" (1893), in *Gesamtausgabe,* vol. 1, esp. pp. 316 ff., 347 ff.; cf. Lindenfeld, *Transformation,* pp. 118–119.

25 Hans Cornelius, *Psychologie als Erfahrungswissenschaft* (1897), pp. 70 ff., 165; idem., "Über Gestaltqualitäten," *ZfP,* 22 (1900): 114 ff., 117 ff.; Felix Krueger, "Die Theorie der Konsonanz, I. Eine psychologische Auseinandersetzung vornehmlich mit C. Stumpf und Th. Lipps," *Psychologische Studien,* 1 (1906): esp. pp. 314, 318, 379, 387; 2 (1907), pp. 211 ff.

26 Theodor Lipps, "Zu den Gestaltqualitäten," *ZfP,* 22 (1900): 384–385; idem., *Einheiten und Relationen* (Leipzig, 1902), pp. 103–104. Cf. Georg Anschütz, *Über Gestaltqualitäten* (Erlangen, 1909), p. 36.

27 Friedrich Schumann, "Beiträge zur Psychologie der Gesichtswahrnehmungen," *ZfP,* 23 (1900): 18 f.

28 Boring, *History,* pp. 445–446; cf. Herrmann, "Ganzheitspsychologie."

29 Schumann, "Beiträge," pp. 7 ff., 13, 27.

30 Ibid., pp. 28 ff.

31 G. E. Müller, *Gesichtspunkte und Tatsachen der psychophysischen Methodik* (Wiesbaden, 1904), pp. 237–238, esp. 238, n. 1.

32 Meinong, "Über Gegenstände höherer Ordnung und deren Verhältnis zur inneren Wahrnehmung," *ZfP,* 11 (1899): esp. pp. 186 f., 198 ff.; idem., "Über Gegenstandstheorie," in Meinong, ed., *Untersuchungen zur Gegenstandstheorie und Psychologie* (Leipzig, 1904), pp. 1–50; cf. Smith, "Gestalt Theory," pp. 22–23.

33 Stefan Witasek, "Über die Natur der geometrisch-optischen Täuschungen," *ZfP,* 19 (1899): esp. pp. 121–126, 131. See also Lindenfeld, *Transformation,* esp. pp. 232 ff., and Smith, "Gestalt Theory," pp. 27 ff.

34 Vittorio Benussi, "Zur Psychologie des Gestalterfassens," in Meinong, ed., *Untersuchungen,* esp. pp. 308–310.

35 Benussi, "Gestalterfassen," pp. 218, 332 n. 1, 403–404, 444 ff.; idem., "Gesetze des inadäquaten Gestaltauffassens," *AfgP,* 32 (1914): 409; see also idem., "Experimentelles über Vorstellungsinadäquatheit," *ZfP,* 42 (1906): 22–55; 45 (1906): 188–230; idem., "Über die Motive der Scheinkörperlichkeit bei umkehrbaren Zeichnungen," *AgP,* 20 (1911): 363–396; idem., *Psychologie der Zeitauffassung* (Heidelberg, 1912); cf. Schumann, "Beiträge," vol. 30, pp. 263–264. For further discussion, see Chapter 9.

36 Benussi, "Gesetze," p. 401.

37 Karl Bühler, *Die Gestaltwahrnehmungen* (Stuttgart, 1913), pp. 28–30; Benussi, "Die Gestaltwahrnehmungen," *ZfP,* 69 (1914): 259–292; Cassirer, *Substance and Function,* p. 335; Anton Marty, *Untersuchungen zur allgemeinen Grammatik und Sprachphilosophie,* vol. 1, pp. 199 ff. Other Brentano students thought differently. See Josef Kreibig, *Die intellektuellen Funktionen. Untersuchungen über Grenzfragen der Logik, Psychologie und Erkenntnistheorie* (Vienna, 1909), pp. 111 ff.; Alois Höfler, "Gestalt und Beziehung, Gestalt und Anschauung," *ZfP,* 60 (1912): 161–228.

38 Adhemar Gelb, "Theoretisches über Gestaltqualitäten," *ZfP,* 58 (1911): 1; Ebbinghaus, *Grundzüge* (cited in Chapter 4), 2nd ed. (1905), p. 462; idem., *Abriss* (cited in Chapter 4), p. 67.

39 Stumpf, "Erscheinungen und psychische Funktionen" (cited in Chapter 2), pp. 28 f.; Gelb, "Theoretisches über Gestaltqualitäten," pp. 51, 56.

40 Quoted in Allen, *Life Sciences* (cited in Chapter 5), p. 94.

41 A. Büttner, *Zweierlei Denken. Ein Beitrag zur Physiologie des Denkens* (Leipzig, 1910), p. 6. On Sherrington, see Allen, *Life Sciences,* pp. 88 ff., and Judith P. Swazey, *Reflexes and Motor Integration* (Cambridge, Mass., 1969).

42 Johannes von Kries, *Über die materiellen Grundlagen der Bewusstseinserscheinungen* (Freiburg, 1901), pp. v, 5, 14 f., 17 ff., 22–24.

43 Von Kries, *Bewusstseinserscheinungen,* pp. 34–36, 44 ff., 51–52.

44 Erich Becher, *Gehirn und Seele* (Heidelberg, 1911), pp. 216 ff., 219 ff., 227.

45 Ibid., pp. 230 f., 249, 283.

46 Ibid., pp. 293, 344, 374.

47 Cassirer, *Substance and Function,* p. 332.

Chapter 7

1 The most extensive available accounts of Wertheimer's early life are Michael Wertheimer, "Max Wertheimer, Gestalt Prophet," *Gestalt Theory,* 2 (1980): 3–17; Abraham S. Luchins and Edith H. Luchins, "An Introduction to the Origins of Wertheimer's Gestalt Psychologie" [sic!], *Gestalt Theory,* 4 (1982): 145–171. The latter must be used with caution. See also Edwin B. Newman, "Max Wertheimer," *American Journal of Psychology,* 57 (1944): 429–435; Abraham S. Luchins, "Wertheimer, Max," in David Sills (ed.), *International Encyclopedia of the Social Sciences,* vol. 16 (New York, 1968), pp. 522–526.

2 Gary B. Cohen, "Jews in German Society: Prague, 1860–1914," in David Bronson (ed.), *Jews and Germans from 1860 to 1933: The Problematic Synthesis* (Heidelberg, 1979), pp. 306–337.

3 Luchins, "Wertheimer," p. 522; Wertheimer, "Max Wertheimer," p. 8; cf. Luchins and Luchins, "Introduction," pp. 146 f., 149 f.

4 Max Brod, *Der Prager Kreis* (Frankfurt a.M., 1979), p. 138. For a vivid portrait of Ehrenfels as teacher and personality, see Brod's autobiography, *Streitbares Leben* (rev. ed., Munich, 1969), pp. 209ff. A transcript of Wertheimer's Charles University record is in the Max Wertheimer Papers (MWPB), Boulder, Colo.; see also Wertheimer, "Max Wertheimer," p. 9.

5 Ehrenfels, "Zuchtwahl und Monogamie," *Politisch-anthropologische Revue,* 1 (1906): Hefte 8–9; idem., "Monogamische Entwicklungsaussichten," ibid., 2 (1907): Heft 9; idem.,

"Die Sexuelle Reform," ibid., 2 (1907): Heft 12. Wertheimer's copies are in MWPB. For accounts of the controversy to which Ehrenfels's views led, see Brod, *Streitbares Leben,* p. 211, and Ann Taylor Allen, "German Radical Feminism and Eugenics, 1900–1918," *German Studies Review,* 11 (1988): 31–56.

6 A record of Wertheimer's courses in Berlin is in MWPB; cf. Wertheimer, "Max Wertheimer," p. 10.

7 Wertheimer, "Max Wertheimer," p. 10.

8 Koffka, "Beginnings of Gestalt Theory," lecture delivered 18 April 1931, Kurt Koffka papers, AHAP, Box M379.

9 Max Wertheimer and Julius Klein, "Psychologische Tatbestandsdiagnostik," *Archiv für Kriminalanthropologie und Kriminalistik,* 15 (1904): 74–75.

10 Ibid., esp. p. 79.

11 Wertheimer, *Experimentelle Untersuchungen zur Tatbestandsdiagnostik* (Leipzig, 1905), and in *AgP,* 6 (1905): 59–131. Citations are from the separate publication, esp. pp. 53 ff., 59, 64.

12 Wertheimer, *Tatbestandsdiagnostik,* p. 6.

13 For later discussions of Wertheimer's research, see August Messer, "Die Bedeutung der Psychologie" (cited in Chapter 3), pp. 203–205; Max Wertheimer, "Tatbestandsdiagnostik," in Emil Abderhalden (ed.), *Handbuch der biologischen Arbeitsmethoden* (Berlin and Vienna, 1933), sec. 6, part C2, pp. 1105–1111.

14 Wertheimer, "Zur Tatbestandsdiagnostik: eine Feststellung," *AgP,* 7 (1906): 139–140; Jung's work is cited in "Untersuchungen," pp. 25, 65, 69. Jung to Wertheimer, 10 October 1906, MWP-NYPL, Box 1, Folder 1. See also Michael Wertheimer, D. Brett King et al., "Carl Jung and Max Wertheimer on a Priority Issue," *JHBS,* 28 (1992), 45–56.

15 Wertheimer, "Experimental-psychologische Analyse hirnpathologischer Erscheinungen," *Bericht über den 5. Kongress für experimentelle Psychologie* (1913); "Über hirnpathologische Erscheinungen und ihre psychologische Analyse," lecture to the Ärztlicher Verein in Frankfurt, 20 October 1913, protocol in *Münchener medizinische Wochenschrift,* 5 November 1913; Luchins and Luchins, "Introduction," pp. 160 f.

16 Wertheimer, "Musik der Wedda," *Sammelbände der Internationalen Musikgesellschaft,* 11 (1910), esp. p. 306; Wertheimer, "Max Wertheimer," p. 12.

17 Wertheimer to "Dear K.," undated, MWPB, cited in Luchins and Luchins, "Max Wertheimer: His Life and Work during 1912–1919," *Gestalt Theory,* 7 (1985): 3, 5. Luchins and Luchins, "Introduction," pp. 158, 159 f., mentions similar letters to others from 1904 to 1905. The authors speculate that the recipient could have been Kurt Koffka, Wolfgang Köhler, or Wertheimer's earlier collaborator Julius Klein; Klein appears to me to be the correct choice.

18 The following account rests primarily on information from Molly Harrower-Erikson, "Kurt Koffka," *American Journal of Psychology,* 55 (1942): 278–281, and Grace M. Heider, "Koffka, Kurt," in David Sills (ed.), *International Encyclopedia of the Social Sciences* (New York, 1968), vol. 8, pp. 435–438. See also Molly Harrower, *Kurt Koffka: An Unwitting Self-Portrait* (Gainesville, Fla., 1984), pp. 253–260, and Wolfgang Metzger, "Koffka, Kurt," *Neue deutsche Biographie,* vol. 12 (Berlin, 1980), pp. 417–418.

19 Kurt Koffka, "The Ontological Status of Value: A Dialogue," in Horace M. Kallen and Sidney Hook (eds.), *American Philosophy Today and Tomorrow* (New York, 1935; repr. 1968), p. 274.

20 Koffka, "Untersuchungen an einem protonomalen System," *Zeitschrift für Sinnesphysiologie* (1908).

21 Koffka, "Experimental-Untersuchungen zur Lehre vom Rhythmus," *ZfP,* 52 (1909): esp. p. 16.

22 Koffka, "Rhythmus," pp. 53 ff.

23 Langfeld, "Stumpf's 'Introduction,'" p. 55; Koffka, "Beginnings of Gestalt Theory," pp. *a* and *b; Chronik der Friedrich-Wilhelm-Universität zu Berlin,* 20 (1906), pp. 65–67; Koffka, "Rhythmus," pp. 57, 104–105.

24 On the Freiburg institute, see Appendix 1, Table 1.

25 Fritz Mauthner, *Die Sprache* (Frankfurt a.M., 1906), pp. 30–31, 86.

26 Koffka to Mauthner, 22 May 1909, Leo Baeck Institute Library, Mauthner Collection.

27 Koffka to Mauthner, 22 November 1909, Mauthner Collection; Koffka, *Zur Analyse der Vorstellungen und ihrer Gesetze* (Leipzig, 1912), pp. vi, 26.

28 Koffka, *Vorstellungen,* p. 26.

29 Koffka, *Vorstellungen,* pp. 21 ff., 191–192 n. 1, 195, 238, 243, 253 ff., 272, 279.

30 Koffka, *Principles,* p. 53. For Koffka's official title in Giessen, see Hans-Georg Burger, "Anfänge und Bedeutung der experimentellen Psychologie in Giessen," *Giessener Universitätsblätter,* 1975, Heft 1, pp. 88–89.

31 The information on Köhler's ancestry comes from Wilhelm Lenz, ed., *Deutschbaltisches biographisches Lexikon 1710–1960* (Cologne and Vienna, 1970), pp. 242–245, 398; Florentine Mütherlich, "Köhler, Wilhelm," and Rudolf Bergius, "Köhler, Wolfgang," *Neue deutsche Biographie,* vol. 12 (Berlin, 1980), pp. 301–302, 302–304 resp. Other useful biographical articles are Mary Henle, "Wolfgang Köhler" (cited in the Introduction), and Solomon E. Asch, "Wolfgang Köhler," *American Journal of Psychology,* 81 (1968): 110–119.

32 Köhler to Geitel, 27 June 1915, in Siegfried Jaeger (ed.), *Briefe Wolfgang Köhlers an Hans Geitel 1907–1920* (Passau, 1989), p. 55.

33 Siegfried Jaeger, "Wolfgang Köhler in Berlin," in Lothar Sprung and Wolfgang Schönpflug (eds.), *Geschichte der Psychologie in Berlin* (Frankfurt a.M., 1992), pp. 163 f. Köhler named Stumpf, Riehl, Nernst, and Planck as his teachers in the autobiography attached to his dissertation, "Akustische Untersuchungen I," Phil. Diss. Berlin, 1909.

34 Köhler to Geitel, 14 November 1907, 20 November 1907, 1 March 1908, 31 May 1908, 23 March 1909, all in Jaeger (ed.), *Briefe Wolfgang Köhlers,* pp. 15 f., 17 f., 21 f., 25 ff.

35 Mach, *Analysis of Sensations,* (cited in Chapter 4), p. 279 n. 1; Köhler, "Akustische Untersuchungen I," *ZfP,* 54 (1909): 243 ff.; Köhler to Geitel, 23 March 1909, in Jaeger (ed.), *Briefe Wolfgang Köhlers;* Hans-Lukas Teuber, "Wolfgang Köhler zum Gedenken," *PsyFo* 31 (1967): vi.

36 Köhler to Geitel, 24 March 1909, in Jaeger (ed.), *Briefe Wolfgang Köhlers,* pp. 27 ff.

37 Köhler to Geitel, 24 April 1909, in Jaeger (ed.), *Briefe Wolfgang Köhlers,* p. 31; Stumpf's assessment is in UAHUB, Philosophische Fakultät, Dekanat, Nr. 469, Bl. 327.

38 Köhler, "Über akustische Prinzipalqualitäten," *Bericht über den 4. Kongress für experimentelle Psychologie* (1911), pp. 229–230; idem., "Akustische Untersuchungen II," *ZfP,* 58 (1910): secs. 2 and 3, esp. p. 130.

39 Köhler, "Akustische Untersuchungen II," pp. 98, 102.

40 Ibid., pp. 98–99.

41 Ibid., p. 111.

42 Ibid., pp. 284, 289–290; "Akustische Untersuchungen," *Bericht über den 5. Kongress für experimentelle Psychologie* (1912), p. 153.

43 For Mach's treatment of "interval colors," see *Analysis,* ch. 13, esp. pp. 286 ff.; cf. Köhler, "Akustische Untersuchungen I," pp. 281–282. Köhler later denied that he had intended to attack Krueger directly with this remark; see "Akustische Untersuchungen II," p. 79 n.

44 Stumpf, "Carl Stumpf," pp. 409–410; idem., discussion contribution to Köhler, "Akustische Untersuchungen," *Bericht* (1912), pp. 155–156; idem., "Über neuere Untersuchungen zur Tonlehre," *Bericht über den 6. Kongress für experimentelle Psychologie* (1914), pp. 305–348; idem, "Die Struktur der Vokale," *Sitzungsberichte der Preussischen Akademie der Wissenschaften,* Phil.-hist. Cl., 1918. For a discussion of subsequent research, see Boring, *Sensation and Perception* (cited in Chapter 2), pp. 372 ff.

45 Köhler, "Akustische Untersuchungen II," p. 59.

46 Max Planck, "The Unity of the Physical World-Picture" (1909), trans. Ann Toulmin, in Stephen Toulmin (ed.), *Physical Reality* (New York, 1970), esp. pp. 6, 17, 24–25, 26–27.

47 Later Planck would say that Spinoza was the philosopher who expressed his views best; see Armin Hermann, *Max Planck* (Reinbek bei Hamburg, 1973), p. 98. For further discussions of Planck's religious and ethical views in their relation to his physics, see Stanley Goldberg, "Max Planck's Philosophy of Nature and His Elaboration of the Special Theory of Relativity," *HSPS,* 7 (1976): 125–160.

48 Köhler to Geitel, 8 May 1909 and 16 February 1910, in Jaeger (ed.), *Briefe Wolfgang Köhlers,* pp. 33 f., 35.

49 Köhler, "Akustische Untersuchungen II," p. 111.

Chapter 8

1 Wertheimer, "Max Wertheimer," p. 13; cf. Newman, "Max Wertheimer," pp. 431–432. In an interview with me on 2 October 1976, Professor Newman could not recall whether he had heard the story from Wertheimer directly or from a third party.

2 On the history of the academy and the psychological institute, see Richard Wachsmuth, *Die Gründung der Universität Frankfurt* (Frankfurt a.M., 1929), esp. pp. 47, 50, 53; for additional background, see Paul Kluke, *Die Stiftungsuniversität Frankfurt am Main 1914– 1932* (Frankfurt a.M., 1972). Information about the institute's initial outlays and budget comes from its account books (*Kassenbücher*) for the years 1905–1906, courtesy of Professor Victor Sarris. The information about the assistantship comes from the Rausch interview (cited in Chapter 3). Cf. Ulfried Geuter (ed.), *Daten Zur Geschichte der deutschen Psychologie,* p. 34.

3 Wachsmuth, *Gründung,* pp. 55 ff.; Cornelius, "Hans Cornelius," in Raymund Schmidt (ed.), *Die Philosophie der Gegenwart in Selbstdarstellungen,* vol. 2 (Leipzig, 1921), p. 86; Koffka to Edwin G. Boring, 22 April 1930, E. G. Boring papers, Harvard University Archives. It is not clear why Koffka failed to mention his first wife in this context.

4 Max Wertheimer, "Über das Denken der Naturvölker I: Zahlen und Zahlgebilde," *ZfP,* 60 (1912): 321–378; reprinted in *Drei Abhandlungen zur Gestalttheorie* (Erlangen, 1925), pp. 106–161. Citations are to the reprint, here p. 107.

5 Wertheimer, "Denken der Naturvölker," pp. 108–109, 127.

6 Ibid., pp. 133–134.

7 Ibid., p. 132.

8 Ibid., p. 144.

9 Ibid., pp. 109, 114.

10 The course titles are taken from *Akademie für Sozial- und Handelswissenschaften zu Frankfurt am Main. Die Vorlesungen* (Frankfurt a.M., 1903 ff.) for the years 1912 to 1914.

11 Gabriele Gräfin von Wartensleben, "Über den Einfluss der Zwischenzeit auf die Reproduktion gelesener Buchstaben," *ZfP*, 64 (1913): 321–385; Köhler and Wertheimer are acknowledged on p. 327. See also Robert S. Harper, Edwin B. Newman, and Frank R. Schab, "Gabriele Gräfin von Wartenslehen and the Birth of *Gestalt Psychologie*" (sic!), *JHBS*, 21 (1985): 118–123.

12 Von Wartensleben, *Die christliche Persönlichkeit im Idealbild. Eine Beschreibung sub specie Psychologica* (Kempten and Munich, 1914), pp. 1–3; repr. in Martin Scheerer, *Die Lehre von der Gestalt* (Berlin, 1931), pp. 84–85. Emphasis in the original. I have used the selective translation by Michael Wertheimer, in "Max Wertheimer," p. 14, with minor alterations; cf. Harper et al., "Gabriele Gräfin von Wartensleben," pp. 121–122.

13 Sigmund Exner, "Über das Sehen von Bewegungen und die Theorie des zusammengesetzten Auges," *Sitzungsberichte der Wiener Akademie der Wissenschaften,* 72 (1875): 156–190; idem., *Entwurf zu einer physiologischen Erklärung der psychischen Erscheinungen* (Leipzig, 1894); Cornelius, *Psychologie* (cited in Chapter 6), p. 132; Mach, *Analysis,* p. 144; Mach, *Grundlinien der Lehre von den Bewegungsempfindungen* (Leipzig, 1875), p. 63; cf. Woodward, "From Association to Gestalt" (cited in Chapter 4), pp. 580–581.

14 Von Kries, suppl. to Helmholtz, *Optics* (cited in Chapter 2), vol. 3, pp. 605–606; Karl Marbe, *Theorie der kinematischen Projektionen* (Leipzig, 1910); Paul Ferdinand Linke, "Die stroboskopischen Täuschungen und das Problem des Sehens von Bewegung," *Psychologische Studien,* 3 (1907): e.g., pp. 544–545.

15 Wertheimer, "Experimentelle Studien über das Sehen von Bewegung," *ZfP*, 61 (1912): 161–265; repr. in *Drei Abhandlungen zur Gestalttheorie,* 1–115; abr. trans. in Thorne Shipley (ed.), *Classics in Psychology* (New York, 1961), pp. 1032–1089. Except where noted, both the translation, designated "Motion," and the German reprint, designated "Bewegung," will be cited. The citation here is: "Motion," pp. 1032, 1036; "Bewegung," pp. 2, 6. Emphasis mine.

16 Wertheimer, "Bewegung," pp. 15–16. See also Viktor Sarris, "Max Wertheimer in Frankfurt – über Beginn und Aufbaukrise der Gestaltpsychologie I. Ausgangsstudien über das Sehen von Bewegung," *ZfP*, 195 (1987): esp. pp. 291–292, 303–304. Professor Sarris and Dr. Gottfried Langenstein reconstructed Wertheimer's apparatus for an exhibition at the University of Frankfurt and kindly demonstrated it to me.

17 Wertheimer, "Bewegung," p. 7.

18 Wertheimer, "Motion," pp. 1035, 1038 f., 1044; "Bewegung," pp. 5, 13 f., 18.

19 Wertheimer, "Motion," p. 1065; "Bewegung," p. 67. Emphasis mine.

20 Wertheimer, "Motion," pp. 1050 ff., 1052; "Bewegung," pp. 57 ff.

21 Wertheimer, "Motion," pp. 1076–1077, 1059 f., 1049; "Bewegung," pp. 79–80, 62–63, 25.

22 Wertheimer, "Bewegung," pp. 30, 78 ff.; cf. "Motion," pp. 1074 ff.

23 Wertheimer, "Motion," pp. 1085–1087, esp. p. 1085, n. 57; "Bewegung," pp. 87 ff., esp. p. 88 n. 1.

24 Wertheimer, "Bewegung," pp. 91 f., p. 92 n. 3.

25 Kurt Koffka, "Beginnings of Gestalt Theory" (Cited in Chapter 7), pp. 1–2. Emphasis in the original; the grammar has been corrected. Koffka later said that this meeting occurred early in 1911; Wertheimer had told Köhler of the results some time before; cf. Koffka, *Principles,* p. 53; Koffka to E. G. Boring, 24 May 1927, Boring papers, Harvard University Archives.

26 Wertheimer, "Motion," p. 1042; "Bewegung," p. 17.

27 Wertheimer, "Motion," p. 1084; "Bewegung," p. 87. Emphasis in the original.

28 Koffka, "Beginnings of Gestalt Theory," p. 4. Emphasis mine.

29 Schumann, "Über einige Probleme der Lehre von den Gesichtswahrnehmungen," *Bericht über den 5. Kongress für experimentelle Psychologie* (1913), pp. 181, 183.

30 Adhemar Gelb, "Versuche auf dem Gebiet der Zeit- und Raumanschauung," *Bericht über den 6. Kongress für experimentelle Psychologie* (1914), pp. 36–42.

31 Vittorio Benussi, "Kinematische Scheinbewegungen und Auffassungsformung," *Bericht über den 6. Kongress für experimentelle Psychologie* (1914), discussion, p. 148.

32 Wertheimer, "Untersuchungen zur Lehre von der Gestalt II," *PsyFo,* 4 (1923): 301–350. On p. 301, Wertheimer stated that these experiments "came essentially from the years 1911–1914." The results are discussed in Chapter 14.

33 Wertheimer to Ehrenfels, 10 April, 25 April, and 25 May 1914, Christian von Ehrenfels papers, Forschungsstelle für österreichische Philosophie, Dokumentationszentrum, Graz. Thanks to Barry Smith for providing me a copy of these documents. It is not clear from this correspondence which professorship was in question.

34 For a previous such incident see Fabian, "Leben und Wirken von Christian von Ehrenfels" (cited in Chapter 6), pp. 42 f.

35 Wertheimer to Ehrenfels, n.d., as above. Emphasis in the original. Brod, *Streitbares Leben* (cited in Chapter 7), p. 211.

Chapter 9

1 Köhler, "Akustische Untersuchungen III and IV. Vorläufige Mitteilung," *ZfP,* 64 (1913): 101 f.; idem., "Akustische Untersuchungen III," *ZfP,* 72 (1915): 121 ff.

2 Köhler, "Über unbemerkte Empfindungen und Urteilstäuschungen," *ZfP,* 66 (1913), 51–80; idem., "On Unnoticed Sensations and Errors of Judgment," trans. Helmut E. Adler, in Mary Henle (ed.), *The Selected Papers of Wolfgang Köhler* (New York, 1971), pp. 13–39. Citations are to the translation, here pp. 35–36.

3 Köhler, "Unnoticed Sensations," esp. pp. 15, 19; Stumpf, "Differenztöne und Konsonanz," *ZfP,* 59 (1911): 175, cited in Köhler, "Unnoticed Sensations," p. 16.

4 Ibid., p. 18.

5 Ibid., p. 23.

6 Ibid., pp. 26–27, 28–29.

7 Ibid., pp. 38–39. Emphasis in the original.

8 Ibid., pp. 17, 36–37, 24 n. 20.

9 Ibid., pp. 29–30, esp. p. 36 n. 37.

10 Koffka, "Beginnings of Gestalt Theory" (cited in Chapter 7).

11 Götz Martius, "Über synthetische und analytische Psychologie," *Bericht über den 5. Kongress für experimentelle Psychologie* (1913), pp. 261–281; Otto Selz, "Experimentelle Untersuchungen über den Verlauf determinierter intellektueller Prozesse," ibid., 229–234 and *Über die Gesetze des geordneten Denkverlaufs* (Stuttgart, 1913), esp. pp. 1 ff., 84 ff., 177 ff., 194–195. Selz's views and their relation to Gestalt theory will be discussed further in Chapter 18.

12 Karl Bühler, "Über die Vergleichung von Raumgestalten," *Bericht über den 5. Kongress der Gesellschaft für experimentelle Psychologie* (1913), pp. 183–185; idem., *Die Gestaltwahrnehmungen* (cited in Chapter 6), esp. pp. 138 ff., 182 ff.

13 Koffka, review of Bühler, "Die Gestaltwahrnehmungen," *Deutsche Literaturzeitung,* 34 (1913): esp. p. 398; idem., "Psychologie der Wahrnehmung," *Die Geisteswissenschaften,* 1 (1914), pp. 798–799; cf. Bühler, *Die Gestaltwahrnehmungen,* esp. pp. 175, 206.

14 Koffka, "Psychologie der Wahrnehmung," pp. 711–713, 716. Emphasis in the original.

15 Koffka, "Beiträge zur Psychologie der Gestalt- und Bewegungserlebnisse. Einleitung" (1913), reprinted in Koffka (ed.), *Beiträge zur Psychologie der Gestalt* (Leipzig, 1919), pp. 2, 6. The formulation Koffka quoted is in Henry Watt's review of Wertheimer's paper, "The Psychology of Visual Motion," *British Journal of Psychology,* 6 (1912): 26–43.

16 Friedrich Kenkel, "Untersuchungen über den Zusammenhang zwischen Erscheinungsgrösse und Erscheinungsbewegung bei einigen sogenannten optischen Täuschungen" (1913), in Koffka (ed.), *Beiträge,* pp. 17, 21 f.

17 Benussi, "Stroboskopische Scheinbewegungen und geometrisch-optische Täuschungen," *AfgP,* 24 (1912): esp. pp. 40, 61; Kenkel, "Untersuchungen," pp. 93–94, 94 n. 1.

18 Benussi, review of Koffka and Kenkel, *AfgP,* 32 (1914): Literatur-Teil, pp. 55–56; idem., "Stroboskopische Scheinbewegungen," p. 41.

19 Koffka, "Psychologie der Wahrnehmung," pp. 797–798.

20 Benussi, "Experimentelles über Vorstellungsinadäquatheit," *ZfP,* 42 (1906): 22–55; 45 (1907): 188–230; idem., "Über die Motive der Scheinkörperlichkeit bei umkehrbaren Zeichnungen," *AfgP,* 20 (1911): 363–396; idem., "Gesetze der inadäquaten Gestaltauffassung," *AfgP,* 32 (1914): esp. p. 400.

21 Benussi, *Psychologie der Zeitauffassung* (Heidelberg, 1913), esp. pp. 102 f., 145 ff.; idem., "Versuche zur Bestimmung der Gestaltzeit," *Bericht über den 6. Kongress für experimentelle Psychologie* (1914), pp. 72–73; idem., "Gesetze," pp. 409–410.

22 Benussi, "Die Gestaltwahrnehmungen," pp. 276–277.

23 Wertheimer to Ehrenfels, 10 April 1914 (cited in Chapter 8).

24 Köhler, "Unnoticed Sensations," p. 30 n. 30; Benussi, "Gesetze," p. 419.

25 Koffka, "Zur Grundlegung der Wahrnehmungspsychologie. Eine Auseinandersetzung mit V. Benussi," *ZfP,* 73 (1915): 11–90; here, pp. 14–15.

26 Ibid., pp. 26 ff.

27 Ibid., pp. 18–19; cf. p. 81.

28 Koffka, "Einleitung," p. 1.

29 Koffka, "Zur Grundlegung," pp. 33–34; Ludwig Edinger, "The Relations of Comparative Anatomy to Comparative Psychology," *Journal of Comparative Neurology and Psychology,* 18 (1908): 437–457.

30 Ibid., pp. 58, 60 n. 2. Emphasis mine.

31 Wertheimer, "Motion," p. 1066; "Bewegung," p. 68.

32 Koffka, "Zur Grundlegung," pp. 36–37, 57. Emphasis in the original.

33 Ibid., p. 36.

34 John Dewey, "The Reflex Arc Concept in Psychology," *Psychological Review,* 3 (1896): 357–370; idem., "Thought and Its Subject-Matter" (1903), in *Essays in Experimental Logic* (New York, 1916), pp. 84, 92, 95. Emphasis mine.

35 Dewey, *Essays,* pp. 93, 95.

36 Koffka, "Zur Grundlegung," pp. 20–21, 41, 49; in *Principles,* pp. 134 ff., Koffka notes that Wundt had used a similar experiment as well.

37 Koffka, "Zur Grundlegung," pp. 25–26, 35–36, 40 n. 1.

Chapter 10

1 For an account of the establishment of the station, see Max Rothmann and Eugen Teuber, "Aus der Anthropoidenstation auf Teneriffa, I. Ziele und Aufgaben der Station," *Abhandlungen der königlich Preussischen Akademie der Wissenschaften,* phys.-math. Kl., 1915, No. 2, esp. pp. 3–5; see also Jaeger (ed.), *Briefe Wolfgang Köhlers* (cited in Chapter 7), pp. 100 ff. Ronald Ley, in *A Whisper of Espionage* (Garden City, N.Y., 1990), repeatedly implies that the station was actually located on Tenerife for espionage purposes because the place was situated on major Atlantic shipping routes. Unfortunately, he provides no direct evidence for this claim and shows no familiarity with the sources cited.

2 Albert Samson Stiftung, Statut vom 19. Juli 1905/7 (sic), September 1914, in AAW, II-XI-121, Bnd. 6, Bl. 5 ff. Cf. Wilhelm Waldeyer, "Ansprache," *Sitzungsberichte der königlich Preussischen Akademie der Wissenschaften,* 1914, Nr. 1, p. 83.

3 For the budget figures, see the reports in *Sitzungsberichte,* 1915, Nr. 1, p. 129; 1916, Nr. 1, pp. 162–163. See also AAW, II-XI-121, Bnd. 13, Bl. 126–128.

4 Teuber to Waldeyer, AAW, II-XI-121, Bnd. 23, Bl. 75–76; cf. Teuber to Yerkes, 30 November 1913, Yerkes to Teuber, 29 December 1913, Robert M. Yerkes papers, Yale Medical Library, New Haven, Conn.

5 The information in this and the following paragraphs comes from Rothmann and Teuber, "Aus der Anthropoidenstation," and from papers of Eugen Teuber cited in Marianne Teuber, "The Founding of the Primate Station, Tenerife, Canary Islands," *American Journal of Psychology,* 107 (1994): 551–581.

6 Rothmann and Teuber, "Aus der Anthropoidenstation," esp. pp. 15–16, 18; Waldeyer, "Ansprache," p. 84.

7 Köhler to Waldeyer, 29 August 1913; contract between Köhler and the Samson Foundation, 20 December 1913; AAW, II-XI-121, Bnd. 13, Bl. 56, 90, 237, 246.

8 Köhler to Waldeyer, 12 March 1914, 29 March 1914; Waldeyer to Köhler, 14 May 1914, 4 April 1914, AAW Berlin, II-XI-121, Bnd. 13, Bl. 112–114, 122–123, 129–131, 139–142.

9 Conwy Lloyd Morgan, *An Introduction to Comparative Psychology* (London, 1895), p. 53; Robert J. Richards, *Darwin and the Emergence of Evolutionary Theories of Mind and Behavior,* (Chicago, 1987), pp. 380–381, 385, 395.

10 Edward L. Thorndike, *Animal Intelligence* (New York, 1911; repr. New York, 1965), esp. pp. 67 ff., 112 ff.; John C. Burnham, "Thorndike's Puzzle Boxes," *JHBS,* 8 (1972): 159–167; Robert Boakes, *From Darwin to Behaviorism: Psychology and the Minds of Animals* (Cambridge, 1984), esp. pp. 69–72.

11 Thorndike, *Animal Intelligence*, pp. 189, 192, 238–239; Burnham, "Thorndike's Puzzle Boxes," p. 165; Phillip Howard Gray, "The Early Animal Behaviorists: Prolegomenon to Ecology," *Isis*, 59 (1968): 372–383. For Thorndike's work in educational psychology, see Geraldine Joncich, *The Sane Positivist: A Biography of Edward Lee Thorndike* (Middletown, Conn., 1968). On technocratic hopes in American psychology, see: O'Donnell, *The Origins of Behaviorism* (cited in Chapter 1); Danziger, *Constructing the Subject* (cited in the Introduction).

12 Oskar Pfungst, *Das Pferd des Herrn von Osten (Der Kluge Hans)* (Leipzig, 1907); Boakes, *From Darwin to Behaviorism*, pp. 78–81.

13 Pfungst, "Zur Psychologie der Affen," *Bericht über den 5. Kongress für experimentelle Psychologie* (1913), pp. 200–205; Rothmann's announcement is on p. 203; cf. Alexander Sokolowsky, *Beobachtungen über die Psyche der Menschenaffen* (Frankfurt a.M., 1908). Ironically, Pfungst deprived himself of the chance to reap fame as director of the station. He was offered the position, but remained indecisive, and Teuber went to Tenerife instead. Stumpf to Waldeyer, 3 December 1912, AAW, II-XI-121, Bnd. 13, Bl. 4.

14 Köhler, discussion contribution in Pfungst, "Zur Psychologie der Affen," p. 204.

15 Köhler to Waldeyer, 23 May 1914, AAW, II-XI-121, Bnd. 13, Bl. 146–147.

16 Köhler, "Intelligenzprüfungen an Anthropoiden," *Abhandlungen der Königlich Preussischen Akademie der Wissenschaften*, phys.-math. Kl., 1917, Nr. 1; 3rd ed., *Intelligenzprüfungen an Menschenaffen* (Berlin, Heidelberg, New York, 1973); idem., *The Mentality of Apes*, translation of second edition by Ella Winter (1925 rev. ed., repr. New York, 1973). Citations are to the translation, hereafter designated *Mentality*, here pp. 13–14, and to the third German edition, designated *Intelligenzprüfungen*, here p. 10. The translation has been altered where appropriate. The following account is arranged systematically rather than chronologically. Boakes, *From Darwin to Behaviorism*, pp. 186 ff., presents Köhler's results in the order he recorded them.

17 Köhler, *Mentality*, p. 16; *Intelligenzprüfungen*, p. 12. Emphasis in the original.

18 Köhler, *Mentality*, pp. 31 ff., 48 f., 55 ff., 69 f.; *Intelligenzprüfungen*, pp. 22 ff., 35 f., 39 ff., 50 f.

19 Köhler, *Mentality*, pp. 39 ff., 135 ff., esp. pp. 40, 136, 149–150; *Intelligenzprüfungen*, pp. 28 ff., 96 ff., esp. pp. 29, 98 ff., 106–107.

20 Köhler, *Mentality*, pp. 103 ff., 109–110, 125–128, 178 f.; *Intelligenzprüfungen*, pp. 73 ff., 78–79, 90–93, 128–129.

21 Köhler, *Mentality*, e.g. pp. 44, 46, 87 ff., 168 f.; *Intelligenzprüfungen*, pp. 32–33, 34, 63 ff., 122.

22 Köhler to Waldeyer, 7 March 1914, AAW, II-XI-121, Bnd. 13, Bl. 111. For discussion of Köhler's films, see: Hermann Kalkofen, *Wolfgang Köhlers Filmaufnahmen der 'Intelligenzprüfungen an Menschenaffen' 1914–1917* (Göttingen, 1975); Helmut E. Lück, "Wolfgang Köhler auf Tenerife," *Gestalt Theory*, 9 (1987): 170–181. Film copies are available through the American Psychological Association in Washington, D.C., and the Institut für den wissenschaftlichen Film in Göttingen.

23 Köhler, *Mentality*, esp. pp. 102 and 204 n. 1; cf. pp. 142 f.; *Intelligenzprüfungen*, pp. 73 and 147 n. 1.

24 Köhler, *Mentality*, pp. 69–72, 157, 221 ff.; *Intelligenzprüfungen*, pp. 50 f., 112, 160 ff.; cf. Köhler, *The Task of Gestalt Psychology* (Princeton, N.J., 1969), pp. 156–157.

25 Köhler, *Mentality*, p. 190; *Intelligenzprüfungen*, pp. 136–137. Emphasis in the original.

26 Köhler, *Mentality,* pp. 156 ff., 192; *Intelligenzprüfungen,* pp. 112 ff., 138.

27 Köhler, *Mentality,* pp. 21 ff.; *Intelligenzprüfungen,* pp. 15 ff.

28 Leonard Trelawney Hobhouse, *Mind in Evolution* (1901), 2nd ed. (London, 1915; repr. New York, 1973), esp. ch. 12, pp. 190 f., 239, 274, 280. Köhler acknowledged Hobhouse's experiments in *Mentality,* p. 30 n. 1; cf. *Intelligenzprüfungen,* p. 22 n. The report to Waldeyer is in Köhler to Waldeyer, 23 May 1914, AAW, II-XI-121, Bnd. 13, Bl. 146.

29 Hobhouse, *Mind in Evolution,* esp. pp. 258 ff., 294–295, 305 ff.; cf. Boakes, *From Darwin to Behaviorism,* pp. 180 ff.

30 Köhler, "Zur Psychologie des Schimpansen" (1921), repr. in *Intelligenzprüfungen,* pp. 195–232, esp. pp. 197, 200–201; idem., "Some Contributions to the Psychology of Chimpanzees," in *Mentality,* esp. pp. 274, 279–280.

31 Köhler, "Methods of Psychological Research with Apes" (1921), trans. Mary Henle, in *Selected Papers* (cited in Chapter 8), pp. 206, 209–210. Emphasis in the original.

32 Köhler, *Mentality,* pp. 198 ff.; *Intelligenzprüfungen,* pp. 143 ff.

33 Köhler, *Mentality,* p. 201; *Intelligenzprüfungen,* p. 145.

34 Köhler, *Mentality,* pp. 19, 193 n. 1; *Intelligenzprüfungen,* pp. 13, 139 n. 1.

35 Köhler, *Mentality,* pp. 198, 268; *Intelligenzprüfungen,* pp. 142, 193.

36 Koffka, "Psychologie," in Max Dessoir (ed.), *Die Philosophie in ihren Einzelgebieten* (Berlin, 1925).

37 Köhler to Waldeyer, 10 August 1914 and 19 October 1914, AAW, II-XI-121, Bnd. 13, Bl. 167, 174. Ley, *A Whisper of Espionage* (cited in note 1), casts doubt on the claim that German men could not easily return home from Tenerife; but even though he quotes other documents from the records cited here, he omits any reference to this one.

38 See, e.g., AAW, II-XI-121, Bnd. 13, Bl. 227.

39 Köhler to Waldeyer, 18 January 1915, 30 June 1915, 5 December 1915, AAW, II-XI-121, Bnd. 13, Bl. 186, 215–216, 230a; cf. Waldeyer, "Die Anthropoidenstation auf Teneriffa," *Sitzungsberichte,* 1917, Nr. 1, 401–402.

40 Köhler to Geitel, 15 July 1914, in Jaeger (ed.), *Briefe Wolfgang Köhlers,* p. 47 (see also p. 99 n. 55); Köhler to Waldeyer, 19 April 1914, 28 August 1915, AAW, II-XI-121, Bnd. 13, Bl. 124–125, 221. Ley, *A Whisper of Espionage,* also reports these and other rumors, and adds that the station's animal keeper, Manuel Gomez y Garcia, told him directly that Köhler "was a spy – that he had a concealed radio, which he operated from the roof of his house," and that the keeper helped warn him of approaching police patrols so he could hide the radio (pp. 8, 14; the words are Ley's). He does not say how Manuel, being ignorant of German, could have known that Köhler used the radio for espionage purposes. This sixty-year-old memory and a second-hand report by Köhler's son of a conversation with a former German resident on the island, who actually did send intelligence by wireless to Germany (pp. 212 f.), are the only evidence provided for Köhler's involvement in espionage that is not purely circumstantial.

41 Köhler to Yerkes, 17 April 1914, 19 May 1916, Yerkes papers, Box 29, folder 548. See also Box 57, Folder 1090.

42 Köhler to Yerkes, 10 June 1915, 13 December 1916, 2 March 1917; Yerkes to Köhler, 20 May 1914, Yerkes papers.

43 Robert M. Yerkes, "The Mental Life of Monkeys and Apes," *Behavior Monographs,* vol. 3, no. 1 (Cambridge and Boston, 1916), pp. 68, 73, 96; see also Boakes, *From Darwin to Behaviorism,* pp. 196–199.

44 Köhler to Yerkes, 19 May 1916, Yerkes papers; cf. Yerkes, "Mental Life," pp. 2, 136–137.

45 Yerkes to Köhler, 17 July 1916 and 20 October 1916, Yerkes papers; Yerkes, "Mental Life," p. 132.

46 Yerkes to Waldeyer, 28 October 1919; Köhler to Waldeyer, 9 December 1919; Sitzungsprotokoll des Sekretariats, 11 December 1919, AAW, II-XI-121, Bnd. 6, Bl. 80, 95, 96. Yerkes to Köhler, 13 April 1921; Köhler to Yerkes, 19 May 1921, Yerkes papers. For further discussion of the Köhler-Yerkes correspondence, see Michael M. Sokal, "The Gestalt Psychologists in Behaviorist America," *American Historical Review,* 89 (1984): 1240–1263. On international relations among scientists during and after the First World War, see Brigitte Schröder-Gudehus, *Deutsche Wissenschaft und internationale Zusammenarbeit 1914–1918* (Geneva, 1966); Paul Forman, "Scientific Internationalism and the Weimar Physicists: The Ideology and Its Manipulation in Germany after World War I," *Isis,* 64 (1973): 151–180; Daniel Kevles, "Into Hostile Camps: The Reorganization of International Science in World War I," *Isis,* 62 (1971): 47–60.

47 Köhler, "Optische Untersuchungen am Schimpansen und am Haushuhn," *Abhandlungen der Königlich Preussischen Akademie der Wissenschaften,* phys.-math. Kl., 1915, No. 3. Martin Uibe's and Thekla Köhler's assistance are acknowledged in a footnote on p. 1. Köhler had wanted to list them as coauthors, and particularly emphasized his wife's role in conceiving as well as executing the experiments. Waldeyer refused to do so, citing the "custom" of the Academy of naming only the principal investigator as author in its reports.

48 Ibid., pp. 36, 55 f., 56 n.

49 Ibid., pp. 59–60, 69–70.

50 Ibid., p. 69; David Katz, review of Köhler, "Optische Untersuchungen," *ZfP,* 75 (1916): esp. pp. 386, 389–390; Köhler, "Die Farbe der Sehdinge beim Schimpansen und beim Haushuhn," *ZfP,* 77 (1917): 248–255.

51 Köhler, "Nachweis einfacher Strukturfunktionen beim Schimpansen und beim Haushuhn. Über eine neue Methode zur Untersuchung des bunten Farbensystems," *Abhandlungen der Königlich Preussischen Akademie der Wissenschaften,* phys.-math. Kl., 1918, Nr. 2, pp. 3–4, 10, 12–13, 33.

52 Ibid., pp. 19, 42–43, 47; Robert M. Yerkes and Ada W. Yerkes, *The Great Apes: A Study of Anthropoid Life* (New Haven, Conn., 1929; repr. New York, 1970), p. 334.

53 Köhler, "Strukturfunktionen," pp. 66 ff., 72, 88; cf. pp. 24, 100.

54 David Katz and Géza Révèsz, "Experimentalpsychologische Untersuchungen mit Hühnern," *ZfP,* 50 (1909): 93–116; Karl S. Lashley, "Visual Discrimination of Size and Form in the Albino Rat," *Journal of Animal Behavior,* 2 (1912): 329; Walter S. Hunter, "The Question of Form-Perception," ibid., 3 (1913): 330.

55 Köhler, "Strukturfunktionen," p. 37; cf. Hans Volkelt, *Über die Vorstellungen der Tiere* (Leipzig, 1914).

56 Ibid., pp. 37–38.

57 Ibid., pp. 40–41.

58 Ibid., p. 35.

59 Köhler to Waldeyer, 2 August 1918, 15 August 1918, and 6 December 1918, AAW, II-XI-121, Bnd. 12, unpaginated. Köhler to Stumpf, 9 May 1919, AAW, II-XI-121, Bnd. 12, Bl. 4. Figures calculated from the financial statements of the station submitted by Köhler, and from the reports of the Samson Foundation, AAW, II-XI-121, Bnd. 13, Bl. 81, pp. 126–128, 160–163, 199–200, 248–250, Bnd. 6, Bl. 4, pp. 23 ff.

60 Köhler to Waldeyer, 18 October 1919; Köhler to Stumpf, 26 April 1920, AAW, II-XI-121, Bnd. 6, Bl. 72, 151. The details of the station's final months, and of the sale of the apes, are in AAW, II-XI-121, Bnd. 6, Bl. 92, pp. 112 ff., 121, 123, 153 ff., 192 ff.; cf. the report in *Sitzungsberichte,* 1921, Nr. 1, p. 166.

61 Johannes von Allesch, "Über die drei ersten Lebensmonate eines Schimpansen," *Sitzungsberichte,* Nr. 2 (1921): 672–685.

62 Köhler to Geitel, 10 December 1915 and 9 September 1919, in Jaeger (ed.), *Briefe Wolfgang Köhlers,* pp. 58, 361.

Chapter 11

1 Köhler to Geitel, 15 July 1914, in Jaeger (ed.), *Briefe Wolfgang Köhlers* (cited in Chapter 7), p. 48.

2 Köhler to Stumpf, 9 May 1919, AAW Berlin, II-XI-121, Bnd. 12, unpaginated.

3 Köhler to Stumpf (as cited); Köhler to Waldeyer, 10 November 1917, AAW Berlin, ibid. Emphasis mine.

4 Köhler, *Mentality,* p. 89, n. 1, 209; *Intelligenzprüfungen,* pp. 65 n. 1, 152.

5 Köhler, *Mentality,* pp. 211–212; *Intelligenzprüfungen,* p. 153. Emphasis mine.

6 Köhler to Geitel, 9 September 1919, in Jaeger (ed.), *Briefe Wolfgang Köhlers,* p. 61.

7 Geitel to Vieweg, 10 August 1919; Geitel to Köhler, 9 September 1919 (draft); Vieweg to Geitel, 17 September 1919; Wiedemann to Vieweg, 20 November 1919 (copy), all in Jaeger (ed.), *Briefe Wolfgang Köhlers,* pp. 62, 64, 66, 76; the text of the original foreword, dated May 1919, is on p. 77.

8 Geitel to Köhler, 26 November 1919 (draft); Köhler to Geitel, 23 December 1919; Vieweg to Köhler, 23 December 1919, all in *Briefe Wolfgang Köhlers,* pp. 82, 84 f., 87.

9 Köhler, *Die physischen Gestalten,* pp. x–xi, xv. Emphasis in the original.

10 Köhler, *Die physischen Gestalten,* p. xx.

11 Köhler, "Gestalt Psychology," *PsyFo,* 31 (1967): xxi.

12 James Clerk Maxwell, *A Treatise on Electricity and Magnetism* (Oxford, 1873), vol. 1, pp. x–xi ff.; William Berkson, *Fields of Force* (New York, 1974), ch. 6, esp. pp. 148 ff.; cf. pp. 175, 188; Max Planck, "James Clerk Maxwell in seiner Bedeutung für die theoretische Physik in Deutschland" (1931), in *Physikalische Abhandlungen,* vol. 3, p. 352.

13 Köhler, *Die physischen Gestalten,* pp. 37–38; cf. p. ix.

14 See Alois Höfler, *Psychologie* (Vienna, 1897), p. 152; cf. Theo Herrmann, "Ganzheitspsychologie und Gestalttheorie" (cited in Chapter 4), p. 609, and Martin Scheerer, *Die Lehre von der Gestalt* (Berlin, 1931), p. 41.

15 Köhler, *Die physischen Gestalten,* pp. 42, 44, 47.

16 Ibid., pp. 32–34.

17 Ibid., pp. 55 ff., 58, 76 ff.

18 Ibid., pp. 60 f., 168. Emphasis in the original.

19 Ibid., pp. 69, 106–109.

20 Ibid., pp. 83, 113, 116 ff.; cf. Maxwell, *Treatise on Electricity and Magnetism,* vol. I, p. 143.

21 Planck, *Eight Lectures in Theoretical Physics* trans. A. P. Willis (New York, 1915), p. 45.

22 Köhler, *Die physischen Gestalten,* pp. 72, 75–76.

23 Ibid., pp. 154, 157–158.

24 Ibid., pp. 47, 168.

25 Ibid., p. 170.

26 Ibid., pp. 49, 170.

27 Ibid., pp. 174, 176–177.

28 Ibid., pp. 192–93. Emphasis in the original. Köhler, *The Task of Gestalt Psychology* (Princeton, N.J., 1966), p. 50.

29 Ibid., p. 193 n. 1.

30 Peter Keiler, "Isomorphie-Konzept und Wertheimer-Problem. Beiträge zu einer historisch-methodologischen Analyse des Köhlerschen Gestaltansatzes, I," *Gestalt Theory,* 2 (1980): 81. Keiler seems not to have considered the possibility mentioned in the text. Köhler cites Müller indirectly in the same paragraph where he set up his postulate, and treats Müller's color theory in much the same radicalizing fashion; see below.

31 Köhler, *Die physischen Gestalten,* pp. 199–200.

32 Köhler, *Mentality,* pp. 267–268; *Intelligenzprüfungen,* p. 193. Emphasis in the original.

33 Köhler, *Die physischen Gestalten,* pp. 242–243. Emphasis in the original. Cf. Koffka, *Principles,* pp. 59, 115, and Pastore, *Selective History* (cited in Chapter 5), p. 304. See also Woodward, "From Association to Gestalt" (cited in Chapter 4).

34 Ibid., pp. 194–195; cf. Pastore, *Selective History* (cited in chapter 5), p. 287. See also Wertheimer, "Bewegung," (cited in Chapter 8), p. 49. Unfortunately, Köhler never defined the term "functional similarity" more precisely. For an example of the resulting confusion, see William H. Rosar, "Visual Space as Physical Geometry," *Perception,* 14 (1985):403–425, esp. pp. 407 ff.

35 Ibid., pp. 26–27, 183. Edgar Rubin, "Die visuelle Wahrnehmung von Figuren," *Bericht über den 6. Kongress für experimentelle Psychologie* (Leipzig, 1914), 60–62; idem., *Synsoplevede Figur* (Copenhagen, 1915), trans. *Visuell wahrgenommene Figuren* (Copenhagen, 1921), esp. pp. 33 ff., 46 ff., 94 of the translation.

36 Köhler, *Die physischen Gestalten,* esp. pp. 3–4, 25.

37 Ibid., pp. 206 ff., 228 ff., esp. p. 207.

38 Ibid., pp. 211–212.

39 Ibid., pp. 212 ff., esp. p. 218.

40 See, e.g., Georg Eckardt, "Wolfgang Köhler, Gestaltpsychologie und 'Naturphilosophie' der Gestalt," *ZfP,* 196 (1988): 14.

41 Ibid., pp. 73–74. Emphasis mine.

42 Ibid., pp, 88, 256.

43 Ibid., pp. 50, 250–251, 256.

44 Ibid., pp. 253–254.

45 Mach, *The Science of Mechanics,* 6th ed. (LaSalle, Ill., 1960), pp. 488–489.

46 Goethe, *Farbenlehre, didaktischer Teil,* Para. 25–26, cited in Köhler, *Die physischen Gestalten,* p. 262; cf. *Goethe's Theory of Colors,* trans. Charles L. Eastlake (1840, repr. Cambridge, Mass., 1970), p. 9.

47 Köhler, *The Place of Value in a World of Facts* (1938, repr. New York, 1976), p. 197.

48 Köhler, *The Task of Gestalt Psychology* (cited in Chapter 10), p. 58.

49 Köhler, *Die physischen Gestalten,* p. 104.

50 Koffka, "Beginnings of Gestalt Theory" (cited in Chapter 7), p. 4; review of *Die physischen Gestalten* in *Die Naturwissenschaften,* 9 (1921): 413, 414.

Chapter 12

1 For the account in this and the following paragraph, see Brod, *Streitbares Leben* (cited in Chapter 7), pp. 136 ff., esp. pp. 142–143.

2 For examples, see Daniel Kevles, *The Physicists: The History of a Scientific Community in Modern America* (New York, 1978), ch. 9; Jeffrey Johnson, *The Kaiser's Chemists* (Chapel Hill, N.C., 1991), ch. 9.

3 Rabinbach, *The Human Motor* (cited in Chapter 4), ch. 9; Ulfried Geuter, "Polemos panton pater – Militär und Psychologie im Deutschen Reich 1914–1945," in M. G. Ash and U. Geuter (eds.), *Geschichte der deutschen Psychologie im 20. Jahrhundert* (Opladen, 1985), esp. pp. 147 ff.

4 The following account is indebted to Christoph Hoffmann, "Wissenschaft und Militär: Das Berliner Psychologische Institut und der Erste Weltkrieg" *Psychologie und Geschichte,* 5 (1994): 261–285.

5 For the original personnel and the beginning date of the project see *Chronik der Friedrich-Wilhelm-Universität zu Berlin,* 29 (1915), p. 51.

6 Wertheimer and von Hornbostel, "Über die Wahrnehmung der Schallrichtung," *Sitzungsberichte der Preussischen Akademie der Wissenschaften,* 20 (1920): 388–396.

7 A copy of the patent, dated 7 July 1915, is in MWPB; for the name "Wertbostel," see Michael Wertheimer, "Max Wertheimer" (cited in Chapter 7), p. 15.

8 Wertheimer provided details of these activities in a handwritten statement, in PAUF, Hauptakte Wertheimer, Bl. 24.

9 Stumpf, "Über den Entwicklungsgang der neueren Psychologie und ihre militärtechnische Verwendung," *Deutsche Militärärztliche Zeitschrift,* 47 (1918): 273.

10 *Verzeichnis der Vorlesungen gehalten an der Friedrich-Wilhelm-Universität zu Berlin,* 1917 ff.; Stumpf to Rektor, 7 July 1916, and Ministry to Wertheimer, 25 August 1916 (copy), UAHUB, Philosophische Fakultät, Dekanat, Nr. 1439, Bl. 151, 155.

11 On Wertheimer's relations with Einstein, see Michael Wertheimer, "Relativity and Gestalt: A Note on Albert Einstein and Max Wertheimer," *JHBS,* 1 (1965): 86–87; Wertheimer states that their conversations about the genesis of relativity theory began in 1916 in *Productive Thinking* (cited in Chapter 7), p. 213; cf. p. 223. For a discussion of the dating of Wertheimer's text and a critical analysis of his account, see Arthur I. Miller, "Albert Einstein and Max Wertheimer: A Gestalt Psychologist's View of the Genesis of Special Relativity Theory," *History of Science,* 13 (1975): 75–103.

12 Wertheimer, "Über Schlussprozesse im produktiven Denken" (1920), reprinted in *Drei Abhandlungen zur Gestalttheorie* (Erlangen, 1925), pp. 164–165.

13 Ibid., pp. 167, 170. Emphasis in the original.

14 Ibid., pp. 165, 169, 181.

15 Ibid., pp. 173–174.

16 Ibid., p. 174.

17 Ibid., pp. 175, 178–179. Emphasis in the original. For a richly detailed report of Wertheimer's thinking as he solved this problem, see *Productive Thinking,* ch. 8, esp. pp. 195 ff.

18 Thomas S. Kuhn, *The Structure of Scientific Revolutions,* rev. ed. (Chicago, 1970), pp. 110 ff.

19 Wertheimer, "Schlussprozesse," pp. 180–181.

20 Ibid., pp. 175, 181. Emphasis in the original. For Wittfogel's recollections of Wertheimer, Kollwitz, and his own politics, see "Die hydraulische Gesellschaft und das Gespenst der asiatischen Restauration: Gespräch mit Karl August Wittfogel," in Matthias Greffrath, *Die Zerstörung einer Zukunft: Gespräche mit emigrierten Sozialwissenschaftlern* (Reinbek, 1979), pp. 304–305.

21 On the political views of German academics in general and psychologists in particular in this period, see Klaus Schwabe, *Wissenschaft und Kriegsmoral: Hochschullehrer und die politischen Grundfragen des Ersten Weltkrieges* (Göttingen, 1969), and Eckart Scheerer, "Kämpfer des Wortes: Die Ideologie deutscher Psychologen im Ersten Weltkrieg und ihr Einfluss auf die Psychologie der Weimarer Zeit," *Psychologie und Geschichte,* 1 (1989): 12–22. For Einstein's views, see, e.g., Ronald Clark, *Einstein: The Life and Times* (New York, 1971). Käthe Kollwitz's most vehement antiwar statement in this period is "An Richard Dehmel!" *Vorwärts,* 30 October 1918, repr. in Hans Kollwitz (ed.), *Ich sah die Welt in liebevollen Blicken. Käthe Kollwitz: Ein Leben in Selbstzeugnissen* (1968, repr. Wiesbaden, n.d.), pp. 189–190.

22 *The Born–Einstein Letters: Correspondence between Albert Einstein and Max and Hedwig Born from 1916 to 1955, with Commentaries by Max Born,* trans. Irene Born (New York, 1971), p. 150.

23 *The Born–Einstein Letters,* p. 151. See also Max Born, *My Life: Recollections of a Nobel Laureate* (New York, 1978), pp. 184ff.

24 Wertheimer, "Vom Geistesleben des Prager Judentums," in *Das jüdische Prag: Eine Sammelschrift* (Prague, 1918; repr. Kronberg/Taunus, 1978), p. 16.

25 Stumpf, "Zum Gedächtnis Lotzes," *Kant Studien,* 22 (1917), p. 26, quoted by Wertheimer in "Feier zu Carl Stumpfs 70. Geburtstag" (cited in Chapter 2). Emphasis mine.

26 Koffka, Review of Köhler, *Die physischen Gestalten* (cited in Chapter 11), p. 414.

Chapter 13

1 On the emergence of applied psychology, see Siegfried Jaeger, "Zur Herausbildung von Praxisfeldern der Psychologie bis 1933," in Ash and Geuter (eds.), *Geschichte der deutschen Psychologie* (cited in Chapter 12); Ulfried Geuter, *The Professionalization of Psychology in Nazi Germany* (cited in the Introduction), pp. 45 ff., 189 ff.

2 Georg Eckardt, "Die Gründung der Psychologischen Anstalt in Jena (1923)," *Wissenschaftliche Zeitschrift der Friedrich-Schiller-Universität Jena,* Gesellschafts- und Sprachwissenschaftliche Reihe, 22 (1973): 517–559; Mitchell G. Ash, "Psychology and Politics in Interwar Vienna: The Vienna Psychological Institute, 1922–1942," in Ash and W. R. Woodward (eds.), *Psychology in Twentieth-Century Thought and Society* (Cambridge 1987), pp. 143–164.

3 "Kundgebung der Deutschen Gesellschaft für Psychologie: Über die Pflege der Psychologie an den deutschen Hochschulen," *Bericht über den 11. Kongress der Deutschen Gesellschaft für Psychologie* (Jena, 1930), pp. vii–viii; for Müller's statement, see Geuter, *Professionalization,* pp. 49 f.

4 Karl Bühler, *Die Krise der Psychologie* (Jena, 1927); Alexandre Métraux, "Die angewandte Psychologie vor und nach 1933 in Deutschland," in C.-F. Graumann (ed.), *Psychologie im Nationalsozialismus* (Berlin, Heidelberg, 1985), pp. 221–262; Helmut Hildebrandt, *Zur Bedeutung des Begriffs der Alltagspsychologie in Theorie und Geschichte der Psychologie: eine psychologiegeschichtliche Studie anhand der Krise der Psychologie in der Weimarer Republik* (Frankfurt a.M., 1991), chs. III–IV. For further discussion, see Chapter 17.

5 William Stern, "Die Stellung der Psychologie an den deutschen Universitäten," *Zeitschrift für pädagogische Psychologie,* 32 (1931): 157.

6 BA Potsdam, REM, 49.01, Nr. 1662, Bl. 7–12, 34–35, 62–77, 78 f.

7 GStA, Rep. 76 Va Sekt. 2 Tit. X Nr. 150, Bnd. 3, Bl. 37ff., 154; Rep. 76 Va, Sekt. 4, Tit. X Nr. 53a, Bnd. 1, Bl. 114, 119; Stumpf, "Carl Stumpf," p. 403; *Minerva, Jahrbuch der gelehrten Welt,* 25 (1921): 73, 522; see also Appendix 1, Table 3. On the Leipzig institute, see Felix Krueger, "Das Psychologische Institut," *Bericht über den 4. Kongress für Heilpädagogik* (Berlin, 1929), pp. 414–415.

8 Samson Foundation, board meeting minutes, 19 February 1920 and 4 November 1920, AAW, II–XI-121, Bnd. 6, Bl. 133, 251.

9 Stumpf to Ministry, 4 August 1920, and Stumpf to Erich Wende, 17 August 1920, GStA (cited in n. 7), Bl. 47, 49–50; UAHUB, Philosophische Fakultät, Dekanat, Nr. 1469, Bl. 159, 160.

10 Ministry to Köhler, 3 August 1921, 10 September 1921, UA Göttingen, Universitätskuratorium, Philosophische Fakultät XIV.IV. A.a. 281; Ministry to University Administration Berlin, 31 August 1921, UAHUB, Philosophische Fakultät, Dekanat, Nr. 1460, Bl. 148.

11 Philosophical Faculty to Ministry, 2 August 1921, GStA, Rep. 76 Va Sekt. 2 Tit. IV Nr. 68A, Bl. 155–156; UAHUB, Philosophische Fakultät, Dekanat, Nr. 1469, Bl. 153 and 439–445, esp. 439–440, 442–443.

12 Ministry to Philosophical Faculty, 12 July 1921; Faculty to Ministry, 2 August 1921, GStA, ibid., Bl. 129, 155–156.

13 Günther Anders interview, 5 October 1980; idem., "'Wenn ich verzweifelt bin, was geht's mich an? Gespräch mit Günther Anders," in Matthias Greffrath (ed.), *Die Zerstörung einer Zukunft* (cited in Chapter 12), pp. 20–21. Anders's statement could neither be confirmed nor disconfirmed in Ministry records.

14 On the new departments and their enrollments, see UAHUB, *Chronik der Friedrich-Wilhelm-Universität zu Berlin,* 1927–1933; course titles are compiled from *Verzeichnis der Vorlesungen an der Friedrich-Wilhelm-Universität zu Berlin,* 1921–1929; cf. Siegfried Jaeger, "Wolfgang Köhler" (cited in Chapter 7); idem., "Entwicklung und struktur des psychologischen Lehrangebotes an der Berliner Universität 1900–1945," in Horst Gundlach (ed.), *Arbeiten zur Psychologie-geschichte* (Göttingen, 1994), pp. 58 ff.

15 Anitra Karsten interview, Frankfurt am Main, 22 February 1978; Köhler's statement on admission requirements is in *Chronik,* 1927–1928, pp. 78–79.

16 Richard Meili, "Richard Meili," in Pongratz et al. (eds.), *Psychologie in Selbstdarstellungen,* vol. 1 (Bern, 1972), p. 173; Kurt Gottschaldt interview, Göttingen, 5 August 1987. On the palace atmosphere, see Wolfgang Metzger, "Verlorenes Paradies. Im Psychologischen Institut in Berlin, 1922–1931," *Schweizerische Zeitschrift für Psychologie,* 29 (1970): 16–25; Gottschaldt interview; Rudolf Arnheim interview, Ann Arbor, Mich., 9 August 1976.

17 Metzger, "Verlorenes Paradies"; Köhler to Wertheimer, 4 August 1929, MWP-NYPL; cf. Geuter, *Professionalisierung,* p. 97. On the changing social composition of the German universities, see Fritz K. Ringer, *Education and Society in Modern Europe* (Bloomington, Ind., 1979).

18 Metzger, "Wolfgang Metzger," p. 197; Arnheim interview; Blyuma Zeigarnik interview, Leipzig, 10 July 1980; Gottschaldt quoted in A. S. Luchins and E. H. Luchins, "Max Wertheimer 1919–1929," *Gestalt Theory,* 8 (1986): 9.

19 Wolfgang Metzger, "Zur Geschichte der Gestalttheorie in Deutschland" (1963), in *Gestaltpsychologie,* eds. Michael Stadler and Heinrich Crabus (Frankfurt a.M., 1986), p. 99; for further recollections of Wertheimer as a teacher, see Luchins and Luchins, "Max Wertheimer: 1919–1929," pp. 9–16.

20 Rudolf Arnheim, quoted in Luchins and Luchins, "Max Wertheimer: 1919–1929."

21 Frank Freeman, "The Beginnings of Gestalt Psychology in the United States," *JHBS,* 13 (1977): 352–353; Mary Henle, "Robert M. Ogden and Gestalt Psychology in America," *JHBS,* 20 (1984): 9–19. Impressions of American visitors' sojourns in Berlin are recorded in J. F. Brown memoir; D. K. Adams report to Yerkes and NRC, Adams papers, AHAP; Edwin Newman interview, Cambridge, Mass., 9 September 1976. Köhler's correspondence with the Lincoln Foundation on behalf of Duncker is in Köhler papers, APS; for Metzger's stay in Iowa, see Michael Stadler and Heinrich Crabus, "Wolfgang Metzger (1899–1979): Leben, Werk und Wirkung," in Metzger, *Gestaltpsychologie,* p. 12.

22 Koffka to University Senate, 6 April 1923, UA Giessen, PrA Nr. 384. In this document, Koffka asked his colleagues for permission to employ a further, unpaid assistant, saying that he had "no aide" other than Hartgenbusch. See also Hans-Georg Burger, "Anfänge und Bedeutung der Universitätspsychologie in Giessen" (cited in Chapter 7).

23 Information on course offerings from *Vorlesungsverzeichnis der Hessischen Landesuniversität zu Giessen,* 1912–1927; for the monographs in the "Contributions" series, see Appendix 2.

24 Walter Eisen, "Kurt Koffka, 1886–1941," *British Journal of Psychology,* 33 (1942): 75.

25 Koffka to Ogden, 4 July 1925, Robert M. Ogden papers, Cornell University Archives, Box 6; Löhlein to University Rector, 27 November 1925, UA Giessen, PrA Nr. 384; Koffka, "Perception: An Introduction to the Gestalt-Theorie," *Psychological Bulletin,* 19 (1922): 531–585. See also Freeman, "Beginnings of Gestalt Psychology in the United States"; Henle, "Robert M. Ogden"; and Michael M. Sokal, "The Gestalt Psychologists in Behaviorist America," *American Historical Review,* 89 (1984): 1240–1263.

26 Koffka to Wakeman, 10 March and 19 March 1927, Kurt Koffka correspondence, Smith College Archives, Northampton, Mass., folder 42; William Allan Nielson to Koffka, 18 March 1927, Nielson papers, Smith College Archives; cf. Henle, "Robert M. Ogden." Sokal, "The Gestalt Psychologists," says Koffka's salary was rumored to be $10,000.

27 David Katz to Messer, 5 December 1927; William Stern to Messer, 5 December 1927; Erich Jaensch to Messer, 6 December 1927, UA Giessen, PrA 24 Phil., Bl. 16, 61, 46. Cf. Report of Philosophical Faculty, 11 December 1927, ibid., Bl. 27–34.

28 Koffka to Messer, 1 November 1927; Köhler to Messer, 2 November 1927, UA Giessen, ibid.

29 Löhlein to Wertheimer, 14 June 1928, 11 July 1928, and 2 August 1928, MWPB. Koffka's salary is inferred from information in UA Giessen, PrA 24 Phil., Bl. 6. A. S. Luchins and E. H. Luchins, "Max Wertheimer: 1919–1929," p. 6, incorrectly assert that Wertheimer was not offered this post.

30 Stumpf to Philosophical Faculty, 4 April 1918, UAHUB, Philosophische Fakultät, Dekanat, Nr. 1236, Bl. 117; Ministry to Wertheimer, 22 March 1922, MWPB; Minister Becker to Dean of Philosophical Faculty, 21 February 1922, UAHUB, Philosophische Fakultät, Dekanat, Nr. 1439, Bl. 429 ff.; Köhler to Einstein, 26 April 1922 and Einstein to Schlick, 28 April 1922, Albert Einstein Papers, Jewish and National University Library, Jerusalem; cf. Luchins and Luchins, "Max Wertheimer: 1919–1929," pp. 5, 7.

31 UAHUB, Philosophische Fakultät, Dekanat, Nr. 152, Bl. 3–4, 8; Nr. 1471, Bl. 285–288.

32 Karl-Dietrich Erdmann, ed., *Kurt Riezler: Tagebücher – Aufsätze – Dokumente* (Göttingen, 1972), esp. pp. 7, 143, 145–146.

33 Philosophical Faculty to Ministry, 22 October 1928 and Ministry to Faculty, 10 November 1928, GStA Rep. 76 Va Sekt V Tit. IV Nr. 4, Bnd. 3, Bl. 78–79, 86.

34 Natural Science Faculty to University Kuratorium, 30 July 1928; Riezler to Dean of Natural Sciences, 23 November 1928; Natural Science Faculty to Ministry, 4 December 1928, PAUF, Personalhauptakte Max Wertheimer.

35 Gelb to Wertheimer, 30 December 1928, MWP-NYPL.

36 Ministry to Wertheimer and attached Agreement, 20 March 1929, MWPB. Riezler to Richter, 20 December 1928; Ministry to Philosophical Faculty, 26 January 1929; Philosophical Faculty to Ministry, 4 February 1929; Riezler to Windelband, 6 February 1929, all in GStA, Rep. 76 Va Sekt. 5 Tit. IV Nr. 4 Bnd. 3, Bl. 116, 118, 125–126.

37 Erdmann, "Kurt Riezler," p. 144; Gertrud Siemsen interview, Berlin, 11 May 1979; Wolfgang Hochheimer interview, Berlin, 26 May 1980; Erika Oppenheimer Fromm interview, Chicago, 15 November 1982. See also Paul Kluke, *Die Stiftungsuniversität Frankfurt am Main* (cited in Chapter 8), and Wilhelm Pauck and Marion Pauck, *Paul Tillich: His Life and Work,* vol. 1 (New York, 1976), pp. 117 ff.; Karl Korn, *Lange Lehrzeit. Ein Deutsches Leben* (Munich, 1979); Wolfgang Schivelbusch, *Intellektuellendämmerung* (Frankfurt a.M., 1982).

38 Newman interview (cited in n. 21); *Johann Wolfgang Goethe-Universität Frankfurt am Main. Verzeichnis der Vorlesungen,* 1929–1933; Gertrud Siemsen memoir, spring 1978, pp. 5–6. On Gelb's departure and Metzger's appointment, cf. Wertheimer to Köhler, 17 November 1930, Köhler papers, APS.

39 Gertrud Siemsen memoir (Berlin, 1978).

40 Wertheimer to Kuratorium, 1932, MWPB; Metzger, "Wolfgang Metzger," p. 200; cf. Wertheimer to Köhler (cited in n. 33). Günther Anders (cited in n. 13) states that Wertheimer was naive about the Nazi threat, and adds that he was "a completely unpolitical person."

41 A copy of the contract is in Köhler papers, APS, correspondence Julius Springer.

42 Verlagsarchiv des Springer-Verlags, Heidelberg, folders G-118, G-177, K-185, K-188.

43 "Vorwort," *PsyFo,* 1 (1921): p. 1; for further discussion of this issue, see Eckart Scheerer, "Fifty Volumes of *Psychological Research/Psychologische Forschung:* The History and Present Status of the Journal," *Psychological Research,* 50 (1988): esp. pp. 71–73.

44 Siemsen memoir.

Chapter 14

1 Ludwik Fleck, *Genesis and Development of a Scientific Fact* (1935), trans. Fred Bradley and Thaddeus Trenn (Chicago, 1979).

2 Kurt Danziger, "Generative Metaphor and the History of Psychological Discourse" (cited in the Introduction).

3 Wertheimer, "Untersuchungen zur Lehre von der Gestalt, I: Prinzipielle Bemerkungen," *PsyFo*, 1 (1921): esp. pp. 48–50, 52, 55. Emphasis in the original.

4 Ibid., pp. 51 n. 1 and 56 n. 1; idem., "Gestaltpsychologische Forschung," in Emil Saupe (ed.), *Einführung in die neuere Psychologie* (Osterwieck, 1928), pp. 47–54.

5 Koffka, "Introspection and the Method of Psychology," *British Journal of Psychology*, 15 (1925): 149, 151–152, 154–155.

6 Schumann, "Über einige Probleme der Lehre von den Gesichtswahrnehmungen," *Bericht über den 5. Kongress für experimentelle Psychologie* (1913) pp. 181, 183; idem., "Untersuchungen über die psychologischen Grundprobleme der Tiefenwahrnehmung. II. Die Dimensionen des Sehraumes," *ZfP*, 86 (1921): esp. pp. 266–267, 270; cf. Wilhelm Fuchs, "Experimentelle Untersuchungen über das simultane Hintereinander auf derselben Sehrichtung," *ZfP*, 91 (1923): 145–235.

7 Fromm interview (cited in Chapter 13).

8 On crucial demonstrations as dramatic, didactic devices in the natural sciences, see Geoffrey Cantor, "The Rhetoric of Experiment," in David Gooding, Trevor Pinch, and Simon Schaffer (eds.), *The Uses of Experiment: Studies in the Natural Sciences* (Cambridge, 1989), esp. p. 176.

9 Kurt Danziger, "Social Context and Investigative Practice in Early Twentieth-Century Psychology," in Ash and Woodward (eds.), *Psychology in Twentieth-Century Thought and Society* (cited in Chapter 13), p. 28.

10 Danziger, "Social Context and Investigative Practice." See also Danziger,"Origins of the Psychological Experiment as a Social Institution" (cited in Chapter 1).

11 Andrew Pickering, "From Science as Knowledge to Science as Practice," in idem. (ed.), *Science as Practice and Culture* (cited in the Introduction), esp. p. 9.

12 For more comprehensive, but systematic rather than historical accounts of research in perception to 1931 and 1935, respectively, see Kurt Koffka, "Die Wahrnehmung von Bewegung," in Albrecht Bethe et al. (eds.), *Handbuch der normalen und pathologischen Physiologie*, vol. 12, part 2 (Berlin, 1931), pp. 1166–1214; idem., "Psychologie der optischen Wahrnehmung," ibid., pp. 1215–1271; and *Principles*, esp. chs. 4–7; see also Pastore, *Selective History* (cited in Chapter 5), ch. 14. For citations of Gestalt theorists' work in discussions of recent research on perception, see Kenneth R. Boff et al. (eds.), *Handbook of Perception and Human Performance*, 2 vols. (New York, 1986).

13 Wertheimer, "Untersuchungen zur Lehre von der Gestalt, II," *PsyFo*, 4 (1923): 301–302; abr. trans. by Michael Wertheimer as "Principles of Perceptual Organization," in David C. Beardslee and Michael Wertheimer (eds.), *Readings in Perception* (Princeton, N.J., 1958), pp. 115–116. Interestingly, Wertheimer referred here to Stumpf's call for experimental research to determine the "laws of summation" in individual sense modalities. Cf. Stumpf, "Erscheinungen und psychische Funktionen" (cited in Chapter 2), p. 24; Wertheimer, "Untersuchungen, II," p. 301.

14 Wertheimer, "Untersuchungen II," pp. 337–347; "Perceptual Organization," pp. 126 f., 132–134.

15 Wertheimer, "Untersuchungen II," e.g. pp. 308, 311, 315–316, 343 ff.; "Perceptual Organization," pp. 118–119, 120–121. On the *Punktarbeit* as a research program and Wertheimer's insistence on quantification, see Viktor Sarris, "Max Wertheimer in Frankfurt–Über Beginn und Aufbaukrise der Gestaltpsychologie, II. Strukturgesetze der Bewegungs- und Raumwahrnehmung," *ZfP*, 195 (1987): 409, 410–411.

16 Friedrich Wulf, "Über die Veränderung von Vorstellungen. Gedächtnis und Gestalt," *PsyFo*, 1 (1921): 333–373; Koffka, *Principles*, pp. 493 ff.

17 Wilhelm Benary, "Beobachtungen zu einem Experiment über Helligkeitskontrast," *PsyFo*, 5 (1924): esp. pp. 131–132; Wolfgang Metzger, "Gestalt und Kontrast," *PsyFo*, 15 (1931): 374–386; cf. Koffka, *Principles*, pp. 136–138.

18 Koffka, "Über Feldbegrenzung und Felderfüllung," *PsyFo*, 4 (1923): esp. pp. 193–194, 202; Koffka, *Principles*, pp. 135 and 245 n. 22.

19 Adhemar Gelb and Ragnar Granit, "Die Bedeutung von 'Figur und 'Grund' für die Farbschwelle," *ZfP*, 93 (1923): 83–118; Susanne Liebmann, "Über das Verhalten farbiger Formen bei Helligkeitsgleichheit von Figur und Grund," *PsyFo*, 9 (1927): 300–353; Koffka, *Principles*, pp. 126 f., 265.

20 Wolfgang Metzger, "Optische Untersuchungen am Ganzfeld, II, Mitteilung: Zur Phänomenologie des homogenen Ganzfeldes," *PsyFo*, 13 (1930): esp. pp. 6–7, 9. The setup is described in Willy Engel, "Optische Untersuchungen am Ganzfeld, I. Mitteilung: Die Ganzfeldanordnung," ibid., pp. 1–5.

21 Kurt Gottschaldt, "Über den Einfluss der Erfahrung auf die Wahrnehmung von Figuren, I," *PsyFo*, 8 (1926): esp. pp. 267–279. On the influence of the experimenter's instruction on the perception of embedded figures, see Gottschaldt, "Über den Einfluss der Erfahrung auf die Wahrnehmumg von Figuren, II," ibid., 12 (1929): 1–60. Cf. Koffka, *Principles*, pp. 155–158, 395–398.

22 Herta Kopfermann, "Psychologische Untersuchungen über die Wirkung zweidimensionaler Darstellungen körperlicher Gebilde," *PsyFo*, 13 (1930): 293–364; cf. Koffka, *Principles*, pp. 152–155, 162 f.

23 Joseph Ternus, "Experimentelle Untersuchungen über phänomenale Identität," *PsyFo*, 7 (1926): esp. pp. 82 ff., 86–88, 101; cf. Koffka, *Principles*, pp. 299–300, 301–303.

24 Wolfgang Metzger, "Beobachtungen über phänomenale Identität," *PsyFo*, 19 (1934): esp. p. 40. The experiments were first carried out in Berlin in the late 1920s.

25 Imre Lakatos, "Falsification and the Methodology of Scientific Research Programmes," in Imre Lakatos and Ian Musgrave (eds.), *Criticism and the Growth of Knowledge* (Cambridge, 1970).

26 Karl Duncker, "Über induzierte Bewegung," *PsyFo*, 12 (1929): 187–192, esp. pp. 188–189; Koffka, *Principles*, pp. 282 ff.

27 Ernst Mach, *Grundlinien der Lehre von den Bewegungsempfindungen* (Leipzig, 1875).

28 Duncker, "Induzierte Bewegung," pp. 207–208, 246, 258. For later work on frames of reference done under Wertheimer's direction in Frankfurt, see Erika Oppenheimer, "Optische Versuche über Ruhe und Bewegung," *PsyFo*, 20 (1934): 1–46; Walter Krolik, "Über Erfahrungswirkungen bei Bewegungssehen," ibid., 47–101; Viktor Sarris, "Max Wertheimer in Frankfurt, III. Weitere Studien über das Sehen von Bewegung (1929–1933)," *ZfP*, 196 (1988): 27–62.

29 J. F. Brown, "Über gesehene Geschwindigkeiten," *PsyFo*, 10 (1928): 84–101; Hans Wallach, "How the Transposition Prinicple in Speed Perception Was Discovered," *Gestalt Theory*, 9 (1987): 242–243.

30 Brown, "The Visual Perception of Velocity," *PsyFo*, 14 (1931): 199–232; "On Time Perception in Visual Movement Fields," ibid., pp. 233–248; "The Thresholds for Visual Movement," ibid., pp. 249–268. Wallach, "Über visuell wahrgenommene Bewegungsrichtung," *PsyFo*, 20 (1935): 325–380. "Transposition Principle," p. 245. Cf.

Koffka, *Principles,* pp. 288–291, 296–297. For extensive discussion of this research and later work on the topic, see Hans Wallach, *On Perception* (New York, 1976).

31 Erich von Hornbostel and Max Wertheimer, "Über die Wahrnehmung der Schallrichtung" (cited in Chapter 12), esp. p. 391; Hornbostel, "Das räumliche Hören," in Albrecht Bethe et al. (eds.), *Handbuch der normalen und pathologischen Physiologie,* vol. 11 (Berlin, 1926), esp. pp. 612 ff.

32 Hornbostel, "Beobachtungen über ein- und zweiohriges Hören," *PsyFo,* 4 (1923): 64–114; idem., "The Psychophysiology of Monotic and Diotic Hearing," *7. International Congress of Psychology* (1923), pp. 377–381. For related work by an American visitor to Berlin in the 1920s, see Paul T. Young, "Auditory Localization with Acoustical Transposition of the Ears." *American Journal of Psychology,* 11 (1928): 399–429.

33 Hornbostel, "Psychologie der Gehörserscheinungen," in Bethe et al. (eds.), *Handbuch,* vol. 11, 701–729 idem., "Die Einheit der Sinne," *Melos,* 4 (1926): 230–297.

34 Duncker, "A Qualitative (Experimental and Theoretical) Study of Productive Thinking (Solving of Comprehensible Problems)," *Pedagogical Seminary,* 33 (1926): esp. pp. 646–647. Cf. Norman R. F. Maier, "Reasoning in Humans I. On Direction," *Journal of Comparative Psychology,* 10 (1930): 115–143; "Reasoning in Humans II. The Solution of a Problem and its Appearance in Consciousness," *Journal of Comparative Psychology,* 12 (1931): 181–194.

35 Duncker, "A Qualitative Study," pp. 658–659, 664 f.; idem., *Zur Psychologie des produktiven Denkens* (1935; repr. Berlin, 1974), pp. 21 ff., ch. 6.

36 Ibid., Duncker, "A Qualitative Study," pp. 669 f.; *Zur Psychologie des produktiven Denkens,* pp. 2–3, 17.

37 Ibid., esp. pp. 693, 703.

38 Duncker, *Zur Psychologie des produktiven Denkens,* pp. 10, 14–15, 36–37.

39 J. Borak, "Über die Empfindlichkeit für Gewichtsunterschiede bei abnehmender Reizstärke," *PsyFo,* 1 (1922): 374–389; Köhler, "Zur Theorie des Sukzessivvergleichs und der Zeitfehler," ibid., 4 (1923): esp. pp. 115–117.

40 Köhler, "Sukzessivvergleich," pp. 121–122, esp. pp. 139 ff., 146, 166 ff.; idem., *Gestalt Psychology,* pp. 271–272.

41 Köhler, *Gestalt Psychology,* pp. 271–272, 274.

42 Otto von Lauenstein, "Ansatz zu einer physiologischen Theorie des Vergleichs und der Zeitfehler," *PsyFo,* 17 (1933): pp. 137 ff., 141.

43 Ibid., pp. 147–148, pp. 153 ff.; Koffka, *Principles,* pp. 468 ff.

44 Köhler, *Gestalt Psychology,* pp. 279–280, 282, 292, 294 f., 347.

45 Hedwig von Restorff, "Über die Wirkung von Bereichsbildungen im Spurenfeld," *PsyFo,* 18 (1933): esp. p. 202; cf. Koffka, *Principles,* p. 483.

46 Restorff, "Bereichsbildung," esp. p. 305; Koffka, *Principles,* pp. 482, 484.

47 Erich Goldmeier, *The Memory Trace* (Hillsdale, N.J., 1982), pp. 3–4, 105.

48 Köhler, *Place of Value* (cited in Chapter 11), pp. 188–189.

49 Koffka, *Principles,* p. 679.

Chapter 15

1 Kurt Koffka, *Die Grundlagen der psychischen Entwicklung. Eine Einführung in die Kinderpsychologie* (Osterwieck, 1921; 2nd ed., 1925), trans. by Robert M. Ogden as *The*

Growth of the Mind: An Introduction to Child Psychology (New York, 1925; 2nd ed., 1928). Quotations are from the first edition of the translation, and have been compared with the original.

2 Ludwig Edinger, "The Relations of Comparative Anatomy to Comparative Psychology" (cited in Chapter 9), pp. 443–444.

3 Edinger, *Vorlesungen über den Bau der nervösen Zentralorgane der Menschen und der Tiere*, 8th ed. (Leipzig, 1911), p. 523. Emphasis in the original.

4 Koffka, *Growth*, pp. 55, 285.

5 Ibid., pp. 38–39, 51, 55, 146, 152, 172, 209.

6 Ibid., pp. 151, 234. Emphasis mine.

7 Ibid., p. 106.

8 Ibid., p. 52.

9 Ibid., pp. 264 ff., esp. p. 272.

10 Ibid., pp. 46–47.

11 Lucien Lévy-Bruhl, *Les fonctions mentales dans les sociétés inférieures*, 2nd ed. (Paris, 1912).

12 Koffka and Stählin, "Zur Einführung," *Zeitschrift für Religionspsychologie*, 1 (1914), pp. 2, 6; Herbert Spencer, *First Principles*, 2nd ed. (London, 1867), p. 369.

13 Koffka, *Growth*, pp. 336–337, 339–340. Emphasis mine.

14 Wertheimer, "Denken der Naturvölker" (cited in Chapter 8), p. 114 n. 1.

15 Ibid., p. 335.

16 Koffka, *Growth*, p. 356.

17 Helene Frank, "Untersuchungen über Sehgrössenkonstanz bei Kindern," *PsyFo*, 7 (1926): 137–145; "Die Sehgrössenkonstanz bei Kindern," *PsyFo*, 10 (1928), 102–106.

18 Eckart Scheerer, "Gestalt Psychology in the Soviet Union. I. The Period of Enthusiasm," *Psychological Research*, 41 (1980): 113–132.

19 Alexander Gurwitsch, "Über den Begriff des embryonalen Feldes," *Archiv für Entwicklungsmechanik der Organismen*, 51 (1922): 383–415; Paul Weiss, *Morphodynamik. Ein Einblick in die Gesetze der organischen Gestaltung an Hand von experimentellen Ergebnissen* (Berlin, 1926); Donna Harraway, *Crystals, Fabrics and Fields: Metaphors of Organicism in Twentieth-Century Developmental Biology* (Baltimore, 1976); T. J. Horder and Paul Weindling, "Hans Spemann and the Organiser," in T. J. Harder, et al. (eds.), *History of Embryology* (Oxford, 1986).

20 Köhler to Hans Geitel, 4 December 1919, in Jaeger (ed.), *Briefe Wolfgang Köhlers* (cited in Chapter 7), p. 80.

21 Köhler, "Gestaltprobleme und Anfänge einer Gestalttheorie," *Jahresbericht für die gesamte Physiologie und experimentelle Pharmakologie*, 3 (1924), Bericht über das Jahr 1922, pp. 512, 534. Cf. Becher, "Kritik der Widerlegung des Parallelismus," (Cited in Chap. 5).

22 Ibid., pp. 535–536.

23 Ibid., pp. 535, 538–539; cf. Oskar Hertwig, *Das Werden der Organismen. Eine Widerlegung von Darwins Zufallstheorie* (Jena, 1916).

24 Hans Driesch, "'Physische Gestalten' und Organismen," *Annalen der Philosophie*, 5 (1925): 1, 6, 11.

25 Köhler, "Zum Problem der Regulation" (1927), trans. Mary Henle and Erich Goldmeier as "On the Problem of Regulation," in Mary Henle (ed.), *Selected Papers of Wolfgang Köhler* (New York, 1971), pp. 305 ff. Emphasis in the original. Further quotations are from the translation.

26 Ibid., p. 324.

27 Ibid., pp. 324–325. Emphasis mine.

28 Richard Woltereck, "Über die Spezifizität des Lebensraumes, der Nahrung und der Körperformen bei pelagischen Cladoceren und über 'Ökologische Gestalt-Systeme'," *Biologisches Zentralblatt*, 48 (1928): 521–551, esp. pp. 541, 544 ff.

29 Ludwig von Bertalanffy, *Theoretische Biologie*, vol. 1 (Berlin, 1932), e.g. pp. viii, 47, 49, 90 ff., 120 f.; vol. 2 (Berlin, 1942), ch. 1; cf. *General Systems Theory* (New York, 1968): esp. pp. 12–13. Von Bertalanffy cited Kurt Koffka, "Die Krisis in der Psychologie. Bemerkungen zu dem Buch gleichen Namens von Hans Driesch," *Die Naturwissenschaften*, 25 (1926): 581–586.

30 Köhler to Geitel, 4 December 1919, in Jaeger, *Briefe Wolfgang Köhlers*, p. 80.

31 Köhler, *Intelligenzprüfungen* (cited in Chapter 10), pp. 227–228; *Mentality* (cited in Chapter 10), pp. 322–323. Cf. Heinz Werner, *Einführung in die Entwicklungspsychologie* (Leipzig, 1926), pp. 45 ff.

32 Fritz Heider, *The Life of a Psychologist: An Autobiography* (Lawrence, Kans., 1983), p. 45; Arnheim interview (cited in Chapter 13). Wertheimer, *Über Gestalttheorie* (Erlangen, 1925), p. 19; trans. N. Nairn-Allison as "Gestalt Theory," *Social Research*, 11 (1944): 94. Arnheim studied these phenomena in his research on expression (see Chapter 17).

33 Köhler, "Ein altes Scheinproblem" (1929), trans. as "An Old Pseudoproblem," in *Selected Papers*, pp. 125–141. For Mach's view, see Chapter 4, this volume.

34 Köhler, *Gestalt Psychology*, pp. 227–228.

35 Ibid., 2nd ed. (1947), pp. 132, 142, 148; cf. Max Scheler, *Wesen und Formen der Sympathie*, 2nd ed. (Bonn, 1923), sec. C.

36 Heinrich Schulte, "Versuch einer Theorie der paranoischen Eigenbeziehung und Wahnbildung," *PsyFo*, 5 (1924): 1–23; "A Gestalt Theory of Paranoia," trans. Ervin Levy, *Gestalt Theory*, 8 (1986): 230–255, esp. pp. 232–233, 234, 237.

37 Köhler, *Gestalt Psychology*, pp. 322–323, 324–325 f., 327–328.

38 Köhler, "Zur Psychophysik des Vergleichs und des Raumes," *PsyFo*, 18 (1933): 358, 359, 360.

39 Stephen Brush, *The Kind of Motion We Call Heat* (Amsterdam, 1976), esp. vol. 1, pp. 80 ff.

40 Planck, *Eight Lectures on Theoretical Physics* (cited in Chapter 11), p. 50.

41 Köhler, "Zur Boltzmannschen Theorie des zweiten Hauptsatzes," *Erkenntnis*, 2 (1931): 336–353; Reichenbach to Köhler, 11 August 1931, Wolfgang Köhler papers, APS.

42 Köhler, "Zur Boltzmannschen Theorie," esp. pp. 343, 349, 353; idem., "On the Problem of Regulation," pp. 309–310.

43 Köhler, *The Place of Value in a World of Facts* (cited in Chap. 11), pp. 141, 146; cf. idem., "Über die gegenwärtige Lage der Psychophysik," *Forschungen und Fortschritte*, 10:13 (1 May 1934): 168, where he makes the same point without referring to Heisenberg.

Chapter 16

1 Alexandre Métraux, "Kurt Lewin: Philosopher-Psychologist," *Science in Context*, 5 (1992): 373–384.

2 For the following, see Alfred J. Marrow, *The Practical Theorist: A Biography of Kurt Lewin* (New York, 1969), esp. pp. 7 ff.

3 *Mitteilungen aus dem freistudentischem Bunde. Beilage zur akademischen Rundschau*, vol. 12 (1910), Heft 13, p. 146; vol. 13 (1911), p. 276; Hans-Ulrich Wipf, "'Es war das Gefühl, dass die Universitätsbildung in irgend einem Punkte versagte' . . . Hans Reichenbach als Freistudent 1910 bis 1916," in Lutz Danneberg et al. (eds.), *Hans Reichenbach und die Berliner Gruppe* (Braunschweig, 1994), p. 162. I am grateful to Hans-Ulrich Wipf for providing me with copies of the documents just cited.

4 Kurt Lewin, *Die Sozialisierung des Taylorsystems. Eine grundsätzliche Untersuchung zur Arbeits- und Betriebspsychologie* (Praktischer Sozialismus, Heft 4, Berlin, 1920); repr. *Gestalt Theory*, 3 (1981), pp. 129–151; citations to the reprint. On Lewin's relationship with Korsch, see Marrow, *The Practical Theorist*, p. 12.

5 Lewin, *Die Sozialisierung des Taylorsystems*, here pp. 129, 138–140, 143, 148.

6 See, e.g., Lewin and Hans Rupp, "Untersuchungen zur Textilindustrie," *Psychotechnische Zeitschrift*, 3 (1928–1929), 51–63; Lewin, "Die Bedeutung der psychischen 'Sättigung' für einige Probleme der Psychotechnik," ibid., pp. 182–188; "Filmaufnahmen über Trieb- und Affektäusserungen psychopathischer Kinder" (1926), in *Kurt-Lewin-Werkausgabe*, vol. 6, ed. Franz E. Weinert and Horst Gundlach (Bern and Stuttgart, 1982). See also Lewin, *Die Entwicklung der experimentellen Willenspsychologie und die Psychotherapie* (Berlin, 1928).

7 Lewin and Rupp, "Untersuchungen zur Textilindustrie," pp. 51–52.

8 Lewin, "Der Begriff der Genese in Physik, Biologie und Entwicklungsgeschichte" (1922), in *Kurt-Lewin-Werkausgabe*, vol. 2, ed. Alexandre Métraux, (Stuttgart and Bern, 1982), esp. pp. 53–57, 277 ff., 303; idem., "Wissenschaftslehre" (1925–1928), in *Kurt-Lewin-Werkausgabe*, vol. 2; for Lewin's acknowledgment of Cassirer's influence, see idem., "Cassirer's Philosophy of Science and the Social Sciences," in Paul A. Schilpp (ed.), *The Philosophy of Ernst Cassirer* (Evanston, Ill., 1949), esp. pp. 271 f., 276.

9 Lewin, "Begriff der Genese," pp. 59, 60 ff., 112 ff., 241 ff.

10 Lewin, "Gibt es einzelne Wissenschaften?" (1925), in *Kurt-Lewin-Werkausgabe*, vol. 6, eds. Franz Weinert and Horst Gundlach (Stuttgart and Bern, 1982), esp. pp. 469, 475; cf. Lewin, "Wissenschaftslehre" (1925–1928), esp. pp. 335 ff.

11 Lewin to Reichenbach, 19 July 1923, 10 October 1923, 19 November 1923, 11 January 1924, and 21 January 1924, all in Hans Reichenbach papers, Archives of Scientific Philosophy, University of Pittsburgh. Thanks to Andreas Kamlah for directing me to these documents.

12 Rudolf Carnap, "Autobiography," in Paul A. Schilpp (ed.), *The Philosophy of Rudolf Carnap* (La Salle, Ill., 1969), pp. 14–15; Hans Reichenbach, review of "Der Begriff der Genese," *PsyFo* 5 (1924): 188–190; idem., *The Philosophy of Space and Time* (1927), trans. Maria R. Freund and John Freund (New York, 1957), pp. 268–271; Lewin, *A Dynamic Theory of Personality* (New York, 1935), p. 1.

13 Lewin, "Gesetz und Experiment in der Psychologie" (1927), in *Kurt-Lewin-Werkausgabe*, vol. 1, ed. Alexandre Métraux (Bern and Stuttgart, 1982), pp. 282–283, 286 ff., 293, 307 ff.; trans. as "Law and Experiment in Psychology," *Science in Context*, 5 (1992): 388, 390 ff., 396, 407 ff.

14 Lewin, "Der Übergang von aristotelischen zu galileischen Begriffen in der Psychologie" (1931), in *Werkausgabe*, vol. 1, esp. pp. 251–252; idem., "The Conflict between Aristotelian and Galilean Modes of Thought in Contemporary Psychology," repr. in Lewin, *A Dynamic Theory of Personality*, (New York, 1935), here p. 20. Cf. Lewin, "Gesetz und Experiment," p. 316; "Law and Experiment," p. 415.

15 Lewin, "Übergang," p. 262; *Dynamic Theory*, p. 31.

16 Cassirer, *Substance and Function*, pp. 254, 256; cf. Kurt Danziger, "The Project of an Experimental Social Psychology: Historical Perspectives," *Science in Context*, 5 (1992): 319.

17 Lewin, "Die psychische Tätigkeit bei der Hemmung von Willensvorgängen und das Grundgesetz der Assoziation," *ZfP*, 77 (1917): 212–247; see also Joseph Schwermer, *Die experimentelle Willenspsychologie Kurt Lewins* (Meisenhein am Glan, 1966), ch. 2.

18 Lewin, "Das Problem der Willensmessung und das Grundgesetz der Assoziation, I.," *PsyFo*, 1 (1921): 301; idem., "Das Problem der Willensmessung und das Grundgesetz der Assoziation, II.," *PsyFo*, 2 (1922): esp. pp. 67 ff., 138–139; cf. Koffka, *Principles*, pp. 578 ff.; Köhler, *Gestalt Psychology*, pp. 373 f.

19 Lewin, "Über die Umkehrung der Raumlage auf dem Kopf stehender Worte und figuren in der Wahrnehmung," *PsyFo*, 4 (1923): 210–261; Lewin and Kanae Sakuma, "Die Sehrichtung monokularer und binokularer Objekte bei Bewegung und das Zustandekommen des Tiefeneffektes," *PsyFo*, 6 (1925): 298–357.

20 Lewin, "Preliminary Remarks on the Structure of the Mind," reprinted in *Dynamic Theory*, esp. pp. 46 ff., 53, 56, 61.

21 Ibid., p. 53.

22 Lewin, "Kriegslandschaft" (1917), in *Kurt-Lewin-Werkausgabe*, vol. 4, ed. Carl-Friedrich Graumann (Bern and Stuttgart, 1982), p. 320; cf. Schwermer, *Die experimentelle Willenspsychologie Kurt Lewins*, pp. 21 f.

23 Lewin, "Preliminary Remarks," p. 46; idem., "Intention, Will and Need," trans. in David Rapaport (ed.), *Organization and Pathology of Thought: Selected Sources* (New York, 1951), pp. 110 ff.; cf. Joseph De Rivera (comp.), *Field Theory as Human Science: Contributions of Lewin's Berlin Group* (Hillsdale, N.J., 1976), pp. 45–46.

24 Karsten interview (cited in Chapter 13). For more extensive accounts, see Alfred Marrow, *The Practical Theorist*, esp. pp. 26–27.

25 Zeigarnik and Karsten interviews (cited in Chapter 13). Marrow, *The Practical Theorist*, p. 27, quotes the same story from Donald Mackinnon, who worked with Lewin in the early 1930s.

26 Bluma Zeigarnik, "Über das Behalten von erledigten und unerledigten Handlungen," *PsyFo*, 9 (1927): 1–85; Maria Rickers-Ovsiankina, "Die wiederaufnahme unterbrochener Handlungen," *PsyFo* 11 (1928): 302–379.

27 Zeigarnik, "Über das Behalten," esp. p. 77; cf. Lewin, *Dynamic Theory*, pp. 243–244, and De Rivera, *Field Theory*, pp. 133 ff.

28 Zeigarnik, "Über das Behalten," pp. 21 ff., 29, 30, 56–58, 63 ff.

29 Lewin, "Die Erziehung der Versuchsperson zur richtigen Selbstbeobachtung und die Kontrolle psychologischer Beschreibungsaufgaben" (1918), in *Kurt-Lewin-Werkausgabe*, vol. 1, e.g., p. 189. For an excellent account of the research style of Lewin's Berlin group, see Kurt Danziger, *Constructing the Subject* (cited in the Introduction), pp. 173 ff.

30 Zeigarnik interview; Zeigarnik, "Über das Behalten," pp. 17–19. For an overview of "Zeigarnik effect" studies, see Annie van Bergen, *Task Integration* (Amsterdam, 1968).

31 Tamara Dembo, "Der Ärger als dynamisches Problem," *PsyFo*, 15 (1931): 1–144; idem., "The Dynamics of Anger," abr. trans. Hedda Korsch, in De Rivera, *Field Theory*, pp. 324–422. Citations to the translation.

32 Dembo, "Dynamics of Anger," pp. 327–333, 340.

33 Dembo, "Dynamics of Anger," pp. 334 ff., 379 f., esp. 384.

34 Lewin, "Übergang," p. 262; *Dynamic Theory*, p. 31, 258 ff. For early statements by Lewin on topology in psychology, see Schwermer, *Die experimentelle Willenspsychologie Kurt Lewins*, p. 52.

35 Lewin, "Education for Reality" (1931), reprinted in *Dynamic Theory*, pp. 171–179.

36 Ibid., pp. 175, 177, 178–179.

37 Lewin, *Die psychologische Situation bei Lohn und Strafe* (Leipzig, 1931); trans. as "The Psychological Situation Attending Reward and Punishment" (1931), repr. in *Dynamic Theory*, here pp. 114, 126 n. 3, 138, 153, 169.

38 For further discussion, see my article, "Cultural Contexts and Scientific Change in Psychology: Kurt Lewin in Iowa," *American Psychologist*, 47 (1992): 198–207.

39 Lewin, *Dynamic Theory*, p. 240.

40 Norman R. F. Maier, quoted in Luchins and Luchins, "Max Wertheimer's Life and Work, 1919–1929" (cited in Chapter 13).

41 Köhler to University Administration, 14 August 1929, UAHUB, Universitätskurator, Nr. 8391, Bl. 214.

42 Kurt Goldstein, *Über Rassenhygiene* (Berlin, 1913), pp. v, 54 f.; idem., *Die Behandlung, Fürsorge und Begutachtung der Hirnverletzten* (Leipzig, 1919), pp. 1–2, 25, 211 ff.

43 On Wernicke's paradigm and its reception, see Marc Jeannerod, *The Brain Machine: The Development of Neurophysiological Thought*, trans. D. Urion (Cambridge, Mass., 1985), esp. pp. 72 ff.; and Anne Harrington, *Medicine, Mind and the Double Brain* (Cambridge, 1986), esp. pp. 71 ff., 155 ff., 164–165, 261 ff. For Goldstein's account of the background, see Goldstein, "Die Lokalisierung in der Grosshirnrinde," in Albrecht Bethe et al. (eds.), *Handbuch der normalen und pathologischen Physiologie*, vol. 10 (Berlin, 1927), esp. pp. 620 ff.; idem., "Notes on the Development of My Concepts," in *Kurt Goldstein: Selected Papers*, eds. Aron Gurwitsch et al. (The Hague, 1971), pp. 1–12.

44 Wertheimer, "Bewegung" (cited in Chapter 8), p. 247; see also the sources cited in Chapter 7, n. 15.

45 Goldstein, *Die Behandlung*, pp. 23–25.

46 Ibid., p. 69.

47 Adhemar Gelb and Kurt Goldstein, "Zur Psychologie des optischen Wahrnehmungs- und Erkennungsvorganges" (1918), reprinted in Gelb and Goldstein, *Psychologische Analysen hirnpathologischer Fälle* (Leipzig, 1920), esp. pp. 24, 96 ff., 101.

48 Wilhelm Fuchs, "Untersuchungen über das Sehen der Hemianopiker und Hemiamblyopiker, I. Verlagerungserscheinungen," *ZfP*, 84 (1920): esp. pp. 79, 81 ff.; idem., "Untersuchungen über das Sehen der Hemianopiker und Hemiamblyopiker, II. Die totalisierende Gestaltauffassung," *ZfP*, 86 (1921): esp. pp. 7–8, 37 f., 46.

49 Fuchs, "Untersuchungen I," pp. 65 ff., 81; Walter Poppelreuter, *Die Schädigungen durch Kopfschuss im Kriege,* vol. 1 (Bonn, 1916), pp. 130 f.; but cf. p. 152.

50 Fuchs, "Untersuchungen I," pp. 131–132, 134–135, 137 f.; idem., "Eine Pseudofovea bei Hemianopikern," *PsyFo,* 1 (1921): esp. pp. 158 ff., 172.

51 Gelb and Goldstein, "Über Farbenamnesie nebst Bemerkungen über das Wesen der amnestiaschen Aphasie überhaupt und die Beziehung zwischen Sprachen und dem Verhalten zur Umwelt," *PsyFo,* 6 (1924): 127–186.

52 Goldstein, "Die Lokalisierung in der Grosshirnrinde," p. 629.

53 Ibid., pp. 646 ff.; *Die Behandlung,* p. 1.

54 Goldstein, "Die Lokalisierung," pp. 645–646, 647, 650 f. Emphasis in the original. For a clear account of Goldstein's philosophy of the organism, see Marjorie Grene, *Toward a Philosophical Biology* (New York, 1969).

55 Goldstein, *Der Aufbau des Organismus* (Amsterdam, 1934), p. 314; cf. *The Organism* (trans. New York, 1938), p. 377; see also Grene, *Philosophical Biology,* p. 280.

56 Goldstein, "Die Lokalisation," pp. 640 f., 652. On the relationship of Cassirer's and Goldstein's views, see John Michael Krois, "Problematik, Eigenart und Aktualität der Cassirschen Philosophie der symbolischen Formen," in Hans-Jürg Braun et al. (eds.), *Ernst Cassirers Philosophie der symbolischen Formen* (Frankfurt a.M., 1988), pp. 24 ff.

57 Goldstein, *Der Aufbau des Organismus,* pp. 206–207; *The Organism,* pp. 369, 374; see also Grene, *Philosophical Biology,* p. 250.

58 Karl Lashley, *Brain Mechanisms and Intelligence* (Chicago, 1929), esp. pp. 25, 122 f., 154 f., 175; Henry Head, *Aphasia and Kindred Disorders of Speech* (New York, 1926); Albrecht Bethe, "Plastizität und Zentrenlehre," in Bethe et al. (eds.), *Handbuch der normalen und pathologischen Physiologie,* vol. 15 (1930).

59 Anne Harrington, "Interwar 'German' Psychobiology: Between Nationalism and the Irrational," *Science in Context,* 4 (1991): 429–447; idem., "'Other Ways of Knowing': The Politics of Knowledge in Interwar German Brain Science," in Anne Harrington (ed.), *So Human a Brain: Knowledge and Values in the Human Sciences* (Boston, 1992).

60 Koffka to Molly Harrower, 17 January 1938, in Molly Harrower, *Kurt Koffka* (cited in Chapter 7), pp. 27–28; B. F. Skinner, *The Shaping of a Behaviorist* (New York, 1979), p. 246.

Chapter 17

1 Ringer, *Mandarins* (cited in the Introduction); Peter Gay, *Weimar Culture: The Outsider as Insider* (New York, 1969), p. 70; Walter Laqueur, *Weimar: A Cultural History 1918–1933* (New York, 1974), esp. Chs. 3 and 6; Detlev J. K. Peukert, *Die Weimarer Republik* (cited in the Introduction); Konrad H. Jarausch, "Die Krise des deutschen Bildungsbürgertums im ersten Drittel des 20. Jahrhunderts," in Jürgen Kocka (ed.), *Bildungsbürgertum im 19. Jahrhundert. Teil IV. Politischer Einfluss und gesellschaftliche Formation* (Stuttgart, 1989), pp. 180–205; Paul Forman, "The Financial Support and Political Alignment of Physicists in Weimar Germany," *Minerva,* 12 (1974): 39–66.

2 Weyl quoted in Herbert Mehrtens, "Anschauungswelt und Papierwelt: Zur historischen Interpretation der Grundlagenkrise der Mathematik," in Heinrich Poser and Hans-Werner Schütt (eds.), *Ontologie und Wissenschaft* (Berlin, 1984), here p. 231; cf. Heinrich Löwy, "Die Krisis in der Mathematik und ihre philosophische Bedeutung," *Die Natur-*

wissenschaften, 14 (1926): 706–708. For a richly detailed and illuminating account of the controversy, see Mehrtens, *Moderne Sprache Mathematik* (Frankfurt a.M., 1990).

3 Paul Forman, "Weimar Culture, Causality and Quantum Theory, 1918–1927: Adaptation by German Physicists and Mathematicians to a Hostile Intellectual Environment," *Historical Studies in the Physical Sciences,* 3 (1971): esp. pp. 104–105; John Heilbron, *The Dilemmas of an Upright Man: Max Planck as Spokeman for German Science* (Berkeley, Calif. 1986), pp. 124, 140; Ludwig von Bertalanffy, *Theoretische Biologie,* vol. 1 (Berlin, 1932), p. 109.

4 Riezler, "Die Krise der 'Wirklichkeit'," *Die Naturwissenschaften,* 16 (1928): esp. pp. 708, 711–712.

5 Theodor von Uexküll, *Staatsbiologie* (Berlin, 1920): p. 42; cf. his *Theoretische Biologie* (1920; repr. Frankfurt a.M., 1973), pp. 330 ff. Othmar Spann, *Der Wahre Staat. Vorlesungen über Abbruch und Neubau der Gesellschaft* (1921), in Spann, *Gesamtausgabe,* vol. 5 (Graz, 1972), esp. pp. 211, 230, 243, 327. On these and other "organic" political theories of the Weimar period, see Eckart Scheerer, "Organische Weltanschauung und Ganzheitspsychologie," in Carl-Friedrich Graumann (ed.), *Psychologie im Nationalsozialismus* (Heidelberg, 1985), esp. pp. 30–33; Anne Harrington, "Interwar 'German' Psychobiology: Between Nationalism and the Irrational," *Science in Context,* 4 (1991): 429–447; idem., *Reenchanted Science* (New York, forthcoming).

6 Oswald Spengler, *Der Untergang des Abendlandes* (Munich, 1923, repr. 1977), vol. 1, pp. 130–131, 135 ff.; vol. 2, pp. 660 ff. Cf. Jeffrey Herf, *Reactionary Modernism: Technology, Culture and Politics in Weimar and the Third Reich* (Princeton, N.J., 1984), pp. 52 ff., 63.

7 George Mosse, *Germans and Jews: The Right, the Left and the Search for a 'Third Force' in Weimar Germany* (New York, 1970); Karl-Dietrich Bracher, *The Age of Ideology: A History of Political Thought in the Twentieth Century,* trans. Ewald Osers (New York, 1984).

8 Gilian Rose, *The Melancholy Science* (New York and London, 1974), p. 7. For a thorough discussion of holistic thought in these and related thinkers, see Martin Jay, *Marxism and Totality* (Berkeley, Calif., 1984).

9 Karl Jaspers, *Allgemeine Psychopathologie* (1913), 6th ed. (Berlin, 1953), Part II, e.g., p. 350; cf. idem., *Philosophische Autobiographie* (Munich, 1977), pp. 23 ff..

10 Jaspers, *Psychologie der Weltanschauungen* (1919), 3rd ed. (Berlin, 1925), pp. v, 1, 30f., and esp. p. 159.

11 Spranger, "Die Frage nach der Einheit der Psychologie," *Sitzungsberichte der Berliner Akademie der Wissenschaften,* 24 (1926): esp. p. 199; idem., *Lebensformen: Geisteswissenschaftliche Psychologie und Ethik der Persönlichkeit* (1913), 3rd ed. (Halle, 1922), Vorwort. On the relation of Spranger's views to the "crisis" mood of the Weimar period, see Ringer, *Mandarins,* esp. pp. 352–366, 380–382, 401–418.

12 Spranger, *Lebensformen,* passim; idem., "Die drei Motive der Schulreform," *Monatsschrift für höhere Schulen,* 20 (1921): 267.

13 Ludwig Klages, *Die Grundlagen der Charakterkunde* (1910), 4th ed. (Leipzig, 1926), p. 1; idem., *Vom Wesen des Bewusstseins* (Leipzig, 1921), esp. pp. 26–27, 28 ff.; idem., *Der Geist als Widersacher der Seele,* 3 vols. (Leipzig, 1929). Robert Saudek propounded a different version of graphology in *Wissenschaftliche Graphologie* (Munich, 1926). Kracauer's article is reprinted in *Das Ornament der Masse* (Frankfurt a.M., 1963).

14 Einstein to Wertheimer, 12 September and 18 September 1922; Wertheimer to Einstein, 17 September 1922, MWP-NYPL. Emphasis in the original. Cf. the edition of the Einstein–Wertheimer correspondence by Abraham S. Luchins and Edith H. Luchins, *Methodology and Science,* 12 (1979): here pp. 178–180. For the context, see Paul Forman, "Scientific Internationalism: The Ideology and its Manipulation in Weimar Germany" (cited in Chapter 10); Heilbron, *Dilemmas,* pp. 100 ff.

15 Köhler to Geitel, 9 March 1920, in Jaeger (ed.), *Briefe Wolfgang Köhlers* (cited in Chapter 15). Emphasis in the original.

16 Köhler to E. B. Titchener, 12 May 1925, E. B. Titchener papers, Olin Library, Cornell University; Hans Wallach and Rudolf Arnheim interviews (cited in Chapter 13); Walter Eisen, "Kurt Koffka" (cited in Chapter 13), p. 76.

17 Zeigarnik interview (cited in Chapter 13).

18 Wolfgang Metzger, "Psychologie," *Sozialistische Monatshefte,* from vol. 60 (1923), 635 ff. to vol. 63 (1926): 869–872; Rudolf Arnheim, "Psychologie," ibid., from vol. 64 (1927): 230–233 to vol. 68 (1929): 450–452.

19 Arnheim, "Geschlechtercharakterologie," *Sozialistische Monatshefte,* 65 (1927): pp. 1018–1019.

20 Wertheimer, "Unterschungen zur Lehre von der Gestalt I" (cited in Chapter 13), p. 57.

21 *Annalen der Philosophie,* 1 (1919): i, et seq.; Hartmut Hecht and Dieter Hoffmann, "Die Berliner 'Gesellschaft für wissenschaftliche Philosophie': Naturwissenschaften und Philosophie zu Beginn des 20. Jahrhunderts in Berlin," NTM, 28 (1991): esp. pp. 44, 49 f., 47 ff. On the Korsch incident, see Reichenbach to Lewin, 23 February 1933, cited in Hoffmann, "Die Berliner 'Gesellschaft für wissenschaftliche Philosophie'," in Friedrich Stadler (ed.), *Wien-Berlin-Prag. Der Aufstieg der wissenschaftlichen Philosophie* (Vienna, 1993), p. 396.

22 Max Wertheimer, *Über Gestalttheorie* (Erlangen, 1925), pp. 3, 9, trans. N. Nairn-Allison as "Gestalt Theory," *Social Research,* 11 (1944): 81, 86. Further quotations are from the translation, which has been corrected where necessary.

23 Wertheimer, "Gestalt Theory," pp. 81–82, 83–84.

24 Ibid., pp. 82, 87–88, 83.

25 Ibid., pp. 85, 88, 91.

26 Ibid., pp. 95–96, 96–97.

27 Ibid., pp. 98–99.

28 Koffka, "Psychologie" (cited in Chapter 10, p. 600.

29 Ringer, *Mandarins,* p. 324.

30 Köhler, "Wesen und Tatsachen," *Forschungen und Fortschritte,* 8 (1932), 152; idem., "Wesen Und Tatsachen: Nachschrift eines am 23. Januar 1932 in der Kant-Gesellschaft (Berlin) gehaltenen Vortrags (Nachschrift von [Otto von] Lauenstein)," *Gestalt Theory,* 9 (1987): 247–250; idem., *Gestalt Psychology,* 2nd ed., pp. 324, 325 f.

31 On the concept of "requiredness," see Köhler, *Place of Value* (cited in Chapter 11).

32 Köhler, "The Nature of Intelligence" (1930), trans. Erich Goldmeier, in *Selected Papers,* esp. pp. 187–188.

33 Ibid., p. 186.

34 Anton Kaes, "The Debate about Cinema: Charting a Controversy (1909–1929)," *New German Critique,* No. 40 (Winter 1987): 7–33.

35 Arnheim interview (cited in Chapter 13).

36 Arnheim, "Experimentalpsychologische Studien zum Ausdrucksproblem," *PsyFo*, 11 (1928): esp. pp. 3–4, 6–7.

37 Arnheim, "Ausdrucksproblem," pp. 51 ff., 90–91, 96, 98–101.

38 Arnheim, "Nachwort," in *Zwischenrufe: Kleine Aufsätze aus den Jahren 1926–1940,* ed. Ursula Madrasch-Groschopp (Leipzig and Weimar, 1985), p. 235.

39 Arnheim, *Film as Art* (1932; trans. L. M. Sieveking and Ian F. D. Morrow, London, 1933), abr. trans. revised by the author in *Film as Art* (Berkeley, calif., 1957). Quotations are from the latter edition.

40 Arnheim, *Film as Art,* pp. 11 ff., 30–31, 59–60; for a more complete account, see J. Dudley Andrew, *The Major Film Theories: An Introduction* (Oxford, 1976), ch. 4.

41 Ibid., pp. 29, 132–133; for early evidence of Pudowkin's influence, see Arnheim, "Filmbücher" (1928), abstracted in *Kritiken und Aufsätze zum Film,* ed. Helmut H. Diederichs (Frankfurt a.M., 1979), pp. 313–314.

42 Arnheim, *Film as Art,* pp. 155–156.

43 Ibid., p. 154; idem., "Garbo und Gassenhauer" (1931) in *Kritiken und Aufsätze,* p. 230; "Beitrag zur Krise der Montage" (1930), ibid., p. 369; "A New Laocoon" (1938), in *Film as Art.*

44 Ibid., pp. 74 ff., 98 ff., 136.

45 Arnheim, "Die Seele in der Silberschicht" (1925), quoted in *Kritiken und Aufsätze,* p. 306; "Alte Chaplinfilme" (1929), ibid., p. 214; "Zwei Filme" (1931), quoted in ibid., p. 320. On "new objectivity" in general, see John Willett, *Art and Politics in the Weimar Period: The New Sobriety 1917–1933* (New York, 1978), esp. chs. 9–11, 15.

46 Arnheim, "Pudowkins 'Mutter'" (1927), in *Kritiken und Aufsätze,* pp. 188–189.

47 Arnheim, "Kuhle Wampe" (1932), in *Kritiken und Aufsätze,* p. 158.

48 Arnheim, "Der Film und seine Stiefmutter" (1929), in *Kritiken und Aufsätze,* p. 156; "Zensur ohne Hemmung" (1932), ibid., p. 159; "Die sogenannte Freiheit," ibid., p. 163.

49 Arnheim, "A Personal Note" (1957), in *Film as Art,* pp. 5–6.

50 Robert Musil, "Das Hilflose Europa, oder: Reise vom Hundertsten ins Tausendste" (1922), in *Gesammelte Werke,* ed. A. Frisé (Reinbek, n.d.), p. 1085.

51 Horkheimer to Naumann (Dean of Natural Sciences Faculty), 24 July 1928, Personalhauptakte Max Wertheimer, PAUF.

52 Ernst Bloch, *Erbschaft dieser Zeit* (1935; repr. Frankfurt a.M., 1962), pp. 303 ff.

53 Käthe Busch, "Die Gestalttheorie vom Standpunkt des dialektischen Materialismus," in *Beiträge der Fachgruppe für Dialektisch-Materialistische Psychologie: Krise der Psychologie – Psychologie der Krise* (Berlin, 1932), p. 102.

Chapter 18

1 Richard Whitley, *The Intellectual and Social Organization of the Sciences* (Oxford, 1984), esp. pp. 88, 91. The reception of Gestalt theory in general philosophy and psychoanalysis cannot be discussed here. For the latter see Bruno Waldvogel, *Psychoanalyse und Gestaltpsychologie: Historische und theoretische Berührungspunkte* (Stuttgart, 1992).

2 Karl Bühler, *Die Krise der Psychologie* (Jena, 1927).

3 G. E. Müller, *Komplextheorie und Gestalttheorie. Ein Beitrag zur Wahrnehmungspsychologie* (Göttingen, 1923), pp. 1, 104. In support of this claim Müller cited his *Gesichtspunkte und Tatsachen der Psychophysik* (cited in Chapter 6), and "Zur Analyse des Vorstellungsverlaufes I," *ZfP,* Beiheft (1911): 302.

4 Ibid., esp. pp. 24 ff., 32 ff., 43.

5 Ibid., pp. 18, 49 ff., 51, 53–54; cf. Gerald M. Edelman, *Neural Darwinism: The Theory of Neuronal Group Selection* (New York, 1987).

6 Ibid., pp. 95, 101, 105.

7 Köhler, "Komplextheorie und Gestalttheorie: Antwort auf G. E. Müllers Schrift gleichen Namens," *PsyFo,* 6 (1925): 365–366.

8 Ibid., pp. 368–369, 371. Köhler cited Karl Bühler, *Die Gestaltwahrnehmungen* (cited in Chapter 6), p. 27.

9 Ibid., esp. p. 390; cf. Köhler, "An Aspect of Gestalt Psychology," in Carl Murchison (ed.), *Psychologies of 1925* (Worcester, Mass., 1925), pp. 163–198.

10 G. E. Müller, "Bemerkungen zu Wolfgang Köhlers Artikel 'Komplextheorie und Gestalttheorie,'" *ZfP,* 99 (1926): 1, 15; Köhler, "Zur Komplextheorie," *PsyFo,* 8 (1926): p. 243.

11 Karl Bühler, "Die 'neue' Psychologie Koffkas," *ZfP,* 99 (1926), esp. pp. 146, 156, 159. Bühler was referring to Koffka, "Psychologie" (cited in Chapter 10).

12 Otto Selz, "Zur Psychologie der Gegenwart," *ZfP,* 99 (1926): esp. pp. 168, 169 f., 172, 181, 182, 186. Cf. Karl Gerhards, "Zur naturwissenschaftlichen Erforschung des Denkens," *Die Naturwissenschaften,* 13 (1925): 471–477, 506–510.

13 Koffka, "Bemerkungen zur Denk-Psychologie," *Psy Fo,* 9 (1927): 177–178, 181, 183.

14 Felix Krueger, *Über Entwicklungspsychologie* (Jena, 1915). On the roots of Krueger's "holistic" psychology, see Theo Herrmann, "Ganzheitspsychologie und Gestalttheorie" (cited in Chapter 6). On Krueger's social philosophy, see Ulfried Geuter, "Das Ganze und die Gemeinschaft. Wissenschaftliches und politisches Denken in der Ganzheitspsychologie Felix Kruegers," in C.-F. Graumann (ed.), *Psychologie im Nationalsozialismus* (cited in Chapter 17), pp. 55–88.

15 Krueger, "Der Strukturbegriff in der Psychologie" (1923), reprinted in *Zur Philosophie und Psychologie der Ganzheit,* ed. E. Heuss (Berlin, 1953), pp. 128, 132–134; idem., "Über psychische Ganzheit" (1926), reprinted in ibid., p. 107.

16 Krueger, "Strukturbegriff," pp. 142, 144; cf. Geuter, "Das Ganze und die Gemeinschaft."

17 Krueger, "Psychische Ganzheit," p. 120; Hans Volkelt, *Über die Forschungsrichtung des Psychologischen Instituts der Universität Leipzig* (Erfurt, 1925), pp. 4–5. For the term "methodological irrationalism," see Eckardt Scheerer, "Organische Weltanschauung und Ganzheitspsychologie" (cited in Chapter 17), p. 23.

18 Krueger, "Strukturbegriff," pp. 139, 145; "Psychische Ganzheit," p. 107; cf. Geuter, "Das Ganze und die Gemeinschaft," p. 59.

19 Jaensch, *Eidetic Imagery* (1925), trans. from 2nd ed. (London, 1930), esp. pp. 9, 14, 127; Hans Henning, *Psychologie der Gegenwart* (1925), 2nd ed. (Leipzig, 1931), p. 38.

20 Erich Jaensch and F. Reich, "Der Aufbau der Wahrnehmungswelt, II. Über die Lokalisation im Sehraum," *ZfP,* 86 (1921): 280, n. 1. Jaensch, "Über den Aufbau der Wahrnehmungswelt und ihre Struktur im Jugendalter. Über Gegenwartsaufgaben der Jugendpsychologie (als Schlusswort)," *ZfP,* 94 (1924): 52, n. 1.

21 Jaensch, "Gegenwartsaufgaben der Jugendpsychologie," p. 51; Erich Jaensch and Lazlo Grünhut, *Über Gestalttheorie* (Osterwieck, 1929), esp. pp. 9, 31, 35, 43, 50 f., 71–72. Ulrich Sieg suggests that personal jealousy of Köhler may account for the vehemence of Jaensch's criticism: see "Psychologie als Wirklichkeitswissenschaft" (cited in Chapter 3), p. 325.

22 Jaensch, *Wirklichkeit und Wert in der Philosophie und Kultur der Neuzeit* (Berlin, 1929).

23 Stern, *Studien zur Personwissenschaft, Teil I. Personalistik als Wissenschaft* (Leipzig, 1930), p. vii; idem., "Personalistische Psychologie," in Emil Saupe (ed.), *Einführung in die neuere Psychologie* (Osterwieck, 1928), p. 169; Martin Scheerer, *Die Lehre von der Gestalt. Ihre Methode und ihr psychologischer Gegenstand* (Berlin, 1931), esp. p. 142.

24 Stern, *Studien,* pp. 14, 18; cf. Geuter, "Das Ganze und die Gemeinschaft," pp. 64–65.

25 Scheerer, *Die Lehre von der Gestalt,* esp. pp. 143 ff., 150, n. 1.

26 Ibid., pp. 156, 395.

27 Karl Bühler, *Die Krise der Psychologie* (cited in note 2), pp. 109, 112–113, 122. In 1921, Becher argued that, although physiological processes must be complex enough to account for psychological phenomena, this does not mean that both must have the same structure. Erich Becher, "W. Köhlers physikalische Theorie der physiologischen Vargänge, die der Gestaltwahrnehmung zugrunde liegen," *ZfP,* 87 (1921): 1–44.

28 On Bühler's theory of language, see "Über den Begriff der sprachlichen Darstellung," *PsyFo,* 3 (1923): 282–294, and *Sprachtheorie. Die Darstellungsfunktion der Sprache* (Jena, 1934). Cf. Achim Eschbach, "Wahrnehmungen und Zeichen: Die sematologischen Grundlagen der Wahrnehmungstheorie Karl Bühlers," *Ars Semiotica,* 4 (1981): 219–235; Robert E. Innis, *Karl Bühler: Semiotic Foundations of Language Theory* (New York, 1982).

29 Bühler, *Krise,* esp. pp. 115, 119, 120–121; idem., "Die 'neue' Psychologie Koffkas," pp. 152–153.

30 Köhler, *Gestalt Psychology,* p. 243.

31 Bruno Latour, *Science in Action,* and Peter Galison, *How Experiments End* (both cited in the Introduction).

32 Joseph Fröbes, *Lehrbuch der experimentellen Psychologie,* vol. 1, 2nd and 3rd ed. (Freiburg, 1923), pp. 405 ff., esp. 412–413; cf. Sarris, "Max Wertheimer in Frankfurt I" (cited in Chapter 8), p. 297. Johannes Wittman, *Über das Sehen von Scheinbewegungen und Scheinkörpern* (Leipzig, 1921); Bruno Petermann, *The Gestalt Theory and the Problem of Configuration* (1929), trans. Meyer Fortes (London, 1932).

33 Wilhelm Neuhaus, "Experimentelle Untersuchung der Schweinbewegung," *AgP,* 75 (1930), 315–458; cf. Viktor Sarris, "Max Wertheimer in Frankfurt . . . III: Weitere Studien über das Sehen von Bewegung (1929–1933)" (cited in Chapter 14).

34 K. Zietz and Heinz Werner, "Über die dynamische Struktur der Bewegung," *ZfP,* 105 (1928): esp. p. 226 f.

35 Friedrich Sander, "Experimentelle Ergebnisse der Gestaltpsychologie," *Bericht über den 10. Kongress für experimentelle Psychologie* (1928), pp. 57 ff.; cf. idem., "Funktionale Struktur, Erlebnisganzheit und Gestalt," *AfgP,* 85 (1932), esp. pp. 244 ff., and Erich Wohlfahrt, "Der Auffassungsvorgang an kleinsten Gestalten," *Neue Psychologische Studien,* 4 (1932): 347–414.

36 Metzger, "Zur Phänomenologie des homogenen Ganzfelds" (cited in Chapter 14), pp. 22 ff.; "Optische Untersuchungen am Ganzfeld, III. Mitteilung: Die Schwelle für plötzliche Helligkeitsänderungen," *PsyFo,* 13 (1930): 30–53.

37 Sander, "Experimentelle Ergebnisse der Gestaltpsychologie," pp. 39–40; cf. pp. 28, 45. See also Günther Ipsen, "Über Gestaltauffassung. Erörterung des Sanderschen Parallelograms," *Neue Psychologische Studien,* 1 (1926): 167–278.

38 For Kardos's basically favorable reception of Gestalt theory, see "Die 'Konstanz' phänomenaler Dingmomente: Problemgeschichtlicher Darstellung," in Egon Brunswik et al., *Beiträge zur Problemgeschichte der Psychologie* (Jena, 1929), pp. 1–77.

39 Egon Brunswik, "Prinzipienfragen der Gestalttheorie," in *Beiträge zur Problemgeschichte der Psychologie*, esp. pp. 92f., 95–96, 106.

40 Egon Brunswik, "Die Zugänglichkeit von Gegenständen für die Wahrnehmung," *AgP*, 88 (1933): esp. p. 382; idem., *Wahrnehmung und Gegenstandwelt* (Vienna, 1934). On the emergence of this view and the influence of logical empiricism, see David Leary, "From Act Psychology to Probabilistic Functionalism: The Place of Egon Brunswik in the History of Psychology," in Ash and Woodward (eds.), *Psychology in Twentieth-Century Thought and Society*, esp. pp. 119, 121 f.

41 Beverly E. Holaday, "Die Grössenkonstanz der Sehdinge bei Variation der inneren und äusseren Wahrnehmungsbedingungen," *AgP*, 88 (1933): 419–486; Kurt Eissler, "Die Gestaltkonstanz der Sehdinge bei Variation der Objekte und ihrer Einwirkungsweise auf den Wahrnehmenden," ibid., 487–550; Sylvia Klimpfinger, "Über den Einfluss von intentionaler Einstellung und Übung auf die Gestaltkonstanz," ibid., 551–598; Brunswik, "Zugänglichkeit von Gegenständen," pp. 409–410, 418.

42 Koffka, *Principles*, esp. p. 231.

43 Ibid., pp. 238–240; cf. W. Burzlaff, "Methodologische Beiträge zum Problem der Farbenkonstanz," *ZfP*, 119 (1931): 177–235; Helene Frank, "Die Sehgrössenkonstanz bei Kindern" (cited in Chapter 15).

44 Duncker, "A Qualitative Study of Productive Thinking" (cited in Chapter 14), p. 697.

45 See, e.g., Henning, *Psychologie der Gegenwart*, pp. 33–34; cf. August Messer, *Einführung in die Psychologie und die psychologische Richtungen der Gegenwart* (Leipzig, 1927), pp. 152–154, 164.

Chapter 19

1 Fritz Stern, "National Socialism as Temptation," in *Dreams and Delusions: National Socialism in the Drama of the German Past* (New York, 1989), p. 151.

2 Alan D. Beyerchen, *Science under Hitler: Politics and the Physics Community in the Third Reich* (New Haven, Conn., 1977); idem., "What We Now Know about Nazism and Science," *Social Research*, 59 (1992): 615–641. Herbert Mehrtens and Stefan Richter (eds.), *Naturwissenschaft, Technik und Nationalsozialistische Ideologie* (Frankfurt a.M., 1980); Ute Deichmann, *Biologen unter Hitler: Vertreibung, Karrieren, Forschung* (Frankfurt A.M., 1992); Kristie Macrakis, *Surviving the Swastika: Scientific Research in Nazi Germany* (New York, 1993), esp. chs. 4–5; Herbert Mehrtens, "Kollaborationsverhältnisse: Natur- und Technikwissenschaften im Nationalsozialismus und ihre Historie," in Christoph Meinel and Peter Voseinckel (eds.), *Medizin, Naturwissenschaft, Technik und Nationalsozialismus: Kontinuitäten und Diskontinuitäten* (Stuttgart, 1994); Monika Renneberg and Mark Walker (eds.), *Science, Technology, and National Socialism* (Cambridge, 1994).

3 See, for example, the essays in Peter Lundgreen (ed.), *Wissenschaft im Nationalsozialismus* (Frankfurt a.M., 1985); and Jerry Z. Muller, *The Other God that Failed: Hans Freyer and the Deradicalization of German Conservatism* (Princeton N.J., 1987).

4 On psychology, see Ulfried Geuter, *Professionalization* (cited in the Introduction); on eugenics, see Sheila F. Weiss, "The Race Hygiene Movement in Germany," *Osiris*, 2nd series, 3 (1987): 193–236; Peter Weingart, Jürgen Kroll, and Kurt Bayertz, *Rasse, Blut und Gene: Geschichte der Eugenik und Rassenhygiene in Deutschland* (Frankfurt a.M., 1988);

and Robert N. Procter, *Racial Hygiene: Medicine under the Nazis* (Cambridge, Mass., 1988). For related analyses, see Otthein Rammstedt, *Deutsche Soziologie 1933–1945: Die Normalität einer Anpassung* (Frankfurt a.M., 1986); Herbert Mehrtens, "Angewandte Mathematik und Anwendungen der Mathematik im nationalsozialistischen Deutschland," *Geschichte und Gesellschaft,* 12 (1986): 317–346. For the "deprofessionalization" argument, see Konrad Jarausch, *The Unfree Professions* (Cited in the Introduction).

5 See, for example, Horst Möller, *Exodus der Kultur: Deutsche Schriftsteller, Wissenschaftler und Künstler im Exil* (Munich, 1984).

6 Geoffrey Cocks, *Psychotherapy in the Third Reich: The Göring Institute* (New York, 1985); idem., "The Professionalization of Psychotherapy in Germany, 1928–1949," in Geoffrey Cocks and Konrad Jarausch (eds.), *German Professions, 1800–1950* (New York, 1990), pp. 308–328; Regine Lockot, *Erinnern und Durcharbeiten. Psychoanalyse und Psychotherapie im Nationalsozialismus* (Frankfurt a.M., 1985).

7 On the dismissals, see Geuter, *Professionalization,* pp. 52 ff., and "German Psychology in the Nazi Period," in Ash and Woodward (eds.), *Psychology in Twentieth-Century Thought and Society* (cited in Chapter 15), pp. 168 f.

8 Wertheimer to Dean of Natural Sciences, handwritten copy dated 7 April 1933; Dean of Natural Sciences to Wertheimer, 8 April 1933; University Chancellery to Wertheimer, 24 April 1933, all in MWPB; *Vossische Zeitung,* 26 April 1933, cited in Ulfried Geuter, "'Gleichschaltung' von Oben? Universitätspolitische Strategien und Verhaltensweisen in der Psychologie während des Nationalsozialismus," *Psychologische Rundschau,* 35 (1984): 213. For the oral tradition, see Michael Wertheimer, "Max Wertheimer: Gestalt Prophet" (cited in Chapter 7). On the firings in Frankfurt, see Gerda Stuchlik, *Goethe im Braunhemd. Die Universität Frankfurt im Nationalsozialismus* (Frankfurt a.M., 1984), pp. 90, 93.

9 Martin Buber to Wertheimer, 24 August 1933; Alvin Johnson to Wertheimer, 5 August 1933; University Chancellery to Wertheimer, 25 September and 12 December 1933, MWPB; Alvin Johnson, *Pioneer's Progress: An Autobiography* (New York, 1952), p. 343.

10 Köhler to University Administration, 11 September 1933, UAHUB, Assistenten 8391, Bl. 131, 127; Ulrich Jahnke, "Zur Entwicklung der Psychologie an der Berliner Universität nach 1933," in A. Thom and H. Spaar (eds.), *Medizin im Faschismus* (Berlin, GDR, 1983), p. 226.

11 Kurt Lewin, "'Everything within Me Rebels': A Letter from Kurt Lewin to Wolfgang Köhler," trans. G. Wickert and M. Lewin, *Journal of Social Issues,* 42 (1986), pp. 40–41, 46–47. The translators note that this letter was never sent. They claim that "mailing it would have been very dangerous to both the sender and the receiver," but offer no evidence for this.

12 Mary Henle, "One Man against the Nazis – Wolfgang Köhler," *American Psychologist,* 33 (1978): 939–944, reprinted in *1879 and All That: Essays in the Theory and History of Psychology* (New York, 1986), pp. 225–237. Citations are to the reprint. See also Curt Weinschenck, "Wolfgang Köhler im Jahre 1933," *Psychologische Beiträge,* 10 (1967): 622–624. But see the rather different view in Geuter, "'Gleichschaltung' von Oben?" (cited in n. 8).

13 Wolfgang Köhler, "Gespräche in Deutschland," *Deutsche Allgemeine Zeitung* (Reichsausgabe), 28 April 1933, p. 2, reprinted in Graumann (ed.), *Psychologie im Nationalsozialismus,* pp. 305–306, excerpts translated as "Germans Who Doubt the Nazi Doctrines," *New York Times,* 11 June 1933, Sect. 8, p. 2. "Köhler Foresees a Liberal Germany" (interview), *New York Times,* 7 July 1933, p. 23. Köhler to Foreign Office, 4 April 1933, copy in Wolfgang Köhler Papers, APS, Correspondence Achelis.

14 Köhler, "Gespräche in Deutschland."

15 Ibid., emphasis mine; "Köhler Foresees a Liberal Germany."

16 Max Planck, "Mein Besuch bei Adolf Hitler," *Physikalische Blätter,* 3 (1947): 143. For the date of this meeting, see Macrakis, *Surviving the Swastika,* p. 57.

17 Köhler to Wertheimer, 14 May 1933, MWPB; Henle, "One Man against the Nazis," p. 228; Fritz Stern, "National Socialism as Temptation," (cited in n. 1), p. 169.

18 Quoted in Molly Harrower, *Kurt Koffka: An Unwitting Self-Portrait* (Gainesville, Fla., 1983), p. 15.

19 Ibid.

20 Richard Hertz to Köhler, 24 April 1933; Max Dessoir to Kohler, 28 April 1933, Köhler papers, APS, folder "Gespräche in Deutschland." Cf. Henle, "One Man against the Nazis," pp. 228 f. For an analysis of the responses to Köhler's article, see Siegfried Jaeger, "Zur 'Widerständigkeit' der Hochschullehrer zu Beginn der NS-Herrschaft," *Psychologie und Geschichte,* 4 (1993): 219–228.

21 Clark Crannell, "Wolfgang Köhler," *JHBS,* 6 (1970): 267–268; Henle, "One Man against the Nazis," pp. 229–230.

22 Köhler to Rektor Fischer, 13 April 1934, copy in Köhler Papers, APS, Correspondence Achelis.

23 Beyerchen, *Scientists under Hitler* (cited in note 2), pp. 15 ff., 64 ff. Macrakis, *Surviving the Swastika,* p. 52, interprets the Haber ceremony as "an act of passive opposition" that was not originally intended as resistance. For a discussion of Laue's ambivalent stance that nicely parallels this one of Köhler's, see Mark Walker, *Nazi Science: Myth, Truth, and the German Atomic Bomb* (New York, 1995), pp. 70 ff. On the behavior of professors see Anselm Faust, "Professoren für die NSDAP. Zum politischen Verhalten der Hochschullehrer 1932–1933," in Manfred Heinemann (ed.), *Erziehung und Schulung im Dritten Reich,* vol. 2 (Stuttgart, 1980), pp. 31–49; and Michael Kater, "Die nationalsozialistische Machtergreifung an den deutschen Hochschulen. Zum politischen Verhalten akademischer Lehrer bis 1939," in Hans-Jochen Vogel (ed.), *Die Freiheit der Anderen. Festschrift für Martin Hirsch* (Baden-Baden, 1981), pp. 49–75.

24 Köhler to Rector Fischer, 13 April and 20 (?) 1934; cf. Henle, "One Man against the Nazis," p. 231.

25 Michael Kater, *Studentenschaft und Rechtsradikalismus in Deutschland 1918–1933. Eine sozialgeschichtliche Studie zur Bildungskrise in der Weimarer Republik* (Hamburg, 1975); H. W. Strätz, "Die geistige SA rückt ein. Die studentische 'Aktion wider dem undeutschen Geist' im Frühjahr 1933," *Vierteljahresschrift für Zeitgeschichte* (1968): 347–372; Geoffrey J. Giles, *Students and National Socialism in Germany* (Princeton, N.J., 1985).

26 "Hat sich das Psychologische Institut gleichgeschaltet?" *Wissen und Dienst,* 1:2 (1933), reprinted in Graumann (ed.), *Psychologie im Nationalsozialismus,* p. 308. On Preuss' authorship, see Jahnke, "Psychologie an der Berliner Universität" (cited in n. 9), p. 230, n. 27. On Duncker's participation in anti-Nazi activities before 1933, see Newman interview. Duncker's father, Hermann Duncker, was a leading Marxist educator; his mother, Käte Duncker, was a KPD member of the Thuringian State Assembly from 1922 to 1924.

27 Köhler to Rector Fischer, quoted in Jahnke, "Psychologie an der Berliner Universität," pp. 226–227. "Zu dem Artikel in der Zeitschrift 'Wissen und Dienst' Nr. 2. Antwort der Studierenden des Psychologischen Instituts," typescript, n.d., reprinted in Graumann (ed.), *Psychologie im Nationalsozialismus,* p. 309. Hans Haar lent me his copies of this document and of the Preuss article cited above.

28 Margarete Jucknat et al., to Leader of the German Students' League of the University of Berlin, 30 May 1934, copy in Köhler Papers, APS, correspondence Achelis. Ottilie Selbach (née Redslob), interview with the author, 20 August 1987.

29 Notice of torchlight parade set for 7 June 1934; "Protokol der Amtshandlung der Deutschen Studentenschaft im Psychologischen Institut," 5 June 1934; protocol of discussion with Amtsleiter Dr. Martin, 6 June 1934, copies in private hands; Selbach interview. Köhler's statement is quoted in Henle, "One Man against the Nazis," p. 230.

30 Protocol of 18 June 1934, p. 2, copy in Köhler Papers, correspondence Achelis.

31 Ottilie Redslob and Ilse Müller to Achelis, 18 June 1934; copy in private hands.

32 Hedwig von Restorff to Ministry Director Vahlen, 15 July 1934, copy with handwritten corrections in Köhler Papers, correspondence Achelis.

33 Köhler to Vahlen, 21 July 1934; Vahlen to Rector, 24 September 1934, copies in Köhler Papers, correspondence Achelis. See also Geuter, "'Gleichschaltung von oben?'" p. 204. On the fate of the student groups, see Kater, *Studentenschaft und Rechtsradikalismus,* pp. 175 ff.

34 Edgar Rubin to Department of Cultural Affairs, German Foreign Office, 9 May 1934, PAAA Bonn, Abt. VI W Hochschulwesen: Deutschland, Bnd. 26.

35 Vertrauliche Aufzeichnung, Berlin, 17 April 1934, PAAA Bonn, Bnd. 25, Journal Nr. 3644. The author may have been Richard Hertz, whose earlier correspondence with Köhler is cited above, n. 20.

36 Vahlen to Foreign Office, 24 September 1934, PAAA Bonn, Bnd. 26.

37 Planck to Köhler, 11 June 1934; Köhler to Planck, 12 June 34, Köhler papers, correspondence Achelis.

38 Köhler to Ministry, 3 February 1935, Köhler Papers, correspondence Achelis; cf. Henle, "One Man against the Nazis," p. 235. Köhler to Lashley, 6 February 1935, Rockefeller Family Archives Center, North Tarrytown, N.Y., RG 1.1, Series 216, Box 11, Folder 2161.

39 Geheime Staatspolizei to Ministry, 6 May 1935; Vahlen to Administrative Director of University, 14 June 1935. GStA, Rep. 76 37 Xa No. 150, Bnd. 4.

40 GStA, Rep. 90 Abt. Q Tit. VI Bnd. V Nr. 1, Bl. 399–402a; *Amtsblatt des Reichsministerium für Wissenschaft, Erziehung und Volksbildung,* 20 September 1935, p. 379. Köhler to Wertheimer, 8 April 1934, MWPB. Köhler to Frank Aydelotte, 13 July 1935; Aydelotte to Köhler, 15 July 1935, copies in Robert MacLeod Papers, in private hands. Köhler folder in Papers of the Emergency Committee in Aid of Displaced Foreign Scholars, NYPL. Cf. Jahnke, "Psychologie an der Berliner Universität," p. 228. For the continued payment of Köhler's pension see Lothar Sprung and Helga Sprung, "Zur Geschichte der Psychologie an der Berliner Universität II (1922–1935)," *Psychologie für die Praxis,* 3 (1987): 202 f.

41 Quoted in Henle, "One Man against the Nazis," p. 236.

42 Matthiat report to Ministry Director Vahlen, GStA (cited in n. 39).

43 Rieffert file, BDC, REM file R 100, here Bl. 2947, 2950. Jaeger, "Zur 'Widerständigkeit' der Hochschullehrer" (cited in note 20), p. 221. On Rieffert's promotion see also UAHUB, *Chronik der Friedrich-Wilhelms-Universität zu Berlin,* April 1932- March 1935, p. 19.

44 Bieberbach to Ministry, 5 July 1935, 6 September 1935; Rieffert file, BDC, Bl. 2926, 2929.

45 Geuter, "'Gleichschaltung' von oben?," pp. 205–206.

46 Rieffert file, BDC, Bl. 2964–2965, 3022–3023, 3027, 3041. On Rieffert's later career see Geuter, *Professionalisierung,* pp. 248 f.

47 Köhler to Wertheimer, 8 April 1934 (cited in n. 40). Kirchenrat Matthiat to Karl Duncker, 1935, GStA, Rep. 76 37 Xa No. 150, Bnd. 4, Bl. 146, 147. Otto von Lauenstein to Administrative Director, 20 April 1934, GStA, Rep. 76 Va Sekt. 2 Tit. X, 150 Bd. 3, p. 483, cit. in Geuter, "'Gleichschaltung' von oben?" p. 213, n. 10.

48 On Duncker's and Lauenstein's subsequent careers, see Jean Matter Mandler and George Mandler, "The Diaspora of Experimental Psychology: The Gestaltists and Others," in Bernard Bailyn and Donald Fleming (eds.), *The Intellectual Migration: Europe and America, 1930–1960* (Cambridge, Mass., 1969), pp. 392 f.; Wolfgang Metzger, "Gestalt-theorie im Exil," in H. Balmer (ed.), *Die Psychologie des 20. Jahrhunderts, Band 1: Die Europäische Tradition* (Zurich, 1976), pp. 659–683.

49 Geuter, "'Gleichschaltung' von Oben?" p. 213; *Professionalization,* p. 63. Rudolf Bergius, interview with the author, Bad Homburg, 11 November 1983. Jaeger, "Ent-wicklung und struktur des psychologischen Lehrangebotes" (cited in Chapter 13), pp. 63 f. For institute personnel in this period, see Appendix 1, Table 4.

50 Bergius interview; Barbara Burks, "Into the Third Reich," Lawrence K. Frank papers, National Library of Medicine; Geuter, *Professionalization,* p. 81. Examples of Berlin dissertations from the period are Gopal Mamdapurkar, "Versuch über die räumliche Wirkung ebener Bilder mit besonderer Berücksichtigung des Films," *ZfP,* 144 (1938): 273–354; Rudolf Bergius, "Die Ablenkung von der Arbeit durch Lärm und Musik und ihre strukturtypischen Zusammenhänge," *Zeitschrift für Arbeitspsychologie,* 12 (1940): Heft 4–6; K. J. Hene, "Über die Entwicklung der kindlichen Persönlichkeit (Erbpsychologische Untersuchungen an Kleinkindzwillingen" (Diss. Berlin, Math.-Nat. Fakultät, 1940); and Kripal Singh Sodhi, "Dynamik des Tiefensehens," *ZfP,* 151 (1941): 81–176. But see also H. J. Tilse, "Zur Psychologie des Antigermanismus in Frankreich" (Diss. Berlin, 1939).

Chapter 20

1 Geuter, *Professionalization* (cited in the Introduction), pp. 55 f.; "German Psychology during the Nazi Period," p. 169.

2 Geuter, "Das Ganze und die Gemeinschaft" (cited in Chapter 17), esp. pp. 68 ff., 77 ff.

3 Friedrich Sander, "Deutsche Psychologie und nationalsozialistische Weltanschauung," *Nationalsozialistische Bildungswesen,* 2 (1937): 642; Geuter, "Das Ganze und die Ge-meinschaft," p. 72; idem., "German Psychology during the Nazi Period," p. 173.

4 On Jaensch and Bieberbach's "German Mathematics," see Helmut Lindner, "'Deutsche und 'Gegentypische' Mathematik: Zur Begrundung einer 'arteigenen' Mathe-matik im 'Dritten Reich' durch Ludwig Bieberbach," in Mehrtens and Richter (eds.), *Naturwissenschaft, Technik und NS-Ideologie* (cited in Chapter 19), esp. pp. 98, 100. For Jaensch's reorientation under Nazism, see Ulfried Geuter, "Nationalsozialistische Ide-ologie und Psychologie," in Ash and Geuter (eds.), *Geschichte der deutschen Psychologie,* pp. 185 ff.; idem., "German Psychology during the Nazi Period," pp. 172–173.

5 Erich R. Jaensch, *Die Lage und Aufgaben der Psychologie. Ihre Sendung in der Deutschen Bewegung und an der Kulturwende* (Leipzig, 1933), pp. 7, 22; idem., *Der Gegentypus. Psychologisch-anthropologische Grundlagen deutscher Kulturphilosophie ausgehend von dem, was wir überwinden wollen* (Leipzig, 1938), p. 75. For a still more direct version of this line by a Jaensch student, see Hans Eilks, Gestalttheorie, "Gestaltpsychologie und Typologie. Teil I," *ZfP,* 136 (1935): 209–261. See also Ulrich Sieg, "Psychologie als Wirklichkeitswissenschaft" (cited in Chapter 3), pp. 329 f.

6 Bruno Petermann, *Grundfragen seelischen Seins* (Leipzig, 1938), esp. pp. 27 ff.; "Beiträge zur Rassenseelenlehre und völkischen Anthropologie: Einleitung," *AfgP,* 97 (1936): 251–256. On Petermann's appointment in Göttingen, see Rainer Paul, "Psychologie unter den Bedingungen der 'Kulturwende': Das Psychologische Institut 1933–1945," in H. Becker et al. (eds.), *Die Universität Göttingen unter dem Nationalsozialismus* (Munich, 1987), pp. 336–337.

7 Ludwig Ferdinand Clauss, *Rasse und Seele* (Munich, 1926), pp. vi, 124 ff., 142 ff.; idem., *Rasse ist Gestalt.* Schriften der Bewegung, Heft 3 (Munich, 1937), pp. 9–10, 13.

8 Ferdinand Weinhandl, *Die Gestaltanalyse* (1928); idem., *Philosophie als Werkzeug und als Waffe* (Neumünster i. Holstein, 1940), p. 5.

9 Oswald Kroh, "Die Psychologie im Dienste völkischer Erziehung," in *Charakter und Erziehung. Bericht über den 16. Kongress der Deutschen Gesellschaft für Psychologie* (1939), p. 43; "Erbpsychologie der Berufsneigung und Berufseignung," in Günther Just (ed.), *Handbuch der Erbbiologie des Menschen,* vol. 5:1 (Berlin, 1939), pp. 638 ff.

10 For detailed discussion of these developments, see Geuter, *Professionalization,* chs. 4, 6–7.

11 Philipp Lersch, *Gesicht und Seele. Grundlinien einer mimischen Diagnostik* (Munich, 1932); idem., *Der Aufbau des Charakters* (Leipzig, 1938), esp. pp. 36, 174. On Lersch's career and Nazi party membership in psychology, see Mitchell G. Ash and Ulfried Geuter, "NSDAP-Mitgliedschaft und Universitätskarriere in der Psychologie," in Graumann (ed.), *Psychologie im Nationalsozialismus,* p. 272.

12 Geuter, *Professionalization,* esp. ch. 3; idem., "German Psychology in the Nazi Period," pp. 176 ff.

13 For the information, though not the interpretation, in the following paragraphs, see the *Lebenslauf* appended to Metzger's dissertation (listed in Appendix 2); Metzger, "Wolfgang Metzger" (cited in Chapter 13); Michael Stadler and Heinrich Crabus, "Wolfgang Metzger (1899–1979)" (cit. chap. 13); and Michael Stadler, "Das Schicksal der nichtemigrierten Gestaltpsychologen im Nationalsozialismus," in Graumann (ed.), *Psychologie im Nationalsozialismus,* pp. 141 ff.

14 Stadler, "Das Schicksal," p. 142.

15 Metzger, "Gestaltpsychologie – ein Ärgernis für die Nazis," *Psychologie heute,* March 1979, p. 85. Metzger's SA membership is recorded in the Berlin Document Center (BDC).

16 Metzger, NSDAP membership card 4702876, BDC.

17 Metzger, "Gesetze des Sehens – angewandt," *Natur und Volk,* 67 (1937): esp. pp. 556–557.

18 Geuter, *Professionalization,* p. 64. See also Geuter, "Der Nationalsozialismus und die Entwicklung der deutschen Psychologie," *Bericht über den 33. Kongress der Deutschen Gesellschaft für Psychologie* (Gottingen, 1983), pp. 99–106.

19 Wolfgang Metzger, "Ganzheit und Gestalt. Ein Blick in die Werkstatt der Psychologie," *Erzieher im Braunhemd,* 6 (1938): esp. p. 92; idem., "Lebendiges Denken. Nach Schopenhauer und v. Clausewitz," *Erzieher im Braunhemd,* 6 (1938): 193–196; idem., *Gesetze des Sehens,* 1st ed., p. 159; Stadler, "Das Schicksal," p. 144.

20 Köhler to Springer, 7 April 1939, Köhler Papers, Correspondence Springer-Verlag. In 1933 and 1934, only Kurt Goldstein's name had disappeared from the masthead, while Wertheimer's and Gelb's remained.

21 Köhler to Metzger, 1 June 1937, copy in Kurt Koffka Papers, AHAP, Box M376.

22 Köhler to Springer, 7 April 1939 (cited in note 20). See also Joachim Wohwill, "German Psychological Journals under National Socialism: A History of Contrasting Paths," *JHBS,* 23 (1987): 177 ff.

23 Geuter, *Professionalization,* p. 65; Dr. Brinck, Gutachten von Professor Dr. von Allesch, 25.2.1938, BDC, REM file A6.

24 Metzger, "Wolfgang Metzger," p. 203; idem., "Gestaltpsychologie – ein Ärgernis für die Nazis," p. 85; cf. Stadler, "Das Schicksal," p. 146.

25 Metzger, "Der Auftrag der Psychologie in der Auseinandersetzung mit dem Geist des Westens," *Volk im Werden,* 10 (1942): esp. pp. 139–140; cf. Wolfgang Prinz, "Ganzheits-und Gestaltpsychologie im Nationalsozialismus" in Lundgreen (ed.), *Wissenschaft im Nationalsozialismus* (cited in Chapter 19), esp. pp. 103–105. For further discussions of this article, see Geuter, "Der Nationalsozialismus und die Entwicklung der deutschen Psychologie" and Stadler, "Das Schicksal," pp. 145 f.

26 Prinz, "Ganzheits- und Gestaltpsychologie," p. 107.

27 Metzger, *Psychologie. Die Entwicklung ihrer Grundannahmen seit der Einführung des Experiments* (Darmstadt, 1941; 2nd ed., 1953; 4th ed., 1964). Citations here are to the first edition.

28 Stadler, "Das Schicksal," pp. 149, 151; cf. Metzger, *Psychologie,* 4th ed., p. viii.

29 Ulfried Geuter, personal communication.

30 Hauptamt Wissenschaft der NSDAP to Party Chancellery, 24 February 1942; Assessment of Reichsdozentenführung des NSD, 21 January 1942, UA Münster, MA 116/10, both cited in Geuter, *Professionalization,* pp. 216 f.; cf. Stadler, "Das Schicksal," p. 150.

31 Reese Conn Kelly, "German Professoriate under Nazism: A Failure of Totalitarian Aspirations," *History of Education Quarterly,* 25 (1985): 273.

32 Kurt Gottschaldt, *Lebenslauf* attached to Berlin dissertation (listed in Appendix 2); Gottschaldt interview (cited in Chapter 13); Köhler to Administrative Director, 10 February 1929, UAHUB, Universitäts-Kuratorium Nr. 8391, Bl. 239–240.

33 Gottschaldt, "Der Aufbau des kindlichen Handelns." *Zeitschrift für angewandte Psychologie,* Sonderheft (Leipzig, 1933), pp. 6–7, 11 f.; Gottschaldt interview.

34 Gottschaldt, *Aufbau,* p. iv.

35 Gottschaldt interview; Geuter, *Daten,* vol. 1 (cited in chapter 13), p. 20. Gottschaldt to Rothacker, 18 April, 6 May, 11 May, 17 June 1933, 7 November, 19 December 1934, 8 January 1935. Gottschaldt papers. Max Planck Institute for Psychological Research, Munich.

36 Adolf von Harnack to KWG Board, 18 June 1926, MPGA, Berlin-Dahlem, KWI Anthropologie, Bnd. 2411. On the background of eugenics and "race hygiene" in the Weimar period, see Weiss, "The Race Hygiene Movement in Germany" (cited in the Introduction), and Paul Weindling, "Weimar Eugenics: The Kaiser Wilhelm Institute for Anthropology, Human Heredity and Eugenics in Social Context," *Annals of Science,* 42 (1985): esp. pp. 312 f. On Britain and the United States, see Daniel Kevles, *In the Name of Eugenics* (New York, 1985).

37 Otmar von Verschuer, "Erbpsychologische Untersuchungen an Zwillingen," *Zeitschrift für induktive Abstammungs- und Vererbungslehre,* 54 (1930): 280–285; idem., "Intellektuelle Entwicklung und Vererbung," in Günther Just (ed.), *Vererbung und Erziehung* (Berlin, 1930), pp. 197 ff.; Ida Frischeisen-Köhler, "Untersuchungen an Schulzeugnissen von Zwillingen," *Zeitschrift für angewandte Psychologie,* 37 (1930): 385–416; idem., "Über die Empfindlichkeit für Schnelligkeitsunterschiede," *PsyFo,* 18 (1933): 286–290; Marie-Therese Lassen, "Zur Frage der Vererbung sozialer und sittlicher Charakteranlagen (auf Grund von Fragebögen von Zwillingen)," *Archiv für Rassen- und Gesellschaftsbiologie,* 25 (1931): 268–278.

38 *Bekenntnis der Professoren an den deutschen Universitäten und Hochschulen zum Adolf Hitler und dem nationalsozialistischen Staat* (Dresden, n.d.), pp. 9–10; Fischer, "Tätigkeitsbericht," 15 June 1933, MPGA, Berlin-Dahlem, KWI Anthropologie, Bnd. 2399; Fischer, "Kaiser-Wilhelm-Institut für Anthropologie, menschliche Erblehre und Eugenik," in M. Hartmann (ed.), *25 Jahre Kaiser-Wilhelm-Gesellschaft zur Förderung der Wissenschaften,* vol. 2 (Berlin, 1936), p. 117; cf. Benno Müller-Hill, *Tödliche Wissenschaft: Die Aussonderung von Juden, Zigeunern und Geisteskranken 1933–1945* (Reinbek b. Hamburg, 1984), p. 24; Weiss, "The Race Hygiene Movement."

39 Fischer to KWG Board, 9 April 1935, MPGA, KWI Anthropologie, Bnd. 2406, Bl. 180; Fischer to KWG Board, 22 and 26 June 1936, MPGA, KWI Anthropologie, Bnd. 2399, Bl. 75, 77; budget figures from MPGA, KWI Anthropologie, Bnd. 2399, Bl. 47, 142. Available documents do not state the exact size of the budget.

40 Fischer, "Das Kaiser-Wilhelm-Institut für Anthropologie, menschliche Erblehre und Eugenik," in M. Planck (ed.), *25. Jahre Kaiser-Wilhelm-Gesellschaft zur Förderung der Wissenschaften,* vol. 1 (Berlin, 1936), p. 355. Emphasis mine.

41 Gottschaldt, "Über die Vererbung von Intelligenz und Charakter," *Fortschritte der Erbpathologie und Rassenhygiene,* 1 (1937): 6–7; idem., "Zur Methodik erbpsychologischer Untersuchungen in einem Zwillingslager, *Zeitschrift fur induktive Abstammungs- und Vererbungslehre,* 73 (1937): pp. 522 f.; idem., *Die Methodik der Persönlichkeitsforschung in der Erbpsychologie* (Leipzig, 1942), pp. 94 ff.

42 Gottschaldt, "Umwelterscheinungen im erbpsychologischen Bild," *Die Naturwissenschaften,* 25 (1937), pp. 433–444; idem., "Über die Vererbung," p. 4. Emphasis mine. Cf. A. Ohm, "Die Entwicklung der sozialen Person während der Untersuchungshaft" (Diss. Berlin, Math.-Nat. Fakultät, 1938); Helmut von Bracken, "Verbundenheit und Ordnung im Binnenleben von Zwillingspaaren," *Zeitschrift für pädagogische Psychologie,* 37 (1936): 65–81. Wilde's SA and NSDAP memberships are documented in BDC files.

43 Gottschaldt, "Erbpsychologie der Elementarfunktionen der Begabung," in Günther Just (ed.), *Handbuch der Erbbiologie,* vol. 5:1 (Berlin, 1939), esp. pp. 461, n. 1, 488, 493. See also Stadler, "Das Schicksal," p. 155.

44 Gottschaldt, "Erbpsychologie," p. 512; idem., "Phänogenetische Fragestellungen im Bereich der Erbpsychologie," *Zeitschrift für induktive Abstammungs- u. Vererbungslehre,* 76 (1939): 118–157.

45 Gottschaldt, "Erbpsychologie," pp. 462, 464, 467.

46 Gottschaldt, *Die Methodik der Persönlichkeitsforschung,* p. 56.

47 Gottschaldt, "Erbpsychologie," esp. pp. 457–459; cf. Kurt Wilde, "Über Intelligenzuntersuchungen an Zwillingen," *Zeitschrift fur induktive Abstammungs- u. Vererbungslehre,* 73 (1937): 512–517.

48 Gottschaldt, "Über die Vererbung"; idem., "Erbpsychologie," pp. 502, 517 f., 519. Cf. Horatio H. Newman, *Twins: A Study of Heredity and Environment* (Chicago, 1937).

49 Gottschaldt, "Erbpsychologie," pp. 449–450.

50 See, e.g., F. Stumpfl, "Kriminalität und Vererbung," in Just (ed.), *Handbuch* (cited in n. 9), p. 1271.

51 Gottschaldt, "Erbpsychologie," p. 514.

52 Gottschaldt interview.

53 Gottschaldt interview; Müller-Hill, *Tödliche Wissenschaft,* pp. 151, 161. Rieffert's attack of 22 January 1936 is noted in BDC, REM file card G 64.

54 Fischer, Gutachten Dr. Gottschaldt betreffend, 14 January 1938, UAHUB, Philosophische Fakultät. Nr. 1441, Bl. 323; Rothacker to Fischer, 6 January 1938, UAHUB, Philosophische Fakultät, K 221, Bnd. 3, Bl. 64; Faculty to Ministry, 14 April 1938, ibid., Bl. 75.

55 Vorschlag ord. Prof. Köhler (sic), 31 May 1939, BDC, REM file card G 64; Harmjanz statement for Gottschaldt file, 29 December 1939, BDC, REM W 48/5.

56 Dekan Leipzig to Ministry, 19 December 1942; Verschuer to Ministerialrat Dr. Frey, 18 May 1943, BDC, REM W 48/5; NS Dozentenbund evaluation, BA Koblenz, NS 15/243, Bl. 89. Cf. Stadler, "Das Schicksal," p. 160.

Chapter 21

1 Erich R. Jaensch, "Der Hühnerhof als Forschungs- und Aufklärungsmittel in menschlichen Rassenfragen," *Zeitschrift für Tierpsychologie,* 2 (1939): 223–258; H. Ermisch, "Psychophysische und Psychologische Untersuchungen an verschiedenen Hühnerrassen," *ZfP,* 137 (1936): 209–244.

2 Gottfried Hausmann, "Zur Aktualgenese räumlicher Gestalten," *AfgP,* 93 (1935): 289–334; Hermann Busse, "Rhythmische Gestaltbildungen bei der Arbeit in der Gruppe," *AfgP,* 99 (1937): 213–259.

3 Paul Wachter, "Über den Zusammenhang der typischen Formen des Gestalterlebens mit den Temperamentskreisen Kretschmers," *AfgP,* 104 (1939): 1–47; Ludwig Schiller, "Ganzheitliche Auffassung und Persönlichkeitstypus," *ZfP,* 153 (1942): 43–80.

4 Gerda Ostermeyer and Franz Lotz, "Erbcharakterkunde, Gestaltpsychologie und Integrationstypologie," *Zeitschrift für angewandte Psychologie und Charakterkunde,* Beiheft 73, ed. Gerhard Pfahler (Leipzig, 1937); Geuter, "German Psychology in the Nazi Period," pp. 171–172. On psychology journals, see Joachim Wohlwill, "German Psychological Journals under National Socialism" (cited in Chapter 20).

5 Prinz, "Ganzheits- und Gestaltpsychologie" (cited in Chapter 20).

6 Edwin Rausch, interviews with the author, 23 February and 22 September 1978; Stadler, "Das Schicksal" (cited in Chapter 20), p. 157; cf. Geuter, *Daten* (cited in Chapter 1), vol. 1, p. 33. Budget figures from the Frankfurt *Kassenbücher* (cited in Chap. 7), pp. 96, 110, 123.

7 See the dissertations by Jacobs, Madelung, Becker and Siemsen listed in Appendix 2.

8 See the dissertations by Bartel, Müller and Ortner listed in Appendix 2. See also Wolfgang Köhler and Hedwig von Restorff, "Analyse von Vorgängen im Spurenfeld, II. Zur Theorie der Reproduktion," *PsyFo,* 21 (1935): 56–112.

9 Ludwig Kardos, "Ding und Schatten," *ZfP,* Ergänzungsband 23 (Leipzig, 1934).

10 Wilhelm Wolff, "Induzierte Helligkeitsveränderung," *PsyFo,* 20 (1935): 159–194, esp. pp. 174–175, 190–191.

11 Mümtaz Turhan, "Über räumliche Wirkungen von Helligkeitsgefällen," *PsyFo,* 21 (1937): esp. pp. 2 ff. Cf. Kardos, "Ding und Schatten," cited in Turhan, "Über räumliche Wirkungen," p. 12, n. 3.

12 Turhan, "Über räumliche Wirkungen," pp. 41, 48–49; cf. Metzger, *Gesetze des Sehens,* pp. 124 ff.

13 Erich Goldmeier, "Über Ähnlichkeit bei gesehenen Figuren," *PsyFo,* 21 (1936): 146–208; translated as chs. 1–6 of "Similarity in Visually Perceived Forms," *Psychological Issues,* 8 (1972), Monograph 29. Citations to the translation, here pp. 40f., 55, 69, 81.

14 Ibid., pp. 25, 70–71.

15 Ibid., pp. 17, 49, 82–83.

16 Siegfried Sorge, "Neue Versuche über die Wiedergabe abstrakter optischer Gebilde," *AgP,* 106 (1940): esp. pp. 10 ff.

17 Ibid., pp. 67, 86–87.

18 Unless otherwise noted, the information in the following paragraphs comes from Rausch interviews (cited in n. 6) and idem., "Edwin Rausch," in Ludwig Pongratz et al. (eds.), *Psychologie in Selbstdarstellungen,* vol. 2 (Bern, 1979), pp. 211–256.

19 Edwin Rausch, "Über Summativität und Nichtsummativität," *PsyFo,* 21 (1937): 209–289; (reprint ed. Darmstadt, 1967).

20 Ibid., pp. 212, 215 f., 229, 238 ff., 249, 258.

21 Ibid., pp. 221, n. 1; 261, 280, 289.

22 Kurt Grelling and Paul Oppenheim, "Der Gestaltbegriff im Lichte der neuen Logik," *Erkenntnis* 7 (1937/1938), esp. pp. 221 ff. and n. 10; "Supplementary Remarks on the Concept of Gestalt," ibid., 357–359.

23 Metzger, "Zur anschaulichen Repräsentation von Rotationsvorgängen und ihre Deutung durch Gestaltkreislehre und Gestalttheorie," *Zeitschrift für Sinnesphysiologie,* 68 (1940): 272, n. 1; cf. Stadler and Crabus, "Wolfgang Metzger," p. 15.

24 Koffka, *Principles* (cited in Chapter 1), p. 684; cf. pp. 10 ff.

25 Metzger, *Psychologie* (cited in Chapter 20), 1st ed., pp. 2–3, 7 f., 221; 4th ed., pp. 2–3, 7 f., 231. The names given are drawn from the citations Metzger added to the second and subsequent editions.

26 Metzger, *Psychologie,* 1st and 4th eds., pp. 2–3, 4–5.

27 Ibid., 1st and 4th eds., pp. 12 f., 23–25.

28 Ibid., 1st and 4th eds., pp. 16, 18, 29 ff.

29 Ibid., 1st and 4th eds., pp. 45 f.

30 Ibid., 1st ed., pp. 199–200; 4th ed., pp. 209–210; cf. p. 208.

31 Ibid., 1st ed., pp. 133 f., 135 f.; 4th ed., pp. 138, 140.

32 Ibid., 1st ed., pp. 135–137; 4th ed., pp. 140–141.

33 Ibid., 1st ed., p. 164; 4th ed., p. 170.

34 Ibid., 1st ed., pp. 136, 155 ff.; 4th ed., pp. 141, 160 ff. Cf. Koffka, *Principles,* pp. 474 ff. Such considerations ultimately led to Harry Helson's *Adaptation Level Theory* (New York, Evanston, London, 1964).

35 Ibid., 1st ed., pp. 140 f., 146; 4th ed., pp. 144, 151.

36 Ibid., 1st ed., p. 271; 4th ed., p. 283.

37 Ibid., 1st ed., p. 287; 4th ed., p. 300.

38 Ibid., 1st ed., pp. 270, 287, 289, 292; 4th ed., pp. 283, 299, 302, 305.

39 Viktor von Weizsäcker, *Der Gestaltkreis: Theorie der Einheit von Wahrnehmen und Bewegen* (1940; 4th ed. (1950), repr. Frankfurt a.M., 1973), esp. p. 4 and ch. V. For Weizsäcker's account of his intellectual background and relationship to Gestalt theory, see his *Natur und Geist: Erinnerungen eines Arztes,* 3rd ed. (Munich, 1977), esp. ch. 5.

40 Von Weizsäcker, *Der Gestaltkreis,* pp. 9 f., 30 ff., 39; cf. Alfred Prinz von Auersperg and Harry C. Buhrmeister, Jr., "Experimenteller Beitrag zur Frage des Bewegtsehens," *Zeitschrift für Sinnesphysiologie,* 66 (1935): 274–309; Paul Christian, "Wirklichkeit und Erscheinung in der Wahrnehmung von Bewegung dargestellt an experimentellen Beispielen," ibid., 68 (1940): 151–184. See also Thomas Henckelmann, *Viktor von Weizsäcker (1886–1957). Materialien zum Leben und Werk* (Berlin, Heidelberg, New York, 1986), chs. 6, 8.

41 Von Weizsäcker, *Der Gestaltkreis,* p. 36.

42 Ibid., pp. 37 ff.

43 Metzger, *Psychologie,* 1st ed., pp. 198 n. 1, 262–263, 281 f.; 4th ed., pp. 208, 275–276, 293 f. See also Metzger, "Zur anschaulichen Representation von Rotationsvorgängen," *Zeitschrift für Sinnesphysiologie,* 68 (1940), 261–279; idem., "Zur Theorie der Rotationserlebnisse," ibid., 69 (1940), 94–96; Paul Christian, "Wirklichkeit und Erscheinung" (cited in n. 43); idem., "Entgegnung zum Aufsatz Metzgers . . . ," ibid., 69 (1940), 91–93.

44 Metzger, *Psychologie,* 1st ed., pp. 105, 117 f., 226; 4th ed., pp. 109, 121, 236; but cf. Sorge (cited in n. 16), pp. 19–20.

45 Rausch interview.

46 J. Deussen, review of Metzger, *Psychologie, Archiv für Rassen- und Gesellschaftsbiologie,* 36 (1942), esp. pp. 481, 484.

47 Metzger, "Psychologie und Menschenkenntnis," *Die Erziehung,* 16 (1941), esp. pp. 63, 68.

Chapter 22

1 For the context, see Hermann Glaser, *The Rubble Years: The Cultural Roots of Postwar Germany* (New York, 1986). On academic disciplines, see, e.g., Christoph Cobet (ed.), *Einführung in Fragen an die Soziologie in Deutschland nach Hitler 1945–1950* (Frankfurt a.M., 1988); Walter Pehle and Peter Sillem (eds.), *Wissenschaft im geteilten Deutschland* (Frankfurt a.M., 1992).

2 For further details, see Ash and Geuter (eds.), *Geschichte der deutschen Psychologie* (cited in Chapter 13), and Ash, "Psychology in Twentieth-Century Germany: Science and Profession," in Konrad Jarausch and Geoffrey Cocks (eds.), *German Professions 1800–1950* (Oxford and New York, 1990), esp. pp. 299 ff.

3 Albert Wellek, *Die Polarität im Aufbau des Charakters* (Bern, 1950); idem., *Die genetische Ganzheitspsychologie* (Munich, 1954); Philipp Lersch, *Der Aufbau der Person,* 3rd ed. (Munich, 1953); idem., *Der Mensch in der Gegenwart* (Munich, 1947), p. 79; cf. Peter Mattes, "Psychologie im westlichen Nachkriegsdeutschland – Fachliche Kontinuität und gesellschaftliche Restauration," in Ash and Geuter (eds.), *Geschichte der deutschen Psychologie,* esp. pp. 205 f., 210 f., 215; idem., "Zur Kontinuität in der deutschen Psychologie über die NS-Zeit hinaus," *Psychologie und Geschichte,* 1 (1989): 1–11.

4 Ulfried Geuter, "The Uses of History for the Shaping of the Field: Observations on German Psychology," in Loren Graham et al. (eds.), *Functions and Uses of Disciplinary Histories* (Dordrecht and Boston, 1983), pp. 206 ff.; Mark Walker, "Legenden um die deutsche Atombombe," *Vierteljahreshefte für Zeitgeschichte,* 38 (1990): 45–74; idem., *Nazi Science* (cited in Chapter 19), ch. 10.

5 Erich Mittenecker, "Die Weiterentwicklung der deutschsprachigen experimentellen Psychologie nach dem Krieg – ein persönlicher Rückblick," in Kurt Pawlik (ed.), *Fortschritte der experimentellen Psychologie* (Berlin, 1984), p. 14; Mattes, "Zur Kontinuität in der deutschen Psychologie," pp. 9–10.

6 Mattes, "Psychologie im westlichen Nachkriegsdeutschland," pp. 206 ff., 217–218.

7 For the broader context, see Volker Berghahn, *The Americanization of West German Industry, 1945–1973* (Cambridge, 1986).

8 Peter R. Hofstätter, *Einführung in die quantitativen Methoden der Psychologie* (Munich, 1953), p. 3; Albert Wellek, *Der Rückfall in die Methodenkrise der Psychologie und ihre Überwindung* (Göttingen, 1959), pp. 6, 28 f.; cf. Alexandre Métraux, "Der Methodenstreit und die Amerikanisierung der Psychologie in der Bundesrepublik 1950–1970," in Ash and Geuter, *Geschichte der deutschen Psychologie,* esp. pp. 236–244.

9 On the early years of psychology in the GDR, see Heinz Gast, "Zur Entwicklung des psychologischen Fachstudiums in den letzten Jahren," *ZfP,* 158 (1955): 294–304. For alliances with congenial Soviet theorists see, e.g., Hans Hiebsch, review of Sergei Rubinstein, *Grundlagen der allgemeinen Psychologie* (Berlin, GDR, 1958), *ZfP,* 163 (1959): 309–310. On the stages in the development of psychology in East Germany and the effort to create a "GDR psychology," see Stefan Busse, "Gab es eine DDR-Psychologie?" *Psychologie und Geschichte,* 5 (1993): 40–62.

10 Brigitte Schunter-Kleemann, "Leipziger Allerlei: Psychologie in der DDR," *Psychologie heute,* June 1980; on Soviet psychologists in this period, see David Joravsky, *Russian Psychology: A Critical History* (Oxford, 1989), esp. pp. 448–449.

11 Stadler, "Das Schicksal der nichtemigrierten Gestaltpsychologen" (cited in Chapter 20), pp. 149–150; cf. Stadler and Crabus, "Wolfgang Metzger" (cited in Chapter 13), p. 16.

12 J. F. Adams, "The Status of Psychology in the Universities of Austria and Germany," *Journal of Psychology,* 63 (1966): 117–134.

13 Metzger, "Psychologie zwischen Natur- und Geisteswissenschaften," in H. Balmer (ed.), *Die Psychologie des 20. Jahrhunderts,* vol. 1 (Zurich, 1976), pp. 28–29; Stadler and Crabus, "Wolfgang Metzger," p. 17. Comparisons to chemistry and biology are in Metzger to Kurator der Universität, 12 May 1942 and 5 January 1946, Psych. Institut/Nachlass Metzger, UA Münster, Nr. 2.

14 Metzger, "Das Bild des Menschen in der neueren Psychologie" (1952), reprinted in *Gestaltpsychologie,* esp. pp. 33–34, 42; idem., "Das Experiment in der Psychologie," *Studium Generale,* 5 (1952): p. 144; see also Wilhelm Witte, "Zur Geschichte des Gestaltbegriffs in der Psychologie," *Studium Generale,* 5 (1952): 455–464.

15 Metzger, "Das Experiment in der Psychologie," p. 142. Emphasis in the original.

16 Ibid., pp. 142–143, 147, 152, 154.

17 Metzger, "Das Experiment," p. 145; idem., "Wolfgang Metzger," pp. 225 f.; cf. Albert Wellek, "Das Experiment in der Psychologie," *Studium generale,* 1 (1947): esp. pp. 30–31.

18 Julian Hochberg, "Effects of the Gestalt Revolution," *Psychological Review,* 64 (1957): 73, n. 1.

19 Wolfgang Köhler, *Dynamics in Psychology* (New York, 1940), ch. 2; idem., "Direct Currents in the Brain," *Science,* 113 (1951): 478; Köhler and Hans Wallach, "Figural Aftereffects: An Investigation of Visual Processes," *Proceedings of the American Philo-*

sophical Society, 88 (1944): 269–357; Köhler and David A. Emery, "Figural After-Effects in the Third Dimension of Visual Space," *American Journal of Psychology,* 60 (1947): 159–201; Karl S. Lashley, K. L. Chow, and J. Semmes, "An Examination of the Electric Field Theory of Cerebral Integration," *Psychological Review,* 58 (1951): 123–136; Roger W. Sperry and N. Miner, "Pattern Perception Following Insertion of Mica Plates into the Visual Cortex," *Journal of Comparative and Physiological Psychology,* 48 (1955): 463–469; Köhler, "Unsolved Problems in the Field of Figural After-Effects," *Psychological Record,* 15 (1965), 63–83. On the outcome, see Mary Henle, "The Influence of Gestalt Psychology in America" (1980), reprinted in *1879 and All That* (cited in Chapter 20), esp. pp. 125–126.

20 Metzger, "Zum gegenwärtigen Stand der Psychophysik" (1950), reprinted in *Gestaltpsychologie,* esp. p. 240.

21 Metzger, "Das psychophysische Problem" (1952), reprinted in *Gestaltpsychologie,* p. 255; idem., "Über Notwendigkeit und Möglichkeit kybernetischer Vorstellungen in der Theorie des Verhaltens" (1965), reprinted ibid., p. 266; "Die Wahrnehmungswelt als zentrales Steuerungsorgan" (1969), reprinted ibid., pp. 273 ff. Stadler and Crabus, "Wolfgang Metzger," pp. 18–19, erroneously date Metzger's "helmsman" analogy to the 1960s.

22 Metzger, "Über Modellvorstellungen in der Psychologie" (1965), reprinted in *Gestaltpsychologie,* esp. pp. 47–48; cf. Karl Pribram, "Towards a Holonomic Theory of Perception," in S. Ertel et al. (eds.), *Gestalttheorie in der modernen Psychologie* (Darmstadt, 1975), pp. 161–186; Michael Stadler, "Feldtheorie heute – von Wolfgang Köhler zu Karl Pribram," *Gestalt theory,* 3 (1981): 185–199.

23 Stadler and Crabus, "Wolfgang Metzger," pp. 20–21; cf. Metzger, "Adler als Autor" (1977), reprinted in *Gestaltpsychologie,* pp. 478–493.

24 Metzger, "Zur Frage der Bildbarkeit schöpferischer Kräfte," *Arbeit und Betrieb,* 12 (1941): 60–70, 118–127; idem., *Die Grundlagen der Erziehung zu schöpferischer Freiheit* (Frankfurt a.M., 1949); 2nd ed., retitled *Schöpferische Freiheit* (Frankfurt a.M., 1962).

25 Metzger, *Schöpfersiche Freiheit,* esp. pp. 15 ff., 44–46; 2nd ed., pp. 21 ff., 45 ff., 96 ff.; Max Wertheimer, *Produktives Denken,* trans. Wolfgang Metzger (Frankfurt a.M., 1957).

26 Ibid., pp. 40, 46 f., 56 ff.; 2nd ed., pp. 84, 115 f., 130–132.

27 Ibid., pp. 1, 13, 24; 2nd ed., pp. 40 f. Metzger to Köhler 20 April 1950, UA Münster, Psych. Inst./Metzger, Nr. 63.

28 Ibid., pp. 19, 32–33, 48–49; 2nd ed., pp. 60, 72, 112 f..

29 Ibid., pp. 37, 73, 84, 86; 2nd ed., pp. 79, 158, 179, 185.

30 Ibid., p. 87; 2nd ed., p. 186; but cf. the passing remarks on p. 49; 2nd ed., pp. 55 ff., 109, 113.

31 Metzger, *Politische Bildung aus der Sicht des Psychologen* (Lüneburg, 1965), pp. 7–8, 12–13.

32 Ibid., pp. 19–20, 25–26, 34–35.

33 Ibid., pp. 25–26, 28.

34 Ibid., pp. 10, 35, 39 f., 45–46, 49, 58–59.

35 Ulfried Geuter, "Institutionelle und professionelle Schranken der Nachkriegsauseinandersetzungen über die Psychologie im Nationalsozialismus," *Psychologie- und Gesellschaftskritik,* 4 (1980); Stadler and Crabus, "Wolfgang Metzger," pp. 24–25.

36 Metzger, "Wolfgang Metzger," p. 224; Metzger to H. Bauer, 21 September 1960, UA Münster, Nachlass Metzger, Nr. 41.

37 Stadler and Crabus, "Wolfgang Metzger," p. 22; Lothar Spillmann, "Zur Lage der psychophysischen Sehforschung in Deutschland: Rückblicke – Ausblicke," unpublished MS (Freiburg i.Br., 1975).

38 Wallach interview (cited in Chapter 13); Metzger to Köhler, 23 June 1949, 16 May 1956, 30 December 1960, 19 September 1961, 30 July 1962; Köhler to Metzger, 22 October 1962, 30 January 1965, all UA Münster, Psych. Inst./Metzger, Nr. 63; cf. Stadler, "Das Schicksal," p. 152.

39 For the following, see Rausch, "Edwin Rausch" (cited in Chapter 21) and Rausch interviews with the author, 23 February and 22 September 1978; cf. Stadler, "Das Schicksal," pp. 157 ff.

40 Rausch, "Variabilität und Konstanz als phänomenologische Kategorien," *PsyFo,* 23 (1949): esp. pp. 73, 78–79, 86, 89 ff., 95, 113.

41 Rausch, "Das Eigenschaftsproblem in der Gestalttheorie der Wahrnehmung," in Wolfgang Metzger (ed.), *Wahrnehmung und Bewusstsein: Handbuch der Psychologie,* vol. 1:1 (Göttingen, 1966), esp. pp. 872 ff., 879, 902 f.; cf. Barry Smith, "Gestalt Theory" (cited in the Introduction), esp. pp. 50–51.

42 Rausch, "Eigenschaftsproblem," p. 899; Smith, "Gestalt Theory," pp. 56–58. This line of thinking originated in Max Wertheimer, "Zu dem Problem der Unterscheidung von Einzelinhalt und Teil," *ZfP,* 129 (1933): 353–357; idem., "On the Distinction between Arbitrary Component and Necessary Part," trans. Michael Wertheimer, reprinted in *Productive Thinking,* enl. ed. (New York, 1959), pp. 260–265.

43 Rausch, "Eigenschaftsproblem," part III; Smith, "Gestalt Theory," pp. 61 ff.; cf. Ehrenfels, *Kosmogonie* (cited in Chapter 6), pp. 93 ff.

44 Rausch, *Struktur und Metrik geometrisch-optischer Täuschungen* (Frankfurt a.M., 1952), esp. pp. 24 ff., 39 ff.; idem., "Probleme der Metrik (Geometrisch-optische Täuschungen)," in Metzger (ed.), *Handbuch der Psychologie,* vol. 1:1 (cited in note 23), pp. 780–781.

45 Ibid., pp. 79 ff., 89 ff., 96 ff; "Probleme der Metrik," Sec. 7–8.

46 Information on Frankfurt institute budget from *Kassenbücher* (cited in Chapter 13). For a list of Frankfurt dissertations, see Appendix 2; see also Rausch, "Zehn Jahre 'Psychologische Arbeiten'," *Psychologische Beiträge,* 10 (1968): 409–430.

47 Rausch interviews.

48 Rausch, "Variabilität und Konstanz," p. 83.

49 Gottschaldt, Bericht über meine Reise nach Berlin, 26 August 1945; Verschuer to Gottschaldt, 28 August 1945; Telschow to Verschuer, 3 September 1945, copies in MPGA, Abt. II, Rep. 1A, X 97.

50 "Handlanger des Verbrechens," *Tägliche Rundschau,* 23 May 1946; Weingart et al., *Rasse, Blut und Gene* (cited in Chapter 19), pp. 574 f.

51 Verschuer to Hahn, 23 May 1946, 30 October 1946; Planck to Verschuer, 29 October 1946; Hahn to Verschuer, 5 November 1946, MPGA, Abt. II, Rep. 1A, X 97. A fuller account of this incident is in preparation. See also my essay, "Denazifying Scientists – and Science," in Matthias Judt and Burghard Ciesla (eds.), *TechnologyTransfer out of Germany* (Chur, 1995), in press.

52 Gottschaldt interview (cited in Chapter 13).

53 Jürgen Meehl, "Das Institut für Psychologie der Humboldt-Universität zu Berlin," *ZfP,* 157 (1954); 149–150, 153; Hans Hiebsch, "Die Ausbildung der wissenschaftlichen Aspiranten des Faches Psychologie," ibid., 157 (1954), 156–162; Metzger to H. Bauer, 21 September 1960, UA Münster, Nachlass Metzger, Nr. 41.

54 Meehl, "Das Institut für Psychologie," pp. 151–152; Gottschaldt, "Geleitwort," *ZfP,* 157 (1954): 1.

55 Gottschaldt, *Probleme der Jugendverwahrlosung,* 2nd ed. (Leipzig, 1950), pp. 5 ff., 17–18.

56 Ibid., pp. 4, 39.

57 Gottschaldt, "Zur Theorie der Persönlichkeit und ihrer Entwicklung," *ZfP,* 157 (1954): 2–3; p. 4, n. 2.

58 Ibid., pp. 5–6, 9–10, 10–11, 14–15; cf. Gottschaldt, "Phänogenetische Fragestellungen im Bereich der Erbpsychologie," *Zeitschrift für induktive Abstammungslehre,* 76 (1939): 118–157; Thomas J. Bouchard et al, "Sources of Human Psychological Differences: The Minnesota Study of Twins Reared Apart," *Science,* 250 (1990): 223–228.

59 Gottschaldt, "Über Persona-Phänomene," *ZfP,* 157 (1954): 163–200, esp. pp. 167, 170 f., 175 ff., 198.

60 Gottschaldt, "Zur Psychologie der Wir-Gruppe," *ZfP,* 163 (1959): esp. pp. 194 f., 198 ff.

61 Cf. Stadler, "Das Schicksal," p. 156.

62 Gottschaldt, *Jugendverwahrlosung,* p. 180.

63 Friedhart Klix, "Die Auffassung Pawlows und Bykows von den Beziehungen des Organismus zur Umwelt und Fragen der neueren Wahrnehmungspsychologie," *ZfP,* 158 (1955): esp. pp. 5, 20, 26.

64 Hans Dieter Schmidt, "Bedingungsgrundlagen der sozialen Betriebsatmosphäre und Probleme der innerbetrieblichen Kooperation," *ZfP,* 163 (1959): esp. p. 157.

65 Alfred Katzenstein, "Gestalt- und klinische Psychologie," *Neurologie, Psychiatrie und medizinische Psychologie,* Heft 7 (1956); Hans-Dieter Schmidt, "Über eine unsaubere Methode der Diskussion in der Psychologie. Eine notwendige Stellungnahme," *ZfP,* 160 (1956–57): 199–200.

66 Hans Hiebsch, "Aufgaben und Situation der Pädagogischen Psychologie in der DDR," *Pädagogik,* (1958), Heft 4, p. 251; Ulrich Ihlefeld, "Zur ideologischen Situation in der pädagogischen Psychologie," Ibid., Heft 7, p. 508.

67 Georg Eckart, "Wissenschaftshistorische und Wissenschaftspolitische Aspekte der Entwicklung der Psychologie in der DDR," in Georg Eckardt et al. (eds.), *Psychologiehistorische Manuskripte.* Band 2 (Berlin, GDR, 1989), p. 37; Walter Mäder, "Zeitgeschichtliche Aspekte und Zusammenhänge der Entstehung und Entwicklung der Gesellschaft der Psychologie der DDR," ibid., p. 46. Since Mäder was the Central Committee official responsible for the Berlin institute at the time, we can assume that this was his own opinion. Higher level officials, notably ideology chief Kurt Hager, appear to have thought differently.

68 Mäder, "Zeitgeschichtliche Aspekte," esp. pp. 47–49. A detailed analytical account of this incident based on the wealth of archival material that has become available since German unification is in preparation.

69 Gottschaldt, "Zwillingsforschung als Lebenslaufforschung - Längsschnittuntersuchungen über Entwicklungsverläufe von Zwillingen, aufgewachsen unter sich verändernden Zeitumständen," *Bericht über den 33. Kongress der Deutschen Gesellschaft für Psychologie . . . 1982* (Göttingen, 1983), 53–64.

70 Norbert Bischof, "Erkenntnistheoretische Grundlagenprobleme der Wahrnehmungspsychologie," in *Handbuch der Psychologie* (cited in note 41), esp. p. 51.

71 Metzger, "Der Geltungsbereich gestalttheoretischer Ansätze," *Bericht über den 25. Kongress der Deutschen Gesellschaft für Psychologie* (Göttingen, 1967), p. 21.

Conclusion

1 Mary Henle, "Wolfgang Köhler (1887–1967)" (cited in the Introduction); Hans-Lukas Teuber, "Wolfgang Köhler" (cited in Chapter 7); Esther Pabst, Solomon E. Asch, and Wilhelm Witte (eds.), "Festschrift zum 75. Geburtstag von Prof. Dr. Dr. h.c. Wolfgang Köhler am 21. Januar 1962," *Psychologische Beiträge,* 6 (1962), Hefte 3/4.

2 Metzger, "Der Ort der Wahrnehmungslehre im Aufbau der Psychologie," in idem. (ed.), *Handbuch der Psychologie, Band 1:1. Wahrnehmung und Bewusstsein* (Göttingen, 1966), pp. 12 f.

3 Henle, "The Influence of Gestalt Psychology in America" (cited in the Introduction); idem., "Rediscovering Gestalt Psychology," in Sigmund Koch and David E. Leary, *A Century of Psychology as Science* (New York, 1985); Molly Harrower, *Kurt Koffka* (cited in Chapter 7), pp. 185 ff.; cf. my essay, "Gestalt Psychology: Origins in Germany and Reception in the United States" (cited in the Preface), pp. 334 ff.

4 Kai von Feiandt, "Current Trends in Perceptual Psychology," in Pabst et al. (eds.), Festschrift (cited in note 1), esp. pp. 651, 659.

5 Gerd Gigerenzer, "Probabilistic Thinking and the Fight against Subjectivity," in Lorenz Krüger, et al. (eds.), *The Probabilistic Revolution,* vol. 2 (Cambridge, Mass., 1987), pp. 11–34. Nikolas Rose, "Engineering the Human Soul: Analyzing Psychological Expertise," *Science in Context,* 5 (1992): 351–372.

6 Hans-Jürgen Walter, *Gestalttheorie und Psychotherapie* (Darmstadt, 1977; 2nd ed., Opladen, 1985).

7 See, e.g., Roman Jakobsen and L. R. Waugh, *The Sound Shape of Language* (Brighton, 1979); for a different Gestalt approach to language and its relation to experience, see G. Lakoff and M. Johnson, *Metaphors We Live By* (Chicago, 1980), esp. ch. 15.

8 See his explication of "requiredness" in *The Place of Value in a World of Facts* (cited in Chapter 11).

9 Solomon E. Asch, *Social Psychology* (New York, 1952), ch. 16.

10 Carl-Friedrich Graumann, "Gestalt in Social Psychology," *Psychological Research,* 51 (1989), esp. pp. 76, 78.

11 Metzger, *Psychologie,* 1st ed., p. 199; 4th ed., p. 209.

12 Koffka, *Principles,* p. 48.

13 Metzger, "Aporien der Psychophysik" (1961), reprinted in *Gestaltpsychologie* (cited in Chapter 22), esp. pp. 260–261.

14 James J. Gibson, "The Legacies of Koffka's 'Principles'," *JHBS,* 7 (1971): 3–9; J. R. Pomerantz, "Perceptual Organization in Information Processing," in M. Kubovy and J. R. Pomerantz (eds.), *Perceptual Organization* (Hillsdale, N.J., 1981), pp. 141–180. For comments on the latter see my "Gestalt Psychology" and Mary Henle, "Some New Gestalt Psychologies," *Psychological Research,* 51 (1989), 81–85.

15 Julian Hochberg, "Organization and the Gestalt Tradition," in Edward C. Cartarette and Morton P. Friedman (eds.), *Handbook of Perception,* vol. 1 (New York, 1974), pp. 179–

210; Gary Hatfield and William Epstein, "The Status of the Minimum Principle in the Theoretical Analysis of Visual Perception," *Psychological Bulletin,* 97 (1985): 155–186; Epstein, "Has the Time Come to 'Rehabilitate' Gestalt Theory?" *Psychological Research,* 50 (1989): 2–6.

16 Manfred Eigen, "Goethe und das Gestaltproblem in der modernen Biologie," in H. Rössner (ed.), *Rückblick in die Zukunft. Beiträge zur Lage in den achtziger Jahre* (Berlin, 1981); cf. David Hubel and Torsten Wiesel, "Brain Mechanisms in Vision," *Scientific American,* 241 (1979): 150–162. For recent work using positron emission tomography (PET) to localize neural activity involved in perception, see Senir Zeki, *A Vision of the Brain* (Oxford, 1993). On Gestalt theory and self-organization, see also Michael Stadler and Peter Kruse, "Gestalttheorie und Theorie der Selbstorganisation," *Gestalt theory,* 8 (1986): 75–98; idem., "The self-organization perspective in cognition research: Historical remarks and new experimental approaches," in Hermann Haken and Michael Stadler (eds.), *Synergetics of Cognition* (Berlin and Heidelberg, 1990), pp. 32–52.

17 Frederic C. Bartlett, "The Way We Think Now," in Pabst, Asch, and Witte (eds.), *Festschrift für Wolfgang Köhler* (cited in note 1), p. 387.

18 Koffka, *Principles,* pp. 9–10.

Index

Entries pertain to items in main text and notes only

abstract and concrete functions, in language disturbances, 280, 281
Ach, Narziss, 39, 78, 269, 271, 317, 342, 344
acoustics, psychology of (*see* hearing)
Adams, Donald K., 210, 338
Adler, Alfred, 268, 389, 391
Allesch, Johannes von, 34, 299, 350, 357, 363, 382
Althoff, Friedrich, 19, 31 ff., 35
analysis, psychological, 139, 144
and-connections (*Und-Verbindungen*), x, 172
and-sums (*Und-Summen*), 220, 256 (*see also* sum)
Angell, James Rowland, 37
anger, dynamics of, 273
animals, psychology of, 150 ff., 165; form discrimination in, 165; perception in, 163 ff. (*see also* anthropoid apes)
anisotropy, of visual space, 65, 177
Annalen der Philosophie, 294 (see also *Erkenntnis*)
Ansbacher, Heinz, 385
Anschütz, Georg, 46
anthropoid apes, behavior, 149; communication, 149; emotions, 56; perception, 153, 156, 163; problem solving, 149 ff., 153 ff.; social behavior, 156; tool-making, 153, 156
anthropomorphism, 116, 156, 286
anti-Semitism, 134, 208, 213 f., 292
aphasia, 276 f.
apparent distance (*see* distance)
apparent form (*see* form, perception of)
apparent motion (*see* motion)
apparent size (*see* size)
apperception, 61, 91
apprehension, in Gestalt perception, 317

Archiv für die gesamte Psychologie, 26, 382
Aristotle, 85
Arleth, Emil, 105
Arndt, Ernst Moritz, 312
Arnheim, Rudolf, 12, 293, 299–304; film theory, 301 ff.; political views, 303 f.
Arnold, Wilhelm, 389
Asch, Solomon E., 409
assimilation, 125
associationism, and animal learning, 151; criticism of, 69 ff., 79, 84 ff., 309 f.; and neurophysiology, 96, 276 f.
atoms, 174 (*see also* corpuscles and continua)
atomism, 70, 98, 135 f. (*see also* elementism)
attention, 91, 127, 280, 308 f.
attitudes, 226 (*see also* cognitive styles; mental set)
Aubert, Hermann, 25
"authoritarian" and "democratic" groups, 391
Avenarius, Richard, 62, 63 f., 68, 71, 74, 88, 144, 211
Aydelotte, Frank, 338

Baeumler, Alfred, 339
Bain, Alexander, 74
Bartlett, Frederick, 411
Becher, Erich, 9, 81 f., 97–98, 146, 207, 253, 315
Becher, Siegfried, 146
Beck, Robert, 340
Becker, Carl Heinrich, 207, 214, 215, 290
Becquerel, Henri, 112
Beethoven symphony, image of, in Weimar culture, 296 f.
behaviorism, x f.
Benary, Wilhelm, 227, 357

Benjamin, Walter, 299
Benussi, Vittorio, 93 ff., 133, 137, 139 ff., 143
    ff., 199, 227, 317, 392
Bergner, Elizabeth, 302
Bergson, Henri, 9, 68, 69 f., 78, 82, 98, 128,
    140, 169, 211
Berkeley, George, 186, 211
Berlin, 35, 40, 45; psychological institute, x, 33
    ff., 188 ff., 205–211, 332–338; students in,
    33, 209, 332, 333 f.; technical academy, 204;
    university, 31, 195 f., students, 333
Berlin school (of Gestalt theory), ix, x, 3, 11,
    203, 217, 326; attacks on under Nazism,
    344; ownership of term "Gestalt theory,"
    310; research methods, 222 f.; reputation
    abroad, 210 f., 336 f. (*see also* Gestalt the-
    ory; phenomenology)
Bertalanffy, Ludwig von, 256, 286
Bethe, Albrecht, 282
Bieberbach, Ludwig, 332, 333, 338, 339, 343
Bildung, 18, 21, 86, 290, 314
Bildungsbürger/Bildungsbürgertum, 24, 43, in
    Weimar culture, 284 f., 290, 298
Bischof, Norbert, 404
Bleuler, Eugen, 107
Bloch, Ernst, 288, 305
Blumenbach, Johann Friedrich, 80, 85
body image, 376
Bogen, Helmut, 238
Boltzmann, Ludwig, 174, 260
Bolzano, Bernhard, 74 f.
Bondy, Curt, 382
Borak, J., 241
Born, Max, 188, 196 f., 286
Bouhler, Philipp, 344
brain events, observability of, 64, 177 (*see also*
    cortical currents; Gestalt theory; Köhler,
    Wolfgang; psychophysical isomorphism;
    Wertheimer, Max)
branched effects, law of, 375
Brecht, Bertolt, 299, 303 f.
Brentano, Franz, 28 ff., 33, 38, 74, 75, 76, 88,
    115, 133, 145, 320 (*see also* intentionality)
brightness, constancy, 59, 135, 136, 320; con-
    trast, 227, 364 ff.
Britsch, Georg, 389
Brod, Max, 105, 134, 187, 196, 197
Brouwer, L. E. J., 285
Brown, Junius F., 211, 235 f.
Brunswig, Alfred, 78
Brunswik, Egon, 318 ff.; probabilistic theory of
    perception, 320
Buber, Martin, 198, 327
Bühler, Charlotte, 249
Bühler, Karl, 12, 78, 95, 110, 138 f., 207, 243,

307, 309, 310, 314, 319, 322, 388; critique
    of Gestalt theory, 315 f.; theory of signs, 315
bundle hypothesis (*Bündelhypothese*), 220
Bunsen, Robert Wilhelm, 22
Burks, Barbara, 340

Cannon, Walter, 378
Carmichael, Walter, 268
Carnap, Rudolf, 267, 371
Cartwright, Dorwin, 406
Carus, Carl Gustav, 86, 290
Cassirer, Ernst, 46 f., 95, 98, 264, 266, 268,
    281
causal explanation, in psychology, 388 f.
centering, 121 f., 123 f., 281 (*see also*
    recentering)
Chamberlain, Huston Stuart, 89
chance, 152 f., 169
Chaplin, Charlie, 303
character, 73 (*see also* person)
characterology, 289 f., 339, 345 f.; criticism of,
    349
chemistry, institutionalization of, 20 f.
Chica (chimpanzee), 164
chronoscope, 24
Chuang Tzu, 390
Claparède, Eduard, 250
Clausewitz, Karl von, 373
Clauss, Ludwig Ferdinand, 340, 344
"Clever Hans," 151 f.
Coghill, G. E., 378
cognitive science, 14,
cognitive styles, 94, 141
Cohen, Hermann, 46 f.
coherence factors (*see* Müller, Georg Elias)
collective representations, 250
color, constancy, 56, 59, 135, 136; in animals,
    163; contrast, 146, 227 ff.; development of,
    249, 320 f.; Goethe's theory of, 86; phe-
    nomenology of, 58 f.; weakness, 108, 110;
    wheel, 59, 165
complexes, 106 f.
complex theory (*see* Müller, Georg Elias)
conduction theory (of nervous transmission), 96
consciousness, experimental study of, 22, 44;
    intentional model of, 93, 319; as process, 61;
    as spatiotemporal continuum, 69; stream of,
    70; as structured whole, 72, 143 ff.;
    Stumpf's conception of, 36 ff.; of time, 78;
    unity of, 29, 61, 170
conscious disposition (*Bewusstseinslage*), 107
    (*see also* attitudes)
constancies, perceptual, development of, 320 f.
    (*see also* brightness, constancy; color cons-
    tancy; size, form, perception of)

constancy hypothesis, 135 ff., 142 f., 166, 175 f.
context of discovery/justification, 30, 184 f.
Cornelius, Hans, 90 f., 116, 120, 125, 135, 214, 386
corpuscles and continua, 171, 174
cortical currents, 180, 388
creative synthesis (resultants), 61, 72, 138
crisis, cultural, in Weimar era, 285 ff., 322; of psychology, 307; of reality, 286; of science, 2, 7, 285
crucial (decisive) demonstrations, 222
Crutchfield, Richard, 406
cybernetics, 388

Darwin, Charles, 66, 70, 150, 254
demand character (*Aufforderungscharakter*) of objects, 270
Dembo, Tamara, 273
deprofessionalization (of education, law, and engineering) under Nazism, 326
depth perception, 221, 232; and motion, 232
Descartes, Réné, 90, 175 f., 352
Dessoir, Max, 332
determining tendencies, 78
Deutsche Allgemeine Zeitung, 328, 330
development, psychical, as differentiation, 251; as maturation, 248; and pedagogy, 251 f.; social, 250 f.; (*see also* Gestalt theory)
developmental mechanics, 80, 247
Dewey, John, 10, 68, 145, 194, 250
Die Naturwissenschaften, 217
Die Weltbühne, 301, 303
Dilthey, Wilhelm, 9, 21 ff., 68, 72–74, 98, 105, 290, 296, 297, 439n.
directional listener (*Richtungshörer*), 190
Döblin, Alexander, 303
Dodge, Raymond, 84 f.
dominance, as Gestalt category, 318, 352
Driesch, Hans, 9, 11, 79–83, 98, 311
Drobisch, Moritz Wilhelm, 22
Du Bois-Reymond, Emil, 54
Düker, Heinrich, 382
Dürr, Ernst, 107
Duncker, Karl, 210, 245, 260, 321, 364, 365, 369, 373, 375, 378, 406; emigration and suicide, 339f.; on induced motion, 234 f.; persecution under Nazism, 332, 334, 337 f., 349; on problem solving, 238 ff., 357 (*see also* relational systems)
Dunlap, Knight, 238
Durkheim, Emile, 24
dynamics, in psychology, 268, 270 f., 273

Ebbinghaus, Hermann, 25, 27, 31, 44, 65 ff., 73 f., 95, 98, 267, 439n.

Ebert, Friedrich, 196, 214
Edinger, Ludwig, 133, 143, 247 f., 276, 280
ego, 91, 269 f.
Ehrenfels, Christian Freiherr von, x, 1, 8, 105, 133, 134, 141, 165, 172, 344, 386, 395, 396, 407; criteria for existence of Gestalten, 171 f., 173; on Gestalt qualities, 87 ff.; on sexual ethics and eugenics, 105
Ehrenstein, Walter, 375
eidetic imagery, 313
Eigen, Manfred, 410 f.
Eilks, Hans, 369
Einstein, Albert, 13, 133, 170, 191, 195 ff., 213, 214, 262, 286, 291 f., 333
Eisenstein, Sergei, 302, 303
Eissler, Kurt, 320
elan vital, 70, 82, 169
Elberfeld horses, 152
elementism, 60 ff.; criticism of, 84 ff., 98, 175 f.
Ellis, Willis D., 216
embryology, 80, 252 f.
Emergency Committee in Aid of Displaced Foreign Scholars, 338
emotion, 12 (see also *Ganzheitspsychologie; Lewin, Kurt*)
empathy, 42
empiricism, 52, 59, 68, 71, 352; and animal learning theory, 151; as basis for evolutionary history, 165; and idealism, 79
entelechy, 82
entropy, 260 f. (*see also* second law of thermodynamics)
epistemology, 9 f., 30, 35 ff., 78 f., 123 (*see also* Gestalt theory, Wertheimer, Max)
equilibrium, 182 ff., 255
Erdmann, Benno, 31, 84 f., 113, 338
Erkenntnis, 261, 267, 294
Erzberger, Matthias, 292
ethnomusicology, 33,
Eucken, Rudolf, 47, 69
eugenics (race hygiene), 312, 326, 348, 356, 361, 380 f.
evolutionary history, 165
Ewe (African tribe), 122
Exner, Sigmund, 96, 125, 130, 133, 141, 277, 364
eye, as camera oscura, 54; as control apparatus, 253

Faiendt, Kai von, 406
Faraday, Michael, 171
Fechner, Gustav Theodor, 22, 48, 60, 64, 67
Fechner's law, 114 (*see also* psychophysics)
feeling, role of, in perception, 311, 318 (see also *Ganzheitspsychologie*)

Feyder, Jacques, 302
Fichte, Johann Gottlieb, 54, 373
field concept, 11; in theoretical biology, 252 f.
field theory, 171 (*see also* Maxwell, James
   Clerk)
figural aftereffects, 388
figure and ground, 1, 179 f., 229
film (motion pictures), 125; of anthropoid behavior, 156; as mass medium, 299; theory of,
   301 ff.
film colors, 58
Fischer, Aloys, 49
Fischer, Emil, 113
Fischer, Eugen, 332, 333, 334, 336, 338, 339,
   355, 356, 359, 360 (*see also* eugenics; Kaiser Wilhelm Institute for Anthropology)
Fischer, Ruth, 355
Fleck, Ludvik, 219
form discrimination, in animals, 165
form, perception of, 58, 87 ff., 95, 131 (*see
   also* Gestalt)
frames of reference (*Bezugssysteme*), 260; in
   general psychology (*see* relational systems);
   in motion perception, 235 f.
Franck, James, 213, 332; resignation in protest
   in 1933, 328, 333
Frank, Helene, 252, 321
Frank, Jerome, 211
Frankfurt, commercial academy, 49, 118, 150;
   Institute for Social Research, 215, 288; psychological institute, 107, 111, 117, 118 f.,
   215, 221, under Nazism, 363 f., in postwar
   period, 394, 398; university, 135, 203, 214
   ff., 327, 347, dismissals from in 1933, 327
Frege, Gottlob, 74, 195
Freiburg, university, 110
Freud, Sigmund, 98, 259, 268, 269, 277
Friedländer, Max, 105
Frischeisen-Köhler, Ida, 355
Fuchs, Wilhelm, 221, 279
Führer principle (*Führerprinzip*), 351
functions, psychical, 37 (*see also* structural
   functions)
functionalism, 37, 145 f.
fusion, of tones (*see* tonal fusion)

Gad, Johannes, 105, 107
Galilei, Galileo, 222, 268, 333
Ganzfeld, 229
Ganzheitspsychologie, x, 12, 217, 311 ff., 317
   f., 369, 383
Garbo, Greta, 302
Geiger, Moritz, 42
Geitel, Hans F. K., 112, 113, 167, 169 f.
Gelb, Adhemar, x, 12, 34, 95, 109, 120, 132,
   215, 216, 217, 221, 229, 263, 305, 336, 347,

348 f.; dismissal in 1933, 337; death, 406;
   research on perception and language of
   brain-damaged patients, 275 ff., 280
generative metaphors, 11, 219
genesis (of ideas, perceptions, etc.), 30
genidentity, 267 f. (*see also* time series)
George, Stefan, 109, 212
Gerhards, Karl, 396
German Labor Front, 3, 14, 345
German National People's Party, 330
German Physical Society, 333
German Society for Psychology, 205, 209, 338
   f., 343, 345, 347, 383, 384, 392, 393, 403,
   404; adaptation under Nazism, 342 (*see also*
   Society for Experimental Psychology)
Germany, Foreign Office, 336 ff.; universities,
   18 ff.
Gestalt (Gestalten), concept, 11, 85 ff., 121,
   175 f., logical analysis of, 371 f., phenomenological analysis of, 395 f., priority claims
   about, 307 ff.; laws or tendencies, 10, 133,
   224 ff.; microgenesis (*Aktualgenese*) of, 317,
   362; in Nazi Germany, 344, 362 f.; neurological basis of, 96 ff., 130 ff., 176 ff.; perception of, 93 ff.; physical, 171 ff.;
   psychophysical thresholds for, 95, 138;
   qualities, 88 ff., 110 f.; racial, 89 f.; strong
   and weak, 173 f., 270; switch, 9, 10, 193 f.;
   systems, ecological, 256
Gestaltkreis theory, 378
Gestalt theory, ix, x, 4, 87, 88; of action and
   emotion, 268 ff.; as answer to challege of
   psychology's situation, 199 f.; and behavior,
   144 f. (*see also* anthropoid apes); and causal
   explanation, 181, 377 f., 409; conception of
   brain action, 178 ff.; conception of consciousness, 143 ff., 176, 220; conception of
   knowledge process (= cognition), 123 f.;
   continuity of under Nazism, 363 ff.; criticisms of, 307–322; defensive position of in
   1960s, 404 f.; and democratic values, 391
   ff.; and (psychical) development, 247–252;
   emergence of, 103, 122 ff.; versus empiricism and nativism, 4; and expression, 299 f.;
   and film theory (*see* film, theory of); versus
   Gestaltkreis theory, 378 ff.; hypothesis testing in, 131 f., 223; institutional situation, 7
   f., 208 ff.; and language, 316, 407; on learning and memory, 240–244; versus mechanism and vitalism, 200; methodological
   commitments, 220 ff.; and National Socialist
   ideology, 348 ff., 360, 363; ontological
   dimension, 146 f. (*see* Ehrenfels, criteria for
   existence of Gestalten); and pedagogy, 389,
   391; and perception, 138 ff., 144, 146, 165
   f., 179 ff., 224–238; and personal agency,

408; phenomenology in, 131 f.; as philosophy of nature, 10, 168–186; physicalism in, 181, 266, criticism of, 312, 313, 321; and positivism, 372; present status of, 410 ff.; and psychotherapy, 407; and quantification, 222 f., 226; as scientific school, 219 ff.; and social perception, 256 ff.; and theoretical biology, 252–256; and thinking, 190 ff., 238–240; as "third way" versus dualism, 295 f.; in Weimar culture, 12, 294–306
Gestapo, 334, 337 f., 338, 360
Gibson, James J., 388, 410
Giese, Fritz, 204
Giessen, psychological institute, 211 f., 226; university, 46, 111, 120, 138, 199, 203
Girgensohn, Wilhelmine, 111
Goethe, Johann Wolfgang, 64, 85 f., 178, 184, 185, 222, 377, 389; theory of color, 86
Goldmeier, Erich, 13, 244, 364, 366 f., 369
Goldschmidt, Richard, 393
Goldstein, Kurt, x, 12, 216, 217, 276–283, 379; critique of Gestalt theory, 282; on the organism, 281 f.; on "racial hygiene," 276
"good" errors, in animal problem solving, 153, 157 f.
"good" phenomena, 222, 268
Göttingen, university, 19, 21, 29, 33, 78, 132 f., 207
Gottschaldt, Kurt, 13 f., 230 ff., 340, 346, 361, 362, 370, 384, 385, 406; biography, 354 ff.; career after 1945, 398 f.; research under Nazism, 356–360; research in Soviet occupation zone and German Democratic Republic, 398–403 (*see also* twin research)
gradients, in hearing, 236; in vision, 228
Granit, Ragnar, 229
graphology, 290, 300
Graz school, ix, 93 ff., 311; Gestalt theorists' criticism of, 139 ff., 145
Grelling, Kurt, 372
Groos, Karl, 211
Gross, Hans, 105, 106
Gross, Walter, 339, 340
grouping, perceptual, 91 f., 109, 141
Gruhle, Hans, 216, 349, 383
Günther, Hans F. K., 340, 362
Gurwitsch, Alexander, 253
Gütt, Arthur, 356

Haar, Hans, 334
Haber, Fritz, 333
habit, influence on perception, 230 f., 308
Hahn, Otto, 398
Haldane, J. B. S., 217
Halle, university, 30
Hallstein doctrine, 403

Hamann, Johann Georg, 373
Hamburg, colonial institute (later university), 21, 204, 382 (*see also* Stern, William)
Harden, Maximilian, 48
Harmjanz, Heinrich, 360
Harrower, Molly, 331
Hartgenbusch, Hans Georg, 211
Hartmann, Eduard von, 80, 169
Hartnacke, Wilhelm, 347 f., 359
Havemann, Robert, 398
Head, Henry, 282
hearing, 52, 113 ff., 187 ff., 190; Berlin school research on, 236 f.
Heckhausen, Heinz, 393
Hegel, Georg Wilhelm Friedrich, 50
Hegelianism, 43
Heidegger, Martin, 214,
Heidelberg, university, 21, 83,
Heider, Fritz, 257, 319
Heisenberg, Werner, 261, 286
Held, Richard, 406
Hellpach, Willy, 46
Helmholtz, Hermann von, 8, 9, 21, 25, 30, 58 f., 61, 62, 70, 72, 86, 96, 115, 132, 135, 175 f., 280, 315, 405; conception of psychology, 53 f.; theory of sensation, 52 f.
Henning, Hans, 313
Herbart, Johann Friedrich, 61, 72, 87
Herbertz, Richard, 161
hereditary character study (*Erbcharakterkunde*), 362
hereditary psychology (*Erbpsychologie*), 355; and "positive" eugenics, 356
heredity and environment, 249, 299, 359, 400 (*see also* eugenics; twin research)
Hering, Ewald, 8, 52, 54 ff., 58 f., 77, 115, 131, 135, 137, 138, 178, 179, 223, 228 f., 236, 241, 258, 280, 320, 364; conception of "seen objects," 55; explanation of sensory processes, 56
Hertwig, Oskar, 254
Hertz, Heinrich, 332, 343
Hertz, Richard, 332
Hiebsch, Hans, 403
Hilbert, David, 286
Hillebrand, Franz, 50, 58
Hindenburg, Paul von, 292 f.
Hitler, Adolf, 331, 342, 346, 356
Hobhouse, Leonard Trelawny, 158
Höffding, 84, 85, 366
Höfler, Alois, 172
Hofstätter, Peter R., 384, 401
Holaday, Beverly, 320
holism, ix, 1, 3, 72 f.; in Weimar social thought, 287 ff. (*see also* Gestalt; Gestalt theory)

holistic psychology (see *Ganzheitspsychologie*)
Hollingworth, Harry, 243, 376
Holst, Erich von, 378
homeostasis, 378
Horkheimer, Max, 215, 288, 305
Hornbostel, Erich von, 34, 39, 105, 107, 109, 113, 114, 188 ff., 204, 236 f., 299, 316 (*see also* directional listener; hearing)
Hull, Clark Leonard, 406
humanistic psychology (*geisteswissenschaftliche Psychologie*) (*see* psychology)
Hume, David, 37, 75, 79, 111, 124, 176, 321
Hunter, Walter, 165
Husserl, Edmund, 9, 19, 37, 44, 47, 68, 72, 74–78, 79, 82, 98, 108, 110 f., 111, 122, 144, 173 f., 195, 286, 297, 344, 373; analysis of whole and part, 76 f.; on the status of psychology, 44 f. (*see also* phenomenology)
Hüter, Ludwig, 293

Ibsen, Henrik, 212
idealism, 9, 35, 75, 296 (*see also* empiricism; positivism)
Ihlefeld, Ulrich, 403
imagery, 110
imitation, in chimpanzee behavior, 156
industrialization, 43 f.
inference, scientific, 53, 191 ff., 443n.; unconscious, 53, 70 (*see also* thinking)
inner eye, 56, 179
inner perception, 28 f.
insight, in animal problem solving, 157 ff., 162
instinct, 28 f., 248 f.
intelligence tests, criticisms of, 298 f., 357
intentionality, 28, 75 f., 93, 314 (*see also* consciousness)
interactionism (*see* mind-body problem)
intersubjectivity, 258
introspection, 39 f. (*see also* inner perception; self-observation)
intuition, 69 f., 291, 297
intuitionism, 286
invariance (invariants), criterion for suprasummativity, 371; directly experienced, 297 f.; in hearing, 190; in nature, 116 f., 182, 222, 223
isomorphism (*see* psychophysical isomorphism)

Jaensch, Erich Rudolf, 47, 58, 212, 215, 342, 362, 369, 373, 383; critique of Gestalt theory, 312 ff., 343 f.; role under Nazism, 343
Jakobsen, Roman, 407
James, William, 9, 37, 68, 70 f., 75, 91, 98, 238, 241, 258, 310 (*see also* consciousness)
Jaspers, Karl, 73, 214, 215, 289
Jena, psychological institute, 204

Jennings, Herbert Spencer, 82
Jerusalem, Hebrew University of, 327
Joachim, Josef, 39
Johnson, Alvin, 327
Journal of Animal Behavior, 162
Jügel, Carl Christian, Foundation, 118
Julius (orangutan), 162
Jung, Carl Gustav, 107, 401
Jützi, Piel, 303
juvenile delinquency, 400

Kafka, Franz, 10, 105
Kafka, Gustav, 342
Kaiser Wilhelm Institute for Anthropology, Human Heredity and Eugenics, 355 ff., 398 (*see also* eugenics; Fischer, Eugen; Gottschaldt, Kurt; Verschuer, Otmar von)
Kaiser Wilhelm Society, 34, 356, 359, 398
Kant, Immanuel, 53, 75, 80, 85, 87, 374
Kappers, C. Ariens, 161
Kardos, Ludwig, 318
Kastil, Alfred, 133
Katz, David, 58 f., 78, 164, 165, 212, 214, 215, 217, 318, 321, 337, 342; dismissal in 1933, 326 f. (*see also* color; phenomenology)
Katzenstein, Alfred, 402
Keller, Hans, 337, 339, 340
Kenkel, Friedrich, 139 ff.
kinesthetic sensations (*see* sensation)
Kirchhoff, Gustav, 22
Klages, Ludwig, 290 f., 300, 312, 316, 345, 373, 390
Klein, Julius, 106 f.
Klimpfinger, Sylvia, 320
Klix, Friedhart, 402, 403
knowledge (*see* epistemology; Gestalt theory)
Koffka, Kurt, ix, x, 1, 2, 8, 9, 11, 34, 38, 42, 46, 49, 50, 57, 69, 81, 103, 106, 108–111, 117, 118, 120, 124, 125, 130, 132, 158, 165, 176, 186, 188, 198, 199, 200, 203, 210, 212, 226, 227 f., 232, 244, 256, 257, 263, 269, 270, 274, 275, 279, 280, 281, 294, 295, 296, 302, 305, 315, 331 f., 343, 350, 354, 364, 376, 379, 404, 409, 410, 411 f.; conception of experience, 143 ff.; controversy with Bühler and Selz, 310 ff.; death in 1941, 406; dissertation, 108 f.; in Giessen, 211 ff.; methodological views, 220 f.; move to the United States, 212; opposition to Graz school, 139, 142–147; political views, 293; on psychical development, 247 ff.; reply to Brunswik, 320 f.; systematic account of Gestalt theory, 372 ff.; testimonial for Wertheimer, 213
Koffka, Mira (née Klein), 109, 110, 118, 120

Köhler, Franz Eduard, 111 f.
Köhler, Thekla, 163, 338
Köhler, Ulrich, 111
Köhler, Wolfgang, ix, x, 1, 2, 6, 9, 10, 11, 13, 34, 42, 46, 49, 50, 57, 64, 69, 71, 81, 103, 111–117, 118, 120, 123, 124, 125, 142, 143, 148, 150, 166 ff., 198, 199 f., 203, 209, 216, 221, 223, 229, 235, 236, 238, 245, 248, 249, 263, 266, 267, 269, 270, 275, 281, 282, 293, 294, 295, 297 f., 300, 305, 313, 314, 326, 327 f., 342, 345, 346, 353, 354, 357, 358, 360, 361, 363, 364, 369, 371, 372, 373, 376, 377, 379, 383, 388, 404; accused of espionage, 161, 461 n.; anthropoid research, 152–160, 163–166; appointment in Berlin, 207 f.; commitment to physicalism, 181, 266; conception of physical law, 182 ff., 261 f.; conflict with Metzger in Nazi period, 349 f.; correspondence with Metzger after 1945, 393 f.; correspondence with Robert Yerkes, 160–163; critique of intelligence tests, 298 f.; death in 1967, 405; dissertation, 113 ff.; honors and awards, 405; on language 316; on learning, 241 ff.; methodological views, 221; natural philosophy, 168 ff.; on open systems, 252 f.; opposition to constancy hypothesis, 135 ff.; opposition to Nazi policies, 328 ff.; on perception of self and others, 258 ff.; political views, 292, 330 f., 335; reply to G. E. Müller, 308 f.; and Society for Scientific Philosophy, 294; struggle for Berlin institute, 332–338; teaching style, 210; testimonial for Wertheimer, 213; theory of brain action, 178 ff.; tests of, 242 f., 388
Kollwitz, Käthe, 105, 195
König, Arthur, 25
Kopfermann, Herta, 232
Korsch, Karl, 265, 294
Kracauer, Siegfried, 291
Kraepelin, Emil, 57, 289
Kraus, Oskar, 133
Krech, David, 406
Kretschmer, Ernst, 345, 362
Krieck, Ernst, 350
Kries, Johannes von, 9, 25, 59, 96 f., 110, 125, 178
Kroh, Oswald, 249, 313, 341, 345, 362, 380, 382, 393
Krolik, Walter, 364
Krueger, Felix, 12, 91, 116, 125, 135, 307, 313, 314, 373, 383; behavior under Nazism, 342 f.; critique of Gestalt theory, 311 ff.; political views, 312; structural psychology, 312
Külpe, Oswald, 7, 9, 25, 34, 45, 47, 49, 50, 65, 78 f., 83, 105, 110, 111, 120, 133, 138, 144, 194, 224, 310; opposition to associationism,

79; views on status of psychology, 45 f. (*see also* Würzburg school)
Künkel, Fritz, 390
Kuhn, Thomas S., 9, 67, 193 f., 223, 408
Kultur, versus civilization, 284, 288, 312, 314
kymograph, 24

laboratories, as subcultures, 6, 278, 316
Ladenburg, Rudolf, 188
Lakatos, Imre, 234
Lamarckism, 249
Lamprecht, Karl, 48
Langfeld, Herbert, 38
language, 12, 110 (*see also* Bühler, Karl; Gestalt theory)
Lao Tzu, 195,
Laplace equation, 175
Lashley, Karl, 165, 282, 337
Lassen, Marie Therese, 355
Laue, Max von, 333
Lauenstein, Otto von, 242 f., 245; emigration, return and death 339 f.; persecution under Nazism, 332, 333, 336, 337 f., 369, 406
Lavater, Johann Caspar, 86, 257
learning, 248 (*see also* Gestalt theory; habit, perceptual learning; trial and error learning)
Lehmann, Alfred, 84
Lehrfreiheit (freedom of teaching), 19
Leipzig, psychological institute, 11, 22 ff., 33, 35, 40, 206, 217 ; model of experimentation, 223
Leipzig school (see *Ganzheitspsychologie*)
lens model, of perception, 319
Lenz, Fritz, 359
Lersch, Philipp, 345, 350, 358, 360, 383, 402 (*see also* characterology)
Levy, Ervin, 364
Levy, Luise, 107
Levy-Bruhl, Lucien, 247, 250 f.
Lewin, Kurt, x, 11, 34, 208, 209, 217, 260, 263–275, 294, 305, 337, 342, 354 f., 357, 358, 373, 391, 401, 402, 408, 409; biography, 263 ff.; conception of the person, 269; conception of psychology as science, 267 f.; conception of psychological research, 271, 272; death in 1947, 406; early research on will, 269, 271 ff.; emigration from Nazi Germany, 327 f.; pedagogical views, 274; philosophy of science, 266 f.; political views, 265, 274, 293; on psychoanalysis, 268, 270, 273; relation to Gestalt theory, 269, 275; theory of action and emotion, 268 ff. (see also ego; Galilei, Galileo; life space; Society for Scientific Philosophy; Taylorism)
Lewin, Leopold, 264
Lewin, Recha, 264

Liebig, Justus, 20
Liebmann, Susanne, 229
Liepmann, Hugo, 277
life space, 274
Lindworsky, Johannes, 241
Linke, Paul Ferdinand, 125, 129, 232
Lippitt, Ronald, 391
Lippmann, Otto, 238
Lipps, Hans, 363
Lipps, Theodor, 25, 32, 44, 91
Lissauer, Hugo, 279
lived experience, 14, 69
local signs, 29, 53, 179 (*see also* space, perception of)
localization, of brain function, 277 (*see also* neural plasticity)
Locke, John, 120
Loeb, Jacques, 80
logic, 75, 195
Lombroso, Gina, 293
Lorentz, Hendryk, 174
Lorenz, Konrad, 404
Lotze, Rudolph Hermann, 19, 29, 30, 53, 75, 198
Löwenstein, Adolf, 354
Ludwig, Karl, 21
Lukacs, Georg, 288
Luria, Alexander, 210
Luz, Melchor, 149

Mach, Ernst, 9, 37, 39, 54, 55 f., 57, 68, 69, 70, 71, 74, 79, 87 f., 90, 91, 96, 110, 113, 115, 116 f., 125 f., 129, 132, 182, 211, 223, 225, 258, 285, 366 f.; on maximum-minimum processes, 184; on physical systems, 175; on the primacy of sensations, 62 ff.; on psychophysical isomorphism, 64, 177 f.
Mach bands, 55, 228
machine theory, 1, 253, 256
McKinnon, Donald, 211
Mann, Heinrich, 21
Mannheim, Karl, 215
Marbe, Karl, 42, 49, 50, 105, 107, 111, 118, 120, 125 f., 129, 204 (*see also* Würzburg school)
Marburg, university, 46 f.; psychological institute, 313
Marie, Pierre, 277
Marr, David, 410
Martius, Götz, 58, 138
Marty, Anton, 95, 105, 133
Marx, Karl, 195
Masaryk, Thomas, 187
mathematics, theory of, 122, 285 f.
Maturana, Humberto, 410

maturation, 248
Mauthner, Fritz, 110
maximum-minimum problems, 171, 182 ff.
Maxwell, James Clerk, 171, 174, 175, 182, 352
meaning, intrinsic (*Sinn*), 1, 290, 309, 373
Meinong, Alexius, ix, 76, 88, 90 f., 93 f., 122, 141, 142, 147, 174 (see also Graz school)
melody, 87 f., 88, 131, 295 (*see also* Gestalt)
memory, 226 f., 241 ff.; for form, 367 f. (see also *Prägnanz*)
memory colors, 56
Mengele, Josef, 398
mental set, 78, 225 f., 231
Messer, August, 78, 111, 211
Metelli, Fabio, 392
Metzger, Wolfgang, 13 f., 209, 210, 211, 216, 229, 340, 363, 382, 405, 406, 408, 409, 410; biography, 346 f.; career under Nazism, 347 ff.; career in postwar period, 385 ff.; conception of scientific psychology, 386 f.; concessions to Leipzig school, 352; conflict with Köhler, 349 f.; *Ganzfeld* research, 232 f., 318; pedagogical views, 389 ff.; political views, 293, 347 ff., 350 ff., 353 f., 391 f.; research in Frankfurt, 364 ff.; students in Münster, 393; systematic account of Gestalt theory, 372–380
Meumann, Ernst, 25 f., 27, 35, 44
microgenesis (*Aktualgenese*), of Gestalten (*see* Gestalt)
Mill, John Stuart, 40, 53, 61, 72, 74, 191, 192, 195
mind, active, 53 (*see also* consciousness; epistemology; thinking)
mind-body problem, 37, 38, 82, 97 (*see also* psychophysics; psychophysical isomorphism)
minimum principle, in perception, 410 (see also *Prägnanz*)
moment (aspect), 76 f.
Monakow, Constantin von, 277
monism, 62 ff.
Montessori education, 274
Morgan, Conwy Lloyd, 150; "canon," 150, 166
Morgenstern, Christian, 212
Morinaga, Shiro, 390
morphology, 80, 85
motion: alpha and beta, 140 f.; apparent and real, 127 f.; and *Gestaltkreis* theory, 378; perception of, 118, 125–132, 317, in brain-injured patients, 277 ff.; and perceptual organization, 232 ff.
Müller, Georg Elias, 42, 47, 67, 132, 180 f., 204, 207, 223, 225, 226, 235, 241, 243, 340; complex theory, 308 f., 310 f.; concept of "coherence factors" in perception, 92 f., 308

f.; critique of Gestalt theory, 308 ff.; psychophysical axioms, 58, 177, 178
Müller, Johannes, 177
Münster, university, 385 ff.; psychological institute, 386
Münsterberg, Hugo, 26 f., 44
music, psychology of, 107, 108 f. (*see also* ethnomusicology; melody)
Musil, Robert, 39, 305

Nagel, Willibald, 108
naive phenomenology (*see* phenomenology)
naive physics, 238
naive realism, 221
National Socialism (Nazism), 13, 325; and Gestalt theory, 13, 343 f., 351 ff., 363, 380 f.
National Socialist German Students' League, 333, 334, 336
National Socialist German Workers' Party, Race Policy Office, 339
National Socialist Professors' League, 347, 349, 350, 360
nativism, 4, 52, 59
Natorp, Paul, 47, 74
Naturphilosophie, 51, 60
Neobehaviorism, 406
Neo-Kantianism, 8, 46, 69, 98, 266, 441n.
Nernst, Walter, 113, 180
Neue Psychologische Studien, 217
Neumann, Franz Ernst, 21
neural plasticity, 255
Neurath, Otto, 267
Newman, Edwin, 216, 406
new objectivity (*Neue Sachlichkeit*), 285, 303
New School for Social Research, 327
Newton, Isaac, 52, 86, 222
Nietzsche, Friedrich, 48, 98, 110
number, concepts of, 120 ff.
Nuremberg Laws, 349

objectivity, 4, 28, 76, 222, 239
objects, theory of, 93
Ogden, Robert M., 111, 210, 212
Ohm, A., 357
ontogeny and phylogeny, 250
ontology, formal, 76 f., 90 f.
open systems, 11, 252 (*see also* Köhler, Wolfgang)
Oppenheim, Paul, 372
Oppenheimer, Erika, 364
optic sector, as physical system, 177, 179
order and freedom, 391
organicism, in Weimar culture, 3, 287 ff.
organism, and environment, 143 f.; as psychophysical unity, 281

parametric procedure, 127

Paulsen, Friedrich, 105
Pavlovism, 384 f.
Peano, Giuseppe, 122, 371
perception; as achievement of the organism, 320; constancy and variability in, 394 f.; of form (*see* Gestalt); in animals 164–165 (*see also* form discrimination); and personality, 362, 369 (*see also* characterology; typology); physiological correlates of (*see* psychophysical isomorphism); primacy of, in consciousness, 75, 139; production theory of, 94, 141 (*see* Benussi, Vittorio; Graz school; Koffka, Kurt); self-organizing tendencies in (*see* Gestalt laws)
perceptual learning, 158
Perry, Ralph Barton, 335
person, 86; as psychophysical whole, 73, 314; perception of, 401
personalism, in philosophy (*see* Stern, William)
personality, diagnostics 357 f., 362, 370, 383 f. (*see also* characterology)
personality, theory of (*see* characterology)
personality, types, and perception, 362, 369
Petermann, Bruno, 317, 344, 369
Peters, Wilhelm, 343; dismissal in 1933, 326
Pfahler, Gerhard, 360, 362
Pfungst, Oskar, 151, 152
phenomenal identity, 232
phenomenalism, 36 f., 63; heuristic, 54, 62 f., 131
phenomenological physics, 182
phenomenology, 36 f.; of animal problem solving, 152 f., 164; and cultural anthropology, 120; as descriptive psychology, 75, 79; and experimental psychology, 77 f.; in Gestalt theory (*see* Gestalt theory); of seen motion, 126 f.; of time consciousness, 78; transcendental, 79
phi phenomenon, 9, 128 f., 130, 132, 144, 159, 317
philosophy, institutional situation of, in Germany, 18; and natural science, 29, 112; as rigorous (human) science, 44 f.; versus psychology in Germany, 42 ff., 47 ff.; as tool and weapon under Nazism, 344
Philosophische Studien, 23, 24
Phonogramm-Archiv, 33, 34, 41, 105, 107, 149
physicalism (*see* Gestalt theory)
physics, institutionalization of, 21 f.
physiognomics, 86, 257, 300
physiology, 8, 21, 25; institutionalization of in Germany, 21; sensory (*see* sensory physiology)
Piaget, Jean, 249
Pick, Arnold, 133, 217, 277
Pikler, Julius, 242

Planck, Max, 112, 113, 116 f., 149, 169, 170, 171, 174, 182, 185, 236, 260 ff., 285, 286, 337, 398; visit to Adolf Hitler, 331
plasticity, neural, 97, 255
Plato, 29
Plaut Foundation, 148
Poincaré, Henri, 133
Poppelreuter, Walter, 58, 279, 342, 389
positivism, and idealism, 68; revolt against, 2, 3, 68
Pötzl, Otto, 107, 277
practice, 248, 249 (*see also* learning; maturation)
Prägnanz, law or principle of, x, 133, 184 f., 224, 240, 366, 379, 395, 396, 410; and brightness contrast, 227; and memory, 226 f.; and motion perception, 232 f., 234
Prague, 103 ff., 187, 198; Charles University, 105, 133
Prague circle, 10, 187, 197 f.
Preuss, Hans, 333 f., 340
Preyer, William Thierry, 25
Pribram, Karl, 389
Prigogine, Ilya, 410
production theory (*see* Graz school)
productive thinking (*see* thinking)
Prussian Academy of Sciences, 34, 148, 169
Prussian Artillery Testing Commission, 188, 190
pseudo-fovea, 279 f.
psychiatry, critique of, 46
psychical causality, 61, 97
psychoanalysis, 107; under Nazism, 326 (*see also* Freud, Sigmund)
Psychologische Forschung, 10, 203, 216–218, 220, 267, 307, 309, 355, 363, 367, 371; founding, 216; under Nazism, 349 f.; re-establishment after 1945, 382 f.
psychologism, critique of, 43, 45, 74 f.
psychology: applied, 35, 44, 50, 204 f., 206, 265, 383 f., 399 f., in the German military, 187 ff., 345 f. (*see also* psychotechnics); comparative (*see* animals; anthropoid apes); continuity and change under Nazism, 362 f.; crisis of (*see* crisis, cultural); as elite field in Berlin, 209; ethnological (see *Völkerpsychologie*); European and American styles in, 406 f.; historiography of, 4 ff.; humanistic, 288 ff.; ideological transformation under Nazism, 342 ff.; institutional situation of, in Germany, 7 f., 17 f., 22 ff., 42–50, 110, 203 ff., 307, 316, 326 f., in West Germany, 382 ff., in East Germany, 384 f., 399 ff., in the United States, 50; laboratories, 5, 18, 25, 43, 118 f.; and medicine, 45 f.; and natural sci-

ence, 6, 7, 42 f., 211, 386, 392 f.; of perception (*see* perception); and philosophy, 6, 35, 47 ff., 199, 204, 208; physiological, 23, 388; professionalization of, in Nazi Germany, 3, 7, 326, 345; of productive thinking (*see* thinking); of testimony (*see* testimony, psychology of); of world views, 289; as young science, 136
psychophysical axioms (*see* Müller, Georg Elias)
psychophysical isomorphism, 64, 71, 176 ff., 244; empirical tests of, 388 (*see also* Mach, Ernst; Köhler, Wolfgang)
psychophysical parallelism, 57, 71, 81
psychophysics, 22, 48, 51, 60, 61, 181, 388 f.; antinomies of, 410; outer and inner, 64, 376 ff.
psychotechnics, 188 (*see also* Münsterberg, Hugo)
psychovitalism, 82, 146
Pudowkin, Wselewod, 302, 303

quantum mechanics, 286 f., 322

race hygiene (*see* eugenics)
race psychology (*see* Günther, Hans F. K.)
race theory, 353
"racial soul studies" (*Rassenseelenkunde*), 340, 344
Ramon y Cajal, Santiago, 96
Rathenau, Walter, 293
Rausch, Edwin, 13 f.; 363 f., 380; analysis of Gestalt category, 371 f., 394 ff.; biography, 370 f.; postwar career, 394; students, 397
reality, psychological, 37, 75
recentering, 10, 192, 240, 408 (*see also* centering; Gestalt, switch; thinking)
Redslob, Erwin, 334
Redslob, Ottilie (*see also* Selbach, Ottilie), 364
Reichenbach, Hans, 3, 30, 214, 217, 256, 261, 265, 267, 286, 294 (*see also* Society for Scientific Philosophy)
Reichswehr, psychological section, 338
relational systems (*Bezugssysteme*), 375 f.
relations, reality (giveness) of, 37
relativity, 286 f., 322
requiredness, 298
research, freedom of, 18 f.
research program, 11
research styles, 11
Restorff, Hedwig von, 243 f., 245, 335, 349, 408
retinal image, 179
retinal interaction, 55 f., 228 f.
Revesz, Geza, 165
reward and punishment, psychology of, 274 f.

rhythm, sensation and perception of, 108 f.,
141
Rickers-Ovsiankina, Maria, 272
Rickert, Heinrich, 42, 47, 110
Rieffert, Johann Baptist, 338 ff., 360
Riehl, Alois, 38, 47, 108, 113, 264, 275
Riehl, Wilhelm, 312
Riezler, Kurt, 214, 286 f.
Romanticism, 81, 291
Rothmann, Max, 148, 149, 152, 161
Rousseau, Jean-Jacques, 251
Roux, Wilhelm, 80, 247, 255
Rubens, Heinrich, 113
Rubin, Edgar, 179, 229, 249, 336, 367
Rubinstein, Sergei, 385
Rüdin, Ernst, 380
Rupp, Hans, 39, 109, 188, 191, 206, 265
Russell, Bertrand, 371
Rust, Bernhard, 333
Rutherford, Ernest, 112

Sader, Manfred, 397
Sakuma, Kanae, 210, 269
Samson, Albert, Foundation, 34, 148, 166, 168,
207
Sander, Friedrich, 317 f., 343, 362, 382, 392;
parallelogram, 318 f., 396
Schapp, Wilhelm, 78
Scheerer, Martin, 314, 322
Scheerer, Wilhelm, 35
Schelderupp-Ebbe, Theodor, 217
Scheler, Max, 258, 305, 378
Schelling, Joseph, 313, 378
Schilder, Paul, 376
Schlick, Moritz, 214, 217
Schmidt, Hans-Dieter, 402
Schmidt-Ott, Friedrich, 337
Schmoller, Gustav, 105
Schole, Heinrich, 349
Schönpflug, Wolfgang, 397
schools, research or scientific, 5, 11 (*see also*
Berlin school; Gestalt theory)
Schrödinger, Erwin, 337
Schulte, Heinrich, 259
Schultz, Joseph, 105
Schumann, Friedrich, 26, 39, 40, 49, 85, 91 f.,
94, 95, 98, 105, 111, 118, 123, 125 f., 127,
129, 132, 133 f., 136, 138, 175, 214 f., 221,
277, 363 f., 370, 387
Schwann, Theodor, 21
science: aesthetic versus technocratic ideals of,
160, 166, 309; contextualist accounts of, 4 f.;
freedom of (see research); and life, 68, 72,
220, 287 f.; under Nazism, 325 f., 361; in
postwar Germany, 385 ff.; social and cultural

relations of, 5 ff.; and technocracy, 151; in
Weimar Germany, 285 ff.
second law (of thermodynamics), 182 f., 248,
254 (*see also* entropy; *Prägnanz*)
Seele, 81, 82
seen objects (*Sehdinge*), 55, 59, 77
Selbach, Ottilie (née Redslob), 334
Selenka Foundation, 148
self-actualization, 282
self-observation, 39 f.; training for, 272 (*see
also* introspection)
self-organization, 1 f., 411
self-regulation, 81, 255
Selz, Otto, 138, 165, 194, 238, 310 f., 358;
dismissal in 1933, 326
sensation, 37, 52 f., 61, 62, 70, 75; of direction,
65; kinesthetic, 109; and perception, 8, 111,
139 (*see also* perception); as products of
analysis, 79; projection of, 258 f., 376; as
signs, 53, 56; unnoticed, 70, 120, 135 ff. (*see
also* Helmholtz, Hermann von; Hering,
Ewald; Mach, Ernst)
sense data, 1
sense (of a sentence), 172; and reference, 194
sensory physiology, styles of reasoning in, 52
ff. (*see also* hearing; vision)
Simmel, Georg, 48
similarity, 179
Skinner, Burrhus Frederick, 282
Society for Experimental Psychology, 26, 42,
45, 49, 66, 132, 138, 141, 151, 204, 205,
275, 311, 317 (*see also* German Society for
Psychology)
Society for Scientific Philosophy, 3, 256, 261,
294
Sokolowsky, Alexander, 151, 152
solipsism, 63
Sommer, Robert, 26, 139, 211
Sorge, Siegfried, 369, 370
space, perception of, 29, 65 (*see also* local
signs)
Spann, Othmar, 287
speed, perception of, 235 f.
Spemann, Hans, 253
Spence, Kenneth, 406
Spencer, Herbert, 251
Spengler, Oswald, 287 f., 297, 299, 312, 313
Speyer Foundation, 120
Spinoza, Baruch, 104, 186, 343
Spranger, Eduard, 73, 290, 296, 298, 373 (*see
also* humanistic psychology)
Springer-Verlag, 216, 349
Stählin, Wilhelm, 111, 250 f.
Stark, Johannes, 286
state of affairs (*Sachverhalt*), 76
Stern, Georg and Lisbeth, 105

Stern, William, 12, 26 f., 35, 39, 44, 66, 78,
    204, 208, 212, 214, 281, 342; critique of
    Gestalt theory, 313 f.; dismissal in 1933,
    326; personalistic philosophy, 314, 317
stimulus gradients (*see* gradients)
Stout, G. F., 158
Stresemann, Gustav, 292 f.
stroboscope, 118
stroboscopic effect, 125
structure, as psychological category, 312, 322
    (*see also* consciousness)
structural functions, 10, 163 ff.
structural psychology (*see* Krueger, Felix;
    Spranger, Eduard)
Studium generale, 386
Stumpf, Carl, x, 3, 7, 9, 19, 25, 28–41, 42,
    44, 50, 54, 58, 65, 70, 71 f., 74, 75, 76, 77,
    79, 81, 90, 91, 95, 103, 105, 106, 108 f.,
    110, 113 ff., 132, 133, 135, 136, 149, 168,
    169, 178, 182, 188, 190, 191, 198, 199, 203,
    205, 207, 213, 263, 264, 266, 275, 315, 338,
    339; appointment in Berlin, 30 ff.; biogra-
    phy, 28 ff.; conception of experience, 36 ff.;
    conception of psychological research, 35,
    39 f.
Stumpf, Felix, 188
Stumpf, Paul, 188
Stuttgart, technical academy, 204
subwholes (*Unterganze*), 369
successive comparison, 241 f.
Sultan (chimpanzee), 149, 153, 157
sum, 172 ff., 256 (*see also* and-connections;
    and-sums)
summativity, and nonsummativity, 371 (*see
    also* Gestalt concept)
suprasummativity, 172, 394 f.
surface colors, 58
symbols, theory of, 281, 315 f.
symmetry, 184
system, 1 f., 11, 165, 170, 171, 175 (*see also*
    open systems, optic sector)

tachistoscope, 39, 118, 127, 139, 277, 317
Taylorism, 265
teleomechanism, 80
Tenerife, 10, 148, 160, 161, 168, 253, 256,
    292; anthropoid research station, 148 ff., 166
    f., in World War I, 160 ff.
tension systems, 272, 273
tensor tympani, 114
Ternus, Joseph, S. J., 232
testimony, psychology of, 106
Teuber, Eugen, 149, 161
Teuber, Hans-Lukas, 388
things: as collections of sensations, 62; as pri-

mary percepts, 137 (*see also* Gestalt theory;
    Wertheimer, Max)
thinking, 10, 78 f., 138, 240, 389 (*see also*
    Duncker, Karl; Wertheimer, Max)
third way, as trope in Weimar discourse, 287,
    288, 295
Thom, Réné, 410
Thorndike, Edward, 151, 152, 158, 159 f., 169,
    221
threshold, of sensation, 60
Tillich, Paul, 215
time series, in physics and biology, 266 (*see
    also* genidentity)
Titchener, Edward Bradford, 60, 144, 292, 372
tonal fusion, 31, 90 f.
tone color, 115
tone psychology (*see* hearing)
tone variator, 29, 115
topology, in psychology, 273
total image (*Gesamtvorstellung*), 121
transposability, 172
trial and error learning, 151
Tübingen, university, 113
Turhan, Mümtaz, 365, 369
Twardowski, Konstantin, 93
twin research, 13, 355, 356 ff., 400 f.
typology, in psychological theory and research,
    313 (*see also* characterology)

Uexküll, Theodor von, 287
Uibe, Martin, 163
unity of consciousness (*see* consciousness)
Usnadze, G., 210, 316

Vahlen, Theodor, 335 f., 337, 338
Vaihinger, Hans, 294
validity, of ideas, 30
Vedda (Ceylonese tribe), 107
Verschuer, Otmar Freiherr von, 355, 356, 360,
    398
Vienna, psychological institute, 204; university,
    107, 118
vision, 29, 52, 364 ff.; and the brain, 55 (*see
    also* Gestalt theory); zone theory of, 178
visual space (*Sehraum*), 56 (*see also*
    anisotropy)
Volkelt, Hans, 165, 312, 349
Völkerpsychologie (ethnological psychology),
    23, 121, 122
volition (*see* will)
vowel colors, 115
Vygotsky, Lev, 252

Wagner, Adolf, 105
Wagner, Richard, 89, 301

Wakeman, Seth, 212
Waldeyer, Wilhelm, 95, 148 f., 150, 152, 156, 158, 161, 166
Wallach, Hans, 235 f., 349 f., 364, 388
Wartenburg, Paul Yorck von, 31
Wartensleben, Gabrielle von, 123
Watson, John B., 50
Watt, Henry, 78, 107
Weber, Ernst Heinrich, 22, 60
Weber, Max, 289, 290
Weber, Wilhelm, 21, 29
Weber-Fechner law (*see* psychophysics)
"we" group, 401 f.
Wehrmacht, 3, 14, 346, 361
Weimar culture, 284 ff.
Weinhandl, Ferdinand, 344
Weiss, Paul, 253
Weizsäcker, Viktor von, 378 (see also *Gestaltkreis* theory)
Weltsch, Felix, 105
Werfel, Franz, 187
Werner, Heinz, 257, 317
Wernicke, Carl, 276 f.
"Wertbostel" (*see* directional listener)
Wertheimer, Max, ix, x, 1, 2, 6, 9, 10, 11, 13, 34, 41, 42, 49, 50, 72, 117, 118, 120–134, 136, 137, 139, 140, 141 f., 144, 145, 147, 158, 159, 160, 168, 171, 172, 176, 184 f., 203, 204, 208, 209, 216, 227, 230, 235, 236, 238, 244, 248, 249, 251, 263, 268, 269, 270, 274, 275, 279, 282, 299, 300, 302, 304, 308, 310, 317, 318, 331, 335, 338, 346, 347, 350, 352, 353, 357, 358, 363, 369, 370, 371, 372, 373, 375, 386, 388, 391, 393, 395, 400, 401, 404, 407, 408; appointment in Frankfurt, 213 ff.; appointment offer in Giessen, 212 f.; biography, 103 ff.; conception of brain events, 130; conception of consciousness, 220; conception of theory construction, 131 f.; death, 406; dismissal in 1933, 326 ff., 337; dissertation, 106; early career, 105–108, 133 f.; Gestalt epistemology, 123 f., 159, 165, 178, 194 f., 220; health problems, 216, 327; on method, 220; military research, 188–190; on motion perception, 125–131; on paranoid delusions and social perception, 259; on perceptual organization, 224 ff.; on person peception, 257 f.; personal philosophy, 104 f., 197 f.; political views, 187, 197, 291 f., 293, 296 f.; on productive thinking, 190–196; research with brain-injured patients, 277; on self and society, 295; teaching style, 210; on thinking of "natural" people, 120–122; in Weimar culture, 294 ff.
Wertheimer, Walter, 104

Wertheimer, Wilhelm, 103 f.
Westermann, Dietrich, 217
Weyl, Hermann, 286, 287
White, Robert, 391
wholes, as experience-correlates of stimuli, 143 f. (*see also* Gestalt concept)
whole and part, 1, 76 f., 82, 85, 90 ff., 95, 98, 108, 121, 138, 295; in field physics, 171, 173; logical analysis of, 371, 395; perception of, in brain-injured patients, 279 (*see also* creative synthesis; Ehrenfels, criteria; Gestalt concept; Gestalt theory; holism)
Wiedemann, Eilhard, 170
Wien, Wilhelm, 113, 286
Wilde, Kurt, 357
will, psychology of, 269, 271 f.
Willmann, Otto, 105
Windelband, Wilhelm, 43, 47, 69, 83
Wissenschaft (systematic scholarship and research), 18, 20, 21
Witasek, Stefan, 93, 145
Wittgenstein, Ludwig, 110
Wittmann, Johannes, 317, 344, 389
Wittfogel, Karl August, 195
Wohlfahrt, Erich, 318
Wolff, Wilhelm, 364, 366
Woltereck, Richard, 256
women, and academic careers in Germany, 20; in Berlin psychological institute, 209
Wulf, Friedrich, 226, 244, 367 f.
Wundt, Wilhelm, 11, 17, 22–25, 26, 27, 32, 40, 44, 46, 50, 60, 61, 62, 65, 67, 72, 74, 79, 84, 85, 87, 121, 125, 149, 177, 208, 215, 241, 266, 289, 296; analysis of whole and part (*see* creative synthesis); conception of experience, 23; conception of psychological research, 23; on status of psychology, 47 ff.
Würzburg, psychological institute, 42, 49, 105, 106 f., 110 f., 117, 120, 144, 204
Würzburg school, 9, 78 f., 83, 194

Yerkes, Robert M., 149, 161 ff.
Young-Helmholtz color theory, 52

Zeigarnik, Blyuma, 271 f.
Zeitschrift für angewandte Psychologie, 355
Zeitschrift für Psychologie, 25 f., 46, 73, 81, 105, 117, 164, 343
Zeller, Karl, 31
Zener, Karl, 211
Zöllner, Carl Friedrich, 22
Zoltobrocki, Josefa, 397
Zürich, university, 107, 120
Die Zukunft, 48
Zwicker, Rosa, 103